Human Growth: Assessment and Interpretation

Many researchers and professionals need to be able to measure, assess, and interpret human growth between birth and adulthood. However, much of the methodology is scattered in diverse literature. *Human Growth: Assessment and Interpretation* provides a complete reference to the field for all those who measure and assess child growth. It emphasizes the interpretation of growth data taking into account the adjusted effects of influences such as genes, hormones, and substance abuse during pregnancy, gives descriptions of normal and abnormal growth patterns, and of variant growth patterns such as failure-to-thrive and catch-up growth. Including methods to measure size and maturity, the judgement and interpretation of recorded data, evaluations of influences on growth, and the significance of abnormal growth, it will be an essential source of information for pediatricians, human biologists, health workers, nutritionists, epidemiologists, and others who are responsible for the health and welfare of children.

ALEX ROCHE is Fels Professor Emeritus of Community Health and of Pediatrics at Wright State University, Dayton, Ohio. His research has focused on child growth and maturation, body composition, and risk factors for disease, and he has written or edited 60 books and monographs in these fields.

SHUMEI S. SUN is Professor of Community Health at Wright State University, Dayton, Ohio. Trained initially in biostatistics, her expertise is in the development of longitudinal statistical models for analysis of data related to growth, overweight, and obesity, and risk factors for cardiovascular disease.

Human Growth: Assessment and Interpretation

Alex F. Roche
Wright State University

and

Shumei S. Sun (formerly Guo)
Wright State University

CAMBRIDGE
UNIVERSITY PRESS

PUBLISHED BY THE PRESS SYNDICATE OF THE UNIVERSITY OF CAMBRIDGE
The Pitt Building, Trumpington Street, Cambridge, United Kingdom

CAMBRIDGE UNIVERSITY PRESS
The Edinburgh Building, Cambridge CB2 2RU, UK
40 West 20th Street, New York, NY 10011-4211, USA
477 Williamstown Road, Port Melbourne, VIC 3207, Australia
Ruiz de Alarcón 13, 28014 Madrid, Spain
Dock House, The Waterfront, Cape Town 8001, South Africa

http://www.cambridge.org

First published 2003

Printed in the United Kingdom at the University Press, Cambridge

Typeface Ehrhardt 10.5/13 pt   *System* LaTeX 2_ε [TB]

A catalogue record for this book is available from the British Library

Library of Congress Cataloguing in Publication data

Roche, Alex F., 1921–
Human Growth: Assessment and Interpretation /
by Alex F. Roche and Shumei Sun.
     p.   cm.
Includes bibliographical references and index.
ISBN 0 521 78245 7   (hardback)
1. Children – Growth.   2. Child development.   I. Sun, Shumei, 1954–   II. Title.

RJ131 .R5915 2003
612.6'54–dc21   2002073462

To Eileen
Mater Familiae

Contents

Preface

This book collates information from many sources that is necessary for the assessment and interpretation of growth data recorded in clinics or research studies. The large writing task will be justified if the book proves beneficial to professionals who measure the growth of infants and children. These professionals include pediatricians, family practice physicians, dieticians, other health workers, epidemiologists, and human biologists. Part of the material presented may be unfamiliar to some readers, but all the topics covered are relevant to the accurate assessment and interpretation of human growth and maturation, which are central to pediatric practice and public health policies relating to children. The literature review is restricted to reports from developed countries, but there should not be a corresponding restriction of the readership. Those working in lesser-developed countries, who may have limited access to the literature, will gain from a more complete knowledge of what is currently known and the analytic methods that have been applied. The text is appropriate for practicing professionals, graduate students, and senior undergraduates, but is not intended for the casual reader.

One impetus to writing this book was the difficulty of collating findings from a multitude of relevant published reports that are scattered in many journals and monographs. With a few exceptions, no more than four recent references are given for any particular statement, which should be sufficient to help the reader evaluate the evidence. Additional references would have broken up the text further, making it difficult to read. Those who wish to review more primary sources should consult electronic databases or the bibliographies in the reports cited. The reader is provided with a considerable amount of numerical data to assist judgements of the sizes of differences between groups and the strengths of correlations. These data are not so numerous that they occlude the general descriptions and conclusions that are presented.

The text has five parts. Chapter 1 describes the equipment needed to make growth measurements and the procedures to be applied. The criteria for growth chart excellence are given and selected growth charts are described. About 50 years ago, all growth charts were for general populations while today many growth charts are for specific groups such as low birth weight infants and untreated children with particular diseases. The shapes of the percentile levels on these charts are noted and their relevance to growth assessment is explained. Methods for the assessment of maturity are given and sources of reference data for growth and maturational status are provided.

Chapter 2 reviews patterns of growth for individuals including irregularities of growth during infancy, prepubertal, and pubescent spurts, tracking, and variations from normal growth patterns, such as decanalization, failure-to-thrive, and catch-up growth. This Chapter ends with a consideration of the prediction of adult stature and the utility of target stature. Chapter 3 reviews the adjusted effects on growth of genetic and family influences, substance abuse during pregnancy, breast-feeding, hormones, ethnic and nutritional influences, high altitude, and maturity. Knowledge of the effects of these influences is of fundamental importance to the interpretation of growth data. In Chapter 4 attention is given to secular changes in growth and maturity with reference to the influences responsible, and the effects of secular changes on the assessment and interpretation of child growth. Knowledge of these changes within countries is needed to assist the selection of appropriate reference data. Chapter 5 describes the significance of child growth and maturity in relation to future size, function, and disease. The place of anthropometry in the assessment of nutritional status and the possible justification of screening programs based on measures of growth are evaluated.

Growth assessment depends on accurate measurements and the use of growth charts. The accurate assessment and interpretation of the growth status and maturity of a child is important because unusual status values may be associated with current pathological conditions. Additionally, they may be risk factors for chronic diseases. For example, overweight in childhood increases the risk of hypertension, non-insulin-dependent diabetes mellitus and other chronic diseases in adulthood. Growth charts are used primarily by health care professionals in clinical settings and by research workers in epidemiological and clinical studies. They assist judgements of the normality of growth status and help screen for current disease and the presence of risk factors for future disease. These charts present sex-specific, smoothed percentile levels for selected variables or the mean and selected standard deviation levels (Z scores). Typically growth charts display reference levels for weight from birth to 18 or 20 years, length, weight-for-length, and head circumference from birth to 36 months, and stature and body mass index (BMI) from 2 to 18 or 20 years. The measured values for these variables should be plotted on charts *and* recorded in a tabular format. Comparisons between the plotted points and the percentile or Z score levels on the charts show the approximate level for the child or children compared with the reference population.

After the recorded data for a child or a group of children have been compared with the reference percentiles on growth charts, the findings must be interpreted. Accurate interpretation depends on knowledge of growth patterns in relation to age and maturity and the effects of influences that affect the rates of growth and maturation. For such interpretation, any earlier data should be considered so that growth progress can be evaluated. It may be desirable to schedule another examination to evaluate future growth progress. The interval to the next examination will be determined by the age of the child and, if increments are to be evaluated, by the length of the intervals in the incremental reference data that will be used.

This text concentrates on the literature of the past 40 years relating to growth, maturation, and body composition. During this time there has been important progress in the development of statistical methods for the analysis of cross-sectional and serial data, the assessment of skeletal age, and the prediction of adult stature. Furthermore, new procedures have increased our knowledge and understanding of body composition. Much remains to be done. New non-invasive procedures may allow improved studies of prenatal growth and changes in organ size that may relate size to function after adjusting for confounding variables. Our understanding of the construct "socioeconomic status" may improve and better study designs in which reported data are replaced by measured data may provide more accurate estimates of the independent effects of specific influences on growth and maturation. The latter will require the measurement of an enlarged set of confounding variables, including specific genes and selected hormones, to remove the effects of all influences other than the one under study. Most of these advances will require lengthy multi-disciplinary collaboration at all stages of the studies, including the development and choice of apparatus, construction of hypotheses, data collection, and data analysis. Genuine complete collaboration is needed, particularly in emerging topic areas, and it must be based on deep interest in particular research areas and mutual respect among research workers. If this can be achieved, significant increases in our understanding of growth and maturation are likely.

We wish to express our gratitude to Donna Menelle and Marianne King for their accurate and professional secretarial support through many drafts. We are grateful to the participants in the Fels Longitudinal Study. Since 1929, they have provided the data that allow us to address topics of great interest concerning child growth and maturation. We wish to acknowledge the help given by Howard Kulin, Peter Lee, and Bradford Towne, who reviewed sections of typescript; they are, of course, not responsible for any remaining errors. We have been assisted in the development of our approaches to the study of growth and maturation by many consultants and colleagues particularly Laurel Beckett, Cameron Chumlea, Robert Kuczmarski, Roger Siervogel, David Thissen, and Howard Wainer. Considerable support was provided by the National Center for Health Statistics and the National Institute of Child Health and Human Development. We are grateful to Dr. Tracey Sanderson and Dr. Katrina Halliday of the Cambridge University Press who guided the manuscript through all stages of the publication process in a charming and effective manner.

Alex F. Roche
Shumei S. Sun

Abbreviations

AGA	Appropriate-(weight)-for-gestational age	LGA	Large-for-gestational age
BMC	Bone mineral content (g)	LMP	Last menstrual period before pregnancy
BMD	Bone mineral density (g per cm^2)	NIDDM	Non-insulin-dependent diabetes mellitus
BMI	Body mass index	PHV	Peak height velocity (cm per year)
BP	Bayley–Pinneau method of stature prediction	PWV	Peak weight velocity (kg per year)
CTRH	Corticotropin-releasing hormone	%BF	Percentage of body weight that is fat
DHEA	Dehydroepiandrosterone	RNA	Ribonucleic acid
DHEA-S	Dehydroepiandrosterone sulfate	RWT	Roche–Wainer–Thissen method for the prediction of adult stature
ELBW	Extremely low birth weight (<750 g)		
FFM	Fat-free mass (kg)	SD	Standard deviation
FTT	Failure-to-thrive	SGA	Small-for-gestational age
GAA	Gestation-adjusted age (weeks)	SES	Socioeconomic status
GHBP	Growth-hormone binding protein	SMI	Sexual maturity index (scale 1 to 5)
GHRH	Growth-hormone releasing hormone	SRIH	Somatotropin release-inhibiting hormone
GnRH	Gonadotropin-releasing hormone		
IDDM	Insulin-dependent diabetes mellitus	TBBM	Total body bone mineral (kg)
IGF-1	Insulin-like growth factor 1	TBF	Total body fat (kg)
IGF-1 BP	Insulin-like growth factor 1 binding protein	TRH	Thyroid-releasing hormone
IHDP	Infant Health and Development Project	TW	Tanner–Whitehouse method for the prediction of adult stature
LBW	Low birth weight (<2500 g)	VLBW	Very low birth weight (<1500 g)

1 · Measurement and assessment

Body size is important at all ages, but particularly at birth because of its relationships to morbidity and mortality in early infancy and its possible relationships to some diseases in adulthood. Reference data for size at any age allow the recognition of infants with unusual values. Measuring the size of infants and children and comparing the values with appropriate reference data are important parts of a pediatric health assessment. Body measurements are primary indicators for monitoring growth, health, and nutritional status and they provide information regarding future health potential. Infants and children with normal body size are probably receiving adequate nutrition and growing appropriately and are less likely than others to have diseases.

The patterns of growth during infancy differ from those in childhood and adolescence. Consequently, separate sets of growth charts are needed for these periods of growth. Following rapid growth during the first two years after birth, growth proceeds at a relatively steady pace until pubescent accelerations bring children close to their adult body size. Body size is important after infancy in relation to current nutritional status and the probability of current and future diseases.

Differences between males and females in growth after infancy are well documented. One major difference is that males, on average, have pubescent growth spurts about one to two years later than females. Growth is more rapid in males than females during pubescence and pubescent growth lasts longer in males than females. Partly because of these differences, separate growth charts are needed for each sex.

Appropriate reference data to determine the normality of growth status permits the assessment of an infant or a child in relation to the size of his or her age-peers. Such status data are essential to identify infants and children with large or small values for body measurements. This chapter describes procedures for measuring the size of infants and children, the equipment needed to make these measurements and its calibration, quality control procedures and the principles underlying the development of growth charts for clinical use. In addition, data for multiple births, and growth charts for low birth weight infants, and children with some specific diseases are described.

MEASUREMENT OF WEIGHT, SIZE, AND TOTAL BODY COMPOSITION

The assessment and interpretation of child growth requires standardized measurement procedures, trained personnel, the use of appropriate, regularly calibrated instruments and the collection of reliability data. The recorded data should be reviewed by a pediatrician or supervisor at each examination and discussed with the caregiver and child before they leave the examination site. Unless otherwise noted, one person can make each of the measurements described. The descriptions of measurement procedures that follow are for weight, length, stature, body mass index (BMI), segment lengths, circumferences and skinfolds at various sites. Further information about other measures, sources for equipment procurement, and guidelines for calibration can be found in Lohman *et al.* (1988). Reliability data for the measurements considered are given in Table 1.1.

Weight

During infancy, an electronic scale is recommended, but a beam scale with movable weights can be used.

Table 1.1. *Means of inter-observer differences for common anthropometric measures of children in the Fels Longitudinal Study*

Measure	Mean (SD)
Length	0.4 (0.3) cm
Stature	0.3 (0.2) cm
Crown–rump length	0.5 (0.3) cm
Sitting height	0.3 (0.2) cm
Head circumference	0.2 (0.1) cm
Arm circumference	0.2 (0.1) cm
Lower abdominal circumference	0.6 (0.4) cm
Hip circumference	0.5 (0.4) cm
Skinfold thicknesses (triceps, subscapular, suprailiac)	1.1 (1.2) mm

An electronic scale is preferred because the measurements are more accurate, and are affected less by infant movement (Engstrom *et al.*, 1995; Kavanaugh *et al.*, 1990; Torrence *et al.*, 1995). The pan of the scale must be at least 100 cm long. A cloth is placed on the pan and the scale is calibrated across the range of expected weights using objects of known weights. All scales should be calibrated each month, and whenever they are moved. When a beam scale is not in use, the beam should be locked in place or the weights shifted from zero. In studies to assess short-term changes, weights must be recorded at times that are fixed relative to ingestion, defecation, and micturition.

An infant should be weighed nude, because most reference data for infants are based on nude weights. If a diaper is worn, its weight is subtracted from the observed weight. The infant is positioned so that its weight is distributed equally on each side of the pan. Weight is recorded, to the nearest 10 g, with the infant lying still. The average of three measurements is recorded after omitting any clearly erroneous value. When an infant is restless, the caregiver is weighed holding the infant and weighed later without the infant. The difference between the paired measurements is equal to the weight of the infant. There will be some loss of accuracy if the caregiver's weight is not recorded to the nearest 10 g.

A subject able to stand without support is weighed using an electronic scale or a beam scale with moveable weights. The beam should be graduated on both sides and the scale placed so that the measurer can face the subject and move the beam weights without reaching around the subject. Whether an electronic scale or a beam scale is used, the subject stands motionless over the center of the platform and looks straight ahead with the body weight evenly distributed between both feet. Subjects can wear light indoor clothing, excluding shoes, long trousers, and sweaters, but it is better to wear disposable paper gowns. The weight of this clothing is not subtracted from the observed weight when comparisons are made with reference data. Handicapped subjects, who cannot stand without support, can be weighed using a chair or bed scale. Weight is recorded to the nearest 10 g with electronic scales and to the nearest 100 g with beam scales.

Length

Length is measured instead of stature during infancy and in older children who are unable to stand. Two measurers are needed to measure length. A caregiver can substitute for one measurer, but this is not ideal. The subject lies on his/her back on a table designed for length measurements. An infant can wear a diaper, older children should be dressed as for the measurement of weight. The table has a metal rule attached to the top, a stationary headboard, and a moveable footboard. The headboard and footboard are vertical to the tabletop. The crown of the head is placed firmly against the headboard with the centerline of the body over the centerline of the table (Figure 1.1A). One measurer positions the head while standing behind the end of the table and checks the positioning of the body. The other measurer places one hand on the knees of the subject to keep the legs flat on the table and applies firm pressure with the other hand to shift the footboard against the heels. The subject's head is held with the lower border of the left orbital margin and the upper border of the left external auditory meatus in a vertical plane. While the arms rest against the

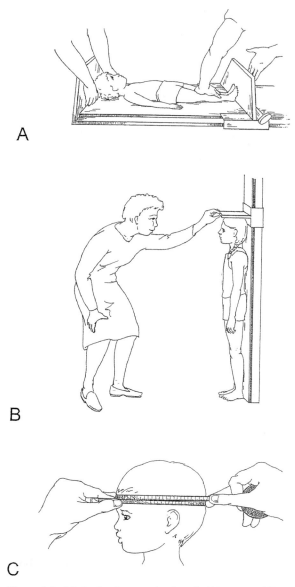

A

B

C

Figure 1.1. Methods of measuring length (A), stature (B), and head circumference (C).

used to describe reliability for this and other measures, which hinders comparisons among reports. In newborn infants, the standard deviations (SDs) of the differences between paired measurements are 0.2 cm for intra-observer data and 0.3 cm for inter-observer data (Chang *et al.*, 1993). Lampl (1993) reported SDs of intra-observer differences of 0.2 cm for infants aged 1 week to 18 months. Median absolute differences for preterm infants, about 3 weeks after birth, are 0.6 cm for intra-observer and 0.9 cm for inter-observer comparisons (Rosenberg *et al.*, 1992). In the Fels Longitudinal Study, the mean absolute inter-observer difference at 1 to 6 months is 0.4 cm (SD 0.3 cm) (Table 1.1).

Stature

A stadiometer is used to measure stature. This apparatus consists of a vertical board with an attached metric rule and a moveable horizontal headboard; the latter can be brought into contact with the most superior point on the head. The subject is barefoot or wears thin socks and minimal clothing so that body positioning can be checked. The subject stands on a flat surface in front of the vertical board with his/her weight distributed evenly on both feet and the arms hanging freely with the palms facing inwards (Figure 1.1B). One measurer ensures that the heels are in contact with each other and touching the base of the vertical board or a heel rest, and that the inner borders of the feet are at an angle of about 60°. If the subject has knock-knees, the feet are separated so that the medial borders of the knees are in contact, but not overlapping. The knees are fully extended and the buttocks, scapulae, and the posterior part of the head are placed in contact with the vertical board. If this is not possible, while maintaining a natural stance, the subject is positioned so that only the buttocks and head are in contact with the vertical board. Two measurers are needed to obtain accurate measurements of stature. The first should face the front of the subject and the second should face the left side of the subject. The second measurer positions the head so that the lowest point on the left orbital margin and the most superior point on the

sides of the trunk, the shoulders and buttocks are kept against the table, and the hips and knees are extended. Length is recorded to the nearest 0.1 cm.

This measurement can be made reliably if attention is given to the details of the recommended procedure. Various summary statistics have been left external auditory meatus are at the same horizontal level. Because this requires palpation of the orbital margin, the subject must remove any spectacles. After the correct positioning is achieved, this measurer keeps her/his left hand under the chin to reduce unwanted movements and exerts slight upward pressure. The first measurer makes sure the head is vertical when viewed from the anterior aspect. The subject inhales deeply and maintains a fully erect position without altering the load on the heels. The movable headboard is brought firmly onto the most superior point on the head by the first measurer. The measurement is recorded to the nearest 0.1 cm. The time of measurement is noted, so that adjustments can be made for possible diurnal changes. In the Fels Longitudinal Study, the mean absolute inter-observer difference is 0.3 cm (SD 0.2 cm).

Stature is measured, instead of length, after 2 or 3 years of age. It is necessary to note whether length or stature was measured because the measures differ systematically with length being larger by 0.4 cm in a recent Dutch national survey, by 0.5 cm in the Euro-Growth Study, and by 0.8 cm in United States national surveys (Fredriks *et al.*, 2000*a*; Haschke *et al.*, 2000; R. J. Kuczmarski, personal communication).

Body mass index

It is useful to adjust an observed weight for length (stature). Therefore, some growth charts include weight-for-length reference data that are almost independent of age during infancy and body mass index (BMI) (weight/stature^2) for older children (Kuczmarski *et al.*, 2000, 2002). There are extremely high correlations ($r = 0.98$–0.99) between weight-for-length and Z scores (calculated by subtracting a mean reference value from an observed value and dividing the difference by the reference standard deviation) for BMI (van't Hof *et al.*, 2000*b*). Values of BMI can be obtained from pairs of recorded weights and statures using a table, a calculator or a nomogram. Tables are available in Himes & Dietz (1994) and at U.S. Department of Health and Human Services (2002). Metric units are used to express

BMI (kg/m^2). If the measurements have been recorded in English units, BMI can be obtained by multiplying the weight in pounds by 703 and dividing the product by the square of stature in inches. To use a nomogram, the child's stature and weight values are located on the left and right scales (Figure 1.2). These locations are joined by a straight line and the BMI value is shown where this line crosses the oblique central scale. Use of a calculator is preferred because tables and nomograms provide BMI values only to the nearest integer, which does not allow adequate differentiation of percentile levels. During childhood, one unit of BMI approximates the difference between the 5th and 10th percentiles. Body mass index has been accepted internationally as an index of overweight and obesity (U.S. Department of Health and Human Services, 2000; World Health Organization, 1998).

Segment lengths

Segment lengths are usually measured as vertical or horizontal distances between bony landmarks. Most of these measurements are made with the subject standing. Arm length is measured on the left side from the tip of the shoulder (acromiale) to the most distal point on the middle finger, excluding the nail. The arm should hang by the side with the palm facing the thigh. A non-stretching tape or a sliding caliper is used. Leg length, which is also called subischial height, is usually obtained as the difference between stature and sitting height. The difference between length and crown–rump length provides a corresponding estimate of leg length in infants. Arm and leg lengths are recorded to the nearest 0.1 cm.

Crown–rump length is a measure of trunk length that is made during infancy. Conceptually, it is similar to sitting height, which is the standard measure of trunk length in older children. Two observers are needed to measure crown–rump length. The infant lies on his/her back upon a table designed for measurement of length with the crown of the head touching the vertical headboard and the body aligned with the long axis of the table. One measurer

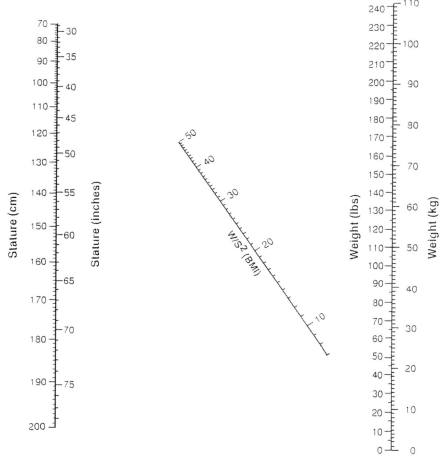

Figure 1.2. A nomogram that provides body mass index from weight and stature. See text for method of use. (Reprinted from *Physical Status: The Use and Interpretation* *of Anthropometry*, World Health Organization, 1995, with permission.)

holds the infant's head with the inferior border of the left orbital margin and the superior border of the left external auditory meatus in the same vertical plane, while ensuring that the long axis of the infant's body coincides with that of the table. The shoulders and hips should be flat against the table. The second measurer raises the legs so that the thighs are vertical to the table surface and moves the sliding footboard firmly against the buttocks with the other hand. The measurement is recorded to the nearest 0.1 cm. The mean absolute inter-observer difference in the Fels Longitudinal Study is 0.5 cm (SD 0.3 cm).

Sitting height is the standard measure of trunk length after infancy. Two measurers are required to obtain sitting height. One procedure is to use a table to which an anthropometer is attached so that the calibrated rod of the anthropometer is vertical and the attached moveable rod is horizontal. Sitting height can also be measured by using a $30 \times 40 \times 50$ cm box, in combination with a vertical anthropometer or a wall-mounted stadiometer. The subject sits on the table or box with the legs hanging unsupported over the edge and the hands resting on the thighs. The midline of the trunk and head must be opposite the vertical rod of the anthropometer or the midline of the stadiometer. The knees are directed straight ahead while the subject sits fully erect, with the head in the same position as for the measurement of stature. An erect position can be ensured by gentle pressure by a measurer simultaneously over the lumbar area with the right hand and over the superior part of the sternum with the left hand. The other measurer should exert gentle upward traction below the mastoid processes and bring the horizontal bar of the anthropometer or stadiometer into contact with the most superior midline point of the head. The subject takes a deep breath and holds it until the measurement is made. The measurement is recorded to the nearest 0.1 cm. The mean absolute difference for inter-observer data is 0.3 cm (SD 0.2 cm) in the Fels Longitudinal Study.

The measurement of armspan requires a non-stretching tape at least 2 m long, a flat surface (usually a wall), and a block with tape spool that is attached to the wall in a way that allows it to be moved vertically. The tape spool is the contact point for the tip of one middle finger when the subject is positioned. The subject stands with feet together and back against the wall. The spool is moved vertically to the shoulder level of the subject and the tape is pulled horizontally along the wall behind the subject. The arms are outstretched laterally at the level of the shoulders, in contact with the wall and the tape, with the palms facing forwards. The tip of the middle finger (excluding the fingernail) is kept in contact with the spool, while the zero end of the tape is at the tip of the other middle finger. One measurer holds the zero end of the tape and the other makes the reading at the spool when the stretch is maximal. The measurement is recorded to the nearest 0.1 cm.

Circumferences

During early infancy, head circumference is measured with the infant seated on the lap of the mother or caretaker. At older ages, head circumference is measured with the subject standing. Added objects, such as combs, are removed from the hair. A non-stretching tape about 0.6 cm wide is used. The measurer faces the left side of the infant or child, passes the tape around the head, and transfers the ends of the tape from one hand to the other so that, on the left aspect of the head, the zero mark on the tape is inferior to the value being recorded. The tape is positioned to exclude any large amounts of cranial hair, such as braids, and placed in a horizontal plane just superior to the eyebrows anteriorly and placed posteriorly so that the maximum circumference is measured (Figure 1.1C). The tape is pulled tightly to compress hair and obtain a measure that approximates cranial circumference. The measurement is recorded to the nearest 0.1 cm. In preterm infants soon after birth, the mean absolute differences are 0.2 cm for intra-observer data and 0.3 cm for inter-observer data (Ifft et al., 1989). Buschan & Paneth (1991) reported a SD of 0.7 cm for inter-observer differences during the neonatal period. The mean absolute inter-observer difference for the measurement of head circumference from birth to 4 years is 0.2 cm (SD 0.1 cm) in the Fels Longitudinal Study.

Arm circumference is measured on the left side of the body. The subject wears loose clothing without sleeves to allow exposure of the shoulder area and stands erect. The measurer stands behind the subject whose left elbow is flexed to 90° with the palm facing superiorly. The lateral tip of the acromial process of the scapula is located by palpating its superior margin and the position of the tip is marked. The most proximal point on the olecranon of the ulna is located and marked. A non-stretching tape is placed over these points and the midpoint between them is marked. With the arm hanging freely, the elbow extended, and the palm facing the thigh, a tape is placed around the arm at the marked midpoint so that it is in contact with the skin, but does not compress the soft tissues. The tape is positioned perpendicular to the long axis of the arm and the circumference is recorded to the nearest 0.1 cm. In the Fels Longitudinal Study, the mean absolute inter-observer difference is 0.2 cm (SD 0.1 cm).

For the measurement of waist circumference, the subject should wear only light underwear or a light gown. The subject stands with the abdomen relaxed, the arms at the sides and the feet together. While facing the subject, one measurer places a non-stretching tape horizontally around the subject at the narrowest part of the abdomen, as seen from the anterior aspect. The other measurer helps position the tape in a horizontal plane without compressing the skin. It may be difficult to identify a waist in overweight subjects. In such cases, the maximum or lower abdominal circumference should be measured and the level of measurement noted. The measurement should be made to the nearest 0.1 cm at the end of a normal expiration. The inter-observer technical error, calculated as the square root of the sum of the squares of differences divided by the number of observations, is 1.6 cm (Malina *et al.*, 1973).

Abdominal circumference is measured with the subject wearing light underwear or a light gown. One measurer faces the subject who stands with the arms by the sides and the feet together. Either maximum or lower abdominal circumference can be measured. For effective evaluation of recorded data, abdominal circumference must be measured at the same level as the reference data used for comparison. To measure maximum abdominal circumference, a non-stretching tape is placed around the subject in a horizontal plane at the level of the greatest anterior extension of the abdomen, which is usually at the level of the umbilicus. Another measurer ensures the tape is horizontal and that it is held snug against the skin without compressing the tissues. The measurement is made at the end of a normal expiration to the nearest 0.1 cm. To measure lower abdominal circumference, the tape is positioned at the level of the anterior superior iliac spines. Except for the difference in level, the procedure matches that for maximum abdominal circumference. Lower abdominal circumference may be more reproducible than maximum abdominal circumference, but its relationships to risk factors for chronic diseases and to body fat may be weaker. The measurement is recorded to the nearest 0.1 cm. The mean absolute inter-observer difference in the Fels Longitudinal Study for lower abdominal circumference is 0.6 cm (SD 0.4 cm).

The subject wears light underwear or a light gown for the measurement of hip circumference and stands with arms at the sides and feet together while one measurer crouches at the side of the subject to identify the level of maximum extension of the buttocks. A non-stretching tape is placed horizontally around the buttocks at this level in contact with the skin, but not compressing subcutaneous tissue. The zero end of the tape is placed under the part of the tape with the measurement value to be recorded. Another measurer helps position the tape correctly on the opposite side of the subject's body. The measurement is recorded to the nearest 0.1 cm. In the Fels Longitudinal Study, the mean absolute inter-observer difference is 0.5 cm (SD 0.4 cm).

Skinfolds

Skinfold thicknesses are measures of double folds of skin and subcutaneous adipose tissue at specific sites most of which are on the left side of the body. Accurate site location is important because small differences in location can significantly alter the measurements. The ease with which the adipose layer can be separated from underlying tissues varies by site and among individuals. Generally, it is more difficult to make reproducible measurements of thick skinfolds. Most skinfold measurement sites need not be marked on the subject, but the triceps skinfold site must be marked if this measurement is to be combined with arm circumference at the same level to estimate cross-sectional areas of arm tissues.

A double fold of skin and subcutaneous adipose tissue is elevated between the thumb and the index finger of the left hand (for a right-handed measurer) about 1 cm from the measurement site. This separation between the elevation site and the measurement site prevents finger pressure from influencing the measured values. To elevate a skinfold, the thumb and index finger are placed on the skin about 6 cm apart, on a line perpendicular to the long axis of the future skinfold. A fold is grasped firmly between the thumb and index finger as they are drawn toward each other. Sufficient tissue must be elevated to form a fold with approximately parallel sides. The amount to be elevated depends on the thickness of the subcutaneous adipose tissue at the site. The thicker the tissue layer, the more the thumb and index finger should be separated when the measurer begins to elevate a skinfold. The fold is raised perpendicular to the surface of the body with the long axis aligned as described for each skinfold and it is kept elevated until the measurement is completed. It may be impossible to elevate a skinfold with parallel sides in the obese, particularly over the abdomen. In these circumstances, the measurement is omitted, unless a two-handed technique produces a satisfactory skinfold (Damon, 1965). In the two-handed technique, one measurer lifts the skinfold using two hands and a second measurer determines its thickness.

Calipers that exert standardized pressure per unit jaw surface are used to measure the thickness of the fold. With young children, it is helpful to demonstrate the caliper on the hand of the measurer and on the hand of the child, simulating the measurement of total palm thickness, before beginning to measure skinfold thicknesses. The skinfold caliper is held in the right hand and pressure is exerted to separate the caliper jaws, which are then slipped over the skinfold. With some plastic calipers, the jaws are apart when the calipers are not in use. To use these calipers, the open jaws are slipped over the skinfold and pressure is exerted to the extent described by the manufacturer. The measurement is made about midway between the surface of the body near the site and the crest of the skinfold. The thickness of the skinfold is measured perpendicular to its long axis about 4 seconds after the pressure on the caliper handles has been fully released. The measurer must be positioned relative to the caliper dial in such a way that parallax errors are avoided. The measurement is repeated three times and the mean recorded. The mean absolute inter-observer differences for triceps, subscapular and suprailiac skinfold thicknesses from the Fels Longitudinal Study are 1.1 mm (SD 1.2 mm).

The triceps skinfold is measured in the midline of the posterior aspect of the arm midway between the lateral projection of the acromial process of the scapula (acromiale) and the inferior margin of the olecranon of the ulna. The method of locating this midpoint is explained in relation to the measurement of arm circumference. During infancy, the measurements of this and other skinfold thicknesses are made with the subject sitting on the lap of the caregiver. At older ages, the subject stands with the arms hanging loosely at the sides. The measurer stands behind the subject holding the caliper in the right hand and places his or her left palm on the subject's left arm proximal to the marked level, with the thumb and index finger directed inferiorly. The triceps skinfold is elevated between the thumb and index finger about 1 cm proximal to the marked site, and the caliper jaws are applied to the skinfold. The measurement is recorded to the nearest 0.1 cm.

The subscapular skinfold is raised on a line, inclined inferolaterally, at approximately 45° to the horizontal plane. The subject stands erect, with the arms relaxed at the sides of the body. To locate the site, the measurer palpates along the vertebral border of the scapula until the inferior angle is identified. The measurement site is slightly lower than the inferior angle of the scapula. For some subjects, especially the obese, gentle placement of the subject's arm behind the back aids the identification of the site. The caliper jaws are applied 1.0 cm inferolateral to the thumb and index finger that are raising the oblique fold, and its thickness is recorded to the nearest 0.1 cm.

The suprailiac skinfold measurement site is in the midaxillary line immediately superior to the iliac crest. The midaxillary line runs vertically through a point midway between the anterior and posterior axillary folds which can be identified easily when the arm is raised. The subject stands with feet together and the arms hanging by the sides or, if necessary, the left arm can be abducted slightly to improve access to the site. A skinfold is grasped just posterior to the midaxillary line and is aligned inferomedially at 45° to the horizontal. The caliper jaws are applied just superior to the iliac crest about 1 cm from the fingers holding the skinfold. The thickness is recorded to the nearest 0.1 cm.

Total body composition

Most of the data for total body composition in infants and children relate to total body fat (TBF), body fat as a percentage of body weight (%BF), fat-free mass (FFM), and total body bone mineral (TBBM). Most procedures to measure body composition, including hydrodensitometry, the measurement of total body potassium and total body water, computer tomography, and magnetic resonance spectroscopy are not used commonly in clinics, but the use of dual energy X-ray absorptiometry is becoming more general (Roche *et al.*, 1996). Research studies that have used these procedures were commonly of non-representative samples. Consequently, they do not provide reference data, but they show age changes for groups and patterns of change within individuals.

THE ASSESSMENT OF MATURITY

The maturity of a child can be assessed at a single examination using sexual, skeletal, or dental criteria. Age at the peak rate of growth in stature during pubescence (age at peak height velocity; PHV) is an index of maturity that is correlated with the timing of sexual and skeletal maturation. Furthermore, pubertal spurts in stature and weight are directly related to many of the differences in growth between rapidly and slowly maturing children. Accurate assessments of maturity are important because misclassification of maturity levels can cause large errors in the interpretation of growth data.

Sexual maturity

Sexual maturity can be assessed from the development of pubic hair in each sex, the genitalia in males and the breasts in females. Usually, a physician makes these assessments with direct inspection. The child is standing, except for the assessment of pubic hair in females. Sexual maturity can be assessed from nude photographs, but this involves even more invasion of privacy and the possibility of misuse. Photographs allow the review of doubtful data, but

it is difficult to recognize from photographs the first stage of pubic hair, the final stage of breast maturation, and thinning of the scrotum. The measurement errors in the grading of sexual maturity in research studies are about 0.4 stages for pubic hair and 0.6 stages for breast and for male genitalia (Voors *et al.*, 1981). Self-assessments of sexual maturity avoid invasion of privacy and the need for a physician. Self-assessed data can be reliable and accurate if made in private using clearly written descriptions, together with line drawings and photographs, and access to a full-length mirror (Brooks-Gunn *et al.*, 1987; Duke *et al.*, 1980; Matsudo & Matsudo, 1994; Neinstein, 1982; Williams *et al.*, 1989).

Sexual maturity assessments are made using the criteria for stages developed by Reynolds and Wines (1948, 1951) and shown in photographs (Herman-Giddens & Bourdony, 1995; Roede & van Wieringen, 1985; van Wieringen *et al.*, 1971). Those of van Wieringen are shown in Figures 1.3–1.6.

The criteria for the assignment of sexual maturity stages are:

PUBIC HAIR (MALES AND FEMALES; FIGURES 1.3 AND 1.4)

Stage 1 The fine non-pigmented hair (vellus) over the pubis is similar to that over the abdominal wall. Pigmented pubic hair is absent.

Stage 2 Sparse growth of long, slightly pigmented downy pubic hair, that is straight or only slightly curled, appearing chiefly at the base of the penis or along the labia.

Stage 3 The pubic hair resembles the adult type, but the area covered is considerably smaller being limited to the pubis or mons veneris. It does not extend to the medial surfaces of the thighs.

Stage 4 The pubic hair resembles the adult type, but the area covered is considerably smaller than the adult and is limited to the pubis or mons veneris. It does not extend to the medial surfaces of the thighs.

Stage 5 The pubic hair is adult in type and is distributed in a classic "male" or "female" pattern with spread to the medial surfaces of the thighs.

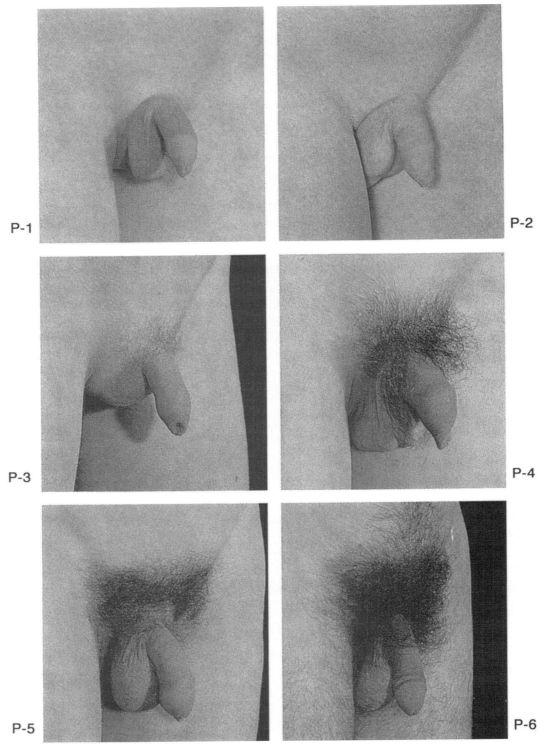

Figure 1.3. Pubic hair maturity stages for males. (From Roede & van Wieringen, 1985, with permission.)

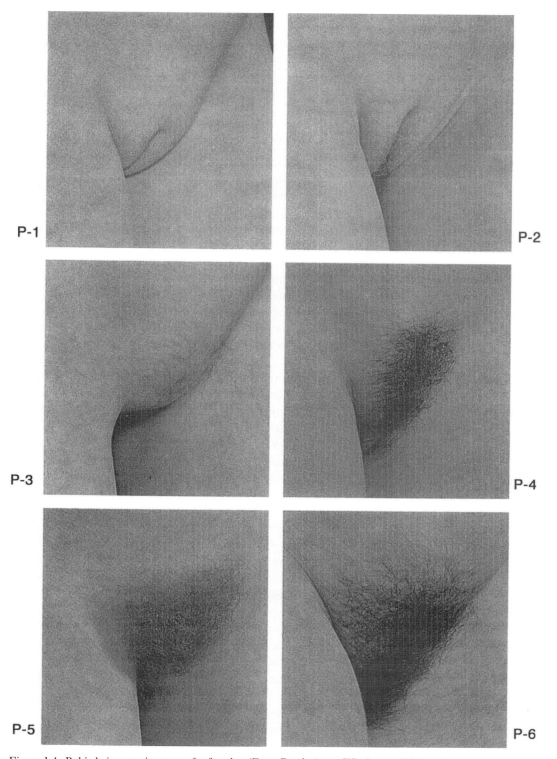

Figure 1.4. Pubic hair maturity stages for females. (From Roede & van Wieringen, 1985, with permission.)

Some have added a stage 6 for pubic hair in which the pubic hair extends along the linea alba and further laterally than in stage 5 (van Wieringen *et al.*, 1971). This stage occurs in about 80% of males and a minority of females and may not be present until about 25 years (Thomas & Ferriman, 1957). This stage is not a valid index of sexual maturity because it does not become universal.

GENITALIA (MALES; FIGURE 1.5)

Stage 1 The testes, scrotum, and penis are about the same size and shape as in early childhood.

Stage 2 There is enlargement of the testes and scrotum. The skin of the scrotum is reddened and the texture is thinner and more wrinkled. There is little or no enlargement of the penis.

Stage 3 The penis is enlarged, at first mainly in length. There is further enlargement of the testes and scrotum.

Stage 4 There is further enlargement of the penis, with increase in its breadth and development of the glans penis. There is also further enlargement of the testes and scrotum and increased darkening of the scrotal skin.

Stage 5 The genitalia are completely mature in size and shape.

When the maturity stages differ between the right and left testes, the less advanced stage is recorded. If one testis is undescended, the stage of the descended testis is recorded.

BREASTS (FEMALES; FIGURE 1.6)

Stage 1 There is elevation of the nipple only.

Stage 2 There is elevation of the nipple and breast as a small mound. A fairly hard area of breast tissue can be palpated. The areola is enlarged, elevated, and pigmented.

Stage 3 There is further enlargement and elevation of the breast and areola without separation of their contours.

Stage 4 The areola and nipple project to form a secondary mound above the level of the breast that is associated with increased deposition of adipose tissue.

Stage 5 Maturity is complete with projection of the nipple only due to recession of the areola to the general contour of the breast.

If the maturity stages differ between the right and left breasts, the less advanced stage is recorded. Stage 4 of breast maturation may be absent from serial data recorded at 6-month intervals. This was unusual in the English group studied by Marshall & Tanner (1969), but it occurred in 33% of the Swiss females studied by Largo & Prader (1983*b*). This stage may be universal in data collected at shorter intervals. Some females do not reach stage 5 of breast maturation when assessments are based on photographs, but recognition of this stage from photographs may be unreliable (Marshall & Tanner, 1969; Nicholson & Hanley, 1953).

The numbers assigned to stages of sexual maturation can be averaged to obtain a sexual maturity index (SMI). With this approach, a female with stage 2 of pubic hair and stage 3 of breasts would be assigned 2.5 as a SMI score. Large variations can occur within individuals between the stages of sexual maturity for different organs (Harlan *et al.*, 1979, 1980; Malina, 1995). Consequently, the use of SMI values leads to the loss of potentially important information available from the stages for separate organs. For example, an adrenal tumor can lead to advanced pubic hair development in the absence of testicular enlargement. Such disharmonic maturation, which is masked by SMI values, reflects unusual relationships among the hormones that control different aspects of sexual maturation (Rosenfeld, 1982). Despite such variations, there are high levels of concordance between maturity stages within groups (Billewicz *et al.*, 1981*a*; Harlan *et al.*, 1979, 1980; Largo & Prader, 1983*a*, *b*). The data of Harlan *et al.* (1979, 1980) from a United States national survey show about 60% have matching stages for pairs of organs, but about 1% of males differ by two stages. Among females aged 12 and 13 years, about 10% have differences of two stages between pubic hair and breasts, but these percentages are much

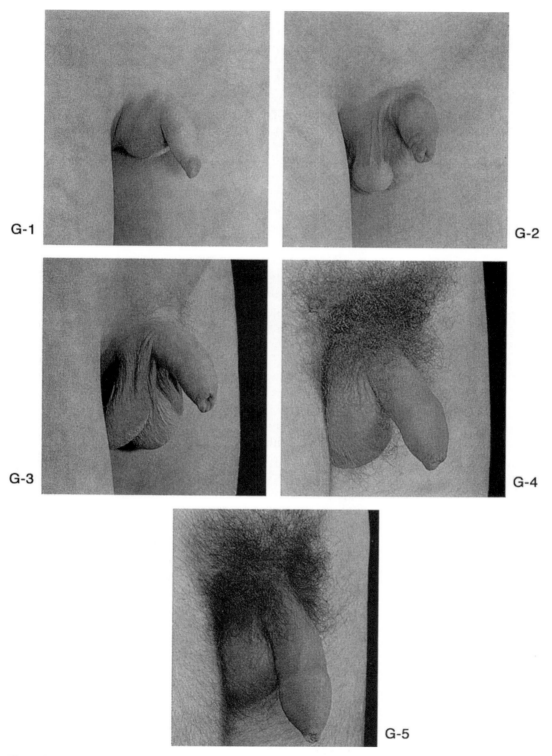

Figure 1.5. Genital maturity stages for males. (From Roede & van Wieringen, 1985, with permission.)

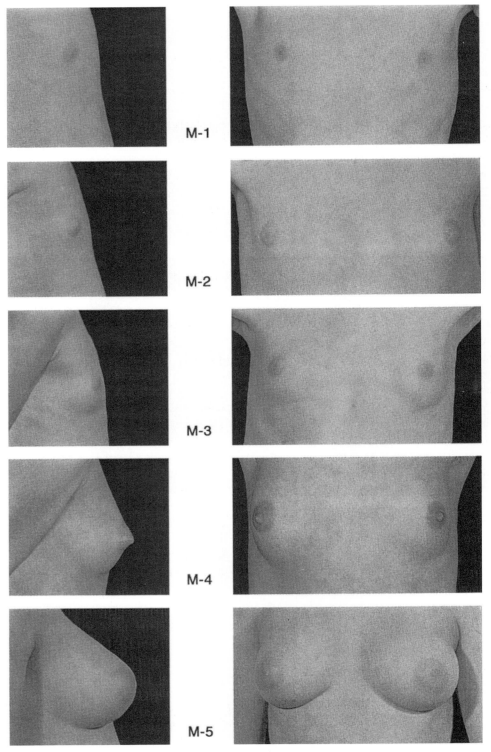

M-1

M-2

M-3

M-4

M-5

Figure 1.6. Breast maturity stages for females. (From Roede & van Wieringen, 1985, with permission.)

smaller at older ages. The extent of concordance between stages is similar for European Americans and African Americans (Harlan *et al.*, 1979, 1980).

Despite the loss of information in some children, it is necessary to average the sexual maturity stages as SMI values, or use the maturity of one organ in each sex, to calculate factors that adjust recorded statures and weights for the level of sexual maturity when comparing groups of children of the same age and when assessing the growth of an individual child using growth charts (Chaning-Pearce & Solomon, 1986; Kulin *et al.*, 1982; Wilson *et al.*, 1987). Without the use of SMI values, sample size constraints would make it difficult to estimate adjustment factors for the many possible combinations of stages for the separate organs and the process of selecting from many adjustment factors in a clinic would be too complex for general use.

Further maturational information can be obtained in males from testicular size. This is graded using an orchidometer, which is a string of 12 elliptical beads ranging in size from 1 to 25 cc. The child stands in front of the seated physician who palpates each testis with the right hand while selecting with the left hand the bead that matches best the size of each testis. The sizes of these beads are recorded. Testicular size is commonly graded by those working with children who are dysmorphic or have endocrine disorders (Daniel *et al.*, 1982; Largo & Prader, 1983*a*; Roede & van Wieringen, 1985; Sempé *et al.*, 1979).

Age at menarche

Age at menarche is an important indicator of the rate of sexual maturation, but is less useful than the grading of secondary sexual characteristics, because it occurs late in pubescence. Age at menarche is recorded from self-reports as the date of the first menstrual period without considering the irregularity of menstruation or the temporary amenorrhea that can follow the first period. Such data are based on recall that may involve errors, but there are high correlations between ages reported close to the time of menarche and those

reported after an interval of several or more years (Bergsten-Brucefors, 1976; Damon *et al.*, 1969; Livson & McNeill, 1962). A major advantage of this approach is that an age at menarche is obtained for each individual. An alternative approach is to obtain group estimates for age at menarche by the use of probit analyses. These analyses estimate the age at which 50% or other percentages of the group reached menarche using information as to whether or not menarche has occurred for each individual.

Skeletal maturity

Skeletal maturity is assessed from radiographs and is recorded as a continuum of skeletal ages (years) that may range from birth to 18 years. Training is needed to ensure high reliability and accuracy. Detailed descriptions of methods to assess skeletal maturity are available elsewhere (Greulich & Pyle, 1959; Roche *et al.*, 1988; Tanner *et al.*, 1983*b*). In these assessments, grades of skeletal maturity indicators are combined to estimate skeletal ages that can be compared with reference data from United States children or English children. Skeletal maturation is more rapid in females than in males with sex differences of about 6 months at 2 years of age, 1 year at 4 years of age and 2 years during pubescence (Roche, 1968). Because of the way in which the skeletal age scales are constructed, a typical 4-year-old child, whether male or female, has a skeletal age of 4 years.

The Fels method of Roche and colleagues, the Greulich–Pyle, and Tanner–Whitehouse methods of assessing skeletal maturity differ in many ways including the standardizing populations, the mathematical bases, and the bones assessed. Consequently, the skeletal ages obtained with these methods are not interchangeable. The reference levels for the Fels method are similar to those of Italian children, but they are more advanced than those of Tanner–Whitehouse by about 1 year at ages 4 to 15 years (Aicardi *et al.*, 2000; Roche *et al.*, 1988). Consequently, Fels skeletal ages are about 1 year less than Tanner–Whitehouse skeletal ages during this period. The Tanner–Whitehouse skeletal maturity levels are less advanced than those of many

European groups after 10 years of age (Beunen *et al.*, 1990; Helm, 1979; Prahl-Andersen & Roede, 1979; Taranger *et al.*, 1976*a*; van Venrooij-Ysselmuiden & van Ipenburg, 1978; Vignolo *et al.*, 1992; Wenzel & Melsen, 1982;Wenzel *et al.*, 1984). The levels are also less advanced than those of Japanese children (Ashizawa, 1994; Murata, 1993). The Tanner–Whitehouse levels have been revised recently for ages 8 to 16 years (Tanner *et al.*, 1997). This new scale matches better to Italian children, but is not sufficiently advanced after 14 years (Vignolo *et al.*, 1999). These 1997 levels may be more advanced than those in some general populations because they were derived from an upper socioeconomic group of European Americans. The Greulich & Pyle *Atlas* (1959) is now used less commonly than in the past to assess skeletal maturity. The standards shown in this *Atlas* are too advanced for the United States population, particularly at 9 to 13 years (Roche *et al.*, 1976).

Age at peak height velocity

Age at peak height velocity (PHV) is commonly used as an index of the rate of maturation. Children in whom PHV occurs early are considered to be maturing rapidly; contrariwise, those in whom it occurs late are considered to be maturing slowly. Recognition of the age of PHV requires serial data extending from at least 2 years before to 2 years after this event. Consequently, this index is not recorded commonly in clinical circumstances or epidemiological studies. Age at PHV is considered here because it is closely related to the variations in statural growth during pubescence that are associated with variations in rates of maturation.

Age at PHV can be recognized as the annual or 6-month interval during pubescence for which the increment in stature is the largest, or by subjectively drawing a curve by hand through plotted serial data (Faust, 1977; Nicholson & Hanley, 1953; Stolz & Stolz, 1951; Tanner *et al.*, 1966*b*). Better estimates can be obtained by fitting a mathematical model to serial data for stature, but the estimates from fitted models differ depending, in part, on their goodness of fit (Guo *et al.*, 1992; Hauspie *et al.*, 1980). The

Preece–Baines model underestimates the velocity of growth at the peak and places PHV about 6 months earlier than the triple logistic model and kernel regression which fit better (Guo *et al.*, 1992).

In each sex, PHV occurs at about stage 3 of sexual maturation, but this relationship is not close enough to allow the prediction of age at PHV for individuals from stages of sexual maturity (Billewicz *et al.*, 1981*a*; Marshall & Tanner, 1969, 1970). On average, PHV in females occurs about 1.4 years after pubic hair stage 2, 1.0 years after breast stage 2, and 1.1 years before menarche (Bielicki, 1975; Lindgren, 1978; Taranger *et al.*, 1976*b*). In males, PHV occurs 0.8 years after pubic hair stage 2 and 1.6 years after genitalia stage 2 (Bielicki *et al.*, 1984). The timing of stages of sexual maturation and of PHV are correlated ($r = 0.3$–0.8 for stage 2 of either pubic hair, genitalia, or breasts, and $r = 0.8$ between ages at PHV and stage 5 of these organs) (Bielicki, 1975; Ellison, 1982; Largo & Prader, 1983*a*, *b*; Taranger *et al.*, 1976*b*). Ages at menarche and at PHV are correlated at about $r = 0.8$ (Deming, 1957; Marshall & Tanner, 1969; Nicholson & Hanley, 1953; Shuttleworth, 1939).

Some have constructed growth charts from serial data in attempts to make them reflect the growth of children in whom PHV occurs at an age that is near average (within ± 2 Z scores of the mean age), or early, or late (Chrzastek-Spruch *et al.*, 1989; Hauspie & Wachholder, 1986). Due to small sample sizes, only the median percentiles for growth rates were reported for early and late groups. A major problem with such charts is the lack of a method to classify children prospectively, preferably from cross-sectional data, in regard to the timing of PHV (Sullivan, 1983).

Age at peak weight velocity

A peak rate of increase in weight (peak weight velocity, PWV) occurs during pubescence that can be identified in the same way as PHV. The timing of PWV and of PHV are closely related, but PWV tends to occur earlier. The mean differences in timing between PWV and PHV are 0.8, 0.5, and

Table 1.2. *Sources of sex-specific reference data for size at birth in singletons born after 37 weeks gestation*

Authors	Country	Number of infants	Age range (weeks)	Variables[a]
Bjerkedal & Skjærven (1980)	Norway	188 226	38–46	W, L
Dombrowski *et al.* (1992)	United States	26 596	38–43	W, L, HC
Lawrence *et al.* (1989)	Sweden	175 453	38–45	W, L, W for L
Overpeck *et al.* (1999)	United States	3 165 339	38–42	W
Roberts & Lancaster (1999)	Australia	343 588	38–44	W
Wilcox *et al.* (1993)	United Kingdom	17 601	38–42	W
Zhang & Bowes (1995)	United States	463 595	38–42	W

[a] HC, head circumference; L, length; W, weight.

0.2 years in males and 0.6, 0.3, and 0.1 years in females for those with early, average or late ages at PHV, respectively (Lindgren, 1978). Although age at PWV has rarely been used as an index of maturity, knowledge of its occurrence assists the interpretation of weight data during pubescence.

GROWTH REFERENCE DATA FOR GENERAL POPULATIONS

Size at birth

Sources of sex-specific data for size at birth of singletons born after 37 weeks gestation are given in Table 1.2. All these reports provide data for weight, but data for other measures are provided only in those of Bjerkedal & Skjærven (1980), Dombrowski *et al.* (1992), and Lawrence *et al.* (1989). Cross-sectional data show decreases in medians for weight and length after 41–42 weeks of gestation (Bjerkedal & Skjærven, 1980; Dombrowski *et al.*, 1992; Roberts & Lancaster, 1999).

Growth charts

The usual growth charts present reference values for growth status at a series of ages. Some of these charts were developed from nationally representative samples to obtain estimates for total populations. Such charts are more general than others in their application and they are derived from large samples, which are needed to provide stable estimates for the outlying percentiles that are important in the assess-

ment of individuals and the surveillance of populations (Guo *et al.*, 2000*b*). Other charts were developed from serial studies or regional surveys. Serial studies are the only source of information about increments and growth patterns, but they usually include relatively small, non-representative samples. Regional surveys can provide useful growth data for local use and for ethnic and local comparisons.

Growth charts consist of a set of smoothed curves for selected percentiles with accompanying tables of means and standard deviations. The curves are usually for the 3rd, 5th, 10th, 25th, 50th, 75th, 90th, 95th, and 97th percentiles. A percentile is a value for which p% of the total sample has a smaller value. For a normal distribution with mean μ and standard deviation σ, the 5th and 95th percentiles are $\mu - 1.645\sigma$ and $\mu + 1.645\sigma$. For a given set of data, percentiles are obtained by replacing μ and σ by x and s. Some growth charts present the mean and selected SD levels (Z scores), which is appropriate when the data are normally distributed. The standard method of developing a growth chart is to calculate the percentiles for each specified age range. The measurement values for each age range are sorted in an ascending order and the percentile estimates are obtained by counting up from the smallest value within each age group. These estimates are plotted against the median age of each group.

The range of ages within each age group affects the precision of the percentiles. If the ranges are very short, the sample size within each age group will be small causing the percentile estimates for each group to be imprecise with a loss of information. For example, when there are 100 children in an age group, the 3rd percentile value depends upon only the smallest three to four observations. These age ranges should be determined also by the age-related patterns of change in the growth variables. Percentile estimates for weight and BMI are less precise during the pubescent period than at other times, but there is little loss of precision for stature during pubescence.

In the revision of the United States growth charts by the National Center for Health Statistics and the Centers for Disease Control and Prevention, infants were grouped to single months of age to 12 months, to three-month groups from 12 through 24 months and to six-month groups at older ages (Kuczmarski *et al.*, 2000, 2002). The estimates obtained were plotted opposite the midpoints of the age ranges. For example, the age group used to estimate percentile values at 2.25 years extended from 2.0 to 2.49 years of age. This pattern was continued to the 19.75-year age group for which the age range was 19.5 to 20 years. In the preparation of the weight-for-length charts, 2–cm length intervals were used. Comparisons among the estimated percentiles in different growth charts must take variations in age grouping into account. In the charts from the Czech Republic and Japan, at least after 6 years, the data relate to the last completed unit of age. Consequently, for example, the estimates for 8 years are derived from data for children aged 8.00 to 8.99 years with a mean age of about 8.5 years. In the charts from Hungary, The Netherlands, Norway and the United States, there are small or zero differences between the ages shown on the charts and the mean ages of the children from whom the estimates were derived.

Weighted univariate statistics

To provide nationally representative reference data, it is necessary to adopt special sampling strategies and to apply statistical sampling weights to individual data. This is possible when there is a complex sampling design, as in the national United States surveys. During the analyses of data from these surveys, sampling weights are applied to the data for individuals to account for individual selection probabilities and adjust for non-response, non-coverage, and post-stratification effects. After incorporating the individual sampling weights, the means, standard deviations and selected empirical percentiles are calculated within each sex and age group. These calculations are made using the following equation:

$$w_i = m_i \frac{n}{\sum_{i=1}^{n} m_i}$$

where w_i is the sampling weight for subject i within a defined age- and sex-specific group, m_i is the sampling weight for subject i in a specific survey, and n is the total number of subjects in each age- and sex-specific group. The values for the weighted means and SDs within age- and sex-specific groups were calculated as follows:

$$\bar{X}_W = \frac{\sum_{i=1}^{n} W_i X_i}{\sum_{i=1}^{n} W_i}$$

$$SD_W = \sqrt{\frac{\sum_{i=1}^{n} W_i (X_i - \bar{X}_W)^2}{\sum_{i=1}^{n} W_i - 1}}$$

where $\bar{X}_W$ is the weighted mean, W_i is the weight for observation x, SD_W is the weighted standard deviation, and n is the total number of observations. Weighted empirical percentile estimates were obtained by representing each subject i "W_i" times and calculating the percentile estimates, where "W_i" is the sampling weight for that subject. The empirical percentiles were plotted at the midpoint of each age group, and these plots form the empirical percentile curves.

Smoothing

Plots of empirical percentile estimates usually show irregular patterns of change across ages, especially for outlying percentiles. Smoothing procedures are used to fit curves to the estimates to remove or reduce the irregularity. These procedures utilize information from measures at nearby ages to estimate a value at an age. Some smoothing procedures describe the relationship between the dependent variable and an independent variable, such as age, by mathematical expressions. These expressions can be used to study the relationship between the dependent and independent variables for group comparisons, and computer-based implementation of the relationships. Prior knowledge about the values of growth variables at various ages and the patterns of change with age greatly assist the choice of smoothing procedures and help avoid under-fitting and over-fitting. In under-fitting, the final curve is smooth, but it does not accurately represent the biological changes with age. Over-fitting occurs when the curve goes through all or nearly all the empirical points and does little to reduce their irregularity. Over-fitted curves describe the aggregate of real changes and of artifacts that lack biological meaning.

Applications of growth charts

Growth measurements should be plotted on age- and sex-specific growth charts and recorded in tables. Comparisons between the plotted data and the reference percentiles will show the approximate levels for the child or children, relative to other children of matching age and sex who belong to the population that provided data for construction of the charts. The accepted normal ranges for growth measurements are determined commonly by the outlying percentiles included in the growth charts. These limits are the 3rd and 97th percentiles in many countries, but the 5th and 95th percentiles are used commonly in the United States and the limits are $\pm 2\,Z$ in some countries. Values outside these limits are more likely to be associated with pathological conditions than values nearer the means or medians. Values that are not within the normal range may be observed in healthy children. Contrariwise, values within the normal range may be found in children with serious diseases. Therefore, a growth assessment, by itself, is not a diagnostic tool.

After the recorded data has been assessed for a child or a group of children, the findings must be interpreted. For such interpretation, any earlier data should be considered so that growth patterns can be evaluated. It may be desirable to schedule another examination to evaluate growth progress. The interval to the next examination should be determined, in part, by the intervals in the reference data for increments that will be used. It is necessary also to consider influences on growth, to estimate future growth, and to consider the likelihood of associated current or future disease.

Healthy children from upper socioeconomic groups differ little in growth among the countries that have been studied (Graitcer & Gentry, 1981; Habicht *et al.*, 1974*a*, *b*). This has led some to conclude that one chart developed from such children could be used worldwide with the assumption that all population groups have the same growth potential, although many do not reach this potential due to adverse environments. Such an international chart would assist epidemiological growth comparisons among countries. If such a chart were applied clinically, the resulting growth assessments could lead to many referrals for future investigation of children particularly for populations in which growth retardation is common. A large number of referrals could be impractical, especially in developing countries where growth retardation is common and services are limited. Clinical judgement could reduce the number of referrals, but such judgements require knowledge of the causes of retarded growth for individuals and groups and the outcomes of various degrees of retardation. The addition of country-specific referral cut-off levels to an international chart could help.

It is reasonable to use an international chart for public health purposes, such as surveys and screening, and to use national or group-specific charts to assess individuals. Ethnic-specific charts will document ethnic differences in growth that may be due to genetic or environmental influences. The use of such charts in clinics would assist the recognition of children whose growth is unusual for their ethnic group. The construction of such charts requires large samples and enrollment into a survey from some ethnic groups may be difficult, especially when the groups are dispersed within countries. When a mean is low for an ethnic group, in comparison with other ethnic groups in the same country, there should not be complacency if the growth of an individual is near the mean for his/her ethnic group. There are growth charts for preterm infants and for infants and children with specific diseases and adjustment factors

Table 1.3. *Characteristics of selected national growth charts*

Authors	Country	Age range (years)[a]	Number of children	Type of reference data	Smoothing procedures[b]	Clinically useful charts
Lhotská et al. (1993, 1995); Vignerová et al. (1997)	Czech Republic	1–18	86 846	percentiles (3rd–97th) from averages, means, SDs	moving averages	yes
Rolland-Cachera et al. (1991)	France	B–21	14 509	percentiles (3rd–97th)	LMS	yes
Eiben et al. (1991)	Hungary	3–18	41 000	empirical percentiles (3rd–97th)	subjective	yes
Hoey et al. (1987)	Ireland	1–18.5	3 284	percentiles (3rd–97th)	subjective	yes
Cortinovis et al. (1993)	Italy	B–3	10 414	percentiles (3rd–97th)	parametric model	yes
Tsuzaki et al. (1987)	Japan	B–18	676 000	percentiles (3rd–97th) from means, SDs	polynomials	yes
Burgemeijer et al. (1998);	The Netherlands	B–20	14 500	means ± 1, 2, 2.5 SDs	LMS	yes
Cole & Roede (1999)		B–20	41 766	percentiles (3rd–97th) from means, SDs	LMS	no
Knudtzon et al. (1989)	Norway	B–4	23 669[c]	percentiles (2.5–97.5th) from means, SDs	none	no
Lindgren et al. (1995)	Sweden	6–16 F 6–19 M	3 647	percentiles (3rd–97th)	LMS	yes
Freeman et al. (1995)	United Kingdom	B–20	20 960	percentiles (3rd–97th)	LMS	yes
Kuczmarski et al. (2000, 2002)	United States	B–20	64 410[d]	empirical percentiles (3rd–97th)	locally weighted regression	yes

[a] B, birth.

[b] LMS, method of Cole & Green (1992).

[c] Each infant examined three times.

[d] Additional data at birth (weight, $n = 83\,587\,123$; length and weight for length, $n = 878\,405$; head circumference, $n = 362$).

are available that make the common growth charts applicable to children with unusual rates of maturation or with parents of unusual stature.

National growth charts

There are growth charts from national surveys in the Czech Republic, Hungary, Japan, The Netherlands, Norway, and the United States and other national charts have been developed by combining data from local or regional surveys (Table 1.3). Most local surveys have been conducted in cities. The study of urban groups has logistic appeal, but the findings may not be applicable to rural populations. Differences in growth between urban and rural children are decreasing in some countries, but they remain large in others (Demoulin, 1998; Floris & Sanna, 1998; Mesa et al., 1996; Sforza et al., 1999).

Van't Hof *et al.* (2000*a*, *b*, *c*) published growth charts using data from eleven European countries with 54 to 131 infants enrolled at each site. These infants were measured at up to 12 ages from birth to 36 months. The numbers of infants measured were 1746 to 12 months, 1205 to 24 months, and 1071 to 36 months. They excluded data from infants of diabetic mothers, multiple births, infants born preterm, and infants with congenital malformations, inherited metabolic diseases, or diseases requiring hospitalization for more than several days. These were unrepresentative samples for which the mothers tended to be older and more educated than local non-participant mothers and more likely to live in urban areas. After the distributions had been normalized, polynomials were fitted to the transformed means and SDs to estimate smoothed percentile curves. They provide clinically useful charts and matching tables of the 3rd to 97th percentiles for weight, length, head circumference, and BMI at 12 ages from 1 to 36 months together with weight-for-length at lengths from 55 to 95 cm.

Growth data were collected from a random national sample of 86 846 Czech children aged 1 to 18 years (Lhotská *et al.*, 1993, 1995; Vignerová *et al.*, 1997). Minority children were included in the survey, but they form only 1% of the population. The empirical means for subjects grouped by last completed unit of age were smoothed by moving averages. These age units were 1 month to 13 months; 3 months to 2 years; 6 months to 6 years; and 1 year at older ages. Clinically useful charts with 3rd–97th percentiles were prepared for length and weight-for-length from birth to 18 months, head circumference from birth to 36 months, and stature and weight-for-stature from 18 months to 18 years (Figures 1.7 and 1.8). The means and SDs are given in tables. Weight data are given in tables, but not in the charts. Because weight-for-stature differs with age after pubescence begins, the charts show separate percentile levels for males older than 15.5 years and females older than 14.5 years. Age-specific percentile levels of weight-for-stature are needed, however, after 12.5 years for males and 11.5 years for females (van Wieringen, 1972).

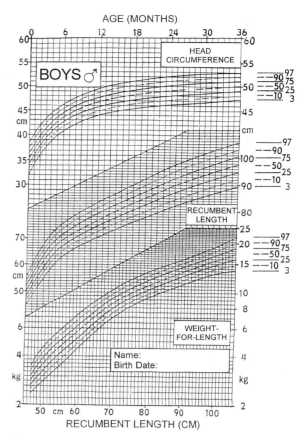

Figure 1.7. Growth charts of head circumference, length, and weight-for-length from birth to 36 months for males from the national data recorded in the Czech Republic. (From Lhotská *et al.*, 1993, with permission.)

A national survey of 14 500 children from birth to 20 years was conducted in The Netherlands in 1997. The sampling and measurement procedures closely matched those of national Dutch surveys in 1955, 1965, and 1980 (Burgemeijer *et al.*, 1998). Low birth weight infants were included in the most recent survey, but excluded from the earlier ones. Their inclusion reduced the means for length by 0.2 cm to 12 months and reduced mean statures by 0.1 cm from infancy to 14 years in males and 18 years in females (Fredriks *et al.*, 2000*a*). In the development of these charts, data from children of non-Dutch parents were excluded unless one parent was Dutch and the other was West European. This limits the applicability of the charts in The Netherlands where 17% of the population are immigrants and

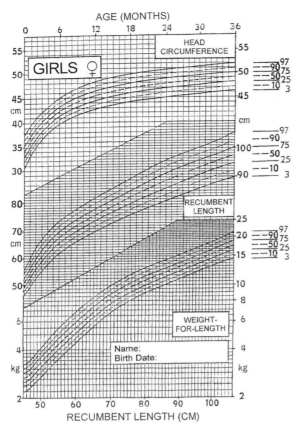

Figure 1.8. Growth charts of head circumference, length, and weight-for-length from birth to 36 months for females from the national data recorded in the Czech Republic. (From Lhotská *et al.*, 1993, with permission.)

about 40% of the children in the large cities have immigrant parents (Fredriks *et al.*, 2000*a*). Clinically useful charts were developed that provide means ± 1, 2, and 2.5 *Z* for weight, length, weight-for-length, and head circumference from birth to 4 years, and for stature and weight-for-stature from 1 to 21 years (Figures 1.9–1.13). The weight-for-stature charts assume the relationship of weight to stature is independent of age, which is not the case after pubescence begins. The charts for children aged 1 to 21 years include percentiles for the timing of secondary sexual characteristics.

Cole & Roede (1999) developed reference data for BMI based on data from 41 766 Dutch children aged from birth to 20 years who were included in a na-

tional survey (Roede & van Wieringen, 1985). Data from children with fewer than two Dutch parents, birth weight <2500 g, or pathological conditions were excluded. The method of Cole & Green (1992) was used to smooth the empirical means and SDs and obtain percentiles (3rd–97th) that are displayed in charts that are not in a clinically applicable format.

Reference data for BMI from 1 month to 16 years were reported by Rolland-Cachera *et al.* (1982) from 494 French children enrolled in a serial study. Later, Rolland-Cachera *et al.* (1991) combined these data with those from three cross-sectional studies (total *n* = 18 363). There were only small differences between these studies for the few ages at which they overlapped. The method of Cole & Green (1992) was applied to the combined data set to develop percentile (3rd–97th) charts in clinically applicable formats.

From 1982 to 1985, the Hungarian National Growth Study recorded data from a random sample of 41 000 children aged 3 to 18 years (Eiben *et al.*, 1991). The children were recruited through 326 nurseries and schools in 114 cities, towns, and villages. All Hungarian counties were represented. Site selection was based on size of place of residence, and location in industrial or agricultural areas. Minority children (5%) were included, but those with chronic diseases or congenital defects were excluded. Empirical percentiles (3rd–97th) were obtained for each year of age from 3 to 18 years, with age groups centered on full years. For example, estimates for 8 years were derived from children aged 7.51 to 8.50 years. These percentile values were smoothed subjectively (Figures 1.14 and 1.15). The variables measured included weight, stature, sitting height, arm length, and triceps, subscapular, and suprailiac skinfolds. Charts of weight-for-stature were published, but these do not take into account the dependence on age after the beginning of pubescence.

Growth charts in clinically useful formats have been developed using cross-sectional data from Irish children aged 5 to 19 years who were living in five locations (Hoey *et al.*, 1987). Means and SDs from the pooled data were used to obtain percentiles (3rd–97th) for weight and stature that were smoothed

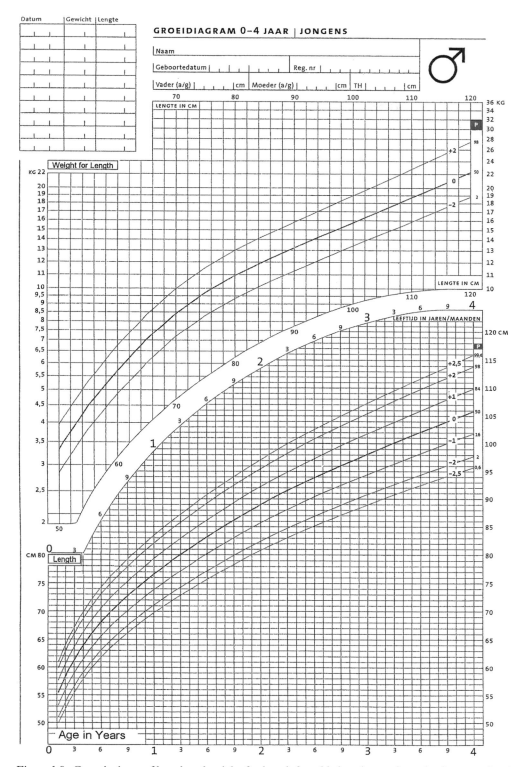

Figure 1.9. Growth charts of length and weight-for-length from birth to 4 years for males from a national survey in The Netherlands. (From Fredriks *et al.*, 2000*a*, with permission.)

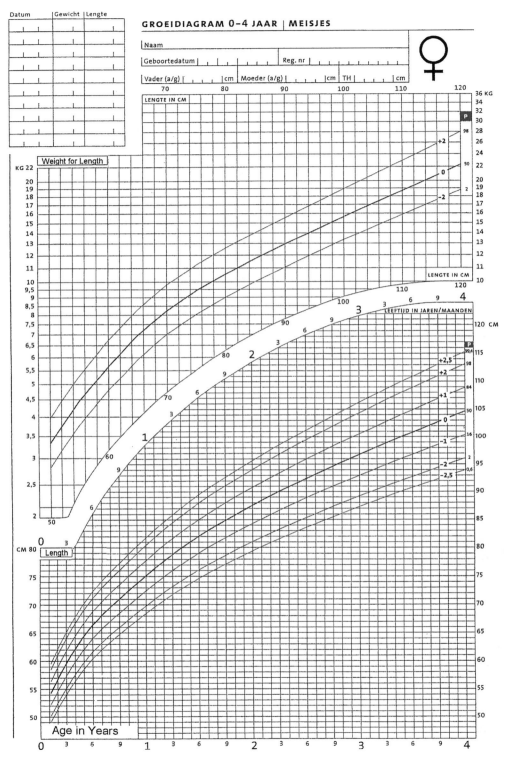

Figure 1.10. Growth charts of length and weight-for-length from birth to 4 years for females from a national survey in The Netherlands. (From Fredriks *et al.*, 2000*a*, with permission.)

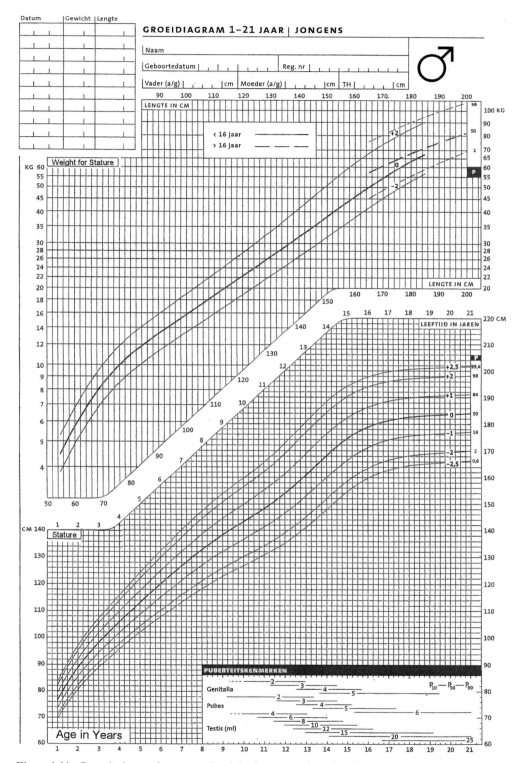

Figure 1.11. Growth charts of stature and weight-for-stature from 1 to 21 years for males from a national survey in The Netherlands. (From Fredriks *et al.*, 2000*a*, with permission.)

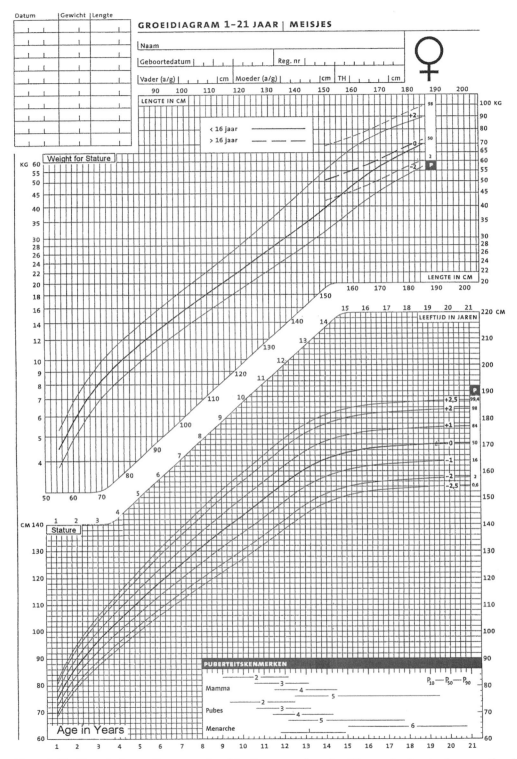

Figure 1.12. Growth charts of stature and weight-for-stature from 1 to 21 years for females from a national survey in The Netherlands. (From Fredriks *et al.*, 2000*a*, with permission.)

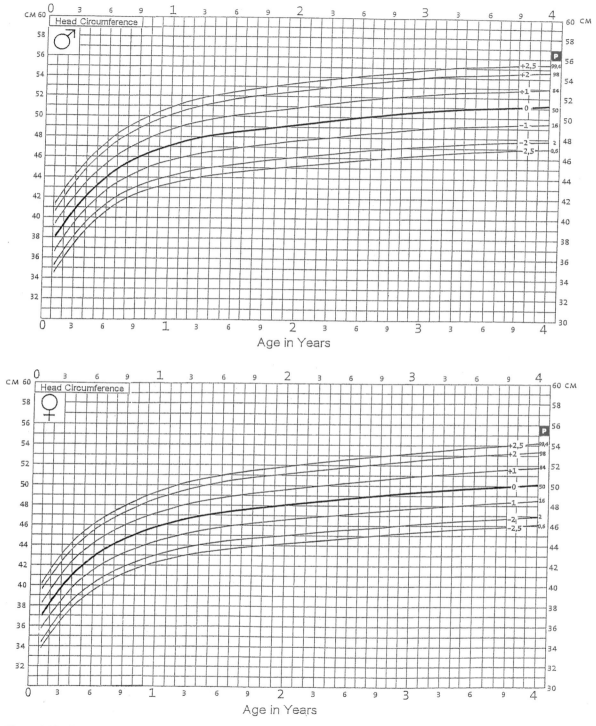

Figure 1.13. Growth chart of head circumference from birth to 4 years for males and females from a national survey in The Netherlands. (From Fredriks *et al.*, 2000*a*, with permission.)

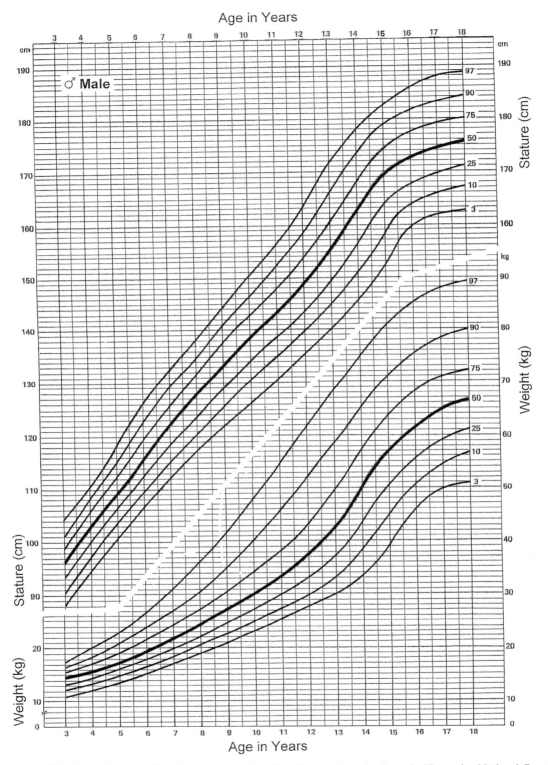

Figure 1.14. Growth charts of weight and stature from 3 to 18 years for males from the Hungarian National Growth Study. (From Eiben *et al.*, 1991, with permission.)

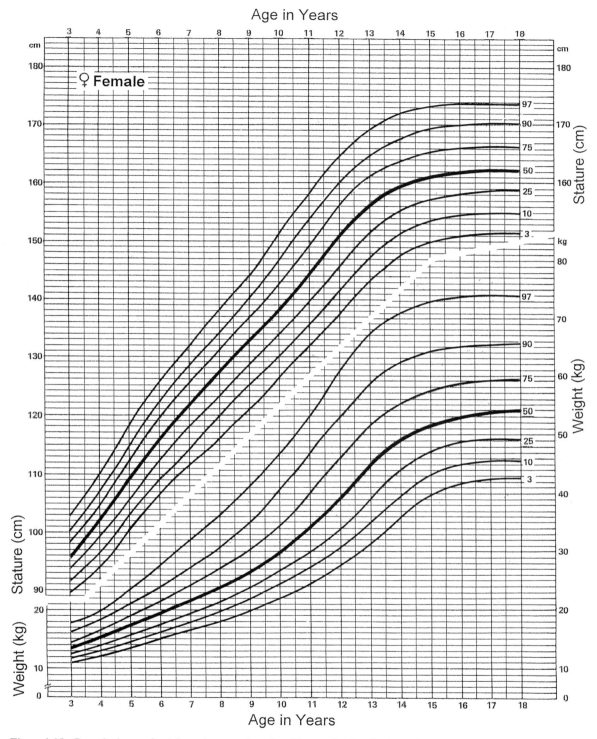

Figure 1.15. Growth charts of weight and stature from 3 to 18 years for females from the Hungarian National Growth Study. (From Eiben *et al.*, 1991, with permission.)

subjectively. Other charts were developed that show velocity medians for those whose timing of PHV was $> +2$ Z or < -2 Z. The differences in growth rates between these maturational groups were based on findings from other countries because the timing of PHV was not known for the children in the study.

Cortinovis et al. (1993) measured 10 414 infants who were born from 1973 to 1981 in Milan, Naples, Parma, Rome, or Trieste. Multiple births and infants with congenital abnormalities were excluded, but preterm infants were included. The infants were measured at nine ages from birth to 36 months. Cortinovis recorded weight to the nearest 100 g and length and head circumference to the nearest 1.0 cm. As a result, the measurements were less precise than usual. These workers used means and SDs to obtain percentiles (3rd–97th) that were smoothed mathematically, before being displayed in tables and in charts that are in a suitable format for clinical use. Because migration was common and differences in growth among the cities were small, Cortinovis concluded the combined data set was suitable for constructing an Italian national reference, although there was no representation of those living in rural areas and in some large parts of the country, such as Sicily.

Kato & Takaishi (1999) analyzed serial data recorded at intervals of two or three months in a multi-site Japanese study of infants aged from birth to 12 months. Data were available for 2620 infants after excluding preterm and multiple births and those not measured on the second or third day after birth. These data were used to construct percentile curves (3rd–97th). Tsuzaki et al. (1987) reported data from a 1980 national survey by the Japanese Ministry of Health and Welfare and the Ministry of Education (Figure 1.16). The sample included 16 000 infants (birth to 4.5 years) and 660 000 children (5 to 18 years). After showing the data were normally distributed, they used means and SDs to calculate percentiles (3rd–97th) that were smoothed within age segments by polynomials and presented in clinically useful formats.

Knudtzon et al. (1989) reported national data for weight, length, and head circumference from 23 669 Norwegian infants. Each infant was measured about three times from birth to 48 months. Infants of non-European immigrants, those with metabolic diseases, malformations, low Apgar scores, low birth weight or prematurity, and those whose mothers might have abused alcohol during pregnancy were excluded. The data were grouped by narrow age ranges (6 weeks $\pm$ 4 days; 3, 4, 6, 12, and 15 months $\pm$ 1 week; 24 months $\pm$ 2 weeks; 48 months $\pm$ 4 weeks). Clinically useful charts were prepared that show percentiles (2.5th–97.5th) for length and weight-for-length.

Freeman et al. (1995) combined data from seven studies in the United Kingdom after excluding children with non-European parents and adjusting the means so that they matched. The data from the seven studies were then combined. The resulting clinically useful growth charts are for length from birth to 2 years and weight and stature from 2 to 20 years. The age-specific empirical percentiles (3rd–97th) were smoothed using the method of Cole & Green (1992). Reference data (3rd–97th percentiles) for BMI from a probability sample of 3357 European-English children aged 4 to 12 years are available in clinically useful formats (Chinn et al., 1992).

The first national United States growth charts were published by Hamill and colleagues (1977, 1979). The charts for infants were developed using data from the Fels Longitudinal Study; the data for older children came from national surveys. The infant growth charts had the same limitations as growth charts developed earlier in the United States: the sample was "limited geographically, culturally, and socioeconomically and genetically" (Hamill et al., 1977). Nevertheless, Fels data were selected to construct the charts because it was considered they constituted the best data set available in the United States at that time and it would have been costly and difficult to collect similar data from a nationally representative sample.

These growth charts have been revised by the National Center for Health Statistics/Centers for Disease Control and Prevention (Kuczmarski et al., 2000, 2002). They are shown in Figures 1.17–1.30

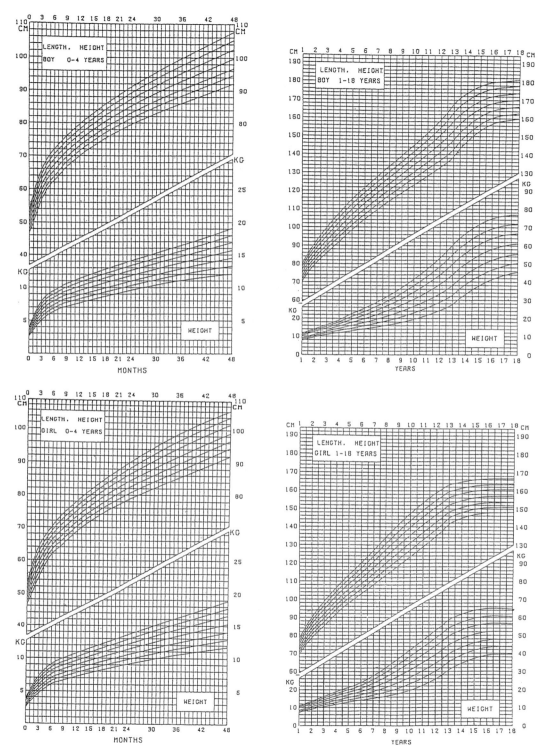

Figure 1.16. Growth charts of length, height, and weight from birth to 4 years, and from 1 to 18 years from a Japanese national survey. (From Tsuzaki *et al.*, 1987, with permission.)

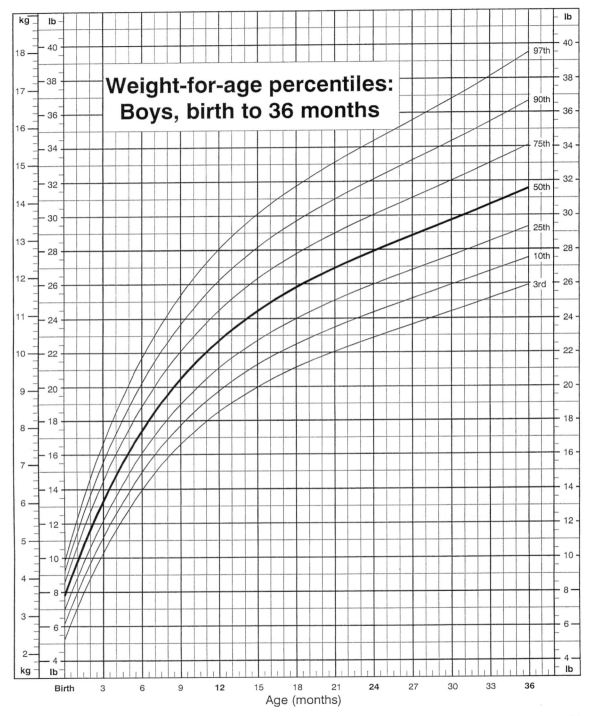

Figure 1.17. Growth chart of weight in males from birth to 36 months from United States national surveys (Kuczmarski *et al.*, 2000, 2002).

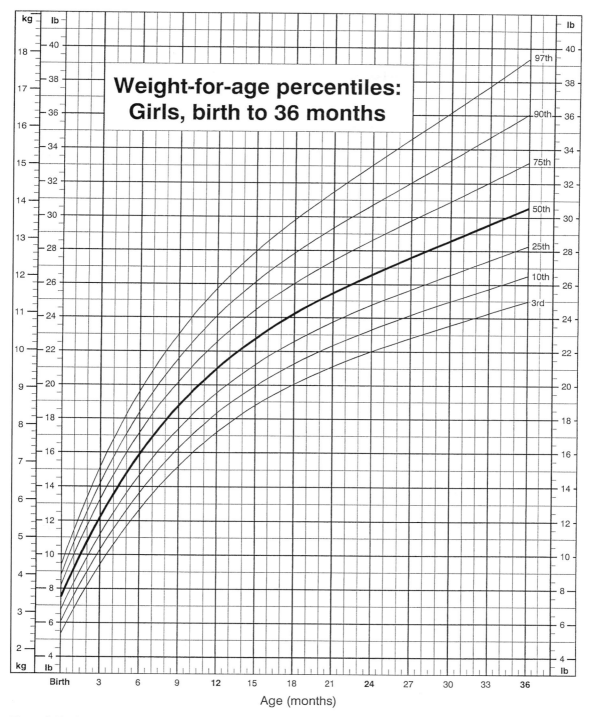

Figure 1.18. Growth chart of weight in females from birth to 36 months from United States national surveys (Kuczmarski *et al.*, 2000, 2002).

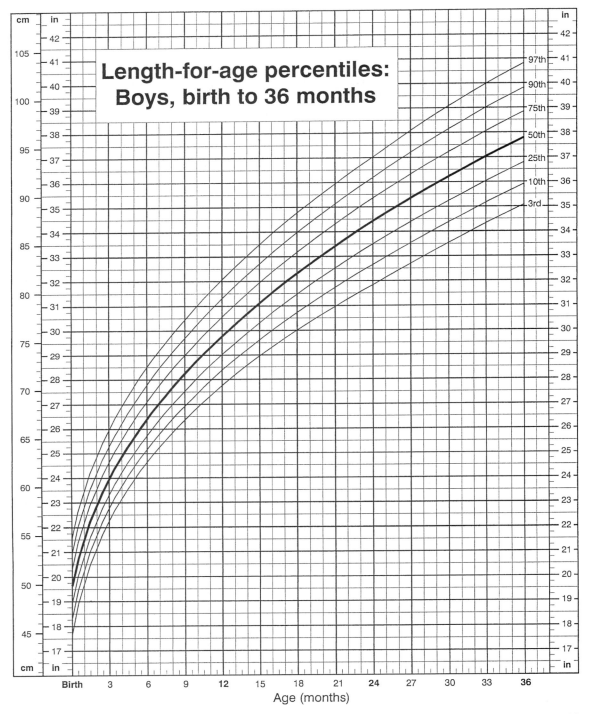

Figure 1.19. Growth chart of length in males from birth to 36 months from United States national surveys (Kuczmarski *et al.*, 2000, 2002).

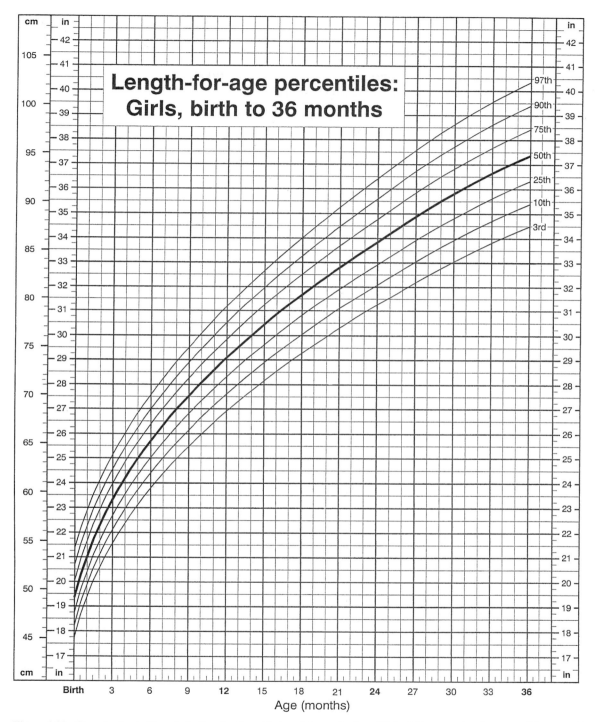

Figure 1.20. Growth chart of length in females from birth to 36 months from United States national surveys (Kuczmarski *et al.*, 2000, 2002).

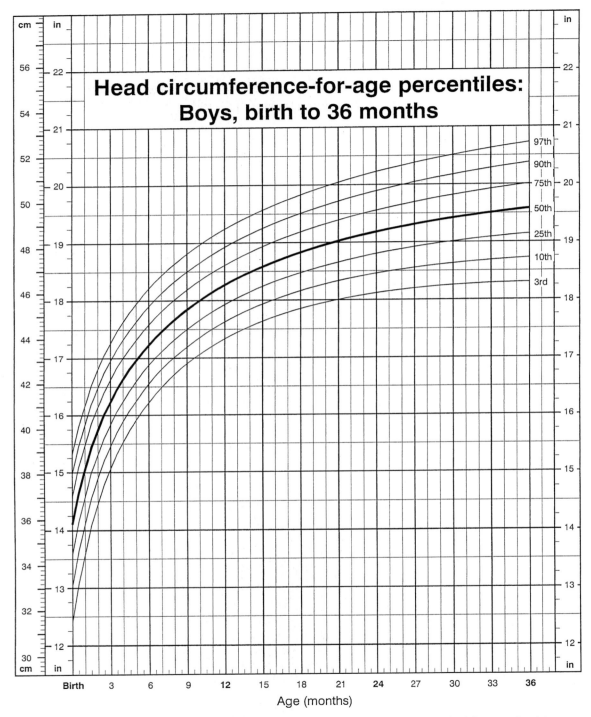

Figure 1.21. Growth chart of head circumference in males from birth to 36 months from United States national surveys (Kuczmarski *et al.*, 2000, 2002).

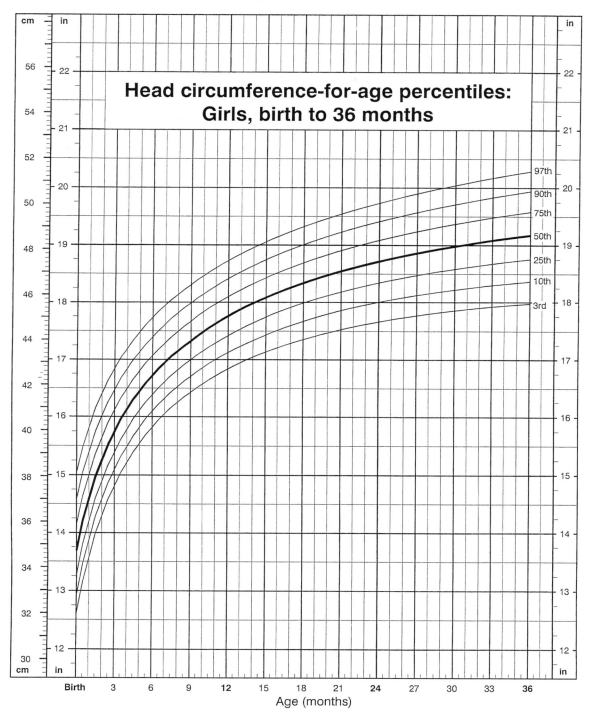

Figure 1.22. Growth chart of head circumference in females from birth to 36 months from United States national surveys (Kuczmarski *et al.*, 2000, 2002).

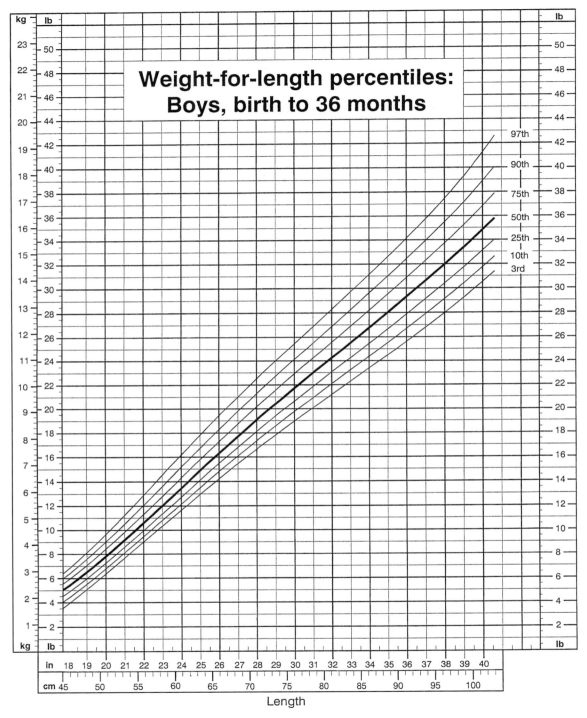

Figure 1.23. Growth chart of weight-for-length from birth to 36 months in males from United States national surveys (Kuczmarski *et al.*, 2000, 2002).

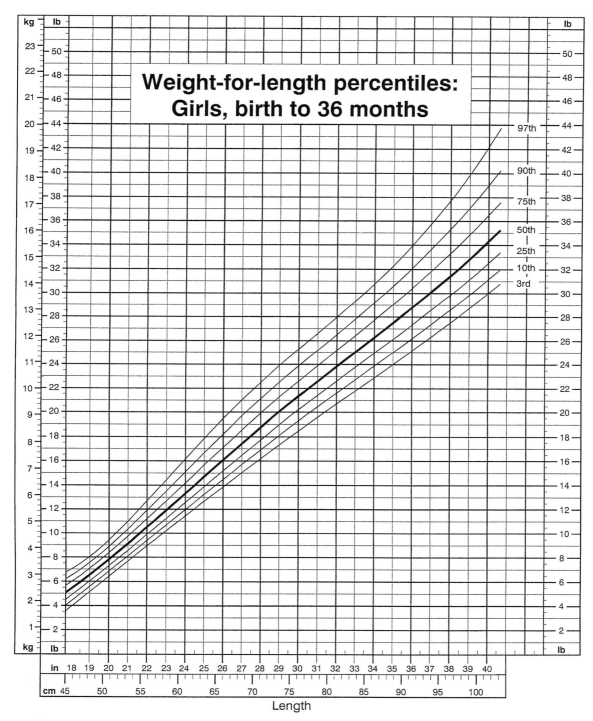

Figure 1.24. Growth chart of weight–for–length from birth to 36 months in females from United States national surveys (Kuczmarski *et al.*, 2000, 2002).

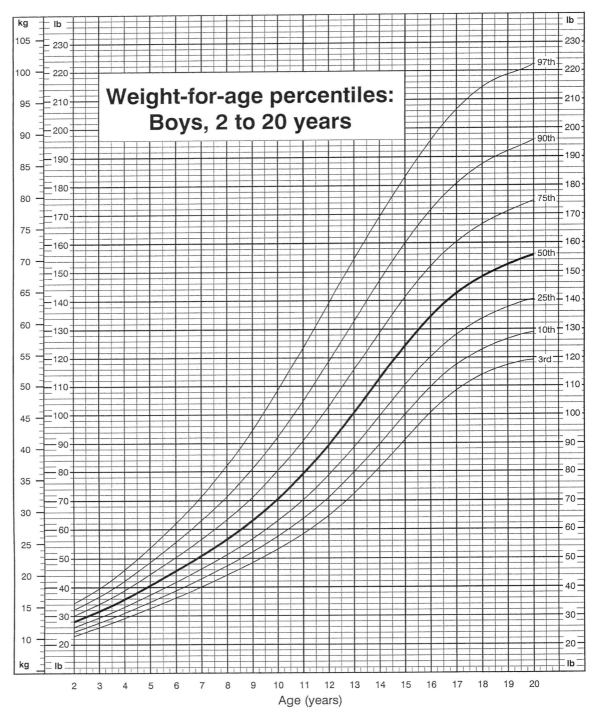

Figure 1.25. Growth chart of weight from 2 to 20 years in males from United States national surveys (Kuczmarski *et al.*, 2000, 2002).

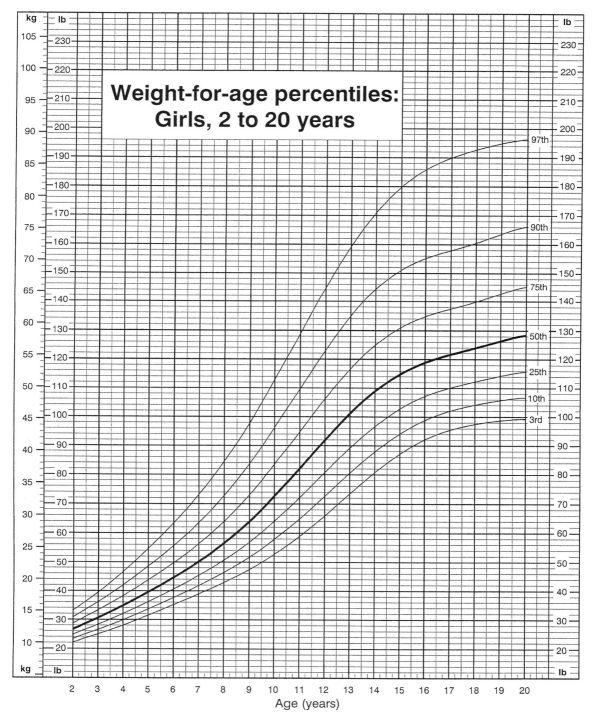

Figure 1.26. Growth chart of weight from 2 to 20 years in females from United States national surveys (Kuczmarski *et al.*, 2000, 2002).

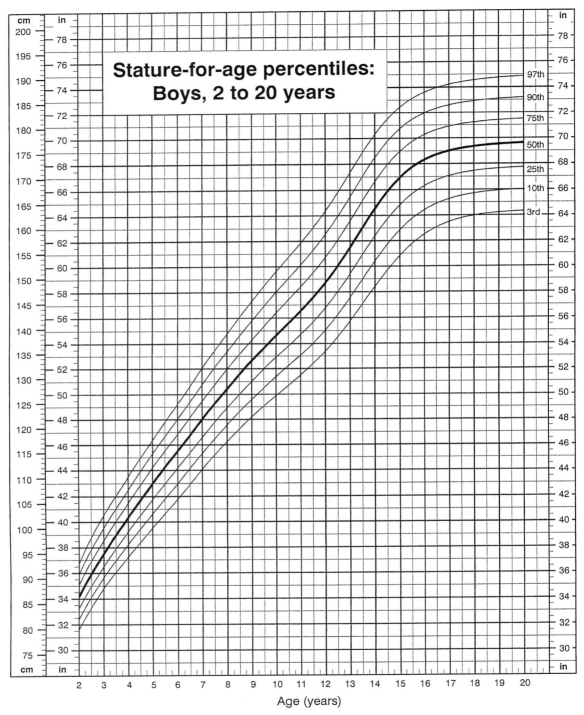

Figure 1.27. Growth chart of stature from 2 to 20 years in males from United States national surveys (Kuczmarski *et al.*, 2000, 2002).

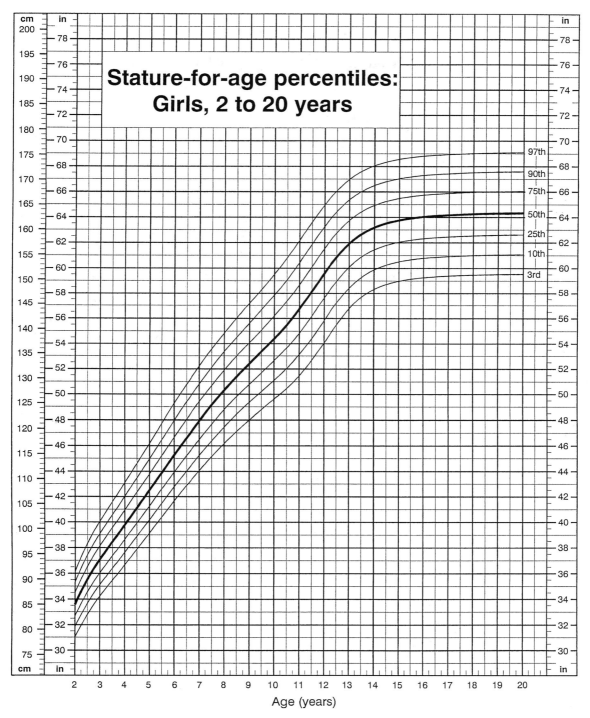

Figure 1.28. Growth chart of stature from 2 to 20 years in females from United States national surveys (Kuczmarski *et al.*, 2000, 2002).

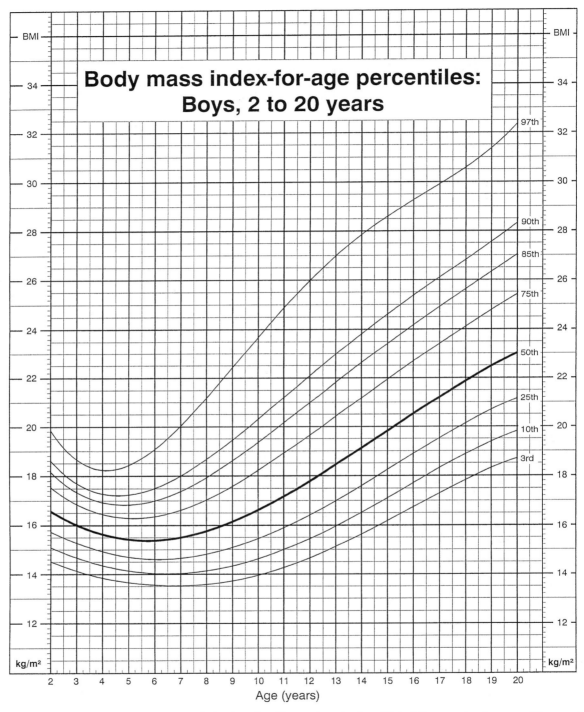

Figure 1.29. Growth chart of body mass index from 2 to 20 years in males from United States national surveys (Kuczmarski *et al.*, 2000, 2002).

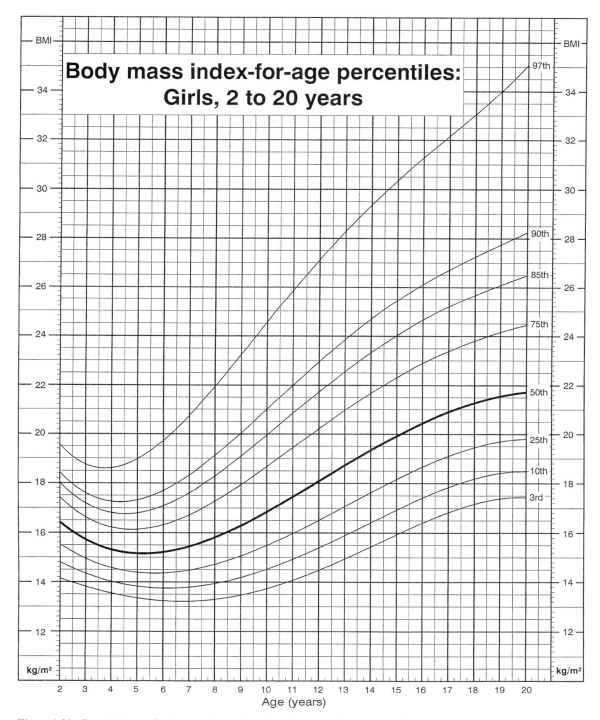

Figure 1.30. Growth chart of body mass index from 2 to 20 years in females from United States national surveys (Kuczmarski *et al.*, 2000, 2002).

and are available at US Department of Health and Human Services (2003). The revised charts were derived, almost entirely, from national probability samples measured from 1963 to 1994 that included 64 410 subjects aged 2 months to 20 years. National estimates were obtained after applying sampling weights to the data for each child. Birth weight data for the total national population, data from repeated surveys in two states for weight and length and data from the Fels Longitudinal Study for head circumference supplemented these data at birth. Data were excluded for infants with birth weights <1500 g and for weight and BMI in children aged more than 6 years who were measured after 1988. These latter exclusions were made because of positive secular trends that were considered undesirable from a public health viewpoint.

The revised charts and the accompanying tables provide reference data from birth to 20 years for weight, and for length, head circumference, and weight-for-length from birth to 36 months and for stature and BMI from 2 to 20 years. As noted earlier, empirical percentiles were obtained for each month of age to 11 months, each 3 months from 12 through 24 months, and then each 6 months. The age-specific percentile estimates were smoothed using a three-parameter model in infancy. Weight and BMI after infancy were smoothed by locally weighted regression followed by a polynomial. The percentiles for stature were smoothed by the triple logistic model (Bock & Thissen, 1976; Guo *et al.*, 1992). Finally, the smoothed percentile curves were again smoothed using a modification of the method of Cole & Green (1992). The ultimate smoothed percentiles, 3rd–97th in one version and 5th–95th in another, are shown in charts suitable for clinical use for two age ranges: birth to 36 months and 2 to 20 years and in tables.

CHANGES IN PERCENTILE LEVELS WITH AGE

Inferences about growth can be made from inspection of reference values for successive ages on growth charts or in tables. Those interpreting growth data should be familiar with the changes in percentile levels with age, although they do not represent the growth rates of individuals that are described in Chapter 2. Individual growth patterns are obscured in the percentile levels on growth charts derived from cross-sectional data. Age changes in percentile levels are described separately for infancy (birth to 36 months) and for childhood to young adulthood (3 to 20 years). Commonly, data for these age ranges are displayed in separate charts with larger scales on the infant charts than on those for older children. The age ranges shown in these separate charts may overlap at 2 to 3 years. During this period, some record length and some record stature as a measure of total body length.

Infancy

The percentile lines on growth charts for weight, length, and head circumference are markedly convex upwards from birth to 18 months. This convexity persists throughout infancy, but becomes less marked with age (Demirjian & Brault-Dubuc, 1985; Hansman, 1970; Karlberg *et al.*, 1976; Prader *et al.*, 1989). Males tend to be slightly larger than females in weight, length, and head circumference; these sex-associated differences increase from birth to 12 months, but change little from then to 36 months (Amini *et al.*, 1994; Guihard-Costa *et al.*, 1997; Kuczmarski *et al.*, 2000, 2002; Voigt *et al.*, 1996).

During infancy, it is helpful to remove effects of length before judging the normality of an observed weight because weight and length are correlated. This has led to the common use of weight-for-length during infancy. This relationship is almost independent of age, which greatly facilitates the presentation of reference data in growth charts and allows them to be applied to infants who lack recorded birth dates. Weight-for-length is not, however, age-independent before three months when weights-for-length increase with age. Due to sampling limitations, weight-for-length reference data may not accurately represent sub-groups of young infants with small lengths and older infants with large lengths (van't Hof *et al.*, 1997).

Weight-for-length percentile lines are less curvilinear than the separate lines for weight and length. Sex differences in weight-for-length are small during infancy, but at corresponding lengths, weights are somewhat larger in males than females by amounts that increase after 4 months. The percentiles of weight-for-length that correspond to lengths from 50 to 65 cm (ages from birth to about 6 months) have slight upward concavities in each sex, but they are slightly upwardly convex at lengths from 65 to 85 cm (ages about 6 to 21 months) and slightly concave upwards for lengths from 85 to 100 cm which correspond to ages about 23 to 36 months (Kuczmarski *et al.*, 2000, 2002; Prader *et al.*, 1989; Sempé *et al.*, 1979; van't Hof *et al.*, 2000*a*). These changes in weight-for-length during early infancy reflect a continuation of rapid increases in length and body fat during the last 4 months of gestation (Fomon *et al.*, 1982; Gampel, 1965; Vaucher *et al.*, 1984). The increases in body fat are also mirrored in rapid increases in reference data for BMI, skinfold thicknesses, and arm circumference from birth to 9 months.

Childhood and adolescence

The differences between successive annual medians for weight and stature, as shown in the percentile levels on growth charts, are fairly constant from 3 to 10 years in males and to 8 years in females when the pubescent spurt in body size begins. As a result of this spurt, the percentile lines for weight and stature become steeper from 9 to 15 years in males and 8 to 13 years in females. At the same time, the canals between the percentile lines widen, because the upper lines accelerate more rapidly than the lower ones. The pubescent spurt in weight has received less attention than the pubescent spurt in stature, but the percentage increase in weight during the spurt is about three times larger than that in stature. These spurts in weight and stature are indistinct in cross-sectional reference data due to variations among individuals in the timing of these spurts. The sex differences in weight are small to 14 years after which the percentile values for males are markedly larger than those for females. Percentile values for stature are similar in the two sexes to 11 years, but the percentile values for females slightly exceed those for males from 11 to 13 years; later the values for males are the larger by amounts that increase with age.

The patterns of change with age in median BMI values are similar to those in skinfold thicknesses, but differ from those in weight and body lengths. The percentile values for BMI decrease rapidly after 1 year until a nadir is reached at about 5 years (Kuczmarski *et al.*, 2000, 2002; Prokopec & Bellisle, 1992; Rolland-Cachera *et al.*, 1984; Siervogel *et al.*, 1991). The increase in BMI values for individuals after the nadir has been called a rebound (Rolland-Cachera *et al.*, 1984). The percentile lines for BMI are almost rectilinear from 6 to 20 years in males and from 6 to 15 years in females except for the 95th and 97th percentile levels, which show accelerations to 15 years in females (Figure 1.18). After 15 years, the percentile levels are near constant in males, but the 75th and lower percentile levels decrease in females. The values at the 90th percentile level and above are larger for females than for males at all ages. The other percentile values differ little between the sexes, except at 12 to 16 years when they are larger for females than for males.

OTHER IMPORTANT MEASURES

Infancy

The most common measures of regional body composition are skinfold thicknesses. Good sets of reference data are available and the measurement techniques are clinically applicable. Skinfold thicknesses are closely related to the amounts of subcutaneous adipose tissue at the sites measured. Although they are not direct measures of either the adipose tissue or fat at these sites, they are closely correlated with percent body fat (%BF) and total body fat (TBF) and are therefore informative (Frerichs *et al.*, 1979; Johnston, 1985; Roche *et al.*, 1981). Ratios between trunk and limb skinfolds are interpreted as indices of central adiposity, but skinfold thicknesses do not allow useful inferences about the amount of visceral adipose tissue.

When skinfold thicknesses of a child are measured, the caliper used must match that employed during collection of the reference data with which comparisons will be made. Major sets of data from soon after birth to adulthood have been obtained for the triceps, subscapular, and suprailiac skinfold thicknesses using the Harpenden and Holtain calipers (Prader *et al.* 1989; Sempé *et al.*, 1979; Karlberg *et al.*, 1976). The applicability of these data is questionable because of secular changes. Other reference data are for term infants measured a few hours after birth (Brooke *et al.*, 1981; Oakley *et al.*, 1977; Vaucher *et al.*, 1984). There are nonsignificant sex and site differences for skinfold thicknesses at birth in term infants except for subscapular skinfold thicknesses which may be significantly larger in females (Amit *et al.*, 1993; Guihard-Costa *et al.*, 1997). Skinfold thicknesses may be used with weight-for-age and weight-for-length to judge nutritional status at birth, but they are of limited value in estimating future progress because they are not correlated with the corresponding values at 3 months or older (Enzi *et al.*, 1982; Roche *et al.*, 1982; Whitelaw, 1977). The percentile levels for skinfold thicknesses are about 0.5 cm higher in females than in males from 9 to 36 months (Karlberg *et al.*, 1976; Prader *et al.*, 1976; Sempé *et al.*, 1979).

The percentile values for triceps skinfold thicknesses are small at birth and increase rapidly during early infancy to reach peaks at about 6 months after which they decrease slowly to 36 months in most studies (Prader *et al.*, 1989; Sann *et al.*, 1988; Schlüter *et al.*, 1976; Sempé *et al.*, 1979). In Canadian infants, however, there are small increases in percentile values from 12 to 24 months (Demirjian & Brault-Dubuc, 1985; Yeung, 1983). The changes with age in percentile values during infancy are largest at the upper percentile levels and some consider the age changes are limited to these percentiles (Cole *et al.*, 1989; Karlberg *et al.*, 1976; Sempé *et al.*, 1979). Percentile levels for subscapular skinfold thickness increase rapidly from birth to about 5 months and then decrease during the remainder of infancy in most studies (Karlberg *et al.*, 1976; Prader *et al.*, 1989; Schlüter *et al.*, 1976; Sempé *et al.*, 1979). In one Canadian study, however, these thicknesses changed only slightly from 3 to 18 months (Yeung, 1983). The percentiles for skinfold thicknesses at the suprailiac site increase rapidly during the first 6 months after birth. These increases continue at a much slower rate to 36 months in Swiss infants and in Belgian females (Gerver & de Bruin, 2001; Hernesniemi *et al.*, 1974; Prader *et al.*, 1989), but do not increase after 9 months in some other studies (Karlberg *et al.*, 1976; Schlüter *et al.*, 1976; Sempé *et al.*, 1979).

The age changes in skinfold thicknesses during infancy have implications for the interpretation of recorded measurements. The rapid increases in skinfold thicknesses until about 6 months and the decreases later in infancy make it difficult to interpret progress during the first year, because the change from rapid increases to decreases varies in timing among infants. A decrease in skinfolds from 5 to 6 months for an infant may occur because of a nutritional deficit or because the infant had an early peak value and a natural decrease began at 5 months. Data from serial examinations should make apparent the true nature of the changes.

Childhood and adolescence

Reports from France, Hungary, and Switzerland show that percentile values for triceps skinfold thicknesses in males change little from 3 to 9 years, but increase rapidly from then to 11 years. This is followed by a decrease to 17 years after which there are only small changes to 20 years (Eiben *et al.*, 1991; Prader *et al.*, 1989; Sempé *et al.*, 1979). Reports from The Netherlands, Norway, and Sweden, however, show gradual decreases from 3 to 8 years in males when increases begin that continue to 18 years (Gerver & de Bruin, 2001; Karlberg *et al.*, 1976; Waaler, 1983). In females, there are only small changes in percentile values from 3 to 11 years when increases begin that continue to 16 years at which age they cease or become small (Gerver & de Bruin, 2001; Sempé *et al.*, 1979; Karlberg *et al.*, 1976).

At all childhood ages, the percentile values for triceps skinfold thicknesses in females exceed those in males. These sex differences increase slowly to 11 years and then increase more rapidly.

In males, percentile levels for subscapular skinfold thicknesses change little from 3 to 9 years and then increase slowly to 20 years (Eiben *et al.*, 1991; Gerver & de Bruin, 2001; Prader *et al.*, 1989; Waaler, 1983). In females, percentile levels for subscapular skinfold thicknesses decrease slightly from 3 to 7 years and then increase rapidly to 14 years after which the changes are small (Prader *et al.*, 1989; Sempé *et al.*, 1979). At all ages, the percentile values for females are larger than those for males by amounts that increase with age, especially from 11 to 14 years.

The percentile values for suprailiac skinfold thicknesses in males and females change little from 3 to 7 years, but they increase rapidly from 7 to 16 years; these increases are larger in females than in males. After 16 years, suprailiac skinfold thicknesses increase slowly to 20 years in males, but they are constant or decrease slightly in females (Gerver & de Bruin, 2001; Karlberg *et al.*, 1976; Prader *et al.*, 1989; Sempé *et al.*, 1979). In Hungarian children, however, the percentile levels for males decrease slightly from 3 to 7 years and then increase at a constant rate to 18 years (Eiben *et al.*, 1991). At all ages, the percentile values for females are larger than those for males by amounts that increase with age. The increase in the sex differences is rapid from 9 to 14 years.

TOTAL BODY COMPOSITION

Infancy

Conclusions about total body composition during infancy are tentative because of the large differences among measurements by different methods (Butte *et al.*, 1999). Some have compared infants with birth weights in the normal range for gestational age (appropriate-for-gestational age; AGA) with infants whose birth weights are less than the 10th percentile for gestational age (small-for-gestational age; SGA).

Table 1.4. *Body composition components during infancy expressed as percentages of body weight*

Age (months)	Total body fat	Fat-free mass (including bone mineral)	Bone mineral
Males			
0.5	11.4	88.6	0.7
3	30.2	69.8	0.5
6	29.1	70.9	0.5
9	25.7	74.3	0.6
12	25.6	74.4	0.5
18	24.5	75.4	0.6
24	25.4	75.6	0.5
Females			
0.5	14.2	85.8	0.7
3	31.5	68.5	0.5
6	32.0	68.0	0.5
9	28.8	71.2	0.5
12	27.6	72.4	0.5
18	26.3	73.7	0.5
24	25.4	74.6	0.5

Source: Data from Butte *et al.* (2000*a*).

At birth, TBF, %BF, FFM, and TBBM are the same in SGA and AGA infants who are matched for birth weight (Lapillonne *et al.*, 1997; Petersen *et al.*, 1989), but females have more fat than males (Fomon *et al.*, 1982). The rapid growth of infants, particularly during the first year, is accompanied by large changes in body composition (Butte *et al.*, 2000*a*) (Table 1.4). The general nature of these changes is well established, but satisfactory reference data are scarce. The values for %BF and skinfold thicknesses are low at term, but increase rapidly to 4 months and then increase slowly with larger values in females than in males (de Bruin *et al.*, 1995; Hurgoiu & Mihetiu, 1993; Lapillonne *et al.*, 1997; Petersen *et al.*, 1988).

The composition of FFM changes during infancy; the percentage water content decreases and the percentage mineral content increases (Fomon *et al.*, 1982). Percentile levels for FFM increase rapidly to 3 months and double from birth to 12 months with larger values in males than in females at most ages (Butte *et al.*, 2000*a*; de Bruin *et al.*, 1995; Fomon *et al.*, 1982; Rutledge *et al.*, 1976). Increases in percentile levels for FFM continue from 12 to 36 months, but are slower than those during early infancy. These age changes are similar in term and preterm infants, but the values are lower in preterm infants at the same chronological ages. These increases in FFM may be mainly in vital organs because there are only small increases from birth to 12 months in the excretion of creatinine, which is an index of muscle mass (Fomon, 1993). As a percentage of body weight, FFM decreases to 6 months and then increases to 36 months (Fomon *et al.*, 1982). The changes reflect alterations in body fatness. Values for FFM at birth, as a percentage of body weight, are larger in SGA term infants than in AGA term infants (Petersen *et al.*, 1988). These variations reflect differences in body fat.

Percentile levels for TBBM increase markedly from birth to 4 months in term infants and to 12 months in preterm infants, after which the increases are smaller (Fomon *et al.*, 1982; Rawlings *et al.*, 1999). As a percentage of body weight, TBBM changes little with age during infancy (Chan & Mileu, 1985; Fomon *et al.*, 1982; Petersen *et al.*, 1989; Zanchetta *et al.*, 1995).

Childhood and adolescence

The values for %BF and TBF tend to be larger for females than for males at all ages throughout childhood (Faulkner *et al.*, 1993; Maynard *et al.*, 2001; Mesa *et al.*, 1996; Ruxton *et al.*, 1999). There are decreases in %BF in each sex from 4 to 8 years (Fomon *et al.*, 1982; Zanchetta *et al.*, 1995). In serial data for males, the means for TBF increase from 8 to 15 years, but change little from 15 to 19 years (Figure 1.31). The means for females change little from 10 to 12 years. They increased markedly during pubescence in the study by Faulkner *et al.* (1993), but only slightly in the study by Guo *et al.* (1997*b*).

Medians for FFM increase with age to at least 19 years in males, but they do not increase after 17 years in females. Estimates of FFM are slightly larger for males than females at 6 to 14 years, later they are markedly larger in males (Boot *et al.*, 1997*b*; Guo *et al.*, 1997*b*). Total body potassium data, which are closely related to FFM, indicate that means for FFM, as a percentage of body weight, are larger in males than in females (Novak *et al.*, 1973). These sex differences are significant at 13 to 15 years. Muscle is a major component of FFM and creatinine excretion is a reliable index of muscle mass. Means for creatinine excretion increase with age, particularly in males from 12 to 14 years, and are larger in males than females after 12 years (Novak, 1963). Similar age changes occur in the cross-sectional area of arm muscle (Kanehisa *et al.*, 1995).

Bone mineral values are usually reported as bone mineral content (BMC, g) or bone mineral density (BMD, g/cm^2). The latter is not a true density because the area, which is available from dual energy X-ray absorptiometry, is used in place of volume, which is rarely measured in the living. The means for total body BMC increase from 4 to 16 years in males and 14 years in females, and then change little to 20 years (Boot *et al.*, 1997*b*; Ogle *et al.*, 1995; Takahashi *et al.*, 1996). The sex differences in these means are small to 15 years, after which the values for males become increasingly larger than those for females (Bachrach *et al.*, 1999; Boot *et al.*, 1997*b*; Maynard *et al.*, 1998; Rico *et al.*, 1992). There is a steady increase in annual means for BMD of the lumbar spine and radius until pubescence when the increases between annual values become larger (Bonjour *et al.*, 1991; Gilsanz *et al.*, 1991; Rubin *et al.*, 1993; Takahashi *et al.*, 1996). The pubescent changes in BMC and BMD of the lumbar spine are similar with larger increases in females than males. The increases between annual means decrease markedly after 15 years in males and 13 years in females (Bonjour *et al.*, 1991; Theintz *et al.*, 1992).

PERCENTAGES OF VALUES AT OLDER AGES

Median values for measurements at young ages during infancy can be expressed as percentages of the

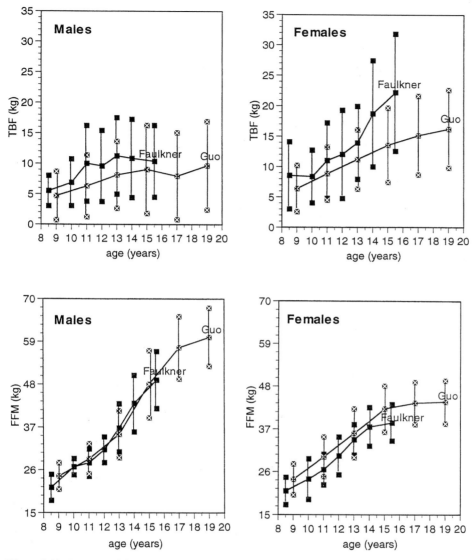

Figure 1.31. Means and SDs of body composition measures (above, total body fat; below, fat-free mass) at 8.5 to 19 years drawn from data of Faulkner *et al.* (1993) and Guo *et al.* (1997*b*).

medians at 36 months (Figure 1.32). These percentages differ only slightly between the sexes for weight, length, and head circumference. Those for length increase from about 50% at birth to about 75% at 9 months and then increase slowly. Median weight at birth is about 26% of median weight at 36 months; the 50% level is reached at about 6 months. The median head circumference at birth is about 72% of the 36-month median and it is about

90% of this value at 9 months. There are large differences among skinfold sites in these percentages at birth. They are about 50% for the triceps site and 73% for the suprailiac site in each sex, but about 100% in males and 113% in females for the subscapular site. The percentages for the triceps site increase markedly to 6 months when they exceed 100%; later there are only small changes. The subscapular and suprailiac percentages increase after

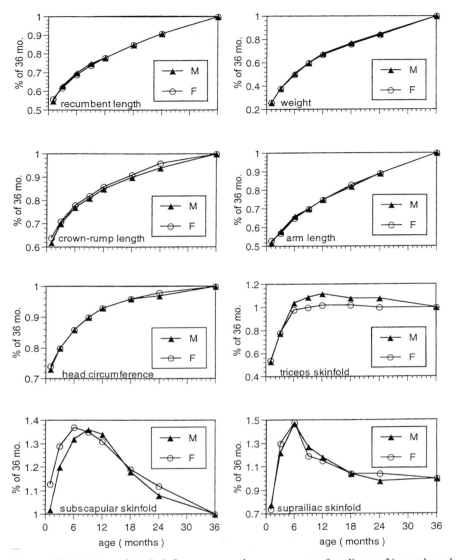

Figure 1.32. Medians of size in infancy expressed as percentages of medians at 36 months, calculated from data of Prader *et al.* (1989).

birth to be 135% to 145% of the 36-month values at 6 months.

Expressing the median values at younger ages during childhood as percentages of the medians at 20 years, using the data of Prader *et al.* (1989), shows that females have achieved larger percentages than males for weight, stature, sitting height, arm length, and arm circumference at matching ages (Figure 1.33). These sex differences are as-sociated with the more rapid maturation of females. The percentage for each of these measures increases markedly during the pubescent age range, with larger changes in males than females, except for weight. Corresponding percentages for head circumference calculated from values at 18 years are almost identical in both sexes (Roche *et al.*, 1987). They show linear trends except for larger increases from birth to 4 years than at older ages.

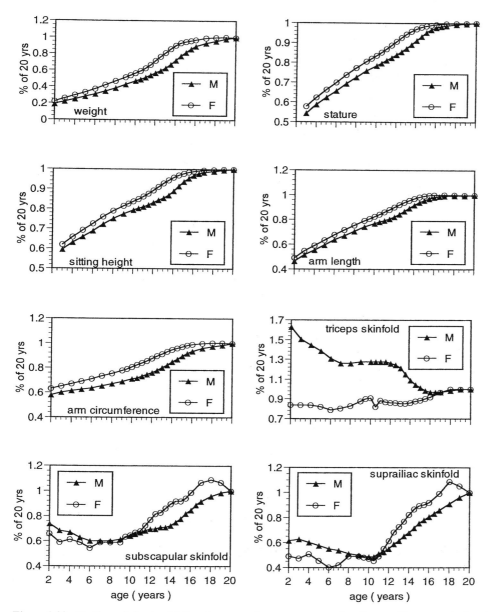

Figure 1.33. Medians of size in childhood expressed as percentages of medians at 20 years, calculated from data of Prader *et al.* (1989).

The patterns of change in percentages of 20-year values for skinfold thicknesses differ markedly from those for the variables just considered. In males, the percentages for triceps skinfold thicknesses are markedly larger at 2 to 16 years than at 20 years because the medians decrease after the pubescent spurt. Indeed, the percentages of the 20-year values are 163% at 2 years and 130% at 6 years. The percentages for triceps skinfold thicknesses in females change little until the pubescent period when increases begin that continue to 18 years. The corresponding percentages for subscapular and suprailiac skinfold thicknesses are similar in the two sexes and change little from 2 to 11 years, when increases begin that are slightly larger in females than in males. After 18 years, the percentages decrease in females, but continue to increase slowly in males.

MULTIPLE BIRTHS

Twins and triplets are generally smaller than singletons at the same gestational age. About 50% of twins weigh less than 2500 g at birth (Ho & Wu, 1975; Keith *et al.*, 1980) and 22–54% are born preterm (Bleker *et al.*, 1979; Koch, 1966; Roberts & Lancaster, 1999). These differences from singletons are probably due to placental and uterine limitations that affect the removal of waste products, blood oxygenation, and the production of essential metabolites. Most reports of size at birth in twins are for combined groups of monozygotic and dizygotic twins. The differences between monozygotic and dizygotic twins are small and not significant (Alfieri *et al.*, 1987; Corney *et al.*, 1979; Ljung *et al.*, 1977; Wilson, 1986). The terms monozygotic and dizygotic refer to the number of fertilized ova involved in each twin pregnancy. The mean birth weights of twins of unlike sex are 70–90 g larger than those of like-sex twins (Corey *et al.*, 1979; Grennert *et al.*, 1980; Hemon *et al.*, 1982).

Deficits in the size of twins and triplets at birth, compared with singletons, become evident at 28–30 weeks of gestation (Bleker *et al.*, 1988; Roberts & Lancaster, 1999; Secher *et al.*, 1985; Simon *et al.*, 1989). These deficits are larger for weight than for length or head circumference (Bossi *et al.*, 1993; Bulmer, 1970; Wilson, 1986). Male twins are 50–130 g heavier than female twins at birth, although female twins may have longer gestation periods (Chen *et al.*, 1993; Cohen *et al.*, 1997; Rydhström, 1992). This sex difference is more marked after 34 weeks gestation than earlier and is larger at the 90th percentile than at lower ones (Cohen *et al.*, 1997). As occurs for singletons, the birth weight of twins is influenced by parity, ethnicity, maternal age and stature, pre-gravid weight, weight gain during pregnancy, socioeconomic status and tobacco use (Asaka *et al.*, 1980; Hemon *et al.*, 1982; Luke *et al.*, 1993; Wilson, 1986).

Selected reference data for size of twins at birth are listed in Table 1.5. Bossi *et al.* (1993) analyzed data from 807 Italian twin pairs in a study that is important because of the strict inclusion criteria and the number of variables for which data are provided. These reference data approximate what can be expected in twins growing without retarding influences. The infants did not have congenital anomalies and their mothers did not have risk factors that limit fetal growth. Bossi and coworkers provide sex-specific empirical percentiles (3rd–97th) for weight, length, and head circumference in relation to gestational age. Other data for the size of twins at birth have been reported from Canada, Israel, and the United Kingdom (Arbuckle & Sherman, 1989; Buckler & Green, 1994; Cohen *et al.*, 1997).

There are data from Israel and the United States for the size of triplets at birth (Table 1.5). Elster *et al.* (1991) presented smoothed percentiles (10th–90th) for sexes combined and provided factors to adjust the medians for infant sex and maternal parity. The adjustments for sex (25 g at 32 weeks, 50 g at 34 weeks, and 100 g at 38 weeks) are to be subtracted from the observed data for males and added to the observed data for females before comparison with the reference data for both sexes combined. Differences associated with parity have been estimated from comparisons between multiparae, who are mothers with previous pregnancies, and primiparae for whom this was the first pregnancy. Triplets born to multiparae are heavier than those born to primiparae. Consequently, adjustment factors (50 g at 28 weeks, 75 g at 30 weeks, 50 g at 32 weeks, and 35 g at 34 weeks) should be added to the observed weights of triplets born to primiparae and subtracted from the observed weights of those born to multiparae before comparisons are made with reference data for both parity groups combined. These adjustments for sex and parity should not be added together in the assessment of an infant because they are not independent estimates.

Data for triplets have been reported from a study in Israel in which gestational ages were obtained

Table 1.5. *Sources of reference data for size at birth in twins and triplets*

Authors	Country	Number of infants	Exclusion of abnormal infants	Exclusion of high-risk mothers	Gestational age method[a]	Sex specific[b]	Variables[c]
Twins							
Arbuckle & Sherman (1989)	Canada	25 330	no	no	A	yes	W
Bossi *et al.* (1993)	Italy	1 614	yes	yes	A	yes	W, L, HC
Buckler & Green (1994)	United Kingdom	38 176	no	no	A	yes	W, HC
Cohen *et al.* (1997)	Israel	1 040	yes	yes	US	yes	W
Triplets							
Elster *et al.* (1991)	United States	3 414	no	no	B	C	W
Jones *et al.* (1991)	United States	580	no	no	US	no	W
Luke *et al.* (1991)	United States	547	yes	no	US + others	yes	W
Yuval *et al.* (1995)	Israel	109	yes	no	US + others	no	W

[a] A, method not stated; B, last menstrual period, pediatric examination and ultrasound; US, ultrasound; "others", days from ovulation induction and from embryo transfer during *in vitro* fertilization.
[b] C, adjustments for sex.
[c] HC, head circumference; L, length; W, weight.

from ultrasound, the day of induced ovulation, and the day of embryo transfer during *in vitro* fertilization (Yuval *et al.*, 1995). The precise gestational ages make this study valuable despite the small sample size. As in the study by Elster, growth in weight is normal for triplets until the third trimester when growth is slow. The mean birth weights of triplets are near the 10th percentile after 37 weeks gestation (Yuval *et al.*, 1995). Head and abdominal circumferences are smaller in triplets than in singletons during the third trimester, when there is also a reduction in total amount of subcutaneous tissue on the face, ribs, thigh, and buttocks (Hata *et al.*, 1991).

Postnatal reference data for twins and triplets are not needed because these infants grow similarly to singletons of the same gestational age and size at birth (Kuno *et al.*, 1999). Most studies of the postnatal growth of twins compare them with general populations of infants without adjusting for prematurity and many of these studies were made before the introduction of high caloric-density feeding.

Wilson (1979, 1986) reported large decreases in the weight deficits to three months followed by slow decreases. In this study, Z scores for length, calculated by comparison with reference data for a general population, increased slightly during infancy for twins with values at birth near the 50th or 90th percentile, but there were larger increases for those with lengths at birth near the 10th percentile. The deficits in weight and length were near zero by 8 years, except that the deficits in weight persisted for twins with birth weights <1750 g. Others have reported marked decreases in weight deficits from birth to 9 months for females, but not males, with slight decreases in the length deficits for each sex (Alfieri & Gatti, 1976). In one study, twins were smaller than singletons from 3 to 11 years in weight, stature, and head circumference without age trends in these differences (Silva & Crosado, 1985). Others have reported an absence of size differences between twins and singletons at 8 to 15 years (Ljung *et al.*, 1977; Sklad, 1975; Wilson, 1986).

LOW BIRTH WEIGHT INFANTS

Infants with low birth weights (<2500 g) differ in postnatal growth from infants of normal birth weight. Due to these differences, data from low birth weight (LBW) or very low birth weight (VLBW) infants (birth weight <1500 g) were omitted when some national data sets were used to construct growth charts for general populations (Cole & Roede, 1999; Knudtzon et al., 1989; Kuczmarski et al., 2000, 2002; Roede & van Wieringen, 1985). Birth weight is used to categorize infants as appropriate-for-gestational age (AGA) or small-for-gestational age (SGA). Infants with birth weights <10th percentile or <−2 Z for gestational age are considered SGA; the remaining infants are considered AGA by many workers, but some distinguish large-for-gestational age infants (LGA) as those with birth weights >4500 g or >+2 Z. Such categorizations are important because these groups of infants differ in neonatal mortality rates, postnatal growth, and body composition at birth. Neonatal mortality rates are particularly high in SGA infants who have lengths <10th percentile (Balcazar & Haas, 1990).

The term intrauterine growth retardation (IUGR) is sometimes applied to SGA infants. The terms IUGR and SGA are not synonymous. Intrauterine growth retardation refers to deficits in intrauterine growth (Owen et al., 1996; Pryor, 1996) whereas SGA refers to weight status at birth. Intrauterine growth rates can be determined from serial ultrasound data that are compared with reference data (3rd–97th percentiles) for weekly increments (Benso et al., 1999). These reference data for biparietal diameter, occipital–frontal diameter, head and abdominal circumferences, and femoral diaphysial length show accelerations from 12 to 18 weeks, with the exception of abdominal circumference for which the acceleration continues to 21 weeks. These accelerations are followed by decelerations that continue to at least 36 weeks. Some LBW infants have asymmetric growth retardation at birth, which can be recognized by a lesser deficit in head circumference or stature than in weight or arm circumference. The growth failure of infants with asymmetrical retardation probably occurred late in gestation during the phase of adipose tissue deposition.

Table 1.6 summarizes sets of reference data for preterm infants measured at birth. Almost all these infants are LBW or VLBW. Only studies that used ultrasound to estimate gestational age are included because this method of estimation will probably become more widespread. It is recommended that the data of Voigt et al. (1996) be used because sex-specific values are given for a wide range of gestational ages. Sex-specific cut-off levels are needed because males are heavier than females at particular gestational ages. Table 1.7 shows selected reference data for the 10th percentiles or −2 Z levels of birth weight at various gestational ages. Therefore, the use of reference data for the two sexes combined would result in more females than males being categorized as SGA infants. Separate 10th percentile values for African Americans and European Americans are available in Amini et al. (1994).

Some charts, that are designed for use in neonatal nurseries, provide reference data for weight at chronological ages up to 105 days (Ehrenkranz et al., 1999; Fenton et al., 1990; Jaworski, 1974; Wright et al., 1993). The chart of Fenton et al. (1990) provides means for weight in five birth weight groups of AGA and SGA infants who were VLBW or ELBW (extremely low birth weight; <750 g). In the ELBW group, the means differ only slightly between AGA and SGA infants, but in other birth weight groups the means for AGA infants increase more slowly than those for SGA infants. Ehrenkranz et al. (1999) provided means for weight, length, and the circumferences of the head and arm for ten birth weight groups (Figure 1.34). These data extend longer after birth for the smaller birth weight groups, which have longer nursery stays. These weight data are presented separately for those with and those without major morbidities and separately for AGA and SGA infants.

With the exception of charts for use in neonatal nurseries, growth charts for LBW and VLBW preterm infants are constructed relative

Table 1.6. *Selected sets of reference data for size at birth of live preterm infants*

Authors	Country	Number of infants[a]	Estimates of gestational age[b]	Exclusion if obs–ped estimate >2 wks	Other exclusions	Gestational age range (wks)	Variables[c]	Sex-specific	Notes
Amini et al. (1994)	United States	8 528	obs, ped	yes	no	24–36	W	yes	percentiles, African Americans and European Americans separately; adjustments for parity
Bjerkedal & Skjærven (1980)	Norway	3 547	Imp	no	no	28–36	W	yes	percentiles
Dombrowski et al. (1992)	United States	3 656	Imp, obs	yes	congenital abnormalities	24–36	W, L, HC	no	percentiles, African Americans and European Americans separately
Högfeldt et al. (1991)	Sweden	43 619[d]	Imp, obs	no	no	28–36	HC	yes	mean; SD
Karlberg et al. (1985)	Sweden	5 695	Imp, obs	obs used if difference ≥2 wks	abnormalities, severe illnesses	28–36	W, L, HC	yes	mean ± 2SD
Milner & Richards (1974)	United Kingdom	16 680	Imp	no	no	28–36	W	yes	mean, SD; primiparae, multiparae
Roberts & Lancaster (1999)	Australia	38 628	Imp	no	outliers	20–36	W	yes	percentiles
Voigt et al. (1996)	Germany	34 935	Imp, obs	no	multiple births, outliers	23–36	W, L, HC	yes for W	percentiles
Wilcox et al. (1993)	United Kingdom	41 718	obs	no	multiple births, congenital abnormalities	24–36	W	no	percentiles
Zhang & Bowes (1995)	United States	310 232	Imp	no	no	25–36	W	no	percentiles, African Americans and European Americans separately; primiparae and multiparae separately

[a] The sample sizes are for the numbers of infants born preterm.
[b] Imp, estimates from last menstrual periods; obs, obstetric estimates; ped, pediatric estimates.
[c] HC, head circumference; L, length; W, weight.
[d] Total sample 3 134 879 for 20–44 weeks; number of preterm not stated.

Table 1.7. *Selected reference criterion weights (g) for the recognition of small-for-gestational-age infants*

| Gestational age (weeks) | 10th percentile | | | −2SD | |
| | Dombrowski *et al.* (1992), United States | Voigt *et al.* (1996), Germany | | Karlberg *et al.* (1985), Sweden | |
	Males plus females	Males	Females	Males	Females
23	–	450	430	–	–
24	475	510	490	–	–
25	540	600	560	–	–
26	630	680	640	–	–
27	670	770	710	–	–
28	810	860	800	690	–
29	835	960	900	850	860
30	1060	1070	990	1020	930
31	1250	1180	1100	1210	1050
32	1320	1340	1260	1410	1210
33	1530	1550	1470	1620	1400
34	1680	1790	1710	1830	1620
35	1870	2060	1980	2030	1840
36	2093	2330	2230	2240	2060
37	2353	2570	2460	2430	2280
38	2540	2780	2660	2600	2470
39	2693	2950	2820	2740	2620
40	2820	3070	2940	2850	2740
41	2948	3160	3020	2910	2790
42	3000	3200	3050	2910	2780
43	3120	3040	2920	–	–

to gestation-adjusted ages (GAA). These are the chronological ages after birth, less the extents that the infants were born preterm. Thus, an infant born at a gestational age of 32 weeks (8 weeks preterm) and measured 12 weeks after delivery will have a GAA of 4 weeks at the time of measurement. The growth of preterm infants should be judged using GAAs to at least 36 months; this practice may need to be continued to 7 years for those born before 32 weeks (Elliman *et al.*, 1992).

The characteristics of the major growth charts for monitoring the growth of preterm LBW and VLBW infants are summarized in Table 1.8. These are for infants born at gestational ages equal to or less than 37 completed weeks. Gestational age can

be estimated as the interval from the first day of the last menstrual period (LMP), from examinations soon after birth (pediatric), or from ultrasound data early in pregnancy (obstetric). Differences between estimates of gestational age by these methods may affect the categorization of an infant as preterm and be important in the evaluation of size at birth. The only truly accurate data are from *in vitro* or artificial insemination (Hall, 1990). The LMP method may systematically overestimate gestational age because ovulation and fertilization may occur more than 14 days after the first day of the last menstrual period (Kramer *et al.*, 1988). Pediatric examinations overestimate short gestations and underestimate long ones (Alexander *et al.*, 1990;

Table 1.8. *Selected characteristics of the Brandt (1978) and Infant Health and Development Program (IHDP) growth charts for low birth weight infants*

Characteristics	Brandt[a]	IHDP
Years of birth	1967–75	1985
Number	65 AGA	985
Birth weight categories	<1500 g	<1500 g, 1501–2500 g
Ethnicity	German	32% European American, 52% African American, 14% Hispanic American
Exclusions	major abnormalities	major abnormalities
Gestational age method	last menstrual period	method of Ballard *et al.* (1979)
Ages corrected for gestation	yes	yes
Ages at examinations	each 1 or 2 wk before term, each mo to 12 mo, then each 3 mo to 24 mo	birth, term, 4, 8, 12, 18, 24, 30 and 36 mo gestation-adjusted age
Growth charts	3rd–97th percentiles after assuming normal distributions	5th–95th empirical
Smoothing	yes	yes
Infants used in charts	weight to term = AGA preterm; 1 to 24 mo = AGA preterm + term length to term = AGA preterm; 0 to 18 mo = term; 18 to 24 mo = AGA preterm + term head circumference birth to 18 mo = AGA preterm + term	all preterm

[a] AGA = appropriate weight >10th percentile for gestational age.
Sources: Casey *et al.* (1991); Guo *et al.* (1996, 1997*a*); Roche *et al.* (1997).

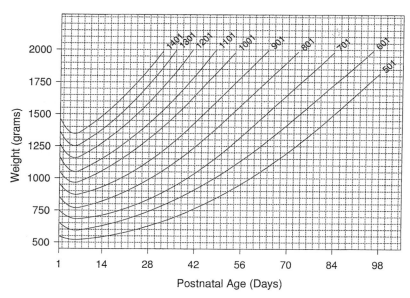

Figure 1.34. Mean postnatal weights of preterm infants grouped by birthweight; the numbers at the upper ends of the curves show mean birth weights (g) for groups. (From Ehrenkrantz *et al.*, 1999, with permission.)

Constantine *et al.*, 1987). Ultrasound examinations early in pregnancy are very useful, although they measure size which varies at specific gestational ages (Hall, 1990). Dombrowski *et al.* (1992) found mean gestational ages from ultrasound tended to be older than those from LMP, but Sanders *et al.* (1991) reported good agreement between these estimates. If hospital records are unavailable, maternal recall data can be used. Recall of gestational age may be less accurate than recall of birth weight, but some have found a high level of accuracy (Cartwright & Smith, 1979; Little, 1986; Seidman *et al.*, 1987).

Preterm and term infants are further categorized by birth weight. Data for birth weight is usually available from hospital records, but maternal reports can be used in the absence of this information. About 95% of mothers accurately identify their infants as LBW or normal birth weight (Burns *et al.*, 1987; Gayle *et al.*, 1988; Tilley *et al.*, 1985; Wilcox *et al.*, 1991). The percentages of recalled birth weights that are within 100 g of the hospital values vary from 47% to 92% among studies (Burns *et al.*, 1987; Gayle *et al.*, 1988; Little, 1986; Wilcox *et al.*, 1991). In some studies, accuracy of recall is not related to maternal age or parity (Querec, 1980; Seidman *et al.*, 1987; Wilcox *et al.*, 1991), but others have reported that recall is less accurate for older multiparae and unmarried mothers and for LBW infants (Burns *et al.*, 1987; Gayle *et al.*, 1988; Hockelman *et al.*, 1976; Querec, 1980).

Some sources of reference data for size at birth in preterm infants are listed in Table 1.6. Each of these studies included many infants, but the samples are small for gestational ages younger than 30 weeks. For example, of the 8528 preterm infants in the study by Amini *et al.* (1994), only 152 and 190 were at gestational ages of 24 and 25 weeks, respectively. In most of these studies, gestational ages were obtained from a combination of LMP and obstetric estimates, but Amini *et al.* (1994) used a combination of LMP and pediatric assessments. The exclusion criteria differed among these studies. They included differences of more than two weeks between estimates of gestational ages by different methods, and the presence of congenital abnormalities, infant illnesses, or

maternal diseases that affect birth weight. Few have excluded induced deliveries, although many such infants have IUGR (Yudkin *et al.*, 1987).

The study of Amini *et al.* (1994) provided sex-specific percentiles for birth weights in African Americans and European Americans. There are only small differences between these ethnic groups until after 38 weeks gestation when the birth weights for African Americans are about 100 to 200 g smaller than those for European Americans. The reference data that are recommended are those of Amini *et al.* (1994) from the United States, Karlberg *et al.* (1985) from Sweden, and Voigt *et al.* (1996) from Germany. These recommendations are based on the ways in which gestational ages were estimated, the criteria for exclusions, provision of sex-specific values and the range of measures for which data are provided. The reports of Milner & Richards (1974) and Roberts & Lancaster (1999) from the United Kingdom and Australia, respectively, are based on less accurate gestational ages, all live infants were included, and the reference data are limited to weight. There are reference data for BMI and for Rohrer's index at birth. Rohrer's index is weight (g) per length^3 (cm^3), and is sometimes called the ponderal index, although the ponderal index was originally defined as stature per weight$^{1/3}$. The best reference data for BMI and ponderal index are given by Lehingue *et al.* (1998) in chart and tabular format by gestational age at birth. These sex-specific data were derived from 88 581 French infants.

There are reference data for the postnatal growth of 65 AGA and 51 SGA preterm German infants who were VLBW and born before the introduction of high caloric-density feeding (Brandt, 1986). A major strength of the study is the large number of examinations from birth to 24 months GAA. Means and SDs were used to calculate the 3rd–97th percentiles, which were published in clinically useful formats. Major aspects of this study are summarized in Table 1.8.

Growth charts were developed for preterm United States infants using data from the Infant Health and Development Program (IHDP) and presented in clinically useful formats (Casey *et al.*, 1991;

Guo *et al.*, 1996, 1997*a*; Roche *et al.*, 1997). The age range covered is birth to 36 months (GAA) for combined groups of AGA and SGA infants (Table 1.8). The sample included 648 moderately LBW infants (birth weight 1501–2500 g) and 219 VLBW infants (birth weight ≤1500 g). The IHDP protocol excluded triplets, quadruplets, infants with serious medical conditions, and those whose mothers abused drugs or alcohol. African Americans were slightly over-represented and there was an over-representation of mothers with less than a high school education, which is common among United States mothers of LBW or VLBW infants. The IHDP charts provide sex-specific 5th–95th percentile values separately for VLBW and LBW infants (Guo *et al.*, 1996). Within birth weight groups, there are very small sex differences in weight and length, but head circumference percentiles are moderately higher in males than in females (Figures 1.35–1.38). The percentile values are higher for LBW infants than for VLBW infants, except for length, which differs only slightly between these groups. The IHDP charts are recommended as reference data, but they are not standards. New growth charts will be needed when the management of such infants changes in ways that affect their growth.

Ancillary reference data from IHDP relate weight and head circumference to length and provide adjustment factors to improve the assessment of SGA infants when reference data from the combined sample of AGA and SGA infants are applied (Guo *et al.*, 1998*b*; Roche *et al.*, 1997). Weight-for-length is smaller from birth to 36 months (GAA) in LBW infants than in VLBW infants. Percentile values of head circumference-for-length are smaller for LBW infants than for VLBW infants at lengths shorter than 60 cm, but they are larger at lengths greater than 80 cm. The differences between SGA infants and the combined data from AGA and SGA infants in the IHDP study vary for LBW and VLBW infants. In LBW infants, the SGA medians are the smaller by amounts that increase with age until they are about 600 g for weight and 0.6 cm for head circumference at 36 months (GAA). The corresponding differences for length change little with age being near zero for males, but the values for females are about 0.5 cm smaller for SGA infants than for AGA infants. In VLBW infants, the differences in median weights between AGA and SGA infants increase with age to 700 g at 36 months (GAA). Those for length and head circumference increase to 0.4 and 0.6 cm respectively in males at 36 months (GAA), but the differences are near zero at all ages in females. The deficits in length for SGA infants may be related to levels of insulin-like growth factor-1 (IGF-1). When IGF-1 levels are normal, catch-up is complete at 12 months, but catch-up is incomplete when these levels are low (Thieriot-Prévost *et al.*, 1988).

The Brandt percentile values for weight and length in VLBW infants are larger than the IHDP values. Indeed, the Brandt medians are about equal to the 90th IHDP percentiles at 12 months; later the differences between these charts decrease. These differences are not entirely due to the inclusion of SGA infants in the IHDP sample. The Brandt percentile values for head circumference are also larger than those from IHDP, but the differences are much smaller than those for weight and length.

If the birth weight is ≤2500 g and the gestational age is unknown, there may be uncertainty as to whether growth charts for preterm infants should be used. These charts cannot be used without gestational age at birth, which is needed to calculate GAA after birth. In these circumstances, gestational age can be estimated as the age at which the observed birth weight is equal to the median value. For example, a singleton infant with a birth weight of 1500 g would be assigned a gestational age of 30 weeks. The way in which gestational age was estimated should be noted on the chart.

DISEASE-SPECIFIC GROWTH CHARTS

There are growth charts for infants and children with some relatively homogeneous diseases that have marked effects on growth. These charts are applicable only to children who have not received specific therapy for the disease. Disease-specific

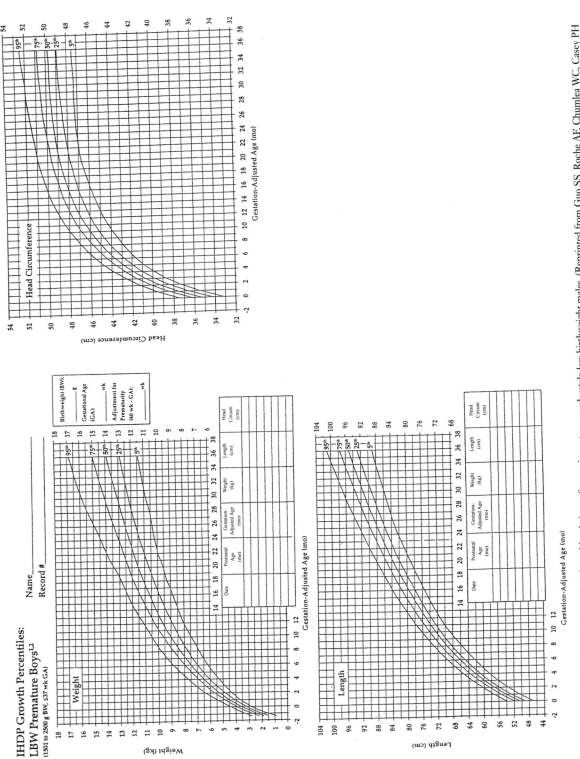

Figure 1.35. Growth charts of weight, length, and head circumference in preterm moderately low birthweight males. (Reprinted from Guo SS, Roche AF, Chumlea WC, Casey PH and Moore WM, Growth in weight, recumbent length, and head circumference for preterm low-birthweight infants during the first three years of life using gestation-adjusted ages, *Early Human Development*, **47**:305–325, 1997; with permission from Elsevier Science and Ross Products Division of Abbott Laboratories.)

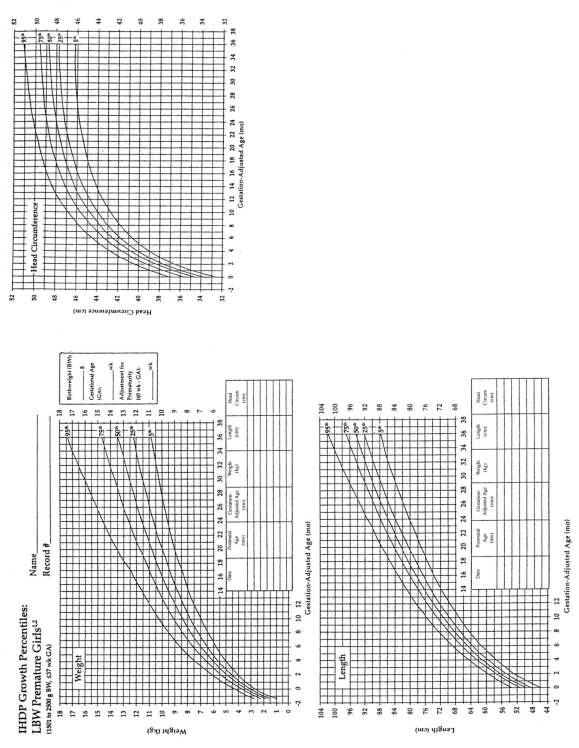

Figure 1.36. Growth charts of weight, length, and head circumference in preterm moderately low birthweight females. (Reprinted from Guo SS, Roche AF, Chumlea WC, Casey PH and Moore WM, Growth in weight, recumbent length, and head circumference for preterm low–birthweight infants during the first three years of life using gestation–adjusted ages, *Early Human Development*, 47:305–325, 1997; with permission from Elsevier Science and Ross Products Division of Abbott Laboratories.)

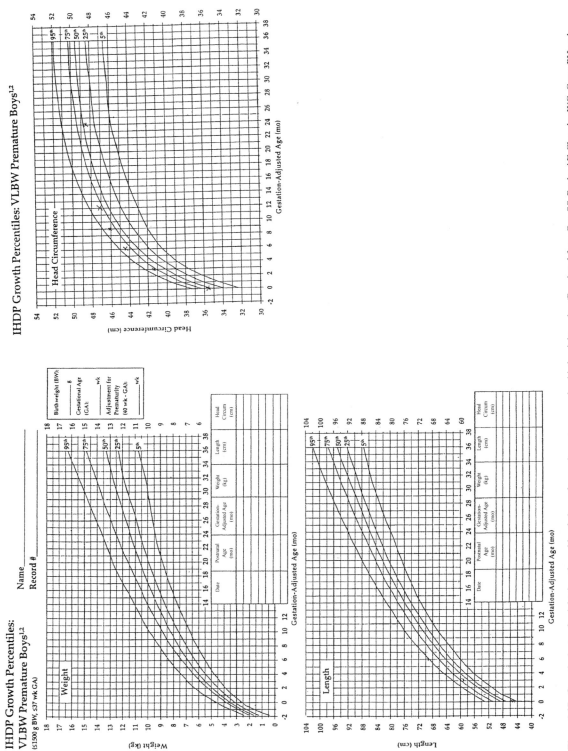

Figure 1.37. Growth charts of weight, length, and head circumference in preterm very low birthweight males. (Reprinted from Guo SS, Roche AF, Chumlea WC, Casey PH and Moore WM, Growth in weight, recumbent length, and head circumference for preterm low–birthweight infants during the first three years of life using gestation-adjusted ages, *Early Human Development*, **47**:305–325, 1997; with permission from Elsevier Science and Ross Products Division of Abbott Laboratories.)

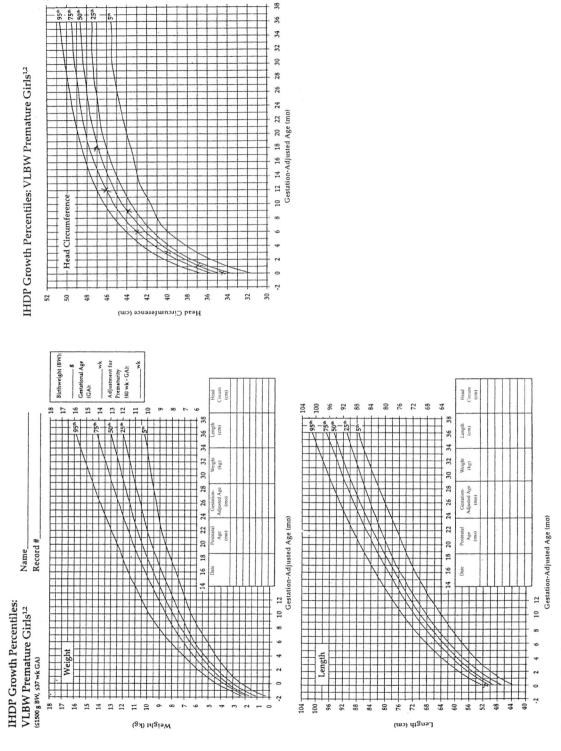

Figure 1.38. Growth charts of weight, length, and head circumference in preterm very low birthweight preterm females. (Reprinted from Guo SS, Roche AF, Chumlea WC, Casey PH and Moore WM, Growth in weight, recumbent length, and head circumference for preterm low-birthweight infants during the first three years of life using gestation-adjusted ages, *Early Human Development*, **47**:305–325, 1997; with permission from Elsevier Science and Ross Products Division of Abbott Laboratories.)

charts allow the growth of an affected child to be judged relative to others with the same disease. Therefore, they assist the counseling of parents and can provide a guide to future growth. Disease-specific growth charts can direct attention to children whose growth is unusual given their primary diagnosis and can assist investigations into the biological basis of the growth disorder. Variations in the growth of children with specific diseases may be due to other conditions that have an increased prevalence in children with the specific disease. For example, children with Turner syndrome have an increased prevalence of Crohn disease, hypothyroidism, and growth hormone deficiency.

Some characteristics of selected disease-specific growth charts are given in Table 1.9 (status values) and Table 1.10 (annual increments). These charts are based on smaller samples than those used to develop growth charts for general populations. As a result, most have reported means and SDs or the 10th, 50th, and 90th percentiles. Many of the samples for these charts are from multiple sites with possible inter-site variations in selection procedures, anthropometric methods, and data analysis. There may be selection bias in some studies due to the type of professionals who conducted them. Obstetricians and neonatologists are likely to reach diagnoses at younger ages than family physicians and the severity of the condition may be related to age at diagnosis.

Children with achondroplasia have extremely small statures. The $+2$ Z stature level for these children is less than the -2 Z level for normal children, but the -1 Z level of head circumference in achondroplastic children is close to the $+2$ Z level for normal children (Horton et $al.$, 1978). At 6 to 12 months, children with celiac disease have median weights less than the 3rd percentile for normal children and they are short.

When chronic renal insufficiency begins in infancy, the deficit in length is marked and may increase at the rate of 0.6 Z per month, but the stature deficit is stable from 3 years to the usual age of pubescence (Rodriguez-Soriano et $al.$, 1992). Commonly, dialysis or renal transplantation does not reduce the growth deficit. Data from a small sample

indicate the rate of growth in stature is normal during the pubertal spurt, which occurs at about the usual time. Nevertheless, adult stature is markedly reduced (-2.4 Z) in females, but not males (Carlucci et $al.$, 2000). Some, but not all, investigators find the growth deficit is related to the reduction in renal function, but peripheral resistance to insulin-like growth factors may be another important determinant of growth in these children (Bettinelli, 1991).

The term "constitutional short stature" (idiopathic short stature) is commonly applied to short children without a recognized disease (Cowell, 1994). When the parents are short, which is common, the term "familial short stature" may be applied. The stature deficits in these children increase after 2 years (Lifshitz & Cervantes, 1990). Adjustments of recorded statures for mid-parental stature assist the diagnosis and management of these children (Himes et $al.$, 1981). Methods to predict adult statures may overestimate in constitutional short stature (Lenko et $al.$, 1982; Zachmann et $al.$, 1978), however some have not found this (La Franchi et $al.$, 1991). Tall children without recognized diseases, who have tall parents, can be categorized as constitutional tall stature or familial tall stature. In these children, length is normal at birth, but becomes larger than normal during infancy. They remain tall throughout childhood and exceed their target statures (Dickerman et $al.$, 1984).

In some children, the major features are slow maturation and short stature, leading to use of the term "constitutional growth delay." In this condition, which is commonly familial, a stature deficit is usually present by 2 years that is more marked in the limbs than in the trunk. There is little change in the deficit from 2 years to the usual age of pubescence when the deficit increases in association with a delayed pubescent growth spurt (Burstein & Rosenfeld, 1987; Lifshitz & Cervantes, 1990). Adult statures tend to be less than the target statures (Crowne et $al.$, 1990, 1991; La Franchi et $al.$, 1991; Rensonnet et $al.$, 1999). Khamis & Roche (1995) analyzed serial data from children free of disease, but with skeletal ages retarded at least 1 Z and statures <10th percentile. These children were less

Table 1.9. *Selected disease-specific growth charts*

Authors	Country	Number of measurements	Criteria for inclusion	Age ranges (years)[a]	Variables[b]	Sex-specific	Reference data
Achondroplasia							
Horton *et al.* (1978)	United States	403	no drug therapy or shunts for hydrocephalus	B–17	S, HC	yes	mean, SD
Hunter *et al.* (1996)	United States + multinational	409	–	infancy–18	W-for-S; W/S	yes	mean, SD
Cystic fibrosis							
Santamaria *et al.* (1991)	Italy	137	–	0.5–18	S, W	yes	percentiles
Down syndrome							
Cremers (1993)	The Netherlands	265	–	B–20	S, W	yes	mean, SD percentiles
Cronk *et al.* (1988)	United States	730	–	1mo–18	S, W	yes	percentiles
Kuroki *et al.* (1995)	Japan	676	–	B–15	S, W, HC	yes	mean, SD
Palmer *et al.* (1992)	United States	421		B–3	HC	yes	mean, SD
Piro *et al.* (1990)	Italy	382	no mosaics or other diseases	B–14	S, W, HC	yes	mean, SD
Marfan syndrome							
Peyritz (1997)	United States	200	–	2–24	S, W	males	mean, SD
Myelomeningocele							
Ekvall (1993)	United States	600	–	2–18	S, W	yes	percentiles
Sickle cell disease							
Phebus *et al.* (1984)	United States	133	–	2–18	S, W	yes	percentiles
Thomas *et al.* (2000)	Jamaica	315	–	8–18	S, W	yes	percentiles
Silver–Russell syndrome							
Wollmann *et al.* (1995)	Germany	386	–	8–18	S, W, HC	–	Z scores
Turner syndrome							
Karlberg *et al.* (1993)	Europe	598	no hormonal therapy	3–21	S	–	mean ± 1, 2 SD
Milani *et al.* (1994)	Italy	772	no hormonal therapy	1–20	S	–	percentiles
Ranke *et al.* (1991)	Europe	617	no hormonal therapy	2–19	S	–	mean, SD
Sempé *et al.* (1996)	France	167	–	B–22	S, SiH, LL, S-inc	–	mean, SD percentiles
Suwa (1993)	Japan	704	–	B–19	S, W	–	mean ± 1, 2 SD

[a] B, birth.

[b] HC, head circumference; LL, leg length; S, stature; SiH, sitting height; S-inc, stature increments; W, weight.

Table 1.10. *Reference data for annual increments in stature for children with selected diseases*

Diseases	Authors	Country	Age range[a]
Achondroplasia	Horton *et al.* (1978)	United States	1–16 years
Constitutional tall stature	Dickerman *et al.* (1984)	Israel	1–10 years
Insulin-dependent diabetes mellitus	Brown *et al.* (1994)	United Kingdom	2–16 years
Down syndrome	Kuroki *et al.* (1995)	Japan	1–17 years
	Toledo *et al.* (1999)	France	4–17 years
Klinefelter syndrome	Schibler *et al.* (1974)	Switzerland	1.5–19.5 years
Noonan syndrome	Ranke *et al.* (1988)	Germany	1–19 years
Rubinstein–Tabyi syndrome	Stevens *et al.* (1990)	The Netherlands, United States	B–18 years
Thalassemia major	de Sanctis *et al.* (1991)	Italy	PHV −2 years to PHV +4 years
Turner syndrome	Massa *et al.* (1990)	Belgium	1–20 years
	Ranke *et al.* (1991)	Germany	B–20 years
	Rongen-Westerlaken *et al.* (1997)	Denmark, Netherlands, Sweden	B–18 years
	Sempé *et al.* (1996)	France	1–22 years
	Suwa (1993)	Japan	1–16 years

[a] B, birth; PHV, peak height velocity.

extreme in maturity and stature than is usual in constitutional growth delay. The mean parental statures were near the 25th percentiles. The growth deficits became more marked until 14 years in males and 12 years in females, but were later reduced when delayed pubescent spurts occurred. The mean deficits in adult stature were 9.5 cm for males and 5.0 cm for females. The errors of adult stature predictions were similar to those for normal children.

In children with cystic fibrosis, mean statures and weights from birth to 5 years are about 0.5 to 1.0 Z less than the means for normal children in males, but there are no stature deficits for females (Stettler *et al.*, 2000). After 5 years, statures and weights are about 0.5 to 1.0 Z less than the means for normal children. The growth deficits in this disease are related to the severity of respiratory impairment (Collins *et al.*, 1999; d'Orazio *et al.*, 1991; Santamaria *et al.*, 1991). Arm circumference, skinfold thicknesses, BMI, and weight-for-stature are near normal (Constantini *et al.*, 1991; Laursen *et al.*,

1999; McNaughton *et al.*, 2000; Taylor *et al.*, 1999), but TBF and FFM values are low (Barera *et al.*, 2000). Normal pubescent spurts occur at the usual age in males, but the spurts are small and delayed in females (Barkhouse *et al.*, 1989). Throughout growth, FFM is reduced in proportion to stature and weight (Shepherd *et al.*, 1989).

Growth charts for children with Down syndrome have been published from Japan, The Netherlands, Sicily, and the United States (Cremers, 1993; Cronk *et al.*, 1988; Kuroki *et al.*, 1995; Palmer *et al.*, 1992; Piro *et al.*, 1990). The means for weight are only slightly smaller than normal, but those for stature and head circumference are less than the 5th percentiles for general populations. These children have small pubescent growth spurts at about the usual age, but growth in stature ceases before adult levels of skeletal maturity are reached (Rarick & Seefeldt, 1974; Roche, 1965). The data of Cronk, Kuroki, and Piro from the United States, Japan, and Sicily, respectively, are similar for stature, but the means of Cremers from The Netherlands are larger by about 1.0 cm to 2 years; the difference increases to about 8 cm in males and 5 cm in females by 10 years and changes little at older ages. The Cremers means for weight are also larger by about 3 kg for males after 10 years, but the differences in weight for females are small except after 13 years when the Cremers values are smaller by about 3 kg. The charts of Cremers are recommended for use in The Netherlands and in other areas such as Scandinavia where children are tall. The data of Kuroki are preferred for use in Japan and those of Cronk and Palmer, which provide percentile levels, are recommended for other countries.

Children with diastrophic dwarfism are short at birth; this shortness becomes more marked with age. The $+1$ Z level of stature in these children is about 10 cm less than the 5th percentile for the general population at 10 years and is 26 cm less at 18 years (Horton et $al.$, 1982). In Duchenne muscular dystrophy, weight and length are normal at birth, but deficits develop during the first 3 years after which the mean weights and lengths are about 1 Z less than those for reference populations (Eiholzer et $al.$, 1998).

Children with insulin-dependent diabetes mellitus (IDDM) are moderately tall at diagnosis, if this occurs before 14 years (Blom et $al.$, 1992; Chiarelli et $al.$, 1988; Domargård et $al.$, 1999; Salardi et $al.$, 1987). The increase in stature is much more marked in males than in females (Blom et $al.$, 1992; Collell et $al.$, 1993; Vicens-Calvet et $al.$, 1977). Brown et $al.$ (1994) found that those diagnosed before 5 years were short, as were their parents. In some other studies, the statures of children with IDDM did not differ from those of controls (Emmerson & Savage, 1988; Holl et $al.$, 1994; Thon et $al.$, 1992). A study of monozygous twin pairs by Leslie et $al.$ (1991) is of marked interest. One member of each pair developed IDDM and growth in stature slowed in the affected twin with a growth velocity nadir about one year before diagnosis. In IDDM, the pubertal spurt is normal in timing, but it is small, especially in females (Brown et $al.$, 1994; Salardi et $al.$, 1987). In females during and after pubescence, %BF and skinfold thicknesses tend to exceed reference values for normal children (Domargård et $al.$, 1999; Gregory et $al.$, 1992).

Children with Klinefelter syndrome (47XXY karyotype) have a normal birth weight, but become tall soon after infancy (Frasier, 1986; Ratcliffe et $al.$, 1979; Schibler et $al.$, 1974). In association with their relatively long limbs, they have small values for BMI and the sitting height/leg length ratio. In Marfan syndrome, the mean length is increased only slightly at birth, but the mean statures are at the 95th percentile for the general population after 6 years. Weight is near normal until 14 years, after which it tends to be larger than reference medians. The limbs are disproportionately long, which tends to reduce BMI and the sitting height/leg length ratio; scoliosis or kyphosis may further reduce the sitting height/leg length ratio (Frasier, 1986; McKusick, 1972).

In myelomeningocele, the medians for weight are near those for normal children, but the 75th percentiles for stature are near the medians for normal children (Ekvall, 1993). The relative deficits in stature change little with age in males, but become larger in females after 12 years. Children with Noonan syndrome are short from birth with mean lengths and statures about -2 Z compared with normal children. The relative deficit in stature changes little with age (Ranke et $al.$, 1988). Children with psychosocial short stature are commonly born preterm with low birth weights (Gohlke et $al.$, 1998). Some consider this condition can occur in infancy as non-organic failure-to-thrive when there are disorganized households, inadequate nutrition, and insufficient attention to the needs of the infants (Krieger, 1973). Growth charts have not been developed that are specific for these children because the severity of the condition varies markedly. Commonly, statures and skeletal ages are about half those expected for chronological age, but weight is large relative to stature.

Prader–Willi children are short; this shortness is more marked after 13 years because they do not have pubescent spurts (Butler & Meaney, 1987; Holm, 1995). Males are increasingly overweight from 2 to 10 years, after which the extent to which they are overweight changes little. Females become increasingly overweight to young adulthood. Children with precocious puberty are tall and BMI is increased until the usual age of pubescence, but adult stature is normal (Bar *et al.*, 1995; Palmert *et al.*, 1999). Head circumference tends to be large relative to weight (Butler & Meaney, 1987). Adult stature is over-predicted by all methods in precocious puberty; these over-predictions are large before 7 years (Bar *et al.*, 1995; Kirkland *et al.*, 1981; Zachmann *et al.*, 1978).

The medians for weight in children with the Rubinstein–Taybi syndrome are near reference medians at birth, but are close to the 3rd percentiles from 4 months to 2 years. Mean weights are near normal from 2 to 10 years, but deficits develop later that are larger in males than females. The stature medians are near the 3rd percentiles of reference data to 10 years; later they are smaller than the 3rd percentile by about 8 cm (Stevens *et al.*, 1990). Weight-for-stature and head circumference are large particularly in males.

In children with sickle cell disease, weight and length are normal at birth, but there are deficits (-1.0 Z) for weight by 3 months and for stature by 9 months (Phebus *et al.*, 1984; Thomas *et al.*, 2000). At 15 months, the Z scores are near -2.0 for weight and stature in males, but are -1.4 for weight and -0.5 for stature in females. About 60% of these children are homozygous and the remainder are variants. In comparison with homozygotes, variant males are short at 2 to 5 years, tall and heavy at 9 to 11 years, and short at 15 to 18 years, while variant females are tall and heavy at 9 to 14 years after which they are similar in stature, but remain heavy. These descriptions of age changes in the size differences between homozygotes and variants are uncertain because they are based on cross-sectional data from 50 variant children (Phebus *et al.*, 1984).

In the Silver–Russell syndrome birth weight is low and stature is <-2 Z (Chatelain & Nicolino, 1994; Tanner *et al.*, 1975). Growth in stature is slow and, if pubescent spurts occur, they are small, but normal in timing (Davies *et al.*, 1988; Tanner *et al.*, 1975). In spondyloepiphysial dysplasia, length is markedly smaller than normal; this deficit increases with age. The $+1$ Z level of stature in these children is less than the 5th percentile for the general population by about 18 cm at 10 years and 31 cm at 18 years (Horton *et al.*, 1982). In thalassemia major (Cooley anemia), the median statures become markedly smaller than the 3rd percentile after 12 years (Iolascon, 1991). This increasing deficit, which is accompanied by slow maturation, is partly due to deficiencies of growth hormone, IGF-1 and zinc.

Most growth charts for children with Turner syndrome are restricted to stature, but Mazzanti *et al.* (1994) provide data for weight and those of Sempé *et al.* (1996) include sitting height and leg length. Lyon *et al.* (1985), using combined data from Finland, France, and Germany, present means, SDs, and percentiles for stature at ages from 1 year to 20+ years. These children are short after 1 year. This shortness becomes more marked until the group means are -2.5 Z compared with reference data for normal children at 4 years and older. Sempé *et al.* (1996) found, however, that the rate of growth in stature was normal to 7 years, but slow from 7 to 14 years. Weight is low at birth, but an increasing deficit develops, partly because these children do not have a pubescent spurt in weight. The means for weight are at about the 10th reference percentile at most ages, but those of weight-for-stature are >90th percentile.

Comparisons among the growth charts for children with Turner syndrome that are listed in Table 1.9 show the values from Sempé are the smallest. This suggests that the Sempé charts, which were developed from French children, should be used in France. There are only small differences among the means reported by Lyon (data from Finland, France, and Germany combined), Milani (data from Italy) and Ranke (data from Austria and Germany combined). Consequently, all these are appropriate for children from Austria, Finland, France, Italy, or Germany. Other data from a combined data set (Denmark, Sweden, The Netherlands) show mean statures at 2 to 8 years that are about 2 cm larger than those of Lyon, Milani, and Ranke (Karlberg *et al.*, 1993). The data of Karlberg are appropriate for the countries from which they were derived.

In this syndrome, adult stature predictions with the Bayley–Pinneau method are more accurate than those with the Roche–Wainer–Thissen or Tanner–Whitehouse methods (Zachmann *et al.*, 1978).

The applicability of growth charts to children with Turner syndrome is somewhat uncertain due to incomplete knowledge of the associations between karyotypes and growth. Information is lacking about karyotypes in all tissues, the parent of origin of the remaining X chromosome, gene expression, and X inactivation mapping of the supposedly inactivated X chromosome in relation to familial stature data (Hall, 1991). Length at birth does not differ among sub-groups with 45XO, X-mosaicism, and X-structural abnormalities, but those with X-mosaicism are shorter than the other sub-groups after 1 year (Mazzanti *et al.*, 1994). Mosaicism is the presence within an individual of two or more cell lines that differ in genotype although they are derived from a single fertilized ovum. Children with 45XO may be smaller than Turner children with other karyotypes, and have a more delayed pubescent spurt (Haeusler *et al.*, 1992). Others, however, did not find associations between karyotype and stature in children with Turner syndrome (Milani *et al.*, 1994; Pelz *et al.*, 1982; Sempé *et al.*, 1996).

In Williams syndrome, there are increasing deficits in weight and length from soon after birth. The means for these measures are smaller than the 5th percentiles for normal children from 1 to 4 years. The stature deficit becomes smaller from 4 to 10 years and then increases to 14 years when the mean is again near the 5th percentile. Adult stature is reached about two years earlier than usual and it is less than target stature (Boscherini *et al.*, 1993; Morris *et al.*, 1988; Pankau *et al.*, 1992).

Table 1.11 provides information about increments and growth patterns in children with selected diseases. Growth is slow in all the diseases listed, except IDDM and Williams syndrome in which the rates of growth are normal. Variations in the occurrence, timing, and size of pubescent spurts affect growth patterns in children with these diseases. Those who lack spurts decrease markedly in per-

centile levels during the usual age of pubescence; this decrease persists to adulthood. Those with small spurts that are normal in timing have a similar, but less marked, pattern of change during the usual age range of pubescence. Those with delayed small spurts decrease in percentile levels early in the pubescent period, and there is incomplete catch-up later. Pubescent spurts occur in the diseases for which information is available, with the exception of Turner syndrome, but spurts may be absent in some children with Down syndrome or thalassemia major. Pubescent spurts are delayed and small in constitutional growth delay, Down syndrome, Noonan syndrome, sickle cell anemia, and in females with cystic fibrosis. In achondroplasia, small spurts occur at a normal age. Pubescent spurts are usually normal in size and timing in familial tall stature, Silver–Russell syndrome and thalassemia major.

MATURITY REFERENCE DATA

Sexual maturity

The timing of stages of sexual maturity for groups can be analyzed by calculating probits, or by the use of cumulative frequencies from cross-sectional data (Michaut *et al.*, 1972; New Zealand Department of Health, 1971; Roede & van Wieringen, 1985; van Wieringen *et al.*, 1971). Alternatively, the percentage prevalences of particular grades can be calculated within age groups. As a result of such analyses, there are many sets of reference data for the timing of stages of sexual maturity in European groups. The timing is similar for English and Swiss groups for most stages (Largo & Prader, 1983*a*, *b*; Marshall & Tanner, 1969, 1970) and reference data for ages at the attainment of stages of secondary sexual maturation differ little among The Netherlands, Poland and Sweden, except that breast stages 3 and 4 and pubic hair stage 4 in females occur later in Poland (Bielicki *et al.*, 1984; Fredriks *et al.*, 2000*a*; Taranger *et al.*, 1976*a*). Data for United States youths have been reported from national surveys (Harlan, 1980; Harlan *et al.*, 1979, 1980; MacMahon, 1973; Villarreal *et al.*, 1989) and from other United States

Table 1.11. *Growth patterns in children with selected diseases*

Diseases	Authors	Rate of growth	Pubescent spurt		
			Occurrence	Timing	Size
Achondroplasia	Horton *et al.* (1978)	slow	yes	normal	small
Chronic renal insufficiency	Carlucci *et al.* (2000)	–	yes	normal	normal
	Rizzoni *et al.* (1984)	normal before pubescence	unusual	normal	small
	Rodriguez-Soriano *et al.* (1992)	–	yes	late	–
Constitutional growth delay	Alp Günöz *et al.* (1989)	–	yes	late	small
	Khamis & Roche (1995)	slow	yes	late	normal
	Rensonnet *et al.* (1999)	–	yes	–	–
Cystic fibrosis	Barkhouse *et al.* (1989)	normal	yes	normal	normal
Down syndrome	Kuroki *et al.* (1995)	slow	yes	early	–
	Rarick & Seefeldt (1974)	slow in males not females	–	–	–
	Roche 1965	slow	yes in 80%	late	small
	Toledo *et al.* (1999)	slow	yes	normal	small
Duchenne muscular dystrophy	Eiholzer *et al.* (1988)	slow	–	–	–
Familial short stature	Cowell (1994)	slow	yes	normal	normal
Hypophosphatemic vitamin D resistant rickets	Hauspie & Steendijk (1993)	–	yes	normal	normal
Insulin dependent diabetes mellitus	Brown *et al.* (1994)	–	yes	normal	small
	Chiarelli *et al.* (1988)	–	yes	–	normal in males; small in females
	Collell *et al.* (1993)	normal	–	–	–
	Domargård *et al.* (1999)	normal	yes	normal	small
	Salardi *et al.* (1987)	normal	yes	normal	small
Klinefelter syndrome	Schibler *et al.* (1974)	rapid	–	–	–
Noonan syndrome	Ranke *et al.* (1988)	–	yes	late	small
Obesity	de Simone *et al.* (1995)	rapid to pubescence	yes	normal	small
Prader–Willi syndrome	Holm & Nugent (1982)	–	no	–	–
Sickle cell anemia	Phebus *et al.* (1984)	slow	yes	late	small
Silver–Russell syndrome	Angehrn *et al.* (1979)	–	–	–	small
	Davies *et al.* (1988)	slow	yes	normal	small
	Tanner *et al.* (1975)	slow	yes	normal	small
	Wollmann *et al.* (1995)	slow	no	–	–
Thalessemia major	de Sanctis *et al.* (1991)	slow	yes	–	small
Turner syndrome	Haeusler & Frisch (1994)	–	yes	late	small
	Massa *et al.* (1990)	slow	no	–	–
	Ranke *et al.* (1991)	slow	no	–	–
	Rochiccioli *et al.* (1991)	slow	no	–	–
	Rongen-Westerlaken (1997)	slow after 6 years	no	–	small
	Sempé *et al.* (1996)	slow	yes	late	small
	Suwa (1993)	slow	no	–	–

Table 1.12. *Ages (years) at which stages of sexual maturity are reached*

	Males		Females	
	Mean	SD	Mean	SD
Genitalia				
Stage 2	11.2	1.5	–	–
Stage 3	12.9	1.2	–	–
Stage 4	13.8	1.1	–	–
Stage 5	14.7	1.1	–	–
Pubic hair				
Stage 2	12.2	1.5	10.4	1.2
Stage 3	13.5	1.2	12.2	1.2
Stage 4	14.2	1.1	13.0	1.1
Stage 5	14.9	1.0	14.0	1.3
Breasts				
Stage 2	–	–	10.9	1.2
Stage 3	–	–	12.2	1.2
Stage 4	–	–	13.2	0.9
Stage 5	–	–	14.0	1.2

Source: Largo & Prader (1983*a*, *b*).

groups (Herman-Giddens *et al.*, 1993; Roche *et al.*, 1995; Wyshak, 1983; Zacharias *et al.*, 1970). Stages 2 and 3 of secondary sexual maturation tend to occur earlier in the United States than in Europe particularly for African Americans (Herman-Giddens *et al.*, 1997).

Although Largo & Prader (1983*a*, *b*) used direct inspection of Swiss children and Marshall & Tanner (1969, 1970) used photographs of English children, there are only small differences between these reports in mean ages at which sexual maturity stages are reached, with the exception of pubic hair stage 2 and breast stage 5 for which the mean ages of Marshall & Tanner (1969) are 1.3 years later than those of Largo & Prader (1983*a*, *b*). The reference data of Largo & Prader (1983*a*, *b*) are recommended for clinical use (Table 1.12). The intervals between stages of sexual maturation are relevant to the interpretation of serial maturational data and to clin-

ical expectations of future progress. The large differences between studies for intervals that include stage 2 of pubic hair reflect methodological differences. Data for testis size show slightly slower maturation in French and Swiss children than in Dutch children (Fredriks *et al.*, 2000*a*; Largo & Prader, 1983*a*; Sempé *et al.*, 1979).

Age at menarche

In developed countries, the mean age at menarche is about 13.2 years (SD about 1.0 year), but it occurs slightly earlier in Italy, Spain, the United States, and the French-speaking part of Belgium (Danker-Hopfe, 1986; Herman-Giddens *et al.*, 1997; Wellens *et al.*, 1990; Zacharias *et al.*, 1970) and slightly later in Ireland (Hoey *et al.*, 1986). Menarche commonly occurs when pubic hair and breasts are at stage 4 (Billewicz *et al.*, 1981*a*; Largo & Prader, 1983*b*; Marshall & Tanner, 1969), but it may occur when breasts and pubic hair are at stage 1 or be delayed until these organs reach stage 5. These are marked differences between studies in the prevalence of menarche when pubic hair or breasts are at stage 5 that may reflect differences in the methods by which sexual maturity was assessed. The correlations between the age at appearance of secondary sexual stages and age at menarche are 0.4 to 0.7 (Bielicki, 1975; Largo & Prader, 1983*b*; Marshall & Tanner, 1969). In males, there are correlations of about 0.7 between the ages at onset of stages for pubic hair and for genitalia; the correlations in females between ages at the onset of pubic hair and of breast stages vary from 0.3 to 0.7 (Largo & Prader, 1983*a*, *b*; Nicholson & Hanley, 1953; Reynolds & Wines, 1948, 1951).

Age at peak height velocity

Estimates of ages at PHV necessarily come from serial studies of unrepresentative groups. Although estimates of the timing of the pubescent spurt reflect the choice of analytic method, there is good agreement among most European studies with reported mean ages at PHV varying from 13.4 to 14.4 years for males and from 11.4 to 12.4 years for females (Beunen & Malina, 1988; Beunen *et al.*, 1988;

Cameron *et al.*, 1982; Gasser *et al.*, 1985; Hauspie & Wachholder, 1986). Marubini *et al.* (1971, 1972), however, reported much younger ages for Italian females (10.2 years with a Gompertz model; 10.5 years with a logistic model). The estimated mean ages for PHV from studies in the United States and Canada range from 13.3 to 14.3 years for males, and from 11.4 to 12.8 years for females (Bock & Thissen, 1980; Demirjian *et al.*, 1985; Guo *et al.*, 1992; Mirwald & Bailey, 1986). Considerably younger ages of 10.8 years have been estimated for African-American and Japanese groups (Berkey *et al.*, 1993; Qin *et al.*, 1996). In a comparative study, the same smoothing procedure was applied to data from the Berkeley and Fels studies in the United States and the Zürich study (Ramsay *et al.*, 1995). The mean estimated ages at PHV in males are 13.6, 13.7, and 14.0 years for Berkeley, Fels, and Zürich respectively. The corresponding estimates in females are 11.5, 11.8, and 12.2 years (SD about 1.0 years). Within studies, the sex difference in mean ages at PHV is 1.7 to 2.2 years.

2 · Patterns of change in size and body composition

Growth is not uniform from infancy to adulthood, it occurs in patterns that occur at different ages and vary in magnitude among children. The patterns of growth of a child or a group of children can be described mathematically through models that summarize the overall pattern of growth across an age range. Knowledge of these patterns is necessary to interpret previous changes in individuals and they are the basis of informed expectations of future changes. This knowledge is fundamental to understanding the mechanisms that influence growth and maturation and the consequences of unusual changes. Considerable information is available about patterns of change in the common growth measures, but knowledge is sparse for patterns of change in total body composition, particularly for infants. This chapter describes normal growth patterns and discusses spurts, increments, diurnal and seasonal variations, tracking, decanalization, failure-to-thrive, catch-up growth, adult stature prediction and target stature. First, the utility of several mathematical models that can be used to describe growth patterns is presented including the subsequent analysis of their parameters.

MATHEMATICAL MODELS FOR DESCRIBING GROWTH PATTERNS

The procedure for describing patterns of change in growth variables begins by developing a model that describes and summarizes the changes in the variable across age (Guo *et al.*, 1997a, 2000a). The selection of a model begins with examination of the patterns of change reported in the literature. If little is known about such changes in the variable to be studied, plots should be made of the serial data for a random sub-set of children. Figures 2.1 and 2.2

are examples of patterns of growth in weight, stature, BMI, and head circumference for a male and a female from the Fels Longitudinal Study. These plots indicate that the patterns of change differ among these variables. The most notable differences are those between BMI, which decreases from about 1 year to 6 years, and the other variables in which decreases are uncommon. The sex differences in patterns of growth are small. This suggests that a single set of models may be appropriate for both sexes. It is also clear that the patterns differ in intensity among these measures. For example, the pubescent spurt is more marked in weight than in stature and it is difficult to detect this spurt in head circumference.

Inspection of these plots should suggest models that are likely to be appropriate. This approach leads to the selection of several functional forms as candidates for the model, with each functional form including a hierarchy of models of increasing complexity. Within each hierarchy, successively more complex models are fitted in order to identify the simplest model that adequately describes the data for individuals. The adequacy of the fit is determined graphically by plotting observed and predicted values and residuals and it is investigated by analyzing the residuals for systematic and random variations. Residuals are the differences between pairs of observed and predicted values.

This model-fitting procedure is summarized as follows:

1. Identify one or more classes of plausible models from the literature and from the examination of individual raw curves of serial data.
2. Fit models of the simplest type to each child's raw data, then calculate residuals.

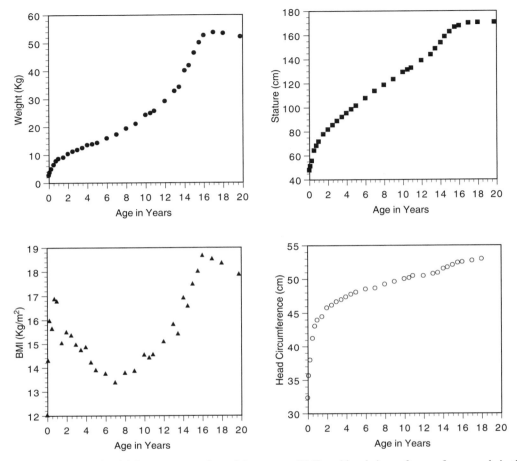

Figure 2.1. Plots of serial data versus age for weight, stature, BMI, and head circumference from a male in the Fels Longitudinal Study.

3. Examine the residuals for systematic deviation, both graphically and analytically, and formally test the next more complex model in the family for improved goodness of fit.
4. If the fit is adequate, ensure that the data are not over-fitted. The distribution of residuals should be consistent with measurement error and biologic variability.
5. If the fit is not adequate, repeat steps 2–4 with the next more complex model.

Weight, length, and head circumference during infancy

A family of mathematical models was fitted to the serial data for each individual for length (cm), weight

(kg), and head circumference (cm). The model was as follows:

$$f(t) = a + b \log(t+1) + c(t+1)^{0.5} + e \qquad (2.1)$$

where $f(t)$ is the length (weight, head circumference) at chronological age t (months), a, b, and c are the parameters to be estimated for each infant, and e is an error term. In addition to these linear growth curve models, non-linear growth models have been used to fit serial data during infancy and early childhood. Jenss & Bayley (1937) developed a model with linear and exponential components. This model is:

$$y = a + bt - e^{c+dt} \qquad (2.2)$$

where y is length or head circumference (cm) or weight (kg), t is age in months, a, b, c, and d are the

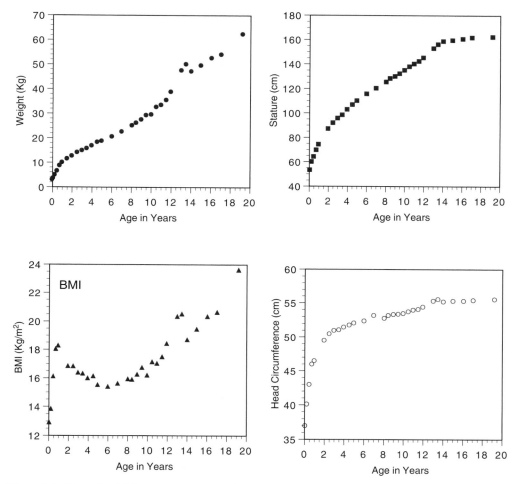

Figure 2.2. Plots of serial data versus age for weight, stature, BMI, and head circumference from a female in the Fels Longitudinal Study.

parameters to be estimated for each infant, and e is an error term. In this model, the linear component is an asymptote that the exponential component approaches over time.

Stature from infancy to adulthood

Numerous mathematical models have been used to describe growth in stature. The Preece–Baines model (1978) originates from a logistic function as the solution of the following differential equations:

$$\frac{ds}{dt} = \gamma(s_1 - s)(s - s_0) \qquad (2.3)$$

$$\frac{dh}{dt} = s(t)(h_1 - h) \qquad (2.4)$$

where h is the stature at an age, h_1 is the adult stature, s_0 and s_1 are rate constants, and γ is the age center of the equation governing the behavior of the rate of growth. The term $s(t)$ is closely related to age at peak height velocity.

The triple-logistic model was developed by Bock & Thissen (1976, 1980) to describe individual stature growth from 1 year to adulthood. In this model, stature growth is governed by separate components for early childhood, middle childhood, and adolescence. Let

$$h(t) = \frac{a_1 q}{1 + e^{-b_1(t-c_1)}} + \frac{a_1 p}{1 + e^{-b_2(t-c_2)}} + \frac{f - a_1}{1 + e^{-b_3(t-c_3)}}$$

$$(2.5)$$

where $h(t)$ is the stature at time t, a_1 is the contribution of prepubertal growth to adult stature, b_1 is the slope of the early-childhood component at maximum velocity, c_1 is the age of maximum velocity of the early-childhood component, b_2 is the slope of the middle-childhood component at maximum velocity, c_2 is the age at maximum velocity of the middle-childhood component, p is the proportion of prepubertal growth attributable to the middle-childhood component, $q = 1 - p$, $f - a_1$ is the contribution of the adolescent component to adult stature, b_3 is the slope of the adolescent component at maximum velocity, and c_3 is the age at maximum velocity of the adolescent component.

Kernel regression is a non-parametric procedure that provides good approximations to individual serial data (Cleveland, 1979). This method is useful and flexible when a parametric method is inappropriate due to restricted assumptions on the shape of the curve. Kernel regression is applicable to variables other than stature, including weight, and to various age ranges, whereas the triple-logistic model is applicable only to stature from 1 to 18 years of age and it may be inefficient when growth is abnormal. The principle of kernel regression is that the estimates are weighted averages of the observations, i.e.,

$$\hat{f}(t) = \sum_{i=1}^{n} f(t_i) p_i(t) \qquad (2.6)$$

The weight function, $p_i(t) = \frac{1}{b(n)} \int_{s_{i-1}}^{s_i} w[\frac{t_i - x}{b(n)}] dx$, is obtained from integration of a kernel function, $w(t)$, where $w(x)$ is a kernel of order k. The kernel function is usually derived from some mathematically optimal conditions such as minimum integrated mean square errors and asymptotic minimum variance.

In our study, the kernel was directly estimated by least squares (Guo *et al.*, 1992). The kernel estimators employed families of polynomials as kernel functions. The kernel estimates were obtained for each participant using 2 years for the bandwidth and a second-degree polynomial as the kernel function. To estimate the value at time t, the smoothing parameter, b (bandwidth), is chosen arbitrarily. The

observations at times t_1, t_2, t_3, and t_4 covered by $t \pm b$ are used to estimate the value at time t. The horizontal axis represents the transformed values of time axes t and S, where S is the midpoint of the t interval. The vertical axis represents the kernel function and the area under the curve is the weight for the corresponding observation. For example, the weight for observation $f(t_3)$ is the integration of the kernel function from $(t - S_3)/b$ to $(t - S_2)/b$. The kernel estimate for $f(t)$ is the summation of the products between observations and weights.

The developed kernel regression was compared with those from two families of mathematical models, one that incorporates the possible existence of the mid-growth spurt (a triple-logistic model) while the other does not (Preece–Baines model). The overall goodness of fit of the Preece–Baines model is less than those of the other two models. A sub-set of the subjects had mid-growth spurts that were apparent with both the kernel regression and the triple-logistic models and both these models performed well in regard to the overall goodness of fit and their ability to quantify the rate of growth, and the timing and duration of growth events.

Body mass index

Serial values for BMI from 2 to 18 years have been fitted by a mathematical model and summarized into a few biological interpretable parameters (Siervogel *et al.*, 1991). These derived parameters were used in an analysis that related childhood adiposity to adulthood adiposity. This model can be expressed as

$$\ln Y = \beta_0 + \beta_1 x + \beta_2 x^2 + \beta_3 x^3 + \varepsilon \qquad (2.7)$$

where Y is BMI at age x, $\beta_0, \beta_1, \beta_2, \beta_3$ are parameters, and ε is an error term. Parameters describing the patterns of growth can be derived from the parameters in the model. These parameters are: estimated value of BMI at 2 years of age (BMI_{2yr}), minimum value of BMI (BMI_{min}), age at minimum value of BMI (Age_{min}), maximum velocity of BMI (V_{max}), age at maximum velocity of BMI ($Age_{V_{max}}$), maximum value of BMI (BMI_{max}), and age at maximum value of BMI (Age_{max}). These estimated values were derived from the fitted four-parameter models:

$$\text{Age}_{\min} = -\frac{\hat{\beta}_2}{3\hat{\beta}_3} + \frac{\sqrt{\hat{\beta}_2^2 - 3\hat{\beta}_1\hat{\beta}_3}}{3\hat{\beta}_3} \quad (2.8)$$

where $\hat{\beta}_2^2 - 3\hat{\beta}_1\hat{\beta}_3 \geq 0$

$$\text{BMI}_{\min} = \exp\left(\hat{\beta}_0 + \hat{\beta}_1\text{Age}_{\min} + \hat{\beta}_2\text{Age}_{\min}^2 + \hat{\beta}_3\text{Age}_{\min}^3\right) \quad (2.9)$$

$\text{Age}_{V_{\max}}$ was calculated by solving a fourth-order equation iteratively using the Newton–Raphson method (Gerald, 1978). The equation is obtained from equating the second derivative of the mathematical function to zero.

$$V_{\max} = \exp\left(\hat{\beta}_0 + \hat{\beta}_1\text{Age}_{V_{\max}} + \hat{\beta}_2\text{Age}_{V_{\max}}^2 + \hat{\beta}_3\text{Age}_{V_{\max}}^3\right) \quad (2.10)$$

$$\text{Age}_{\max} = -\frac{\hat{\beta}_2}{3\hat{\beta}_3} - \frac{\sqrt{\hat{\beta}_2^2 - 3\hat{\beta}_1\hat{\beta}_3}}{3\hat{\beta}_3} \quad (2.11)$$

where $\hat{\beta}_2^2 - 3\hat{\beta}_1\hat{\beta}_3 \geq 0$

$$\text{BMI}_{\max} = \exp\left(\hat{\beta}_0 + \hat{\beta}_1\text{Age}_{\max} + \hat{\beta}_2\text{Age}_{\max}^2 + \hat{\beta}_3\text{Age}_{\max}^3\right) \quad (2.12)$$

Individual serial data for BMI from 2 to 25 years of age have also been summarized by a third-degree polynomial (Guo et al., 2000a). This model described the patterns of change very well. In males, the mean root mean square error (RMSE) was 0.58 kg/m^2 and the SD was about 0.18 kg/m^2. The corresponding mean RMSE for females was 0.63 kg/m^2 and the SD was 0.21 kg/m^2. The individual childhood BMI parameters describing the pattern of change in BMI were derived from the fitted model.

There was a general trend in the BMI patterns for the group. After approximately 2 years of age BMI decreased and reached a minimum at about 5 years, then increased to reach a maximum at approximately 20 to 22 years of age. Three sets of BMI parameters, representing three "critical periods," were developed from the fitted data for each individual. The parameters for age at rebound, BMI minimum (BMI$_{\min}$) and age at BMI minimum (Age$_{\min}$) represent the changes during the early childhood

period. The pubescent period is represented by the maximum velocity of BMI ($V_{\max}$), BMI at maximum velocity (BMI$_{V_{\max}}$) and age at maximum velocity of BMI (Age$_{V_{\max}}$). The post-pubescent period is represented by BMI maximum (BMI$_{\max}$) and age at BMI maximum (Age$_{\max}$).

The derivation of BMI parameters from the third-degree polynomial is as follows:

$$\text{Age}_{\min} = -\frac{\hat{\beta}_2}{3\hat{\beta}_3} + \frac{\sqrt{\hat{\beta}_2^2 - 3\hat{\beta}_1\hat{\beta}_3}}{3\hat{\beta}_3} \quad (2.13)$$

where $\hat{\beta}_2^2 - 3\hat{\beta}_1\hat{\beta}_3 \geq 0$, and

$$\text{BMI}_{\min} = \hat{\beta}_0 + \hat{\beta}_1\text{Age}_{\min} + \hat{\beta}_2\text{Age}_{\min}^2 + \hat{\beta}_3\text{Age}_{\min}^3 \quad (2.14)$$

$\text{Age}_{V_{\max}}$ is calculated by equating the second derivative of the mathematical function to zero.

$$\text{Age}_{V_{\max}} = -\frac{\hat{\beta}_2}{3\hat{\beta}_3} \quad (2.15)$$

The BMI value at maximum velocity of BMI (BMI$_{V_{\max}}$) is calculated as

$$\text{BMI}_{V_{\max}} = \hat{\beta}_0 + \hat{\beta}_1\text{Age}_{V_{\max}} + \hat{\beta}_2\text{Age}_{V_{\max}}^2 + \hat{\beta}_3\text{Age}_{V_{\max}}^3 \quad (2.16)$$

$$V_{\max} = \hat{\beta}_1 + 2\hat{\beta}_2\text{Age}_{V_{\max}} + 3\hat{\beta}_3\text{Age}_{V_{\max}}^2 \quad (2.17)$$

$$\text{Age}_{\max} = -\frac{\hat{\beta}_2}{3\hat{\beta}_3} - \frac{\sqrt{\hat{\beta}_2^2 - 3\hat{\beta}_1\hat{\beta}_3}}{3\hat{\beta}_3} \quad (2.18)$$

where $\hat{\beta}_2^2 - 3\hat{\beta}_1\hat{\beta}_3 \geq 0$, and

$$\text{BMI}_{\max} = \hat{\beta}_0 + \hat{\beta}_1\text{Age}_{\max} + \hat{\beta}_2\text{Age}_{\max}^2 + \hat{\beta}_3\text{Age}_{\max}^3 \quad (2.19)$$

The derived parameters are summarized in Table 2.1. The average BMI$_{\min}$ at the BMI rebound is 15.33 kg/m^2 for males at an average Age$_{\min}$ of 5.4 years. The corresponding mean values for females are 14.87 kg/m^2 at 5.3 years of age. There is a significant difference ($p < 0.05$) between males and females in their average BMI$_{\min}$ values, but not in their Age$_{\min}$ values during the BMI rebound. After the BMI rebound, BMI values increase. The

Table 2.1. *Means and standard deviations of the derived BMI parameters for males and females*

BMI parameters	Males ($n = 180$)		Females ($n = 158$)	
	Mean	SD	Mean	SD
BMI rebound				
BMI_{min} (kg/m^2)	15.33[a]	1.01	14.87	1.06
Age_{min} (years)	5.40	1.36	5.28	1.30
Pubescence				
V_{max} (kg/m^2 per year)	0.71	0.20	0.68	0.20
V_{max} (kg/m^2)	19.57[a]	1.74	18.40	1.60
$Age_{V_{max}}$ (years)	14.30[a]	1.82	13.04	1.58
Post-pubescence				
BMI_{max} (kg/m^2)	23.80[a]	3.01	21.92	2.58
Age_{max} (years)	23.19[a]	3.21	20.80	2.47

[a] Sex differences significant at $p < 0.05$.
Source: Guo *et al.* (2000*b*).

maximum rate of this increase (V_{max}) does not differ between the sexes. $Age_{V_{max}}$ occurs at significantly younger ages for females (13.04 years) than for males (14.3 years), but the $BMI_{V_{max}}$ value is significantly larger in males than in females ($p < 0.05$). After pubescence, BMI values reach their maximum at younger ages in females than in males with a sex difference in timing of about 2.4 years, but BMI_{max} is significantly larger in males than in females ($p < 0.05$). Through childhood into young adulthood, males tend to have larger BMI values than females, but the timing of adiposity events after the BMI rebound is 1–2 years earlier in females than in males.

GROWTH AND BODY COMPOSITION DURING INFANCY

During infancy, body dimensions increase at a faster rate than at any other time in postnatal life. Because an infant is small, the proportional increases are less obvious than those during the pubescent spurt. In the first year after birth, weight increases 200%, body length 55%, and head circumference 30%. Between 1 and 2 years of age, an average child grows

about 12 cm in length and gains about 3.5 kg in weight. An infant's head is disproportionately large at birth. Head circumference is larger than chest circumference at birth, and the length of the head is equal to about a quarter of the body's total length. At birth, head circumference is about 35 cm and increases about 12 cm during the first year and 5 cm during the second year.

Incremental reference data and mathematical functions fitted to serial data show these rapid increases in weight, length, and head circumference during the first year after birth, and that the rate of growth decreases markedly during this period (Guo *et al.*, 1988, 1991; Roche *et al.*, 1989*a*, *b*; Roche & Himes, 1980). Some describe small spurts in length at 3 to 8 months and at 9 months (Karlberg, 1987; van den Broeck *et al.*, 2000), but these findings do not match those from other analyses. Skinfold thicknesses increase rapidly to 5 months after which they decrease slowly at the triceps and subscapular sites, but not at the suprailiac site (Karlberg *et al.*, 1976).

Even brief interruptions in growth from birth to 1 year can have important consequences because very rapid growth is normal during this period (Reinken *et al.*, 1979). Slow growth of head circumference in early infancy can have serious effects because growth in head circumference during the first year is synchronous with increases in the number of glial cells, myelination, growth of dendrites, and the establishment of synapses (Rabinowicz, 1986). Permanent functional deficits may occur when the increases in head circumference are smaller than usual.

Saltations and mini-spurts

There may be short-term changes in length during infancy that are described as mini-spurts by some and as alternating saltations and stasis by others (Hermanussen & Burmeister, 1993; Lampl & Johnson, 1999). Mini-spurts are small undulations in growth rates. Saltations are sudden rapid increases in length which are said to be separated by periods of stasis when length does not increase. The choice between these descriptions could influence the search for underlying physiological mechanisms.

Measurements of infants at intervals of about 1 week may show saltations in length of 0.5 to 2.0 cm separated by periods of 2 to 60 days during which length does not increase or increases slowly (Gibson & Wales, 1994; Heinrichs et al., 1995; Lampl, 1993). Similar, but less marked, changes occur during pubescence (Lampl & Johnson, 1993; Lampl et al., 1998). The recognition of saltations or mini-spurts requires careful measurements at short intervals and may require application of a mathematical curve-fitting procedure. These variations are unlikely to be noted clinically and they are not displayed in reference data. Other studies describe an undulating pattern of mini-spurts (Hermanussen & Burmeister, 1993; Hermanussen et al., 1988). Some of the differences between reports are due to variations in methods of analysis. A possible spurt in weight may occur at about 2 months that is larger in males than females, but this finding has not been replicated (Giani et al., 1996). The common irregularity of weekly weight data during infancy should be interpreted in relation to measurement errors, variations in the timing of measurements relative to feeding, and possible real changes in growth rates.

Increments

Growth progress of an individual is usually judged from serial data points plotted on charts for growth status. This procedure shows whether the trend of the serial status values is parallel to nearby percentile lines, but when the trend deviates from that of the nearby percentiles interpretation is difficult because the distributions of these deviations are unknown. This simple procedure is less sensitive than the use of increments, which are recommended for infants and children at risk of growth failure. Abnormal growth progress can be detected earlier and more precisely from increments than from serial status values (Roche et al., 1989a; Waterlow, 1988).

The use of increments to assess growth has strong appeal, but some conditions must be met before increments can provide accurate and useful information. The errors of measurement must be small because those at each end of the interval are included in the calculated increment. When two successive increments are calculated, an error at the common age point will have opposite effects on the two recorded increments. If the value for the common age point is erroneously large the first increment will be increased and the second decreased. Measurement errors are partly responsible for the low correlations between successive increments, but these also reflect variations in growth rates and regression to the mean (Roche & Himes, 1980).

Incremental reference data are for intervals of specified lengths, but it is logistically difficult to measure children at exactly these intervals. An adjustment should be made when the interval between examinations differs slightly from the specified interval. This can be done by dividing the observed increment by the number of days in the actual interval and multiplying the result by the number of days in the reference interval. Alternatively, this calculation can be made after expressing the actual interval and the reference interval in decimals of a year. An adjustment should not be made if the difference between the actual and reference intervals exceeds one-quarter of the reference interval. Such adjustments would be particularly misleading during infancy and pubescence when the rates of

Table 2.2. *Selected reference data for increments during infancy*

Authors	Country	Variables[a]	Intervals	Age range
Guo *et al.* (1988)	United States	HC	1 mo	birth – 12 mo
Guo *et al.* (1991)	United States	W, L	various	birth – 24 mo
Hansman (1970)	United States	W, L	various	birth – 12 mo
		W, L	6 mo	18–24 mo
Prader *et al.* (1989)	Switzerland	W, L, HC, CRL, LL	various	1–36 mo
Prader & Budliger (1977)	Switzerland	W, L	2 and 6 mo	1–36 mo
Roche & Himes (1980)[b]	United States	W, L, HC	6 mo	birth – 36 mo
Roche *et al.* (1989*b*)	Canada	W, L, HC	1 mo	birth – 12 mo
Russo & Zaccagni (1993)	Italy	W, L	1 mo	1–36 mo
Sempé *et al.* (1979)	France	W, L, HC, CRL, LL	6 mo	birth – 36 mo
Karlberg *et al.* (1976)	Sweden	W, L, HC, AC, SKF	various	1–36 mo
van't Hof *et al.* (2000*a*)	Europe	W, L, HC	various	birth – 36 mo

[a] AC, arm circumference; CRL, crown–rump length; HC, head circumference; L, length; LL, leg length; S, stature; SKF, skinfolds; W, weight.

[b] Charts only; corresponding tabular data are in Baumgartner *et al.* (1986).

growth change rapidly. Some data are expressed as increments per day (Guo *et al.*, 1991). Corresponding restrictions apply to the application of these data for intervals that differ considerably from those for which the reference data were calculated.

In the context of calculating and interpreting increments, the intervals between examinations should be long enough that the 5th percentiles of the increments are larger than the technical errors (TE):

$$\text{TE} = \sqrt{\frac{\sum_{i=1}^{I} \sum_{j=1}^{n_i} (x_{ij} - \bar{x}_i)^2}{\sum_{i=1}^{I} n_i}} \qquad (2.20)$$

where x_{ij} is the jth measurement of the ith participant for $i = 1, 2, \ldots, I$ and $j = 1, 2, \ldots, n_i$ (Roche & Guo, 1998). The TE for weight in infancy is about 10g, which could lead to a suggestion that weight be measured daily to 6 months and at 2- or 3-day intervals for the remainder of infancy. There is, however a biological variation in weight of about 100–250 g associated with feeding, defecation, and urination indicating that intervals of 1 to 3 months, depending on age, are appropriate. The time required for the increment to be twice the standard error, after adjusting for measurement errors, is another guide to the minimum intervals of increments (Hermanussen & Burmeister, 1989; Himes, 1999*a*). With this approach, the estimated minimum intervals for length and stature are about 1 month from birth to 12 months, 6 weeks at 2 years, 3 months at 3 years, 5 months at 6 to 8 years, 4 months from 1 year before to 1 year after PHV, and 6 months from 1 year after PHV to 16 years.

The intervals of reference data for increments must be as short as is practical, particularly during the first year after birth. Increments for long intervals necessitate extensive delays before the rate of growth can be assessed and they are affected by the rapid changes in growth rates during infancy. Short-term increments are, however, more vulnerable to measurement errors than long-term increments, because the errors are large relative to observed increments. When two successive increments are both either small or large, measurement errors are unlikely to be responsible.

Sources of reference data for increments during infancy are given in Table 2.2. Most of these data

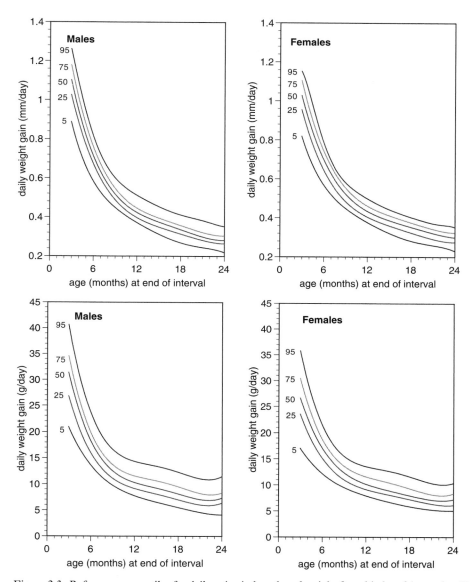

Figure 2.3. Reference percentiles for daily gains in length and weight from birth to 24 months. (Data from Guo *et al.*, 1991.)

sets are for intervals shorter than 6 months and most extend from birth to 36 months. Incremental reference data for normal infants are available in tables or clinically useful charts for weight, length, and head circumference (Figures 2.3 and 2.4). Those from Europe are shown in relation to the midpoint of each interval; those from the United States are related to the end of each interval. The percentile levels for increments during infancy in weight, length, and head circumference decrease rapidly to 12 months, after which they decrease slowly to 36 months (Guo *et al.*, 1991; Prader *et al.*, 1989; Roche & Himes, 1980; Sempé *et al.*, 1979).

Guo *et al.* (1991) developed incremental reference data for 1- 2- or 3-month intervals from a mathematical model fitted to the serial data for each

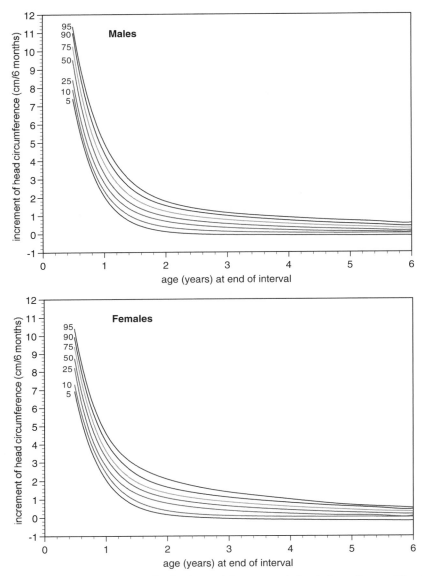

Figure 2.4. Reference percentiles for 6-month increments in head circumference. The observed increments should be plotted opposite the end of each interval. (From Roche & Himes, 1980; with permission from Ross Products Division of Abbott Laboratories.)

infant. One-month increments are recommended for weight from birth to 6 months. At older ages, 2- and 3-month intervals are recommended for weight increments because the differences between the 5th and 50th percentiles of 1-month increments are smaller than the changes expected due to feeding and similar to those due to defecation. For length,

2-month increments are recommended to 4 months and 3-month increments are recommended from 4 to 24 months.

Incremental reference data for triceps and subscapular skinfold thicknesses during 3-month intervals to 12 months and 6-month intervals from 12 to 24 months are available (Karlberg et al., 1976). It is difficult to apply these data because the measurement errors are large relative to the expected increments. The increments from birth to 3 months are large for triceps, subscapular, and suprailiac skinfold thicknesses. The median increments decrease to near zero or become negative for triceps skinfold thicknesses from 24 to 36 months, for subscapular skinfold thicknesses from 12 to 36 months, and for suprailiac skinfold thicknesses from 6 to 36 months. The increments in triceps skinfold thicknesses are larger in males than females to 9 months, but there is a reverse sex difference from 12 to 24 months. Increments in subscapular skinfold thicknesses tend to be larger in males than females, but increments in suprailiac skinfold thicknesses do not show sex differences.

Low birth weight infants

Preterm LBW and VLBW infants grow more rapidly than term infants of normal birth weight to 3 months (GAA). This rapid growth is more marked for weight than for length and head circumference (Guo *et al.*, 1997*a*). The weights of moderately LBW preterm infants (birth weight 1501–2500 g) increase more rapidly for males than females to 9 months (GAA), but there is a reverse sex difference in the rate of growth from 9 to 36 months (GAA). In moderately LBW and in VLBW (birth weight <1500 g) preterm infants, length increases more rapidly in males than females to 6 months, but the increases from 6 to 36 months (GAA) do not differ between the sexes. Head circumference increases more rapidly in males than females for moderately LBW and VLBW preterm infants to 16 months (GAA), but there is an opposite sex difference from 29 to 36 months (GAA).

Incremental reference data for weight, length, and head circumference in LBW preterm infants are for intervals between gestation-adjusted ages (GAA). The data of Brandt (1986) from term to 60 months (GAA) are from VLBW infants who were AGA. Guo *et al.* (1997*a*) presented data from birth to 36 months (GAA) for combinations of AGA and SGA infants in two birth weight groups (VLBW

≤1500 g; LBW 1501–2500g). These charts are restricted to the 5th, 50th, and 95th percentiles because intermediate percentiles would be spaced too closely for accurate plotting and the differences between adjacent lines would be only slightly larger than the measurement errors. These sex-specific charts of Guo are for 1-month increments from birth to 7 months, for 2-month increments from birth to 14 months (GAA) and for 3-month increments from birth to 36 months in moderately LBW and VLBW infants (Figures 2.5–2.10). The percentile values for these increments decrease rapidly with age to 7 months and then decrease slowly to 36 months (GAA).

Preterm LBW infants who are SGA at birth have small weights to 36 months (GAA) compared with general reference data. The median weights are near the 25th percentile throughout infancy for those moderately LBW at birth and near the 10th percentile for VLBW infants. After 18 months (GAA), the median levels for length in SGA infants who are moderately LBW at birth increase from near the 25th percentile to about the 35th percentile and those for infants born VLBW increase from the 10th to the 25th percentile (D'Souza *et al.*, 1985; Guo *et al.*, 1998*b*; Hokken-Koelega *et al.*, 1995). In preterm infants, the means for %BF and TBF increase rapidly to term, but change little during the next 5 months. The rate of increase in FFM becomes slightly slower from birth to 5 months (GAA) in preterm infants, but total body BMC doubles from term to 3 months (GAA) and doubles again by 9 months (Rawlings *et al.*, 1999). In term infants, %BF increases steadily to 24 months (Butte *et al.*, 1999).

Term SGA infants

Term SGA infants (birth weight <2500 g) have small weights, lengths, and head circumferences at birth. The deficits in weight and length decrease markedly from birth to 2 months, but change only slightly from 2 months to 2 years; those in head circumference decrease to near zero by 2 years (Albertsson-Wikland *et al.*, 1993; Fujimora

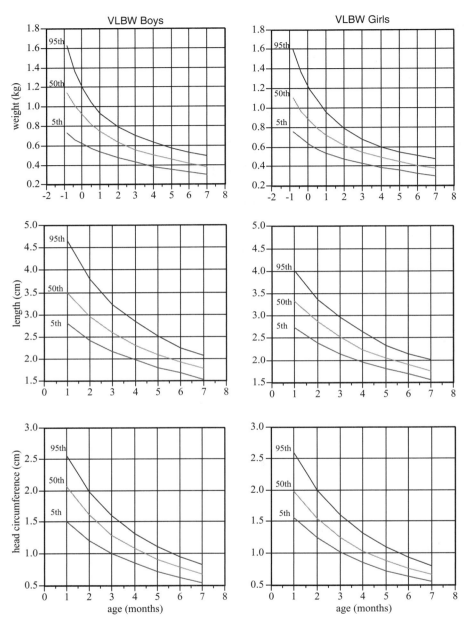

Figure 2.5. Reference percentiles for 1-month increments in weight, length, and head circumference for very low birth weight (VLBW) preterm infants from zero to 7 months GAA. The observed increments should be plotted opposite the end of each interval. (Reprinted from Guo SS, Roche AF, Chumlea WC, Casey PH and Moore WM, Growth in weight, length and head circumference for preterm low-birth weight infants during the first three years of life using gestation-adjusted ages. *Early Human Development* 47: 305–325, 1997 with permission from Elsevier Science and Ross Products Division of Abbott Laboratories.)

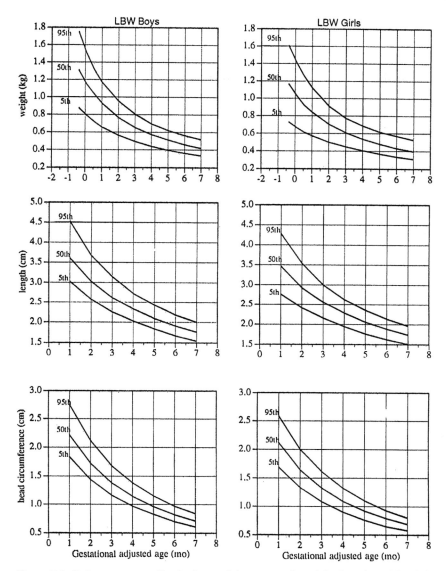

Figure 2.6. Reference percentiles for 1-month increments in weight, length, and head circumference for low birth weight (LBW) preterm infants from 1 to 7 months GAA. The observed increments should be plotted opposite the end of each interval. (Reprinted from Guo SS, Roche AF, Chumlea WC, Casey PH and Moore WM, Growth in weight, length and head circumference for preterm low-birth weight infants during the first three years of life using gestation-adjusted ages. *Early Human Development* 47:305–325, 1997 with permission from Elsevier Science and Ross Products Division of Abbott Laboratories.)

& Seryu, 1977; Hokken-Koelega *et al.*, 1995; Leger *et al.*, 1998). Persistent deficits in stature are commonly associated with low levels of growth hormone (Ackland *et al.*, 1988; Boguszewski *et al.*, 1998; Rochiccioli *et al.*, 1989; Stanhope *et al.*, 1989).

Although the size of the growth deficits in term SGA infants decrease after birth, there are increased prevalences of unusually small values to 5 years for weight and stature, but not head circumference (Albertsson-Wikland *et al.*, 1998; Fitzhardinge

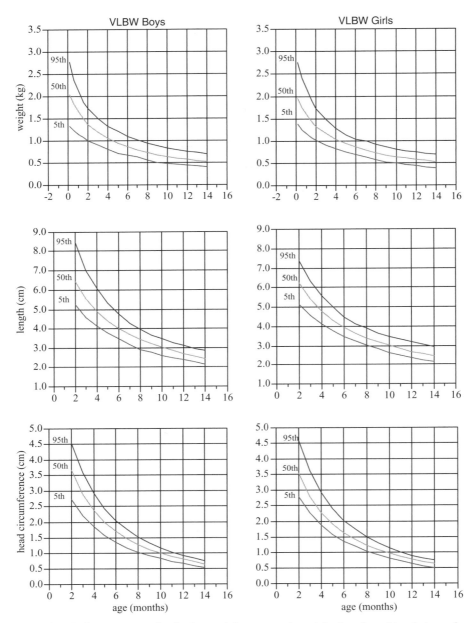

Figure 2.7. Reference percentiles for 1-month increments in weight, length, and head circumference for very low birth weight (VLBW) preterm infants from 2 to 14 months GAA. The observed increments should be plotted opposite the end of each interval. (Reprinted from Guo SS, Roche AF, Chumlea WC, Casey PH and Moore WM, Growth in weight, length and head circumference for preterm low-birth weight infants during the first three years of life using gestation-adjusted ages. *Early Human Development* 47:305–325, 1997 with permission from Elsevier Science and Ross Products Division of Abbott Laboratories.)

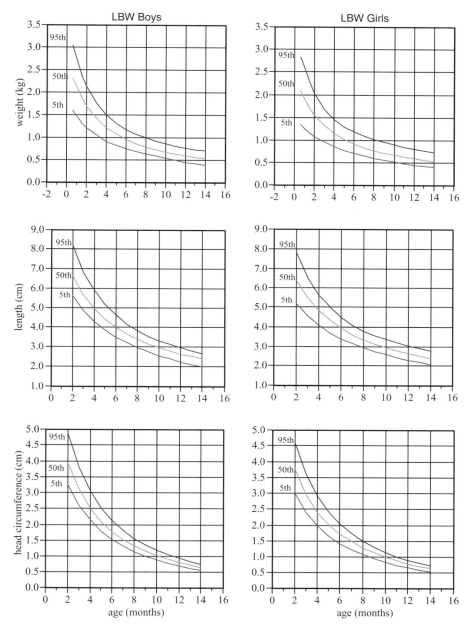

Figure 2.8. Reference percentiles for 2-month increments in weight, length, and head circumference for low birth weight (LBW) preterm infants from zero to 14 months GAA. The observed increments should be plotted opposite the end of each interval. (Reprinted from Guo SS, Roche AF, Chumlea WC, Casey PH and Moore WM, Growth in weight, length and head circumference for preterm low-birth weight infants during the first three years of life using gestation-adjusted ages. *Early Human Development* 47: 305–325, 1997 with permission from Elsevier Science and Ross Products Division of Abbott Laboratories.)

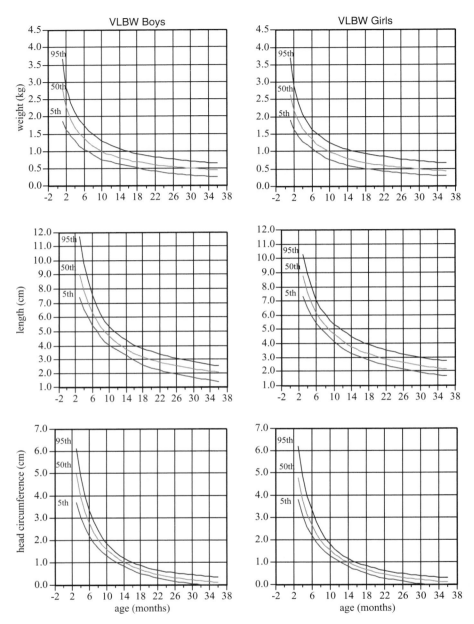

Figure 2.9. Reference percentiles for 3-month increments in weight, length, and head circumference for very low birth weight (VLBW) preterm infants from 3 to 36 months GAA. The observed increments should be plotted opposite the end of each interval. (Reprinted from Guo SS, Roche AF, Chumlea WC, Casey PH and Moore WM, Growth in weight, length and head circumference for preterm low-birth weight infants during the first three years of life using gestation-adjusted ages. *Early Human Development* 47: 305–325, 1997 with permission from Elsevier Science and Ross Products Division of Abbott Laboratories.)

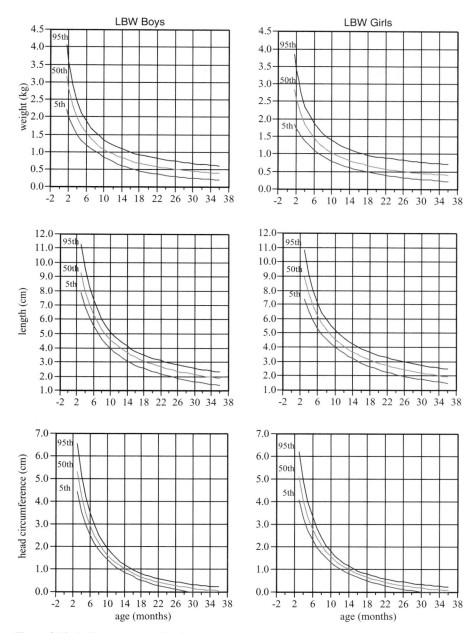

Figure 2.10. Reference percentiles for 3-month increments in weight, length, and head circumference for low birth weight (LBW) preterm infants from 3 to 36 months GAA. The observed increments should be plotted opposite the end of each interval. (Reprinted from Guo SS, Roche AF, Chumlea WC, Casey PH and Moore WM, Growth in weight, length and head circumference for preterm low-birth weight infants during the first three years of life using gestation-adjusted ages. *Early Human Development* 47: 305–325, 1997 with permission from Elsevier Science and Ross Products Division of Abbott Laboratories.)

& Inwood, 1989; Karlberg & Albertsson-Wikland, 1995; Scott *et al.*, 1982*a*). Despite further reductions in deficits during pubescence, the means for stature, but not those for BMI, remain small through early adulthood (Albertsson-Wikland *et al.*, 1998; Leger *et al.*, 1997, 1998; Luo *et al.*, 1998; Persson *et al.*, 1999).

Due to head-sparing, head circumference/length and head circumference/chest circumference ratios are larger in SGA than in AGA term infants, but the crown–rump length/total length ratio does not differ between these groups (Albertsson-Wikland & Karlberg, 1997; Ounsted & Moar, 1986; Ounsted *et al.*, 1986). There is marked variability at birth within groups of AGA and SGA infants, especially in the head circumference/length ratio, which may be positively related to growth during infancy (Lapillonne *et al.*, 1997). In SGA term infants, the means for head circumference and the crown–rump length/total length ratio decrease linearly from 1 to 12 months, but those for head circumference/chest circumference decrease rapidly to 4 months and slowly to 12 months (Lapillonne *et al.*, 1997). From 3 to 6 years, mean skinfold thicknesses increase much more rapidly in term SGA infants than in those who are AGA (Hediger *et al.*, 1998, 1999). Correspondingly, SGA term infants have only slight deficits in arm adipose tissue areas at 4 years, but they have small arm muscle areas (Hediger *et al.*, 1998). Among term infants, those who are SGA have less TBF, FFM and total body BMC than AGA infants, but these groups do not differ when these body composition measurements are expressed as percentages of body weight (Lapillonne *et al.*, 1997).

LGA infants

The literature relating to LGA infants is limited because the care of LGA infants (birth weight >4000 g) causes less concern than that of SGA and LBW infants. In LGA infants, weight, length, and head circumference are larger at birth than in AGA infants. The differences in mean size between these groups decrease rapidly during the first few months after birth and then change little to 4 years (Hediger *et al.*, 1999; Paz *et al.*, 1993; Pryor, 1992). Term LGA infants have slightly larger skinfold thicknesses and mid-arm muscle areas than AGA term infants to 3 years, but they do not differ in arm adipose tissue areas (Hediger *et al.*, 1998, 1999). Mean differences between LGA and SGA infants in weight and stature of about 0.5 to 1.0 Z persist to 17 years (Lhotská *et al.*, 1995; Persson *et al.*, 1999).

GROWTH AND BODY COMPOSITION AFTER INFANCY

During the preschool years, the rate of growth slows from being rapid during infancy to a near constant rate that begins at 4 to 5 years of age when the average annual increments are about 2 kg for weight and 6 cm for stature. Sex differences in stature and weight during the preschool years are slight. Middle childhood (7 to 10 years) is a period of continued steady, but slightly increasing growth in body size for most children. On average, children increase in weight by about 3 kg per year at 7 years and by about 4 kg per year at 10 years and they increase in stature by about 5 cm per year from 7 to 10 years. At 7 years, males are, on average, about 2 cm taller than females, but there is only a small sex difference in weight. By 10 years, the average female is about 1 cm taller than the average male and about 1 kg heavier. The sex difference in growth rate during middle childhood contributes to the earlier appearance of significant sex differences in size at the start of pubescence. During childhood, the lower extremities grow more rapidly than the trunk. The trunk accounts for about 55% of stature at 7 years, but only about 45% at 10 years.

The pubescent and post-pubescent periods extend from about 11 to 20 years of age. At 10 to 13 years, females tend to be taller than males because pubescent spurts occur earlier in females than in males. Within each sex, however, there is a wide range of ages during which normal children become pubescent. In general, children who start their pubescent development early pass through sexual maturation stages quickly; children starting at a later age may take a longer period of time to reach maturity. The pubescent growth spurt lasts longer and the amount of growth during the spurt is larger in males than females. In males, the average maximum rate of growth in stature during the spurt ranges from 8.5 to 11.1 cm per year, while in females the maximum velocity is 7.1 to 9.3 cm per year (Guo *et al.*, 1992). These sex differences in growth contribute to the larger average body size of men than of women. Females reach their adult statures at younger ages than males.

Prepubescent spurts

An inconstant small spurt in the rate of growth in stature, known as the mid-growth spurt, occurs at about 7 years in males and 6 years in females (Butler *et al.*, 1990; Gasser *et al.*, 1991, 1993; Guo *et al.*, 1992; Tanner & Cameron, 1980). At the peak of this spurt, the rate of growth in stature is increased by about 0.5 cm per year in males and 0.3 cm per year in females (Guo *et al.*, 1992; Molinari *et al.*, 1980). Recognition of the mid-growth spurt requires careful serial measurements no more than 6 months apart. These spurts are not evident in reference percentiles on growth charts based on cross-sectional data, but they may be present in increment charts (Reinken & van Oost, 1992; Roche & Himes, 1980). Mid-growth spurts are too small and inconsistent to be important in the evaluation of growth for individuals.

There may be inconsistent spurts in stature at about 9 years in each sex with mean increases in growth rates of about 1.0 to 1.5 cm per year for males and 1.0 cm per year for females (Butler *et al.*, 1990; Tillmann *et al.*, 1998; Wales & Milner, 1987; Wit *et al.*, 1987). These spurts have received little attention, but they could cause misleading conclusions about the effects of therapy that begins just before a natural spurt (Polychronakos *et al.*, 1988).

Pubescent spurt

Pubescence is a period of rapid increases in stature and weight for each sex and in body fat for fe-

males and muscle mass for males. Additionally, pubescence is the time when secondary sexual characteristics mature. The pubescent spurt in stature may be the best-known feature of the human growth curve. This spurt begins in males at about 11 years and the peak rate of growth is reached at about 14 years (Buckler, 1990; Gasser *et al.*, 1984; Guo *et al.*, 1992; Ramsay *et al.*, 1995). The timing is about 2 years earlier in females with SDs of the ages at the beginning and at the peak that are about 1 year in each sex. Peak height velocity (PHV) occurs at stages 2 to 4 of pubic hair in each sex and at stages 3 to 5 of genitalia in males and stages 2 or 3 of breasts in females (Largo & Prader, 1983*a*, *b*; Marshall & Tanner, 1969, 1970). Stages 1 or 2 of each organ are present when the spurt begins at about 2.5 years before PHV. Menarche is not useful in determining whether a child is having a pubescent spurt because it occurs about a year after PHV. Reports of the timing of the peak rate of growth during the spurt (age at PHV) vary due to differences among populations, in study designs, methods of analysis and curve-fitting procedures. The Preece–Baines (1978) curve-fitting model gives ages for PHV that are about 0.5 years younger than other models, which fit the data better (Guo *et al.*, 1992). The mean additional increase in stature due to the spurt is about 2.3 cm for males and 0.8 cm for females. During the spurt, the rate of growth in stature doubles in 70% of males and 11% of females (Largo *et al.*, 1978).

Hormones have major effects on the timing, duration, and rate of pubescent changes in growth and maturity. During pubescence, there are increased levels of growth hormone, and IGF-1, which are probably essential for normal growth. Sex steroids and thyroid hormones accelerate sexual and skeletal maturation and increase the release of growth hormone, and IGF-1. Estradiol may be essential for the occurrence and timing of normal pubescent spurts in females and testosterone may be essential in males. Estradiol may also be responsible for pubescent increases in the body fat of females, despite increases in growth hormone and IGF-1, which tend to reduce body fat. The reduction in the body fat of males during pubescence is due to testosterone, growth hormone, and IGF-1, which jointly increase FFM and muscle mass. Increased levels of sex steroids are largely responsible for the maturation of secondary sex characteristics and the acceleration of skeletal maturation during pubescence. Leptin increases gonadotrophin-releasing hormone and thus increases estradiol, but it is negatively related to testosterone levels and FFM. Leptin levels are positively related to body fatness, but this is an effect of fatness rather than a cause. Estradiol, testosterone, growth hormone, and IGF-1 have positive influences on the pubescent increases in bone mineral.

Changes in a child's percentile levels for stature and weight after 10 years for males and 9 years for females are commonly due to the pubescent growth spurt. This occurs because most growth charts are developed using cross-sectional data from children who varied in the timing of their pubescent spurts. When growth is assessed using these charts, a male with an early spurt will increase in percentile levels at about 10 to 14 years, at which time most males from whom the growth chart was constructed were less advanced in their pubescent changes, but his percentile levels are likely to decrease at 15 to 18 years. Similar percentile changes will occur about 2 years earlier for females who have an early spurt. Children with early pubescent spurts tend to be tall before pubescence, but their adult statures are not increased. Therefore, when children are matched for prepubescent stature, those with early pubescent spurts tend to be shorter in adulthood (Qin et al., 1996; Tanaka, 1996).

Pubescent spurts can be recognized from serial data at 6-month intervals during a period of 18 months or at annual intervals during a period of 3 years. Data from only two examinations may show growth is more rapid than usual, but evidence of a spurt requires a documented increase in the rate of growth. Measures of maturity may assist the recognition of a spurt by suggesting its occurrence is likely. A child with advanced sexual or skeletal maturity for age is likely to have an early pubescent spurt (Largo & Prader, 1983a, b; Marshall & Tanner, 1969, 1970). Data for groups

that differ from the average timing of PHV by about one year are less useful in practice than comparisons based on sexual maturity because the timing of PHV is known only retrospectively in research studies (Buckler, 1990). Reference data for growth status from groups matched for age, but differing in sexual maturity, can provide adjustment factors that are applicable to the observed stature data before comparisons are made with reference values from general populations. This approach is described in Chapter 3.

The pubescence increase is relatively larger for weight than for stature. When pubescence is delayed, the decrease in percentile levels is more marked for weight than for stature and the intervals from the early stages of sexual maturity to PHV are longer, but those from PHV to the cessation of growth in stature are shorter. The peak rate of growth in stature is negatively correlated ($r = -0.5$) with the age at which PHV occurs, but there are positive correlations ($r = 0.4$) between age at PHV and stature at PHV (Billewicz et al., 1981a; Bock & Thissen, 1980; Gasser et al., 1985; Koziel et al., 2001). The timing of the spurt is not closely related to adult stature (Bielicki & Hauspie, 1994; Georgiadis et al., 1997; Qin et al., 1996; Sumiya et al., 1999).

The pubescent growth spurt is not limited to stature and weight. It occurs in many other measures including segment lengths, circumferences, skinfold thicknesses, BMI, FFM, and BMD (Beunen et al., 1988; Geithner et al., 1999; Pařizková, 1976; Roche, 1974; Slemenda et al., 1994). The spurt is more marked in males than females for many of these measures, but not for skinfold thicknesses. The mean ages at the peak rates of growth differ little among measures of body size, but those for stature and bone lengths tend to be early despite marked variability in the sequence of peak rates among various dimensions within individuals (Marshall & Tanner, 1986; Roche, 1974; Stolz & Stolz, 1951). The pubescent peak velocity of FFM occurs at about the same age as PHV, but peak velocities for weight and total body BMC occur about 6 months after PHV. Age at PHV is positively correlated ($r = 0.2$) with peak pubescent velocity of TBF, but is negatively correlated ($r = -0.3$) with peak pubescent velocity of total body BMC in females (Iuliano-Burns et al., 2001).

The medians for total growth in stature after PHV are 17.8 cm for males and 15.8 cm for females and the median total increment after menarche is 7.4 cm for United States females and 5.7 cm for Japanese females (Roche & Davila, 1972, Roche, 1989; Sumiya et al., 1999). The increases in stature after PHV and menarche are negatively correlated with the ages at these events (Hulanicka & Kotlarz, 1983; Roche, 1989). There is a median increment in stature of 1.5 cm for males and 0.9 cm for females after the femur or tibia is fully mature that is due to elongation of the trunk (Roche & Davila, 1972).

The 90th percentile increase in stature after 18 years exceeds 1.0 cm in each sex with median increases of 0.8 cm for males and 0.2 cm for females. Larger increases in stature after 18 years have been reported from studies in Poland and Sweden (Gworys, 1978; Hulanicka & Kotlarz, 1983; Lubicka, 1994; Nylind et al., 1978), suggesting that the rates of maturation were slower in these groups than in the group studied in the United States. The median ages at which growth in stature ceases are 21.2 years for males and 17.3 years for females in a United States study; the ranges between the 10th and 90th percentiles are about 5 years (Roche & Davila, 1972).

It has been claimed that adult stature is reached when the annual increment is <1.0 cm (Tanner et al., 1983b). There are, however, considerable increases in stature after such slowing, particularly when such an annual increment occurs at a young age. Four successive 6-month increments each <0.5 cm are a good indication that growth in stature has ceased (Roche & Davila, 1972). Kato et al. (1998) suggested that the largest stature recorded for an individual should be considered the adult value. In their series, half the males were not measured after 18 years and some of the females were not measured after 14 years. Consequently, valid measures of adult stature were not available for all the individuals studied.

Diurnal variations

Changes in sitting height during the day affect recorded statures. These diurnal variations are due to loss of water from each nucleus pulposus, which reduces the thicknesses of intervertebral disks (Böös et al., 1993; Roberts et al., 1998). In children, there is a decrease of 1.5 cm from rising in the morning to 4:00 p.m. (Strickland & Shearin, 1972). When pressure is exerted under the mandible during the measurement of stature to obtain a stretched value, the mean decrease in stature for children is 0.2 cm from 9:30 a.m. to 2:00 p.m. and 0.6 cm. from 6:00 p.m. to 9:00 p.m. (Buckler, 1978; Whitehouse et al., 1974). These findings suggest that diurnal variations of about 0.3 cm can be expected from early morning to late afternoon. Such variations are not of general concern, but examinations should be scheduled within a time range of a few hours when a research protocol includes analyses of serial data.

Body mass index

Serial data for individuals show BMI decreases from 1 year to about 5 or 6 years and then increases. The timing of the smallest value, known as the nadir, and the rebound that follows it are estimated from curve-fitting procedures in research studies. Clinically, they may be recognized from plots of three or more successive BMI values during the appropriate age range. The timing of the nadir is positively correlated with skeletal age and does not differ between the sexes (Guo et al., 2000a; Rolland-Cachera et al., 1984; Siervogel et al., 1991; Williams et al., 1999). The increase in BMI from the nadir to post-pubescence is negatively related to the age at which the nadir occurs. During pubescence, BMI increases rapidly and has its maximum velocity at about 14 years in males and 13 years in females (Guo et al., 2000a; Siervogel et al., 1991). Because of these pubescent changes, rapidly maturing children tend to increase in BMI percentile levels during the early part of the pubescent age range while opposite changes occur in those who are maturing slowly (Buckler, 1995). Consequently, it could be useful to relate BMI to sexual maturity instead of

Table 2.3. *Selected reference data for increments after infancy*

Authors	Country	Variables[a]	Intervals	Age range (years)	Charts	Tables
Brandt & Reinken (1988)	Germany	S	annual	2–16	yes	yes
Prader et al. (1989)	Switzerland	W, S, SiH, LL	annual	2–19.5	S only	yes
Roche & Himes (1980)[b]	United States	W, S	6 mo	2–18	yes	yes
Sempé et al. (1979)	France	W, S, SiH, HC	6 mo	2–20	yes	yes
Karlberg et al. (1976)	Sweden	W, S, SiH, HC, SKF	annual	2–16	no	yes

[a] HC = head circumference; LL = leg length; S = stature; SiH = sitting height; SKF = skinfolds; W = weight.
[b] Charts only; corresponding tabular data are in Baumgartner et al. (1986).

chronological age during pubescence, but there is a lack of such reference data. The pubescent increases in BMI are mainly due to increases in FFM for males and in TBF for females (Maynard et al., 2001). Small increases in BMI continue after pubescence to at least 20 years in most studies (Prokopec & Bellisle, 1992; Rolland-Cachera et al., 1991; Siervogel et al., 1991), but Luciano et al. (1997) reported decreases in upper percentile levels for females from 15 to 18 years that may be due to social pressures.

In serial data, the sex differences in total body BMC are small to 11 years. At older ages the values are consistently larger in males than in females except perhaps at 15 to 17 years (Bachrach et al., 1999). At PHV, about 90% of adult stature has been reached, but only 60% of adult total body BMC (Bailey, 1997). The peak rate of increase in total body BMC occurs about 1 year after PHV (Bailey, 1997; Lloyd et al., 1996; Lu et al., 1994). This leads to an increased fracture risk during pubescence. As a percentage of body weight, total body BMC increases from 4 to 18 years; this increase is rapid from 12 to 16 years, particularly in males (Fomon et al., 1982; Maynard et al., 1998; Rico et al., 1993). In other study groups, this percentage increases gradually with age in each sex (Faulkner et al., 1993; Haschke, 1983, 1989).

Increments

Sources of reference data for increments after infancy are given in Table 2.3. Roche & Himes (1980) provide reference percentiles for weight and stature during 6-month intervals from 2 to 18 years (Figure 2.11). These intervals are appropriate for their sample in which seasonal effects are absent. Matching tabular data have been published (Baumgartner et al., 1986) and large-scale versions of these charts are available (Ross Products Division of Abbott Laboratories, Columbus, Ohio,). These charts, which are constructed relative to age at the end of each interval, allow judgements of the rate of growth beginning at any age. In each sex, the percentile values for 6-month increments in weight change little from 2 to 5 years when increases begin that continue to about 14.5 years in males and 12.5 years in females. After these ages, there are marked decreases in the reference percentiles with negative values at the lower percentile levels after 12 years in males and 9 years in females.

Percentiles for 6-month increments in stature decrease from 2 years to 10 years in males and 9 years in females. During this period, the mid-growth spurt is indistinct in reference percentiles. Due to the pubescent spurt, the percentile values increase from about 9 years until 13.5 years in males and 12 years in females, after which there are marked decreases. Smoothed empirical percentiles for annual increments in weight and stature beginning at 6-month intervals from 3 to 19.5 years were reported for Swiss children by Prader and colleagues (1989). There are similar data for Swedish children and for children in southeast England (Karlberg et al., 1976; Tanner et al., 1966b). The applicability of

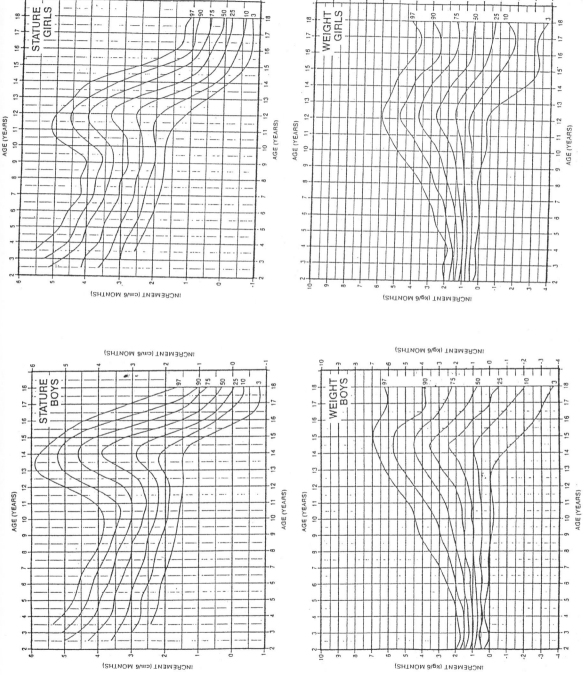

Figure 2.11. Reference percentiles for 6-month increments in stature and weight from 2 to 18 years. The observed increments should be plotted opposite the end of each increment. (From Roche & Himes, 1980; with permission from Ross Products Division of Abbott Laboratories.)

these European reference data has been reduced by secular changes (Cernerud & Lindgren, 1991; Eiholzer *et al.*, 1998; Freeman *et al.*, 1995).

Reference data for increments in crown–rump length and sitting height show the percentile levels for males decrease to 8.5 years after which they are approximately constant until increases occur at 13.25 to 14.75 years. After these ages, the increments decrease to near zero at 19.5 years. The changes for females are similar, but they occur about 2 years earlier than the changes in males (Karlberg *et al.*, 1976; Prader *et al.*, 1989). Medians for annual increments in arm circumference are similar in males and females except that those for males are much smaller than those for females from 12 to 13 years, but larger from 14 to 17 years (van Venrooij-Ysselmuiden, 1978). The median increments for the cross-sectional areas of "muscle plus bone" and adipose tissue in the arm do not differ between the sexes until 12 to 13 years, after which those for "muscle plus bone" become much larger in males and those for adipose tissue become much larger in females.

In any incremental reference data, the intervals between measurements must be long enough for the expected changes to exceed the sum of two measurement errors. Consequently, few other than Karlberg *et al.* (1976) and van Venrooij-Ysselmuiden (1978) have developed reference data for increments in skinfold thicknesses. The medians for annual increments in triceps skinfold thicknesses change little with age except for increases after 10 years in females that are larger for upper percentiles than for lower ones (Figure 2.12). The median increments in subscapular skinfold thicknesses increase from 2 to 16 years in each sex, except for decreases at the 10th percentile level from 9 to 14 years in males. The median increments for suprailiac skinfold thicknesses decrease from 2 years to 8 years in males and from 2 years to 5 years in females. Later, there are moderate increases to 16 years that tend to be larger for females than males.

If seasonal effects are expected, annual increments are indicated, although they are of limited clinical use because of the long delays needed for their application. The ideal reference data in such a situation would be season-specific reference data for 6-month increments, but these are not available. Seasonal effects are extremely small or absent during infancy in developed countries. Seasonal effects occur in about 30% of children aged 8 to 14 years, and they vary markedly in size (Cole, 1993, 1998). Consequently, it is not possible to apply systematic corrections. When seasonal effects occur, growth in stature is typically most rapid in the spring (Gelander *et al.*, 1994; Marshall, 1971; Mirwald & Bailey, 1997), but an additional difficulty is introduced by differences in the timing of seasonal effects between sub-groups from the same population (Neyzi *et al.*, 1993).

Reynolds & Sontag (1944), using a sub-set of the data analyzed by Roche & Himes (1980), found weight gain was maximal from October to December and that stature gain was maximal from April to June. Seasonal effects are not evident when the data available to Reynolds & Sontag are pooled with more recent Fels data. It is not clear why the recent Fels data differ in seasonal effects from some other data sets. The Fels Longitudinal Study (Ohio, United States) is at a latitude of 9° 41′ N, which is not as far north as the studies in which seasonal effects have been observed recently. The Marshall study (1971) in London (United Kingdom) was at 50° 31′ N and that of Mirwald & Bailey (1997) in Saskatoon, Saskatchewan, Canada was at 52° 10′ N. These latitudinal differences may have been associated with differences in exposure to light (Bogin, 1978; Marshall, 1975). Differences in temperature would not be responsible (Lee, 1980) and the availability of vitamin D is unlikely to be a factor because of food fortification.

In clinical practice, 6-month increments have major advantages over annual increments because a growth rate can be established more quickly. Six-month increments should be used when seasonal variations in growth are expected to be small or unusual in the population being served. This is likely to be the case for United States children living

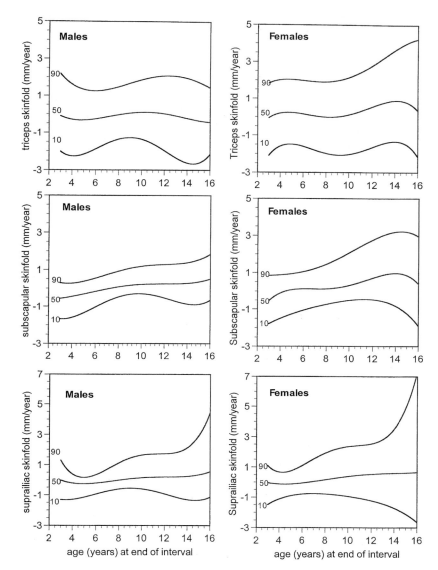

Figure 2.12. Reference percentiles for annual increments in skinfold thicknesses from 2 to 16 years. (Data from Karlberg *et al.*, 1976.)

in states with climates similar to that of Ohio. Annual increments may be needed for populations living in climates similar to those of London (United Kingdom) and Saskatchewan (Canada). During infancy when seasonal effects are small or absent and the rate of growth is changing rapidly, increments for intervals shorter than 6 months are desirable.

TRACKING

Tracking is the tendency for members of a group to retain the rank order of values for a variable across time. Analyses of tracking require serial data, therefore the findings may not be generalizable because the data are unlikely to come from random representative samples, but generalizability of the findings is

Table 2.4. *Age-to-age correlations during infancy*

Authors	Number of subjects[a]	Age intervals (months)[b]	Correlation coefficient
Weight			
Bayley & Davis (1935)	61 M + F	1–3	0.82
		1–6	0.54
		1–24	0.48
		1–36	0.40
Boryslawski (1988)	200 M + F	B–1	0.87
	200 M + F	B–3	0.62
	200 M + F	B–9	0.46
	200 M + F	B–12	0.40
Boulton (1981)	181 M	B–3	0.60
	154 M	3–6	0.79
	143 M	6–12	0.83
	102 M	12–24	0.84
	184 F	B–3	0.46
	153 F	3–6	0.81
	138 F	6–12	0.70
	96 F	12–24	0.52
Cole (1995)	223 M + F	B–3	0.82
	223 M + F	B–9	0.53
	223 M + F	B–12	0.47
	223 M + F	B–24	0.48
	223 M + F	12–24	0.85
Kantero et al. (1971)	57 M	B–3	0.68
		B–12	0.59
		B–24	0.58
		B–36	0.52
	76 F	B–3	0.59
		B–12	0.33
		B–24	0.27
		B–36	0.23
Miller et al. (1972)	327 M + F	B–36	0.30
Tanner et al. (1956)	65 M	B–12	0.48
		B–24	0.43
		B–36	0.49
	59 F	B–12	0.03
		B–24	0.16
		B–36	0.17
Thomson (1955)	130 M	B–6	0.34
	106 F	B–6	0.46
Wright et al. (1994)	1347 M + F	B–3	0.54
	1331 M + F	B–6	0.45

Table 2.4. (*cont.*)

Authors	Number of subjects[a]	Age intervals (months)[b]	Correlation coefficient
	1062 M + F	B–9	0.41
	1084 M + F	B–12	0.45
	848 M + F	B–18	0.43
Length			
Bayley & Davis (1935)	61 M + F	1–3	0.87
		1–6	0.77
		1–24	0.68
		1–36	0.68
Boryslawski (1988)	200 M + F	B–1	0.92
		B–3	0.76
		B–9	0.50
		B–12	0.42
Boulton (1981)	154 M	3–6	0.82
	143 M	6–12	0.78
	102 M	12–24	0.73
	153 F	3–6	0.78
	138 F	6–12	0.76
	96 F	12–24	0.75
Buschang et al. (1985)	104 M	B–3	0.57
		B–6	0.62
		B–9	0.55
		B–12	0.53
		B–18	0.41
		B–24	0.44
		B–36	0.35
	82 F	B–3	0.59
		B–6	0.57
		B–9	0.51
		B–12	0.46
		B–18	0.44
		B–24	0.42
		B–36	0.41
Kantero et al. (1971)	57 M	B–3	0.68
		B–12	0.59
		B–24	0.58
		B–36	0.52
	76 F	B–3	0.59
		B–12	0.33
		B–24	0.27
		B–36	0.23
Miller et al. (1972)	323 M + F	B–36	0.36

Table 2.4. (*cont.*)

Authors	Number of subjects[a]	Age intervals (months)[b]	Correlation coefficient
Smith *et al.*	44 M	B–12	0.13
(1976)	44 M	B–24	0.15
	46 F	B–12	0.39
	46 F	B–24	0.33
Tanner *et al.*	65 M	B–12	0.64
(1956)		B–24	0.48
		B–36	0.52
	59 F	B–12	0.39
		B–24	0.41
		B–36	0.46
Head circumference			
Boryslawski	200 M + F	1–3	0.65
(1988)		1–6	0.45
		1–9	0.34
		1–12	0.27
Body mass index			
Cardon (1996)	260 M + F	B–12	0.09
	260 M + F	B–24	0.14
	260 M + F	B–36	0.11
	260 M + F	12–24	0.52
	260 M + F	12–36	0.46
	260 M + F	24–36	0.52
Skinfold thicknesses (sum of four)			
Boulton (1981)	309 M + F	3–6	0.51
	281 M + F	6–12	0.54
	198 M + F	12–24	0.55
Whitelaw (1977)	74 M + F	B–12	−0.13

[a] M, males; F, females.
[b] B, birth.

increased by replication. Furthermore, analyses of tracking are necessarily retrospective, and changes may have occurred since the beginning of data collection in the determinants of tracking.

Usually tracking is analyzed by calculating age-to-age correlations. These provide estimates of the percentage of the variance at the older age that is accounted for by the values at the younger age. Age-to-age correlations can show whether tracking is more marked for some variables than for others and the degree to which tracking is related to the initial age of measurement, the length of the interval, and sex. Knowledge about tracking can be applied clinically in a non-mathematical manner. For example, less marked tracking is expected for intervals that begin in infancy and for weight than for length or stature.

High age-to-age correlations could justify the construction of conditional reference data that estimate the distribution of values at an older age dependent on the values at a younger age (Berkey *et al.*, 1983; Cameron, 1980; Cole, 1995; Wright *et al.*, 1994). These correlations may differ depending on whether the initial values are large or small, but there is little relevant information, except for weight which has a marked tendency to regress towards the mean during the first few months after birth (Fergusson *et al.*, 1980; Wright *et al.*, 1994). Conditional charts are rarely used because of logistic considerations and because the time intervals between examinations must match those of the conditional charts.

Birth weight is negatively correlated with the increments in weight during infancy (Kato & Takaishi, 1999). This is reflected in the rapid weight gains of LBW infants and the slow gains in LGA infants. Nevertheless, age-to-age correlations for weight during intervals beginning at birth or one month and ending in the first year are positive. These correlations decrease with the length of the interval, but increase with age at the younger measurement (Boryslawski, 1988; Boulton, 1981; Cole, 1995; Fergusson *et al.*, 1980) (Table 2.4). The correlations between weights at birth or 1 month with weights at 24 or 36 months are markedly lower for females than for males in most studies (Boulton, 1981; Kantero *et al.*, 1971).

During the first year after birth, age-to-age correlations for length in AGA term infants decrease with the length of the intervals, as is common for correlations generally (Boryslawski, 1988; Buschang *et al.*, 1985; Kantero *et al.*, 1971), but the correlations for intervals from birth to 12, 24 or 36 months are about 0.4 and differ little with the lengths of the intervals. The reported sex differences in these correlations are inconsistent and generally small (Buschang *et al.*, 1985; Kantero *et al.*, 1971; Smith *et al.*, 1976). In term SGA and LGA infants, there are negative correlations between weight at birth or 1 month and weight at 12 months, but the correlations between values at birth and those at 12 months are positive for length and head circumference (Moar & Ounsted, 1982). During the first year after birth, age-to-age correlations for head circumference decrease with the lengths of the intervals. These correlations are lower than those for length or weight (Boryslawski, 1988).

Age-to-age correlations for BMI are very low for intervals beginning at birth, but those for intervals beginning at 12 or 24 months are about 0.5 (Cardon, 1996). Age-to-age correlations for the sums of four skinfold thicknesses for the interval from birth to 12 months are low ($r = -0.1$), but about 0.5 for the intervals 3 to 6, 6 to 12 and 12 to 24 months (Boulton, 1981; Whitelaw, 1971). These low correlations reflect modest levels of tracking and the effects of large measurement errors relative to the recorded measurements.

Correlations between weights during infancy and weights during childhood or adolescence decrease with the lengths of the intervals and are lower for intervals that begin at birth than for those beginning at 1 year (Table 2.5). Correlations between length at birth and stature at 5 or 10 years are about 0.4 for males and 0.2 for females, but those between statures at 1 year and statures at 5 or 10 years are about 0.7 for each sex.

Age-to-age correlations for weight, stature, and BMI during childhood and adolescence decrease markedly as longer intervals are considered (Table 2.6). These correlations increase for intervals beginning at older ages, except that there are small decreases for intervals beginning during pubescence (Molinari *et al.*, 1995). The correlations for weight tend to be higher for males than for females in some studies (Kantero *et al.*, 1971; Molinari *et al.*, 1995), but Tuddenham & Snyder (1954) reported an opposite sex difference. Age-to-age correlations for BMI are considerably lower for males than for females and differ little between 6 vs. 10 years and 6 vs.

Table 2.5. *Age-to-age correlations from infancy to childhood and adolescence*

Authors	Number of subjects[a]	Age intervals (years)[b]	Correlation coefficient
Weight			
Dine *et al.*	234 M + F	B–5	0.28
(1979)	286 M + F	B–5.5	0.53
Kantero *et al.*	57 M	B–5	0.53
(1971)		B–10	0.37
		1–5	0.65
		1–10	0.60
	76 F	B–5	0.25
		B–10	0.04
		1–5	0.50
		1–10	0.29
Malina *et al.*	72 M	B–7	0.11
(1999)	55 M	B–10	0.37
	57 F	B–7	0.29
	48 F	B–10	0.11
Miller *et al.*	375 M + F	B–5	0.30
(1972)	423 M + F	B–9	0.19
	423 M + F	B–13	0.17
	375 M + F	B–15	0.15
Tanner *et al.*	54 M	B–4	0.51
(1956)		B–5	0.37
	59 F	B–4	0.22
		B–5	0.16
Stature			
Kantero *et al.*	57 M	B–5	0.48
(1971)		B–10	0.37
		1–5	0.85
		1–10	0.69
	76 F	B–5	0.29
		B–10	0.14
		1–5	0.66
		1–10	0.23
Miller *et al.*	347 M + F	B–5	0.18
(1972)	423 M + F	B–9	0.20
	423 M + F	B–13	0.20
	375 M + F	B–15	0.20
Tanner *et al.*	65 M	B–4	0.46
(1956)		B–5	0.43
	59 F	B–4	0.40
		B–5	0.34

[a] M, males; F, females.
[b] B, birth.

Table 2.6. *Age-to-age correlations during childhood and adolescence*

Authors	Number of subjects[a]	Age intervals (years)	Correlation coefficient
Weight			
Kantero *et al.* (1971)	57 M	4–6	0.83
		4–8	0.82
		4–10	0.74
	76 F	4–6	0.69
		4–8	0.58
		4–10	0.44
Tuddenham & Snyder (1954)	66 M	4–6	0.87
		4–8	0.72
		4–10	0.60
		4–12	0.50
		4–14	0.40
		4–16	0.43
	70 F	4–6	0.92
		4–8	0.86
		4–10	0.40
		4–12	0.69
		4–14	0.58
		4–16	0.52
Zacharias & Rand (1986)	207 F	6–9	0.88
		6–12	0.76
		6–15	0.67
Stature			
Kantero *et al.* (1971)	57 M	4–6	0.85
		4–8	0.83
		4–10	0.77
	76 F	4–6	0.80
		4–8	0.65
		4–10	0.49
Tuddenham & Snyder (1954)	66 M	4–6	0.93
		4–8	0.91
		4–10	0.88
		4–12	0.82
		4–14	0.73
		4–16	0.74
	70 F	4–6	0.93
		4–8	0.90
		4–10	0.84
		4–12	0.77
		4–14	0.79
		4–16	0.74

Table 2.6. (*cont.*)

Authors	Number of subjects[a]	Age intervals (years)	Correlation coefficient
Body mass index			
Casey *et al.* (1992)	64 M	6–12	0.64
	60 M	6–14	0.62
	63 M	6–16	0.59
	54 M	6–18	0.44
	64 F	6–10	0.81
	60 F	6–12	0.76
	66 F	6–14	0.70
	54 F	6–18	0.53
Cardon (1996)[b]	260 M + F	3–4	0.56
	260 M + F	3–5	0.47
	260 M + F	3–6	0.44
	260 M + F	3–7	0.47
	260 M + F	3–9	0.34

[a] M, males; F, females.
[b] May be less due to attrition; also has correlations beginning at 4, 5, 6, 7, 8 years.

12 years, but are markedly lower for correlations between 6 years and 14 or 18 years (Casey *et al.*, 1992; Guo *et al.*, 1994). Age-to-age correlations with BMI at 18 years are 0.4 for values at 2 and 4 years, but about 0.7 for values at 10 years (He & Karlberg, 1999). There are low non-significant correlations between triceps, subscapular, and suprailiac skinfold thicknesses at 39 weeks and at 15 years (Prader *et al.*, 1976). The correlations between skinfold thicknesses at 16 years and thicknesses at younger ages are about 0.3 to 6 years and about 0.6 after 6 years; they are considerably higher for females than for males to 6 years (Hernesniemi *et al.*, 1974; Roche *et al.*, 1982).

The correlations of weight and length during infancy with weight and stature in adulthood are about 0.2 (Table 2.7). Age-to-age correlations with adult values for length and stature increase during infancy, but then change little. They tend to be higher for males than for females (Molinari *et al.*, 1995). Correlations of childhood values for stature, sitting height, and arm length with those at 18 years are

Table 2.7. *Age-to-age correlations from younger ages to adulthood*

Authors	Number of subjects[a]	Younger ages (years)	Correlation coefficient
Weight			
Kouchi *et al.*	139 M	birth	0.21
(1985*a*)	127 F	birth	0.23
Miller *et al.*	341 M + F	birth	0.25
(1972)			
Molinari *et al.*	117 M	0.5	0.46
(1995)		1	0.51
		2	0.54
		3	0.56
		6	0.68
		8	0.73
		10	0.78
		12	0.80
		14	0.75
	111 F	0.5	0.45
		1	0.43
		2	0.44
		3	0.48
		6	0.56
		8	0.61
		10	0.59
		12	0.51
		14	0.66
Zacharias &	207 F	6	0.60
Rand (1986)		9	0.65
		12	0.63
		15	0.92
Length and stature			
Kouchi *et al.*	141 M	birth	0.33
(1985*b*)	144 F	birth	0.35
Miller *et al.*	429 M + F	birth	0.26
(1972)			
Molinari *et al.*	117 M	0.5	0.64
(1995)		1	0.73
		2	0.75
		3	0.74
		6	0.80
		8	0.83
		10	0.85
		12	0.82
		14	0.66
	111 F	0.5	0.51
		1	0.52
		2	0.64
		3	0.66
		6	0.68
		8	0.72
		10	0.70

Table 2.7. (*cont.*)

Authors	Number of subjects[a]	Younger ages (years)	Correlation coefficient
		12	0.58
		14	0.82
Body mass index			
Casey *et al.* (1992)	54 M	5–7	0.44
	54 F	5–7	0.53
Molinari *et al.*	117 M	0.5	0.27
(1995)		1	0.28
		2	0.36
		3	0.46
		6	0.60
		8	0.68
		10	0.75
		12	0.76
		14	0.79
	111 F	0.5	0.41
		1	0.38
		2	0.42
		3	0.41
		6	0.53
		8	0.58
		10	0.62
		12	0.62
		14	0.72
Rolland-Cachera	60 M	3	0.39
et al. (1989)	60 M	6	0.60
	59 M	9	0.70
	55 M	12	0.73
	39 M	15	0.77
	57 F	3	0.43
	59 F	6	0.52
	51 F	9	0.54
	57 F	12	0.50
	40 F	15	0.70
Predicted %BF			
Twisk *et al.* (1995)	80 M	13	0.22
	97 F	13	0.56
Triceps skinfold			
Molinari *et al.*	119 M	6	0.58
(1995)	119 M	8	0.68
	119 M	10	0.70
	119 M	12	0.72
	119 M	14	0.74
	111 F	6	0.49
	111 F	8	0.59
	111 F	10	0.65
	111 F	12	0.68
	111 F	14	0.73

[a] M, male; F, female.

generally higher for males than for females and they increase as the intervals begin at progressively older ages. The only exception is a decrease in these correlations for some measures when the interval includes part of the pubescent period, but pubescent decreases do not occur in age-to-age correlations for head and arm circumferences, skinfold thicknesses, and the sitting height/leg length ratio (Molinari et al., 1995).

The correlations of BMI values at 21 years with values at younger ages increase from 0.4 for values at 3 years to 0.6 for values at 6 years, but there are only small increases for correlations at 9, 12, and 15 years (Casey et al., 1992; Rolland-Cachera et al., 1989). Correlations for BMI in childhood and adolescence with BMI at 21 years are higher for males than for females, but those with BMI at 35 years are higher for females ($r = 0.6$) than males ($r = 0.4$) (Guo et al., 1994; Trudeau et al., 2001). The correlations between skinfold thicknesses during pubescence and during early adulthood are about 0.5 (Campbell et al., 2001; Trudeau et al., 2001; Twisk et al., 1995). Similarly, TBF and FFM have tracking coefficients of about 0.6 for values at 8 to 18 years with values 12 years later (Campbell et al., 2001).

Other evidence of tracking comes from studies in which children were categorized into groups at a young age. Kato & Takaishi (1999) placed infants in five birth weight groups and found a strong tendency to maintain group membership during the first year. When term infants are separated into groups with large, average, and small ponderal indices, the group means maintain the same order through the first year for weight, but those for length differ little among the groups at 1 year (Holmes et al., 1977). These authors also found that short infants with small head circumferences at birth retain these deficits throughout the first year. Children placed in three groups by weight and length at birth retain some of the differences between group means at 5 and 18 years (Loesch et al., 1999).

For children in three groups categorized by percentiles of skinfold thicknesses at 0.5 to 5.5 years and examined two decades later, group membership was retained by 72% of those originally in

the 15th–85th percentile group, but by only 20% of those in the other groups (<15th >85th) (Garn & LaVelle, 1985). This difference partly reflects the narrower percentile range in the marginal groups compared with the central group. Quintile membership for stature and relative weight changed from 6 to 17 years in 43% to 83% of Japanese females; these changes were common when the initial quintile levels were near the means. Children with relative weights >140% at 6 years do not return to the normal range (<120%) at 11 years, but 20% of those with relative weights of 120–140% at 6 years have relative weights <120% at 11 years and 80% of those with relative weights 120–140% at 12 years have relative weights >120% in adulthood (Shirai et al., 1990; Yukawa et al., 1985). A reverse analysis shows those with relative weights >120% at 18 to 21 years have significantly larger mean relative weights from 9 to 12 years, but not at 6 to 9 years (Kelly et al., 1984). In children assigned to three groups based on age at rebound in BMI (<5.5 years, 6 to 6.5 years, >7.0 years), there are significant group differences in BMI at 21 years with BMI being largest in the early rebound group and smallest in the late rebound group (Rolland-Cachera et al., 1987). There are high probabilities (0.7 to 0.9) for the retention of tertile membership for %BF, TBF, and FFM from 8 to 13 years and from 13 to 18 years and probabilities of 0.4 to 0.8 for retention of corresponding tertile memberships from 8 to 18 years. These probabilities are higher for males than for females for % BF and TBF, but not for FFM (Guo et al., 1997b).

Some have set cut-off levels for BMI in adulthood beyond which overweight or obesity is considered to be present. These have been used to calculate odds ratios for adult overweight or obesity based on BMI group membership during childhood. Guo et al. (1994) reported odds ratios for BMI >28 kg/m^2 for males and >26 kg/m^2 for females at 35 years that are significantly increased for children with BMI values at the 95th or 75th percentiles from 3 to 18 years, compared with children with BMI values at the 50th percentile. Whitaker et al. (1997) found the odds ratios for BMI >85th percentile in young adulthood were significantly increased for children aged 3 years or more with BMI >85th percentile, but not for younger children. There are similar risks of large relative weights at older ages dependent on relative weight at 9 to 14 years (Kotani *et al.*, 1997).

An assumption that tracking for an individual child indicates normal growth is justified only if the levels and rates of growth are normal before the period for which tracking is assessed. A lack of tracking for an individual, as shown by serial changes that differ from those in adjacent percentile lines, has limited clinical significance if the serial data points remain between the 25th and 75th percentiles. Concern is justified, however, when there is a lack of tracking near the margins of the distributions, particularly if the data points move from the median. A marked lack of tracking for an individual may be categorized as decanalization, failure-to-thrive, or catch-up growth. These phenomena will now be considered.

DECANALIZATION

The term decanalization describes an unusual growth pattern in which the serial points for an individual cross two major percentile lines on a growth chart for status. When this occurs, the points shift from one canal between percentile lines to a non-contiguous canal. Others have regarded decanalization as any change in canals, but some of these changes are due to small deviations in growth rates (Cole, 1995; Wright *et al.*, 1994). A change of >2 Z scores has also been used as the criterion for decanalization. The recognition of decanalization is an important tool in clinical growth assessment because it allows easy identification of major changes. For the recognition of decanalization to be meaningful, the children being assessed must belong to a group similar in growth to those from whom the reference percentile levels or Z scores were derived. This was not the case in some studies (Smith *et al.*, 1976; Terada & Hoshi, 1965*a*, *b*; Tsuzaki *et al.*, 1987; Wright *et al.*, 1994, 1996).

During early infancy, decanalization is more common for weight than for length and is un-

common for head circumference (Berkey *et al.*, 1983; Park *et al.*, 1997; Terada & Hoshi, 1965*b*). The prevalence of decanalization decreases with age. Decanalization of length occurs in 34% of infants from birth to 12 months, in 8% from 12 to 24 months, and in 3% from 24 to 36 months (Buschang *et al.*, 1985; Park *et al.*, 1997). Infants with decanalizations that are increases in relative levels for length tend to be short at birth or have tall parents. This reflects the negative correlations between length at birth and increments in length during infancy and increasing parent–offspring correlations (Boryslawski, 1988; Fergusson *et al.*, 1980). Decanalization before 9 months usually involves regression toward the median (Berkey *et al.*, 1983; Persson, 1985; Wright *et al.*, 1994), but there is only a slight tendency to regress towards the median later in infancy in most studies (Buschang *et al.*, 1985; Elwood *et al.*, 1987; Park *et al.*, 1997). The prevalence of decanalization differs little between the sexes, but may be more common in males than in females for head circumference (Berkey *et al.*, 1983; Terada & Hoshi, 1965*a*, *b*).

Decanalization of weight and stature during 2-year intervals is unusual after the age of 3 years (p <0.05), although decanalizations from the median for weight become more common during pubescence, especially in the upper parts of the distributions (Li *et al.*, 1998). During pubescence, decanalizations in stature that are gains in percentile levels are significantly related to rapid skeletal maturation, while losses in percentile levels are significantly related to slow maturation (Li *et al.*, 1998). These patterns are reversed after pubescence.

FAILURE-TO-THRIVE

The term failure-to-thrive (FTT) refers to a growth deficit during infancy, particularly in weight, that may be due to disease (organic FTT). It commonly occurs in the absence of disease (non-organic FTT), or may be due to a combination of organic and non-organic causes (Casey, 1983; Leung *et al.*, 1993; Ramsay *et al.*, 1993; Singer, 1987). Consequently, like decanalization, FTT refers to an unusual growth pattern; it is not a diagnosis. Unlike decanalization, FTT is restricted to infancy and relates predominantly to levels and rates of change in weight. There should not be confusion between FTT and psychosocial dwarfism because the latter occurs in children older than 3 years and the major deficit is in stature.

Non-organic FTT is commonly due to an unsatisfactory social and emotional environment for the child that leads to malnutrition (Bithoney & Rathbun, 1983; Casey, 1983; Frank, 1986; Skuse, 1985). It is common in low-income families and in those with small birth weights (Homer & Ludwig, 1981; Kotelchuck, 1980; Mitchell *et al.*, 1980; Singer, 1987). The primary cause may be impaired mother–infant interactions, maternal health beliefs, or socioeconomic hardship, which may lead to malnutrition due to low nutrient intakes or impaired absorption and utilization (Holmes, 1979; Pugliese *et al.*, 1987). The common and rapid improvement in growth with hospitalization, but without specific therapy, suggests the importance of non-organic factors (Barbero & Shaheen, 1967; English, 1978).

There is a lack of consensus about criteria for the definition of FTT. Those suggested require the use of reference data that must be appropriate for the population to which the infant belongs (Wright *et al.*, 1991, 1996). Typically the recommended criteria are low weight status *and* slow weight gains; such infants are unlikely to be part of the normal population. Low weight status has been defined as <3rd percentile (Black *et al.*, 1995; Casey *et al.*, 1995; Ellestein & Ostrov, 1985; Singer, 1986) and as weight-for-length <5th percentile (Fomon, 1993). In FTT, the deficit in weight is more marked than that in length or head circumference (Casey & Arnold, 1983; Holmes, 1979; Kirkland, 1990), but some consider FTT is present only when both weight and length are <3rd percentiles or <5th percentiles (Bithoney *et al.*, 1992; Peterson *et al.*, 1985; Reif *et al.*, 1995; Rudolph, 1985).

A slow weight gain, sufficient for categorization as FTT, has been set as a decanalization, which is a decrease in level to a non-adjacent canal between major percentile lines (Berkowitz & Sklaren, 1984; Edwards *et al.*, 1990; Ellestein & Ostrov, 1985; Wright *et al.*, 1998a) or a rate of weight gain that is <-2 Z for at least 2 months in infants less than 6 months old and for at least 3 months in older infants (Fomon, 1993). Age-dependent metric limits have also been used to recognize slow growth rates (Altemeier *et al.*, 1985; Fomon, 1993). Wright *et al.* (1994, 1998a) developed a method for the recognition of FTT that requires normalized reference data, the transformation of serial weights from birth to 2 months to Z scores, and estimations of expected Z scores at 9 to 18 months. Differences between the actual and the expected Z scores that exceeded two of their ten canals were accepted as evidence of failure-to-thrive. This method is not applicable at other ages.

The criteria used to recognize FTT are generally appropriate, but variations among them hinder comparisons among studies. Infants with lesser deficits in weight status and weight gains than those set as criteria for FTT may also require careful clinical attention. These lesser deficits, which are sometimes called "growth faltering," are common during the second and third months of infancy, particularly in developing countries (Martorell & Shekar, 1994; Waterlow, 1981).

Most studies of FTT have been restricted to term infants of normal birth weight. If LBW infants had been included, the inclusion of weight status relative to reference data for general populations would not be appropriate. Criteria for the recognition of FTT in preterm moderately LBW and VLBW infants should be applied that correspond to those for term infants, but are based on reference data specific to preterm infants (Brandt, 1986; Frank & Zeisel, 1988; Guo *et al.*, 1997a; Mitchell *et al.*, 1980).

The onset of FTT is usually during the first year after birth, commonly before 2 months, and almost always before 2 years (Glaser *et al.*, 1968; Reif *et al.*, 1995). Estimates of the prevalence and outcomes of FTT differ widely due to variations in the criteria applied, in the populations studied, and in the influences responsible. Furthermore, changes may occur in these influences during the follow-up period. Usually there is slow incomplete improvement in growth status even without treatment (Wright, 2000). Considerable improvement in growth status by 8 years is common in non-organic FTT, despite a persistent excess of low values for weight and stature (Black *et al.*, 1995; Casey & Arnold, 1985; Oates *et al.*, 1985; Reif *et al.*, 1995). Intensive multidisciplinary intervention may be ineffective, although health monitoring and dietary advice may result in improved weight gains (Bithoney *et al.*, 1989; Casey *et al.*, 1994; Wright *et al.*, 1998*b*). The prognosis is better when FTT occurs after the first year and is due to a low caloric intake (Chase & Martin, 1970; Eid, 1971; Holmes, 1979). Improved growth is to be expected in organic FTT when the disease causing the condition is treated or subsides spontaneously (Kristiansson & Fällström, 1987).

CATCH-UP GROWTH

The term catch-up growth refers to the acceleration seen in many children during recovery from serious illnesses or from environments that retard growth (Barr *et al.*, 1972; Bate *et al.*, 1984; Curtis *et al.*, 1982; Rogers, 1984). Two conditions are necessary for this designation: (1) growth retardation as shown by previous low percentile levels of the measures, and (2) subsequent increases in these percentile levels. The term catch-up growth is often applied, however, when there is improvement in growth with intervention in the absence of data recorded before the apparent retardation began. In these circumstances, estimates of the completeness of catch-up growth are conjectural.

Catch-up growth is complete if the growth data for a child return to the percentile level present before the deficit occurred. This basis for the evaluation of completeness is preferable to an arbitrary target expressed as a particular change in percentile level because it takes genetic potential into account. Complete catch-up is more common in infants than in older children (Chase & Martin, 1970; Prader, 1978), but it may be delayed (Garrow & Pike, 1967; Hansen, 1965; Kato & Takaishi, 1999; Satgé *et al.*, 1970). Catch-up can occur in weight at any age, in stature until the time of epiphysial fusion, and in

head circumference until the sutures of the cranial vault interlock at about 5 years. Catch-up growth can be dramatic in psychosocial dwarfism and other pathological conditions and may be associated with large rapid increases in head circumference with sutural separation indicating increases in intracranial contents (Capitanio & Kirkpatrick, 1969; Roche, 1980). Catch-up growth is commonly incomplete because treatment is inadequate in duration or type, or an unfavorable environment persists (Ashworth, 1969; Krueger, 1969).

The deficit in growth associated with illness, measured in Z scores, is usually larger for weight than for stature, and catch-up growth after an illness occurs earlier and is more complete in weight than in stature (Barr *et al.*, 1972; Prader, 1978). Consequently, there is a tendency for infants to have large weights relative to length for 6 to 12 months after the initiation of treatment for celiac disease, despite rapid growth in stature during the first month of treatment (Barr *et al.*, 1972; Fabiani *et al.*, 1991). Marked catch-up can occur during the treatment of many diseases including celiac disease, Crohn's disease, and hypopituitarism (Barr *et al.*, 1972; Markowitz *et al.*, 1993; van der Werff ten Bosch & Bot, 1986). Catch-up growth in weight, at least during the early stage of catch-up growth, is due more to increases in muscle than in adipose tissue (Golden, 1998; Jackson, 1990; MacLean & Graham, 1980). Malnourished children may catch-up in weight and stature spontaneously or in response to therapy. Similar changes occur when the socioeconomic status of the whole community improves (Adair, 1999; Kulin *et al.*, 1982). There are corresponding changes in malnourished children who migrate to developed countries as refugees or adoptees, but this catch-up is incomplete for many (Proos, 1993; Proos *et al.*, 1991).

The pattern of change in size of LBW infants after birth is sometimes called catch-up growth; this is appropriate if they are SGA at birth. Catch-up may be more complete in preterm SGA infants, than in term SGA infants (Albertsson-Wikland *et al.*, 1993; Job & Rolland, 1986; Sann *et al.*, 1986; Scott *et al.*, 1982*a*). There is rapid partial catch-up of head circumference in preterm SGA infants beginning at 2 to 6 months after birth (Brandt, 1981; Brandt & Hansmann, 1977; Marks *et al.*, 1979; Weber *et al.*, 1976), but many preterm SGA infants do not catch-up completely in weight, stature, or head circumference by 5 years (Hadders-Algra & Touwen, 1990; Karniski *et al.*, 1987; Sann *et al.*, 1988) and stature deficits may remain to 17 years (Lhotská *et al.*, 1995). Most term infants with small weights or lengths at birth show partial catch-up in length by 12 months, but there is still an increased risk of short stature at 18 years (Karlberg & Albertsson-Wikland, 1995).

The mechanisms responsible for catch-up growth are unknown. Circulating levels of growth hormone may increase during catch-up growth, but this is not essential (Mosier & Jansons, 1976; Mosier *et al.*, 1977; Sinha *et al.*, 1973). It has been postulated that each individual has an age-specific set point for body size, which implies central control; this hypothesis is unproven (Prader *et al.*, 1963).

PREDICTION OF ADULT STATURE

When the stature of a child is unusual for age, e.g., <3rd percentile or >97th percentile, there may be concern about the individual's adult stature. In these circumstances, it is helpful to predict his or her adult stature. The prediction may reassure the family or indicate a need for laboratory tests to establish the cause of the unusual growth. All methods for the prediction of adult stature estimate stature at 18 years. Predictions of adult stature have random errors and there may be systematic errors related to ethnicity, overweight, and disease. Events that occur between the time the predictions are made and the cessation of growth can reduce the accuracy of these predictions. The timing and intensity of the pubescent spurt in stature are significantly correlated with the errors of predictions made at 8 to 13 years, which suggests that failure to predict the timing and size of the pubescent spurt contributes to the prediction errors (Preece, 1988).

The Bayley–Pinneau (BP) method predicts the percentage of adult stature achieved (Bayley & Pinneau, 1952). These percentages are provided in tables for chronological age groups of children categorized by whether the Greulich–Pyle (1950) skeletal age differs from the chronological age by more or less than 1 year. This percentage is used with present stature to calculate a predicted adult stature. The BP method is applicable to males from 7 to 18.5 years and to females from 6 to 18 years. Cross-validation studies show median errors of about 4.0 cm (Bayley & Pinneau, 1952; Roche *et al.*, 1975*a*).

The Roche–Wainer–Thissen (RWT) method employs regression equations in which the length and weight of the child, mid-parent stature and Greulich–Pyle (1959) skeletal age, obtained as the median of bone-specific skeletal ages, are used to predict adult stature (Roche *et al.*, 1975*a*). Weight has only small effects on the predictions and its inclusion may have been undesirable, because it leads to under-predictions in overweight adolescent females (Barbaglia *et al.*, 1989). When length is not measured, stature can be adjusted to length by adding 0.8 cm. (R. J. Kuczmarski, personal communication). The RWT method is applicable from 1 to 16 years in males and from 1 to 14 years in females, but only if fewer than half the hand–wrist bones are adult in maturity. Cross-validation shows the median errors are 2.5 to 3.0 cm (Roche *et al.*, 1975*a*). The coefficients in the original RWT regression equations have been re-smoothed across age leading to some reductions in the errors (Khamis & Guo, 1993). The median errors with the RWT method are reduced by about 25% when Greulich–Pyle skeletal ages are replaced by Fels skeletal ages (Roche *et al.*, 1988). The mean changes in serial predicted adult statures for individuals from the RWT method are near zero, but the SDs of these changes are about 1.0 cm for annual intervals, 1.5 cm for 2-year intervals, and 2.5 cm for 5-year intervals (Roche, 1980).

The Tanner–Whitehouse (TW) prediction method uses Tanner–Whitehouse skeletal ages of the hand–wrist and present stature as predictors and is applicable to children 6 years and older (Tanner *et al.*, 1983*a*, *b*). Mid-parent stature is used as a correction factor not a predictor. The error of prediction from data available at one age, which is the common clinical situation, is about 4 cm. There are separate equations for pre- and post-menarcheal females for which the errors are about 3.5 cm before menarche and 1.5 to 2.0 cm after menarche. Post-menarcheal equations are of limited practical value because little growth potential remains. The inclusion of stature measured one year earlier as a predictor reduces the errors at 12 and 13 years in males and after 8 years in females. An earlier skeletal age, used together with the current skeletal age, does not improve the predictions for males, but it reduces the errors for females at a few ages (Onat, 1995; Tanner et al., 1983a).

In children free of organic disease, the RWT and TW methods have similar errors that are smaller than those with the BP method (Brämswig et al., 1990; Harris et al., 1980; Zachmann et al., 1978; Zarów, 1992), except in a Finnish study where the errors with the TW method were larger than those with the BP and RWT methods (Lenko, 1979). The BP, RWT, and TW methods are applicable to tall or short children who are free of disease (Joss et al., 1992; Kopczynska-Sikorska & Miesowicz, 1980; Maes et al., 1997; von Hinkel & Shambach, 1980). The BP method is generally more accurate than the RWT and TW methods in children with pathological conditions.

Other methods predict adult stature without using skeletal age. Beunen et al. (1997a) predicted adult stature in males from childhood anthropometric data. The errors in their sample were 3.5 to 4.0 cm. Similar results were reported by Green & Buckler (2001) who developed a method that requires serial data for stature. Bock & Thissen (1980) used stature at an age, together with detailed knowledge of the growth patterns of the group to which the child belongs. This approach is not generally applicable due to lack of the data needed and it is not as accurate as the RWT and TW methods. The RWT method has been modified for use when

skeletal age is not available (Khamis & Roche, 1994; Wainer et al., 1978). With this modification, there are small increases in the prediction errors, except during pubescence when the increases are about 2 mm at the median level and 8 mm at the 90th percentile level.

TARGET STATURE

Target stature is used by some clinicians to indicate the likely adult stature of a child. The target stature for an individual, with its confidence limits, can be added to the sex-appropriate chart for stature and may help distinguish familial short stature from other conditions. To calculate target stature, the mid-parent stature is adjusted for the sex of the child by adding 6.5 cm for males and subtracting 6.5 cm for females. For 95% of normal children, stature at 18 years is within 19 to 25 cm of the target stature calculated by this method (Karlberg & Fryer, 1990; Mahoney, 1987). The errors may be larger when the children or parents have pathological conditions. The wide limits are not surprising given the inaccuracy of adult stature predictions from mid-parent statures (Gray, 1948; Welch, 1970). An alternative equation has been proposed in which target stature = 0.718 mid-parent stature (cm) + x where $x = 57.6$ cm for males and 44.6 cm for females (Molinari et al., 1995).

Some of the inaccuracy of target stature is due to regression to the mean when the statures of children and their parents are compared. Equations that take this regression into account provide more accurate target statures (Luo et al., 1998; Wright & Cheetham, 1999). Other equations adjust for the generational difference in adult stature between parents and their children by assuming the past generational difference will be repeated and will not vary with the percentile levels of parental statures (Burgemeijer et al., 1998).

3 · Determinants of growth

GENETIC INFLUENCES

Most of the literature considered here relates to continuous quantitative phenotypic traits that are likely to be influenced by a number of genes acting through hormones and their receptors and through effects on metabolic processes. Most of the genetic control of each of these traits is probably due to a few genes while other genes have only small effects. These genetic influences may differ with sex and age and be subject to environmental interactions. The emphasis in this section is on genetic influences that are associated with normal variation in growth. The importance of genetic influences on growth is shown vividly, however, by some chromosomal variations that result in abnormal growth. In Turner syndrome, females lack one X chromosome and they are very short. In Klinefelter syndrome, males have an additional Y chromosome and are very tall. In each of these conditions, the variations from normal growth are most marked in the limbs. In Turner syndrome, mother–daughter correlations for stature are similar to those for normal children indicating that many genes influencing stature are located on autosomes (Brook *et al.*, 1974).

Intra-familial correlations provide initial information about genetic influences on phenotypic traits, although they reflect both genetic and environmental influences. Phenotypic traits are specific observable characteristics of an individual or an organism. These correlations are used to estimate heritability (h^2), which indicates the extent of the genetic control of a trait, usually on a scale from zero to 1.0, but a scale from zero to 100 may be used. These estimates show the proportion of the variance of a phenotypic trait that is attributable to the additive effects of genes that are transmitted from parents to offspring. Less commonly, other sources of genetic variance, such as dominance and interactions between genes, may be included in the model. Moderate positive correlations between traits for measures of size during childhood are common. These could reflect joint environmental influences or pleiotropy, which refers to the ability of genes to influence more than one trait. When two phenotypic traits are moderately or highly correlated, and the h^2 estimates are high for each trait, it is likely that some genes influence both traits. The same genes may increase both traits or may increase one and decrease the other. When pleiotropy is incomplete, the same genes influence both traits, but other genes are involved that influence only one of the traits. Complex traits, such as stature, that show marked tracking, are only slightly sensitive to environmental influences, and the h^2 estimates for them are high. The environment has large effects on soft tissues; consequently, heritability estimates for soft tissue measurements are low to moderate. Heritability estimates are very numerous in the literature, but current genetic research extends to the effects of specific genes and their environmental interactions.

Knowledge of environmental effects is limited mainly because the various elements that constitute the external and internal environments of individuals are measured poorly. Children may differ in their responses to particular environmental variations because of inter-individual genetic variations. Children who differ in the presence or absence of a disease may differ in the genetic variance of a growth measure. This would suggest a genetic × environment interaction in which genes contribute to the differences in growth among children with the disease. Similarly, there may be genotype × sex interactions perhaps due to sex differences in internal hormonal environments. Studies of these possibilities attempt to determine whether the sexual dimorphism of a trait is inherited. The occurrence of large sex differences in stature at an age among children in some families, but not in others, would suggest the presence of gene × sex interactions. The effects of genes on a particular trait may differ with age. Analyses of genotype × age interactions require serial data from related individuals so that the additive genetic variance can be modeled as a function of age. Systematic changes with age in these variances show the presence of genotype × age interactions.

Heritability estimates are specific for particular populations and environments, but generally there is marked concordance among studies regarding the heritabilities of complex traits. Many h^2 estimates have been derived from studies of twins. Studies of differences in phenotypic measures between monozygotic and dizygotic twins assume the environmental effects are the same within monozygotic pairs, but differ within dizygotic pairs. These h^2 estimates are inflated to an unknown degree by the effects of shared pre- and postnatal environments. Prenatally, many twins share a placenta and chorion and, therefore their blood supplies and growth may not be independent (Bulmer, 1970; Corney et al., 1979). After birth, twins typically share the same household environment. Some have tried to remove the latter confounding effect by studying twins adopted into different households. A design based on monozygotic twins was used by Bouchard et al. (1990, 1994) to study genetic influences on the effects of overfeeding and underfeeding. The changes in weight and TBF as a result of dietary manipulation were three times larger between twin pairs than within twin pairs.

Other heritability estimates are made using data from nuclear families that provide correlations between pairs of first-degree relatives, for example, parent–offspring, sibling–sibling, and spouse–spouse pairs. These estimates tend to be increased by shared environments. Secular changes and age differences may influence parent–offspring correlations. Secular changes in the environment may affect the phenotypic trait differentially among individuals and may lower h^2 estimates for groups due to increases in the non-genetic variance of a trait. Age differences between parents and their offspring can also lower h^2 estimates, if some genes are influential only during particular age ranges. More appropriate parent–offspring analyses use data at matching ages, but such data are scarce.

Genetic influences on complex phenotypic traits can be understood more fully from studies of extended families partly because the effects of shared environments are markedly reduced and genotype × environment interactions can be studied. These studies are logistically difficult and expensive and they require intensive computations. The latter need is now met more easily due to improvements in computer science. New methods can also analyze models that include age, sex, and genotype × age and genotype × sex interactions. This is important because the sets of genes that influence growth may vary with age and sex. For example, the genes that influence the rate of growth in weight, length, and head circumference from birth to 2 months are not the same as those that influence growth in these measures from 2 months to 2 years (Kobyliansky et al., 1987).

In addition to determining the importance of genetic influences through h^2 estimates and elucidating their interactions with age and sex, there is marked interest in identifying the genes responsible. Most complex traits are controlled by multiple genes, each of which may influence one or several biological processes. For example, growth in stature can be separated to elongation of the vertebral column and elongation of the inferior extremities. The histological processes differ between these anatomical areas and the enzymes and genes involved may differ also. The search for specific genes with moderate or small effects has been facilitated by knowledge of the sequence of genes in the human genome and by advances in statistical genetic methods. Two basic approaches are used. The first is the association approach in which unrelated individuals are grouped depending on whether or not they have a polymorphism in or near a candidate gene for a trait. A polymorphism is the joint occurrence in a population of two or more genetically determined alternative phenotypes. Relationships between the polymorphism and the trait are then examined. Problems arise when there is linkage disequilibrium, which occurs when the alleles at two loci are not randomly associated due to new mutations, genetic drift or selection. The level of disequilibrium between alleles cannot be predicted from the distance between loci. Other analytic difficulties occur when sub–groups of the population studied differ in allele frequencies or the phenotypic trait. The second approach uses linkage analysis, which is based on the concept that loci near each other on the set of chromosomes (genome) are likely to be linked and inherited together. Studies of the frequency of recombination among genetic markers can identify the chromosomal regions that influence a trait. Linkage is said to be present between a locus that influences a phenotype and a marker in a chromosomal region when pairs of relatives who are more similar in the phenotype have more similar alleles at the marker locus than other pairs of relatives. The linkage approach requires strong genetic effects and large samples that include extended families with a variety of family relationships.

Weight

Most h^2 values for birth weight vary from 0.5 to 0.8 (Clausson et al., 2000; Magnus, 1984; Rao & Morton, 1974). Studies of specific genes are sparse, but there are significant associations between phosphoglucomutase locus 1 genotypes and birth weight in females (Gloria-Bottini et al., 2001). This enzyme is important in carbohydrate metabolism. Estimates of genetic effects on weight at birth and in early infancy are influenced by non-genetic maternal influences. This is shown by the strong relationship of the birth weight of infants born after ovum donation with the weights of the recipient mothers, but not with the weights of the donors (Brooks et al., 1995). The h^2 values for weight may decrease somewhat during infancy due to increases in the effects of environmental influences (Chrzastek-Spruch, 1979; Livshits, 1986; Ounsted et al., 1982), but change

little after 6 years (Clark, 1956; Tanner & Israelsohn, 1963; Tiisala & Kantero, 1971). The h^2 values for all parameters of curves fitted to serial weight data from 10 to 16 years, considered simultaneously, are 0.50 for males and 0.78 for females (Fischbein & Nordqvist, 1978).

Length and stature

The estimates of h^2 for length at birth vary markedly, perhaps due to the study of small samples and variations in the accuracy of the measurements. Some report small h^2 values (Dupae et al., 1982; Feinleib et al., 1977; Langinvainio et al., 1984; Pedersen et al., 1984). Towne et al. (1993), however, found large h^2 values for length at birth (0.83), growth rate from birth to 2 years (0.67) and changes in growth rate (0.76). Baker et al. (1992) reported similar h^2 estimates to those of Towne for length at birth and rate of growth to 2 years, but near zero h^2 values (0.12, males; 0.01, females) for changes in these rates. Both Towne and Baker found indications of genotype × sex interactions for growth rate and change in growth rate during infancy.

Studies of childhood stature have shown h^2 values ranging from 0.38 to 0.95 (Corey et al., 1986; Kuh & Wadsworth, 1989; Phillips & Matheny, 1990; Smith et al., 1973). The h^2 estimates for stature, at the onset of the pubescent spurt, at PHV and in adulthood, range from 0.39 to 0.96 (Bergman & Goracy, 1984; Beunen et al., 2000; Hauspie et al., 1994; Towne et al., 1995). There is low to moderate pleiotropy among the genes controlling stature at an age, rate of growth in stature, and age at PHV (Towne et al., 1995). Simultaneous analyses of all the parameters of curves fitted to serial statures from 10 to 16 years give a heritability estimate of 0.62 (Fischbein & Nordqvist, 1978).

Genetic influences on length and stature may be mediated through the formation and release of hormones that affect bone elongation and through the interactions of these hormones with specific receptors. The h^2 value for growth hormone secretion during wakefulness is 0.74 and that for 24–hour secretion is 0.50 (Mendlewicz et al., 1999). Growth hormone levels are low in some single gene disorders in which the hGH gene is abnormal or mutations alter the hypothalamic control of the pituitary (Fleisher et al., 1980; Nishi, 1990; Phillips et al., 1981; Rimoin, 1976). The h^2 values for testosterone levels during adolescence are 0.6 for males and 0.4 for females (Harris et al., 1998). The absence of a close father–son resemblance in testosterone levels suggests that the genes responsible may differ from adolescence to adulthood. Variations in the genes that control luteinizing hormone production are associated with stature differences that increase from 2 to 18 years (Towne et al., 2000). The secretion of IGF-1 is almost entirely under genetic control (Hong et al., 1996; Kao et al., 1994), but h^2 estimates for insulin levels are only 0.21 to 0.53 (Hong et al., 1998; Iselius et al., 1982; Mitchell et al., 1996; Schumacher et al., 1992). Other genetic influences may affect hormone receptors. For example, variations in estrogen receptor genes influence the rate of growth in stature and the timing of PHV (Parks et al., 2000).

Other measures

During infancy, correlations between mid-parent and offspring head circumferences are 0.2 to 0.5 (Poosha et al., 1984; Tanner & Israelsohn, 1963; van't Hof et al., 2000a), but the h^2 estimate at 12 to 20 years is 0.74 (Clark, 1956). This suggests that the environmental control of head circumference becomes less marked after infancy. Correlations between parents and offspring aged 18 years for BMI, TBF, and FFM are low ($r^2 = 0.0001$ to 0.09), but those relating BMI and TBF to paternal values and those relating FFM to maternal values are significant ($p < 0.05$) in males (Campbell et al., 2001). Estimates of h^2 for BMI range from 0.05 to 0.64 (Allison et al., 1994; Bouchard & Pérusse, 1988; Moll et al., 1991) suggesting marked heterogeneity among studies in genetic and environmental influences or in the accuracy of zygosity determination. Some adoptive studies suggest environmental effects on BMI are very small (Price et al., 1987; Sørensen et al., 1992a, b; Stunkard et al., 1990).

This is puzzling because the large secular increases in childhood BMI must be due to environmental changes. There may be a major gene for BMI combined with polygenic control (Borecki et al., 1993; Lecomte et al., 1997; Moll et al., 1991; Price et al., 1990). Moll concluded a major gene was responsible for about 35% of the variance in BMI and that polygenic loci accounted for a further 42% of the variance. It has been estimated that 6% of individuals have two copies of the major gene, and 37% have one copy (Price et al., 1990; Province et al., 1990; Tiret et al., 1992). Differences in BMI are linked to genetic variations that affect apolipoprotein A-IV, the β_3-adrenergic receptor gene, and leptin receptors (Chagnon et al., 2000; Fujisawa et al., 1998; Lefevre et al., 2000; van der Kallen et al., 2000). Genetic variations may also explain why BMI responds to dietary manipulation in some subjects, but not all (Beynen et al., 1987; Heitmann et al., 1995; Quivers et al., 1992).

The h^2 estimates for skinfold thicknesses range from 0.36 to 0.98 and they increase markedly after PHV (Brook et al., 1975; Katzmarzyk et al., 2000; Thomis et al., 1998). Some of these genetic influences may be due to variations in the leptin receptor gene (Chagnon et al., 2000). Rare variations in the leptin receptor gene are associated with increased %BF in males (Chagnon et al., 2000; Clément et al., 1998, 1999). Major genes with a frequency of 3% may affect %BF and TBF (Comuzzie et al., 1995; Rice et al., 1993a). Genes may influence the strong familial correlations for energy intake and energy expenditure (Bouchard et al., 1994; Pérusse & Bouchard, 1994). Serotonin has a major influence on energy intake, but variations in the gene for its receptor are not associated with variations in body fat (Aubert et al., 2000).

Some genes may influence body fatness by actions on β_2 and β_3 adenoreceptors, hormone-sensitive lipase, tumor necrosis factor α and receptors for low-density lipoprotein cholesterol and leptin. Some β_2 and β_3 adrenergic receptor gene mutations are associated with altered receptor functions in adipocytes and with obesity in females (Clément et al., 1996; Endo et al., 2000; Large et al.,

1997). In some obese individuals, lipolysis may be stimulated by genetic variations that decrease the expression of hormone-sensitive lipase, which is the final limiting step in the breakdown of triglycerides in response to catecholamine stimulation (Pérusse *et al.*, 2001). Variations in the lipoprotein lipase gene are associated with variations in %BF (Bouchard *et al.*, 1994). Tumor necrosis factor α is likely to be controlled by a recessive gene and may influence insulin sensitivity in adipocytes by effects on insulin signaling and may induce apoptosis of adipocytes and stimulate lipolysis by reducing the expression of lipolysis-inhibiting gastrointestinal protein (Hoffstedt *et al.*, 1999, 2000).

Estimates of h^2 for FFM are about 0.5 (Arden & Spector, 1997; Bouchard *et al.*, 1988; Pérusse *et al.*, 1987), but a larger value (0.80) was reported by Seeman *et al.* (1996). Evidence of a major gene for FFM has not been found in regression analyses (Comuzzie *et al.*, 1993; Rice *et al.*, 1993*b*). Cortical thicknesses of metacarpals in adults are correlated within families with r-values of 0.28 for parent–offspring comparisons and 0.39 for sibling comparisons. A major gene may be responsible for about 45% of the variance in cortical thickness (Ginsburg *et al.*, 2001). The h^2 estimates for regional BMD range from 0.30 to 0.66 and those for regional BMC range from 0.62 to 0.80 (Arden *et al.*, 1996; Ferrari *et al.*, 1998; François *et al.*, 1999; Magarey *et al.*, 1999). Much of the variance in lumbar spine BMD can be explained by variations in alleles for the vitamin D receptor (Kelly *et al.*, 1991; Morrison *et al.*, 1992; Pocock *et al.*, 1987) and by variations in stature, weight, FFM, and muscle strength, all of which are under considerable genetic control (Seeman *et al.*, 1996).

Kaprio *et al.* (1995) estimated h^2 for age at menarche at 0.32 and found common genetic influences on age at menarche and BMI accounted for 37% of the variance in age at menarche. The estimated h^2 for the timing of onset of epiphyseal ossification is 0.82 (Reynolds, 1943) and h^2 estimates for skeletal maturity assessed with the method of Roche, Wainer, and Thissen decrease from 0.71 at 3 years to 0.11 at 15 years (Towne *et al.*, 1999). Estimated h^2 values

for TW skeletal ages also decrease with chronological age; these decreases are larger in males than in females (Kimura, 1981). The decreases with age in h^2 estimates for skeletal ages suggest gene × age interactions or that environmental influences on the rate of skeletal maturation become more important after infancy.

FAMILY INFLUENCES

Family influences on growth and maturation are reviewed for developed countries with an emphasis on singletons born at term, because the findings for preterm and multiple births are less certain. The maternal influences considered are age, parity, socioeconomic status (SES), pre-pregnancy weight, weight gain during pregnancy, stature, and prenatal care; the paternal influences considered are weight, stature, and SES. Accurate and valid data can be obtained for each of these measures with the possible exception of SES for which reported data are analyzed using categorical constructs. These constructs differ among studies and commonly have been developed from inadequate sets of descriptors. The influence of ethnicity is considered in a separate section.

Some reports show the effects of selected influences are significant, but do not estimate the size of these effects. Such findings are important for biological understanding, but it is difficult to apply them to the interpretation of growth data. The estimated independent effect of a specific family influence can be used to adjust recorded values if the estimate is considered to be accurate. In theory, after this adjustment, the recorded value will be independent of the family influence. This independence is not complete because the established confounding variables, in combination, explain only about half the variance in the outcome variables (Showstack *et al.*, 1984). Consequently, the estimated effects from such studies should be called adjusted effects rather than independent effects. Few unadjusted estimates of family influences on growth are considered here because they can be applied to the interpretation of growth data only when the child belongs to a group similar

Table 3.1. *Adjusted effects of maternal age on birth weight*

Author	Number	Effect	Notes
Differences between groups			
Dougherty & Jones (1982)	50	−24 g: 18–20 yrs	vs. 21–25 yrs
	381	0: 26–30 yrs	
	249	+9 g: 31–35 yrs	vs. 21–25 yrs
	76	+48 g: ≥36 yrs	vs. 21–25 yrs
Ellard *et al.* (1996)	987	+20 g: <25 yrs	vs. 25–29 yrs
	872	−8 g: >29 yrs	vs. 25–29 yrs
Friedman *et al.* (1993)	4 425	+46.3 g: <20 yrs	vs. 25–29 yrs
	45 023	+14.2 g: 20–24 yrs	vs. 25–29 yrs
	61 029	−6.3 g: 30–34 yrs	vs. 25–29 yrs
	30 260	−18.4 g: ≥35 years	vs. 25–29 yrs
Ogston & Parry (1992)	8 452	+29 g: 20–29 yrs	vs. <20 yrs
		+56.4 g: 30–34 yrs	vs. <20 yrs
		+79.8 g: >35 yrs	vs. <20 yrs
Strobino *et al.* (1995)	227	−54.5 g: 14–17 yrs	vs. 23–25 yrs
	515	+4.1 g: 18–19 yrs	vs. 23–25 yrs
	724	−60.2 g: 20–22yrs	vs. 23–25 yrs
Effect/year			
Hulsey *et al.* (1991)[a]	10 159	+2.4 g per yr	
Marbury *et al.* (1984)	536	−24.8 g per yr: <20 yrs	
	2 175	+15.6 g per yr: >30 yrs	
O'Sullivan *et al.* (1965)[b]	436	+9.9 g per yr	
Rush & Cassano (1983)	16 688	+1.4 g per yr	

[a] No adjustment for socioeconomic status.

[b] No adjustment for substance abuse.

in other influences to the population from which the effect was estimated. It is difficult to establish this multivariate similarity in clinical circumstances.

Many adjusted effects are small, but they indicate shifts in the distributions of values that may be important for infants and children who, in the absence of these effects, would be near the extremes of the distributions. The considerable variation among reported estimates partly reflects differences in study design. Some have performed multivariate analyses while others have compared groups matched for many characteristics. Furthermore, populations differ in the mean values for the recorded measures and in the relationships among them, and studies differ in the methods by which SES is categorized and analyzed.

Size at birth

MATERNAL AGE

Birth weight is significantly smaller for teenage mothers than for older mothers (Amini *et al.*, 1994; Marbury *et al.*, 1984; Ogston & Parry, 1992; Sukanich *et al.*, 1986). After the teenage years, the effects of maternal age on birth weight are small (Dougherty & Jones, 1982; O'Sullivan *et al.*, 1965). Estimates of the effects of maternal age on weight range from +1.4 g per year to +9.9 g per year (Hulsey *et al.*, 1991; O'Sullivan *et al.*, 1965; Rush & Cassano, 1983), except for the much larger estimates by Marbury *et al.* (1984) who adjusted for an incomplete set of confounding variables (Table 3.1). Given the variation among studies and the likely small size

Table 3.2. *Adjusted effects of parity on birth weight*

Author	Number	Effect	Notes
Effects by categories			
Dougherty & Jones (1982)	964	+104 g	multiparae[a]
Ellard *et al.* (1996)	1 477	+150 g	multiparae[a]
Marbury *et al.* (1984)	11 173	+56 g	multiparae[a]
Martin & Bracken (1986)	3 891	+148 g	multiparae[a]
Ogston & Parry (1992)	8 453	+117 g	multiparae[a]
Peters *et al.* (1983): 1958/1970 cohorts[b]	16 989/16 762	−116 g/− 111 g	for parity 1
		+7 g/+ 28 g	for parity 2
		+32 g/+ 40 g	for parity 3 or 4
		+77 g/+ 40 g	for parity 5 or more
Rush & Cassano (1983)	16 688	+23 g	multiparae[a]
Shiono *et al.* (1986*a*)	29 415	+116 g	multiparae[a]
Effect per parity order			
Abrams & Selvin (1995)	2 420	+64 g	–
Fenster & Coye (1990)	1 040	+30 g	–
Hediger *et al.* (1989)	1 767	−76 g	adolescents
O'Sullivan *et al.* (1965)[c]	436	+25 g	–

[a] Vs. primiparae.

[b] 1958 effects/1970 effects.

[c] No adjustment for substance abuse or weight gain in pregnancy.

of the effects, it is recommended that these estimates not be applied clinically. Despite these modest effects, the risk of LBW and preterm infants is significantly increased for teenage mothers and for those >30 years old (Ancel *et al.*, 1999; Fang *et al.*, 1999; Miller *et al.*, 1996; Parker & Schoendorf, 1992). Maternal age is not significantly related to length at birth (Wingerd & Schoen, 1974).

PARITY

Associations of size at birth with parity tend to be confounded with those of maternal age. Mean birth weight is 23 g to 150 g larger for the infants of multiparae than for those of primiparae (Ellard *et al.*, 1996; Marbury *et al.*, 1984; Martin & Bracken, 1986; Ogston & Parry, 1992; Peters *et al.*, 1983). Consequently, multiparae are less likely than primiparae to have LBW or preterm infants (Ancel *et al.*, 1999; Fang *et al.*, 1999; Marouka *et al.*, 1998; Parker &

Schoendorf, 1992), but are more likely to have LGA infants (Scott *et al.*, 1982*b*). The effect of parity on birth weight increases with maternal age (Selvin & Janerich, 1971). The reported effects of parity on birth weight are large from parity 1 to parity 2 (+123 and +139 g in two groups), but increase only slightly for larger parities (Peters *et al.*, 1983) (Table 3.2). In adolescent mothers, however, there is a decrease in birth weight of 76 g per parity order (Hediger *et al.*, 1989). Parity may have a very small positive adjusted effect on length at birth that is about 0.1 cm per parity order (Rush & Cassano, 1983; Wingerd & Schoen, 1974).

SOCIOECONOMIC STATUS

The construct socioeconomic status (SES) commonly includes paternal occupation, maternal education, and family income. These items are partial surrogates for medical, biological, and behavioral influences. The term "social class," as used in many

Table 3.3. *Adjusted effects of employment during pregnancy on birth weight*

Author	Number	Effect	Notes
Hatch et al. (1997)	575	−27 g	standing work, ≥40 h/wk, 1st trimester[a]
		+20 g	standing work, ≥40 h/wk, 2nd trimester[a]
		−25 g	standing work, ≥40 h/wk, 3rd trimester[a]
		+71 g	lifting work, high vs. low frequency, 1st trimester
		+47 g	lifting work, high vs. low frequency, 2nd trimester
		−59 g	lifting work, high vs. low frequency, 3rd trimester
Marbury et al. (1984)	11 173	−43.1 g	employment to 8 months[b]
		−11.5 g	employment to 9 months[b]
Rabkin et al. (1990)	1 507	+12 g	full-time employment[b]

[a] Vs. <40 h/wk.

[b] Vs. no employment.

studies from the United Kingdom, is based on paternal occupation only. The effects of SES on birth weight are significant, but estimates of the size of these effects vary (Butler & Alberman, 1969a; Peters et al., 1983; Vågerö et al., 1999; Weiner & Milton, 1970). Some of the dissimilarity among studies is due to variations in the method of categorizing SES and whether the estimated differences are those between contrasting SES groups or between adjacent groups on a SES scale. The relationships between SES and either length or the ponderal index at birth may not be significant (Hauspie et al., 1996; Vågerö et al., 1999), but there are significant effects of father's occupation on length at birth in a group of German infants (Kromeyer et al., 1997). There is a significant relationship between SES and the risk of preterm and LBW infants (Ancel et al., 1999; Spencer et al., 1999).

Several aspects of SES have received separate attention. Employment during pregnancy is associated with small deficits in birth weight (Hatch et al., 1997; Marbury et al., 1984; Rabkin et al., 1990) (Table 3.3). Others have found significant differences with maternal employment only when mothers with professional spouses are compared with mothers whose spouses are industrial workers or unemployed (Ancel et al., 1999). Employed mothers do not have an altered prevalence of LBW infants,

after adjusting for parity and tobacco use (Gofin, 1979), although physical activity during pregnancy may increase the risk of preterm infants (Mamelle et al., 1984; Meyer & Daling, 1985; Ramirez et al., 1990; Teitelman et al., 1990).

It is uncertain whether leisure physical activity reduces birth weight. Some have found such reductions, which may be associated with an increased prevalence of preterm births (Berkowitz et al., 1983; Clapp & Capeless, 1990; Clapp & Dickstein, 1984), but they were not noted in other studies (Hall & Kaufmann, 1987; Sternfeld et al., 1995). Some of this uncertainty relates to variations in the amounts of physical activity, and inadequate control groups.

The amount and type of work and its timing in relation to gestation may be important. Those working >40 hours per week deliver lighter babies and the deficits in birth weight are larger with employment in the third trimester than with employment in earlier trimesters (Hatch et al., 1997; Peoples-Sheps et al., 1991). Henriksen et al. (1995) studied 4259 women with singleton pregnancies who were working at the 16th week of gestation. They concluded that standing and walking for more than 5 hours per day was associated with a significantly increased risk of preterm delivery. Both the number of hours worked per week and a fatigue score devised from data for standing, physical exertion,

Table 3.4. *Adjusted effects of pre-pregnancy weight on birth weight per kg variation from mean*

Author	Number	Effect	Notes
Abel *et al.* (1991)	772	1.6 g	–
Abrams & Laros (1986)	2 948	15.9 g	all term pregnancies
		22.1 g	uncomplicated term pregnancies
Anderson *et al.* (1984)	1 537	9.1 g	primiparae
		8.8 g	multiparae
Fenster & Coye (1990)	1 040	4.2 g	–
Kuzma & Sokol (1982)	5 093	5.0 g	–
Marbury *et al.* (1984)	11 173	3.6 g	–
O'Sullivan *et al.* (1965)[a]	436	3.0 g	–

[a] No adjustment for substance abuse or weight gain in pregnancy.

mental stress, and work environment have significant positive effects on the risk of preterm delivery (Luke *et al.*, 1995). Similar conclusions have been reached by others (Klebanoff *et al.*, 1990; Launer *et al.*, 1990; Teitelman *et al.*, 1990; Wohlert, 1989), but associations between maternal employment and the prevalence of LBW infants were not found in several studies (Gofin, 1979; Marbury *et al.*, 1984; Meyer & Daling, 1985).

Marital status can be considered an aspect of SES although its significance differs among cultural groups. The relationship between marital status and birth weight was non-significant in the study by Vågerö *et al.* (1999) and the estimated deficits in birth weight associated with the mother being single do not exceed 56 g (Friedman *et al.*, 1993; Hulsey *et al.*, 1991). Single mothers have increased risks of preterm and LBW infants (Ancel *et al.*, 1999; Fang *et al.*, 1999; Hartikainen-Sorri & Sorri, 1989), but the risk of LBW at term is not increased (Lang *et al.*, 1996).

Maternal educational level has small effects on birth weight (grade 9 or lower, −44 g; grade 10 or 11, −27 g; grade 13 or higher, +28 g), but has significant negative relationships to the prevalence of preterm births (Ancel *et al.*, 1999; Basso *et al.*, 1997; Friedman *et al.*, 1993). Maternal educational level is not related to the risk of LBW for African-American or Mexican-American women born in the United States or for Hispanic-American mothers born outside the United States (Collins & Shay, 1994; Fuentes-Afflick *et al.*, 1998; Scribner & Dwyer, 1989). It is uncertain whether maternal educational level is related to the risk of term LBW in European Americans (Alberman *et al.*, 1992; Lang *et al.*, 1996). Maternal educational level may not be related to head circumference at birth and its relationship to length is uncertain (Kromeyer *et al.*, 1997; Lhotská *et al.*, 1995; Villamor *et al.*, 1998).

PRE-PREGNANCY WEIGHT

Pre-pregnancy weight is negatively related to the risk of LBW (Lang *et al.*, 1996). There is an estimated deficit of 231 g in birth weight and an increased prevalence of LBW infants in mothers with weight for stature ≤90% of the expected value (Edwards *et al.* 1979). These effects increase with gestational age, but the corresponding effects on length at birth are not closely related to gestational age (Brown *et al.*, 1980). The effects of pre-pregnancy weight are similar for primiparae and multiparae, but they may be larger for female infants than for male infants (Table 3.4). The estimated effects range from 1.6 to 22.1 g for each kg by which the pre-pregnancy weight varies from the population mean. Given the balance of the evidence and taking sample sizes into account, it is reasonable to assume the effect is about 10 g per kg. This small effect, applied to United States data, would reduce the expected birth weight by 114 g for a mother with a pre-pregnancy weight at the 5th percentile and increase the expected birth weight by 250 g for a mother with a pre-pregnancy weight at the 95th percentile in comparison with mothers whose weights are near the mean (Najjar & Rowland, 1987). The effects of maternal pre-pregnancy weight on birth weight may vary slightly with maternal age. In data that were incompletely adjusted for confounding variables, the effects of pre-pregnancy weights on birth weight decreased slightly with age among teenage mothers being 11.4 g per kg for those younger than 16 years and 9.9 g per kg for those aged 16 to 19 years (Stevens-Simon et al., 1993). There is a slight tendency to smaller effects at young maternal ages for mothers <45 kg but a slight reverse effect of age for mothers weighing 45 to 54 kg (Horon et al., 1983).

Other estimates of the effects of pre-pregnancy weight on size at birth have come from comparisons between groups (Table 3.5). These estimates are somewhat larger than those reported for effects per kg of pre-pregnancy weight. The reported deficits in birth weight for mothers with pre-pregnancy weights <45 kg are 171 to 286 g and there is an excess of 85 to 165 g for those with pre-pregnancy weights >68 kg (Cnattingius et al., 1985; Dougherty & Jones, 1982; Ellard et al., 1996; Horon et al., 1983). Some of these estimates are uncertain because they were derived from small samples, but deficits in birth weight of about 230 g and 135 g can be attributed to pre-pregnancy weights that are <45 kg and 45 to 54 kg, respectively.

The effects of pre-pregnancy weight on length at birth are about 0.3 cm per kg and may increase with parity (Anderson et al., 1984; Hediger et al., 1989; Hoffmans et al., 1988; Wingerd, 1970). Neggers et al. (1995) reported that infants born to mothers with mean pre-pregnancy weights of 86.4 kg had larger lengths (+0.9 cm) and head circumferences (+0.5 cm) than infants born to mothers with mean pre-pregnancy weights of 46.3 kg. These findings are not in agreement with those of Rummler & Woit (1990).

Table 3.5. *Adjusted effects of pre-pregnancy weight on size at birth from group comparisons*

Author	Number	Weight
Cnattingius et al. (1985)	636	−171 g if <45 kg vs. ≥45 kg
Dougherty & Jones (1982)	206	−116 g if <54 kg vs. 55–67 kg
	181	+ 85 g if >68 kg vs. 55–67 kg
Edwards et al. (1979)	708	−231 g if ≤90% of Metropolitan Life Insurance Co. ideal weight (1943)
Ellard et al. (1996)	655	−172 g if <55 kg vs. 55–68 kg
	777	+165 g if >68 kg vs. 55–68 kg
Horon et al. (1983)	33	−210 g if <45 kg[a]
	155	−158 g if 45–54 kg[a]
	38	−44 g if 63–72 kg[a]
	19	−286 g if <45 kg[b]
	122	−111 g if 60–72 kg[b]
	54	+150 g if 63–72 kg[b]
	44	+230 g if >72 kg[b]
Neggers et al. (1995)	1205	+295 g[c]

[a] Age <15 years and vs. 54–63 kg.
[b] Age 20–24 years and vs. 54–63 kg.
[c] For difference between 10th and 90th percentile (46.3 and 86.4 kg, respectively).

Maternal BMI is positively related to the risk of preterm births in primiparae. In multiparae, the risks of preterm births and SGA infants are negatively related to maternal BMI (Cnattingius et al., 1998; Kramer et al., 1999). Mothers with weight-for-stature ≥50% larger than reference values have an increased risk of LGA infants (Edwards et al., 1978; Gross et al., 1980; Harrison et al., 1980). In term infants born to African-American mothers, maternal pre-pregnancy BMI was the best predictor of weight, length, and head circumference at birth.

Table 3.6. *Adjusted effects of maternal weight–stature relationships on weight at birth*

Author	Number	Effect	Notes
BMI			
Abrams & Selvin (1995)	2994	+20.1 g per kg/m^2	–
Goldenberg *et al.* (1992)	1205	+17.7 g per kg/m^2	
Hediger *et al.* (1989)	1767	−99 g if BMI <19.5 kg/m^2	adolescents
Neggers *et al.* (1995)	1205	+267 g	10th vs. 90th percentile
Weight-for-stature percentile			
Shiono *et al.* (1986*a*)	13 420	−106 g^a	≤25th percentile
		+108 g^a	≥75th percentile

^{a} In comparison with 26th–74th percentile range.

For a change from the 10th to the 90th percentile of maternal pre-pregnancy BMI, there are estimated increases of 267 g in weight, 0.7 cm in length, and 0.4 cm in head circumference at birth (Neggers *et al.*, 1995). These findings are in general agreement with those from other reports (Table 3.6). Similarly a low pre-pregnancy BMI (≤19.5 kg/m^2) in teenage mothers is related to a deficit in birth weight of 99 g (Hediger *et al.*, 1989). Nevertheless, the combination of maternal pre-pregnancy weight, weight gain, and BMI during pregnancy explains only about 10% of the variance in weight and length and 4% of the variance in head circumference at birth. Maternal FFM and TBF are correlated with birth weight ($r = 0.5$–0.6). Each 1.0 kg change in maternal FFM is associated with a change of 20 g in birth weight and each 1.0 kg change in TBF is associated with a change in birth weight of 18 g (Mardones-Santander *et al.*, 1998). The infants of mothers at the 10th or 90th percentile for FFM differ by 188 g in mean adjusted birth weight (Neggers *et al.*, 1995).

WEIGHT GAIN DURING PREGNANCY

Birth weight is significantly related to weight gain during pregnancy (Anderson *et al.*, 1984; Hickey *et al.*, 1997). The reported effects are 16 to 27 g per kg of weight gain (Table 3.7), which provides a clinical guide to the expected effect of a small or large weight gain. These expectations are uncertain for

Table 3.7. *Adjusted effects of weight gain during pregnancy on birth weight per kg deviation from mean*

Author	Number	Effect	Notes
Abel *et al.* (1991)	772	+15.8 g	–
Abrams & Laros (1986)	2 948	+15.9 g	–
Abrams & Selvin (1995)	2 994	+18.7 g	–
Anderson *et al.* (1984)	760	+27.3 g	primiparae
	1 215	+21.3 g	multiparae
Kuzma & Sokol (1982)	5 093	+23.4 g	–
Marbury *et al.* (1984)	11 173	+23.5 g	–

individuals because the correlation between weight gain during pregnancy and birth weight in primiparae is only 0.3, after excluding preterm births (Kerr, 1943; Thorsdottir & Birgisdottir, 1998). An increase of only 10 g per kg has been reported for twins that may be explained by their shorter gestation periods (Magnus *et al.*, 1984). The effect of weight gain during pregnancy on birth weight appears to be independent of parity (Cogswell *et al.*, 1995). Weight gain during pregnancy is negatively related to the prevalence of LBW and SGA infants (Eskenazi *et al.*, 1995; Lang *et al.*, 1996; Rantakallio *et al.*, 1995; Stein & Susser, 1984).

Table 3.8. *Adjusted effects of weight gain during pregnancy on birth weight from group comparisons*

Author	Number	Effect	Notes
Hediger et al. (1989)[a]	1767	−187.0 g	small early gain(<4.3 kg by 24 wk GA)
		−154.0 g	small late gain (<4.0 kg from 24 wk to delivery)
		−299.0 g	small early and late gains
Horon et al. (1983)	77	−140.0 g	if ≤9.1 kg and age ≤15 yrs
	50	−92.0 g	if ≤9.1 kg and age 20–24 yrs
	36	+231.0 g	if ≥14.1 kg and age ≤15 yrs
	28	+162.0 g	if ≥14.1 kg and age 20–24 yrs
Scholl et al. (1995)	59	−159.0 g	small gain (<0.34 kg per wk from 24–36 wk)[b]
	77	−17.0 g	large gain (>0.68 kg per wk from 24–36 wk)[b]

[a] Adolescents.

[b] Compared with moderate gains.

Effects on birth weight for categories of weight gain during pregnancy are given in Table 3.8. When the maternal weight gain is ≤9.1 kg, there are decreases in birth weight of 140 and 92 g for mothers ≤15 years and 20 to 24 years, respectively, and increases of 231 and 162 g, respectively, for the same age groups when the maternal gain is ≥14.1 kg (Horon et al., 1983). The effects of weight gain during pregnancy on weight at birth are larger (19 g per kg) for moderately overweight mothers than for those of normal weight (Abrams & Laros, 1986; Cogswell et al., 1995). Weight gain during pregnancy has only small effects on length and head circumference at birth (Brown et al., 1980).

The timing of the weight gain is important. In adolescent mothers, a gain <4.3 kg during the first 6 months of gestation is associated with a deficit in birth weight of 187 g that is not diminished when later gains are compensatory. An inadequate weight gain after 6 months gestation (<6.4 kg) is associated with a deficit in birth weight of 154 g (Hediger et al., 1989). Birth weights were reduced and the length of gestation was less for infants born after pregnancies in which the mothers were exposed to the Dutch famine of 1944 to 1945 during the first and second trimesters, but not for exposure during the third trimester (Lumey, 1992). These effects may have been mediated through reduced weight gain during pregnancy.

STATURE

Maternal stature is significantly related to birth weight after 35 weeks of gestation (Butler & Alberman, 1969a; Witter & Luke, 1991). Such findings led Voigt et al. (1997) to publish reference data from German mothers and infants that relate birth weight to gestational age conditional on the weight and stature of the mother. In singletons, there is a change in birth weight of about 14 g for each cm by which maternal stature differs from the mean (Table 3.9). In addition, significant effects have been shown by the study of groups contrasted by maternal stature (Dougherty & Jones, 1982; Neggers et al., 1995). Others have found that maternal stature is not closely related to size at birth (Emanuel et al., 1992; Hoffmans et al., 1988; Langhoff-Roos et al., 1987; Roche et al., 1993). There are, however, important differences between extreme groups. Neggers et al. (1995) compared the infants of mothers with statures at the 90th percentile and the 10th percentile. At birth, the infants of the tall mothers were heavier by 163 g, longer by 0.9 cm, and their head circumferences were larger by 0.3 cm. The effects of maternal stature on birth weight are larger than those of paternal stature, but maternal and paternal statures have similar correlations with length at birth (Cawley et al., 1954; Wingerd, 1970). Mid-parent stature, which is the mean of the statures of the two parents, has a low correlation ($r = 0.2$)

Table 3.9. *Adjusted effects (change per cm) of maternal stature on size at birth*

Author	Number	Weight	Length	Head circumference
Effects per cm deviation from mean				
Abrams & Selvin (1995)	2 994	12.8 g	–	–
Peters *et al.* (1983)[a]	16 989/16 792	15.2/15.9 g	–	–
Rush & Cassano (1983)	16 688	15 g	–	–
Group comparisons				
Dougherty & Jones (1982)	191	−132 g[b]	–	–
	159	+81 g[c]	–	–
Neggers *et al.* (1995)[d]	1 205	+163 g	+0.9 cm	+0.3 cm

[a] 1958 data/1970 data.
[b] ≤155 cm vs. 156–164 cm.
[c] ≥165 cm vs. 156–164 cm.
[d] 10th vs. 90th percentile.

with length at birth (Himes *et al.*, 1981; Hoffmans *et al.*, 1988; Smith *et al.*, 1976; Tiisala & Kantero, 1971).

PRENATAL CARE

Commonly, prenatal care has been judged from the number and timing of prenatal visits (Kotelchuck, 1994; McDermott *et al.*, 1999). The prevalence of LBW infants is increased when there are either few or many prenatal visits (Harris, 1982; Kotelchuck, 1994). The positive relationship with a large number of visits may be due to their association with high-risk pregnancies. Ideally the quality of care would be considered, including the extent to which there is a focus on substance abuse, weight gain, and specific medical issues and the amount and quality of pre-natal help received from relatives and friends would be taken into account. The prevalence of LBW in-fants is halved when prenatal visits begin in the first trimester, unless the mothers had previous LBW infants, but prenatal care may not reduce the preva-lence of SGA infants (Fang *et al.*, 1999; Parker & Schoendorf, 1992; Raine *et al.*, 1994; Ventura & Martin, 1993).

The effect of adequate prenatal care on birth weight, in comparison with an intermediate level of care, is 21 to 157 g in various studies, but this effect is statistically significant (Hoff *et al.*, 1985; Scholl *et al.*, 1987; Shiono *et al.*, 1986a). The effect on birth weight for adolescent mothers is almost entirely due to an increased weight gain during pregnancy and a decreased prevalence of preterm births (Scholl *et al.*, 1987). There may be no effect for Hispanic-American mothers born outside the United States and the effects are smaller for European Americans than for African Americans (Collins & Shay, 1994; Murray & Bernfield, 1988).

FAMILY BIRTH WEIGHTS

Maternal and paternal birth weights, in combina-tion, account for only 2% to 5% of the variance in birth weight of the offspring. This significant rela-tionship is stronger for mother–daughter pairs than for other pairs (Alberman *et al.*, 1992; Coutinho *et al.*, 1997; Little, 1987). The adjusted effects are 16 to 27 g per 100 g variation in maternal birth weight and 9 to 15 g per 100 g variation in pater-nal birth weight (Coutinho *et al.*, 1997; Goldenberg *et al.*, 1992; Klebanoff *et al.*, 1998b; Ramakrishnan *et al.*, 1999). Mothers who were preterm, LBW or SGA at birth tend to have infants in the same cate-gories (Coutinho *et al.*, 1997; Klebanoff *et al.*, 1989;

Leff *et al.*, 1992). The risk of LBW is also significantly related to the birth weights of older siblings (Klebanoff & Yip, 1987; Magnus *et al.*, 1984; Wang *et al.*, 1995).

Size in infancy

SOCIOECONOMIC STATUS

A large United States study shows that SES has a significant effect on weight, length, and head circumference during infancy after adjustments are made for pre-pregnancy weight and birth weight (Garn *et al.*, 1984). Furthermore, a national United States survey shows that a poverty index derived from income, adjusted for the number of people in the household, is related to growth during infancy (Yip *et al.*, 1993). In the latter data, mean length from birth to 2 years for the lowest of five poverty index groups is significantly reduced by about 6 percentile levels and mean weight-for-length is increased about six percentile levels. Other studies in the United Kingdom and Canada do not show significant relationships of SES with weight, length, and head circumference during infancy (Elwood *et al.*, 1987; Kramer *et al.*, 1985*a*).

In the United Kingdom, SGA infants with fathers in non-manual occupations are heavier and taller and have larger head circumferences from 1 to 4 years than those with fathers in manual occupations. There are similar differences for LGA infants in head circumference, but not in weight or length. Paternal education is positively related to weight, length, and head circumference from birth to 4 years, except for weight and length in LGA infants (Ounsted *et al.*, 1982). In Dutch infants, the mean differences in weight during infancy between groups categorized by paternal occupation are small, being about 20 to 60 g for males and 50 to 160 g for females; the corresponding differences in length are near zero (Roede, 1990; Roede & van Wieringen, 1985). Herngreen *et al.* (1994) placed Dutch infants into four SES groups based on the educational attainment of the mothers. The two highest SES groups had significantly larger increments in length and significantly smaller increments in weight from birth to 2 years than the other SES groups. This difference in weight may be related to the preference of higher SES mothers for leaner infants (Kramer *et al.*, 1983). There are low significant relationships ($r = 0.1$–0.3) between SES and length in Polish infants that become more marked with age (Hauspie *et al.*, 1996). In this study, SES was categorized from the occupations and educational levels of the parents. In Finland, BMI at 1 year is larger in low SES groups than in high SES groups (Laitinen *et al.*, 2001).

The effects of SES on growth during infancy may be larger for males than females. Studies in Brussels, London, Paris, Stockholm, and Zürich show low SES, based on a combination of paternal occupation, parental educational level, source of income, and quality of residence is associated with significant deficits in weight and length for males from birth to 5 years (Graffar, 1956). The corresponding differences for females are smaller and few of them are significant. The differences at 5 years between the mean statures for the combined data from the upper two of five classes and the means for combined data from the lower two classes are 3.4 cm for males and 2.1 cm for females. Adjustments were not made for the associated differences in parental statures (3.6 cm, fathers; 4.0 cm, mothers) between these SES groups (Graffar & Corbier, 1972). In Belgian infants, the differences between high and low SES groups are significant in males for weight and length at 18 and 36 months and for head circumference at 9 and 36 months, but the corresponding differences are not significant for females (Hauspie *et al.*, 1992).

Observed associations between SES and growth could be mediated, in part, by differences in the prevalence of illnesses, but diarrhea and respiratory illnesses are not related to long-term growth rates (Hewitt *et al.*, 1955; Kahn & Harrison, 1989; Rogers, 1984; Rona, 1981) and a health index (0 = healthy, 1 = minor ailments, 2 = major illnesses) is not related to growth patterns in Swedish infants (Karlberg, 1987).

MATERNAL WEIGHT

Maternal weight may influence weights during infancy through similarity of eating styles including overeating and eating in response to emotional and environmental influences (Johnson & Birch, 1994; Ruther & Richman, 1993). Maternal pre-pregnancy weight is not related to infant weight after 3 months in some studies (Eliot & Deniel, 1977; Stunkard *et al.*, 1999), but others have found low to moderate correlations ($r = 0.2$–0.4) with the weights and lengths of infants from birth to 2 years (Edwards *et al.*, 1978; Smith *et al.*, 1976; Wingerd, 1970). Maternal relative weight is significantly correlated with infant weight at 1 and 2 years, although it explains only 14% of the variance even when combined with the birth weight of the infant and the duration of breast-feeding (Garn & Pesick, 1982; Kramer *et al.*, 1985*a*, *b*; Rummler & Woit, 1990).

PARENTAL STATURE

Maternal stature has small but significant effects on weight, length, and head circumference at 12 months (Dewey *et al.*, 1990*a*). Maternal and parental statures are correlated about equally with weight and length at 6 months ($r = 0.2$–0.3) in the data of Wingerd (1970), but others have reported the effect on infant weight is larger for maternal stature than for paternal stature (Cawley *et al.*, 1954). Maternal stature is correlated with the increments in weight from birth to 2 years for females ($r = 0.3$), but not for males, and with increments in length ($r = 0.5$) in each sex (Chrzastek-Spruch & Wolánski, 1969; Drillien, 1958). Paternal stature is also correlated ($r = 0.3$) with the increment in weight from birth to 2 years (Chrzastek-Spruch & Wolánski, 1969). The correlations between mid-parent stature and length during infancy increase from about $r = 0.3$ at 3 months to $r = 0.5$ at 1 to 3 years (Bouchalová *et al.*, 1978; Harvey *et al.*, 1979; Himes *et al.*, 1981; Salmenperä *et al.*, 1985). A 10 cm increase in mid-parent stature is associated with an estimated increase of 0.6 kg in weight, 1.5 cm in length, and 0.5 cm in head circumference at 12 months (Himes *et al.*, 1981; Roche *et al.*, 1993).

Table 3.10. *Adjusted effects of maternal age on stature in childhood*

Author	Number	Age (years)	Effect
Goldstein (1971)	5538	7	−0.4 cm for <25 yrs +0.2 cm for ≥25 yrs
Wingerd & Schoen (1974)	3704	5	+0.3 cm for 25–34 years vs. <25 yrs +0.9 cm for >35 yrs vs. <25 yrs

Size in childhood and adolescence

MATERNAL AGE

Maternal age has significant correlations with childhood statures after adjusting for confounding variables (Fox *et al.*, 1981; Rona *et al.*, 1978; Rona & Chinn, 1995; Wingerd & Schoen, 1974). Children born to mothers younger than 25 years are shorter at 5 to 7 years than those born to older mothers (Davie, 1972; Goldstein, 1971; Wingerd & Schoen, 1974), but these effects are small being about 0.5 cm per decade of maternal age (Table 3.10). Maternal age at the birth of an infant is not significantly related to stature at 16 years (Fogelman, 1980).

PARITY

First-born children are taller than the children of multiparae, but estimates of the size of these differences vary among studies. The effects of parity on stature during childhood and adolescence reported by Fogelman (1980) and by Goldstein (1971) are in fair agreement, but those reported by Wingerd & Schoen (1974) are smaller (Table 3.11). There is a decrease in adult stature of 0.8 cm for each unit of parity (Kuh & Wadsworth, 1989).

FAMILY SIZE

Only children tend to be overweight in most studies (Patterson *et al.*, 1997; Richter, 1980; Rona & Chinn, 1982; Stettler *et al.*, 2000). These variations may reflect cultural and SES differences, but do not seem to be explained by the availability of food (Cook

Table 3.11. *Adjusted effects of parity on stature in childhood and adolescence*

Author	Number	Age (years)	Parity	Effect
Fogelman (1980)	3023	16	>3[a]	−1.2 cm (males)
	2923	16	>3[a]	−1.0 cm (females)
Goldstein (1971)	5538 (total)	7	2 or 3[b]	−1.24 cm
			≥4[b]	−2.84 cm
Wingerd & Schoen (1974)	3707 (total)	5	2 or 3[b]	−0.51cm
			≥4[b]	−0.69 cm

[a] Vs. parity 1 and 2 combined.
[b] Vs. parity 1.

et al., 1973; Rona & Chinn, 1982; Whitelaw, 1971). The number of siblings is associated with reductions in stature of about 0.3 cm for one sibling, 0.7 cm for two siblings, and 1.1 cm for three or more siblings in comparison with those who do not have siblings (Douglas & Simpson, 1964). The deficits associated with multiple siblings change little after 6 years and are present at 19 years (Kolodziej et al., 2001; Miller et al., 1974; Prokopec, 1984; Rona & Chinn, 1995).

SOCIOECONOMIC STATUS

There are few reported estimates of the adjusted effects of SES on growth after infancy. Dietary intakes do not vary with the educational level or occupation of the father, but the educational level of the mother has significant positive relationships with the intakes of calcium, thiamin, and fiber by the child (Magarey & Boulton, 1997). Data from national United States surveys include poverty index values based on family income adjusted for household size. Growth deficits, associated with low poverty index values and low associated educational levels of the parents, were present in children included in national United States surveys conducted from 1971 to 1980 (Hamill et al., 1972; Jones et al., 1985). The mean stature deficits, which may be more marked in leg length than in sitting height (Frisancho et al., 2001), were 1.0 cm at 1 to 5 years, 2 cm at 6 to 11 years, and 2.5 cm at 12 to 17 years except for African-American females who had a deficit of 1.0 cm. The deficits in weight were 0.4 kg at 1

to 5 years and 1.0 kg at 6 to 11 years. At 12 to 17 years, there was no deficit in weight for European-American females but a deficit of 2.5 kg for males and in African Americans there were deficits of 1.5 kg in each sex. The differences in stature between poverty index groups in the survey made in 1976–80 tend to be smaller than those in the 1971–4 survey. The prevalence of large weights for those below the poverty level increased significantly between these surveys for European-American females aged 6 and 7 years and African-American females aged 12 to 17 years (Jones et al., 1985). Yip et al. (1993), analyzing data from the survey made in 1976–80, found the lowest of five poverty index groups, in comparison with the remaining groups, had a significant deficit in stature of about $0.2\ Z$ and an increased prevalence of shortness (stature $< -2.0\ Z$).

The adjusted effects of paternal education and occupation on stature at 5 years are not significant (Wingerd & Schoen, 1974) and there are only small adjusted differences in stature between social classes at 7 and 16 years (Table 3.12). Weights may tend to be small in low SES groups before pubescence and large in females during pubescence, but the evidence is inconclusive (Garn et al., 1984; Kimm et al., 1996; Patterson et al., 1997; Stettler et al., 2000). Combined data from the United States national survey made in 1971–4 and that made in 1976–80 show significant positive associations of the poverty index with stature and BMI in African Americans, European Americans, and Mexican Americans at

Table 3.12. *Adjusted effects of socioeconomic status on stature in childhood*

Author	Number	Age (years)	Effect
Fogelman (1980)	5946	16	class III:[a] −0.3 cm[b]
			classes IV+V: +0.8 cm[b]
			class III: +1.02 cm[c]
			classes IV+V: −0.55 cm[c]
Goldstein (1971)	5538	7	class III: −0.6 cm
			class IV: −0.8 cm
			class V: −1.3 cm
Wingerd & Schoen (1974)	3707	5	high school: −0.3 cm
			maternal college education:[d] −0.5 cm
			family income $6000–10 000:[e] +0.3 cm
			family income >$10 000:[c,e] +0.4 cm

[a] Five classes based on paternal occupation (V is lowest); comparisons with classes I and II combined.
[b] SES at birth.
[c] SES at 16 years.
[d] Vs. high school.
[e] Vs. <$6000 per yr.

1 to 11 years, except for Mexican Americans for whom this relationship was not significant from 6 to 11 years. At 12 to 17 years, poverty index values were significantly correlated with stature in European Americans, but not in the other two ethnic groups (Martorell *et al.*, 1987). In data from the Hispanic Health and Nutrition Examination Survey that was made in the United States in 1982–4, there are small deficits in stature, but slightly larger values for weight and BMI in Mexican-American children compared with European-American and African-American children when poverty index values are held constant (Ryan *et al.*, 1990). Low poverty index groups of Mexican-American children have a high prevalence of short statures (Mendoza & Castillo, 1986).

The separate influences of family income and maternal educational level are about equal for European-American children, but family income has a larger effect than maternal educational level on weight and stature, but not head circumference, in African-American children older than 3 years (Garn & Clark, 1975; Garn *et al.*, 1978, 1984). Kimm *et al.*

(1996) found low SES, based on parental education, household income, and the number of parents living at home, was associated with an increased prevalence of BMI values >85th percentile at 9 to 10 years for European Americans, but not African Americans.

In low SES Canadian children aged 4 to 5 years, unadjusted data show stature is normal, but weight-for-stature and BMI tend to be large and there are deficits in muscle mass (Evers & Hooper, 1995, 1996). The reported nutrient intakes of these children were low, and the actual intakes may have been about 20% smaller than those reported (Bandini *et al.*, 1990; Mertz *et al.*, 1991). A national Canadian survey shows the distributions of BMI values do not vary systematically with family income or occupation except for a negative relationship in European Canadians aged 12 to 17 years (Brunner *et al.*, 1999).

The many studies from the United Kingdom are considered separately from those for the remainder of Europe because, in almost all these studies, SES is based on the occupation of the head of the household using five classes (I–V) with social class

I being the professional group and social class V being unskilled manual workers. These studies have shown only a small part of the variation in stature is explained by paternal occupation or by a combination of crowding, family size, maternal and paternal education, receipt of free school meals, and maternal employment and that SES effects are smaller than those of parental stature, ethnicity, and birth weight (Goldstein, 1971; Rona et al., 1978; Rona & Chinn, 1995). Weight and weight-for-stature values tend to be large in some low social class groups (Fox et al., 1981; Rona, 1981), but others did not find this tendency (Duran-Tauleria et al., 1995; Rona & Chinn, 1987). In English children, a weight-stature index is not significantly related to maternal employment or the number of parents at home (Duran-Tauleria et al., 1995). After adjusting for confounding variables, stature at 7 years is 1.3 cm smaller for social classes I and II combined than for social class V (Goldstein, 1971). There are similar effects on adult stature (Kuh & Wadsworth, 1989).

Important information has come from national surveys in The Netherlands. Roede (1990) compared children in the lower and upper of three groups based on paternal occupation. The deficits in the lower group for weights of males were 50 g during infancy, 100 to 300 g from 3 to 10 years, and 0.5 to 2.0 kg at older ages. The corresponding deficits for females were about 100 g during infancy, 0.5 to 1.5 kg from 3 to 10 years, and 1.0 to 3.0 kg at older ages. The deficits in stature between lower and upper SES groups were small during infancy, but increased to 0.9 cm at 3 to 10 years and 1.3 cm at older ages for males and were 1.3 cm at ages older than 6 years for females. In these Dutch children, weight-for-stature was larger in low SES groups.

Deficits in size during childhood and adolescence, in association with low SES, have been reported from Belgium, Denmark, and France, but the prevalence of large BMI values may be increased in low SES females after 12 years (de Spiegelaere et al., 1998; Lissau-Lund-Sørensen & Sørensen, 1992; Olivier & Devigne, 1983; Susanne, 1980). In Swedish preschool children, the differences in

stature and weight between the highest and lowest SES groups are not significant, but the lowest SES group has a deficit in head circumference (Lindgren, 1994). This is surprising since most influences on growth affect weight more than they affect head circumference. Male Swedish school children with more than one sibling and mothers who are less educated are 1.1 cm shorter than a more privileged group, but stature does not differ between corresponding groups of females (Cernerud, 1994). In Swedish school children born in 1967, females in the lowest of three groups based on paternal occupation and family income were shorter by 1.0 cm and had larger weights-for-stature than other SES groups, but growth did not differ between corresponding groups of males (Lindgren, 1998).

The statures and weights of German children are not related to parental occupation (Kromeyer et al., 1997; Walter, 1977; Walter et al., 1975), but in Hungarian children levels of parental education are positively correlated with the statures of children (Eiben, 1989; Eiben & Pantó, 1988). In Polish children, the deficits in weight and stature for low SES groups, categorized by parental education and occupation and the number of persons per room, compared with high SES groups, decrease with age from about 0.5 Z score at 6.5 years to 0.2 Z score at 18.5 years with larger deficits for males than females (Golab, 1992).

PARENTAL STATURE, WEIGHT, AND BMI
The effects of maternal stature on childhood statures are about 0.4 cm per cm deviation from the population mean. The estimate for paternal stature is slightly smaller (Table 3.13). The correlations between mid-parent stature and the statures of offspring during childhood and adolescence are about 0.4 to 0.5 (Himes et al., 1981; Lasker & Mascie-Taylor, 1996; Prokopec, 1984; Sorva et al., 1989).

Parent–child correlations for stature differ little with social class for European-English groups (Lasker & Mascie-Taylor, 1996; Rona et al., 1978) which implies that similar intergenerational changes have occurred in all social classes. Mother–son, mother–daughter, father–son, and

Table 3.13. *Adjusted effects of parental stature on stature in childhood and adolescence*

Author	Number	Age (years)	Effect per cm deviation from the mean	Notes
Maternal stature				
Alberman *et al.* (1991)	7 710	≥18	0.3 cm	sexes combined
Elwood *et al.* (1987)	513	5	0.4 cm	males
	438	5	0.5 cm	females
Goldstein (1971)	5538	7	0.3 cm	sexes combined
Paternal stature				
Alberman *et al.* (1991)	7 710	18	0.3 cm	sexes combined
Mid-parent stature				
Kuh & Wadsworth (1989)	1667	36	0.5 cm	sexes combined

father–daughter correlations for stature, which reflect genetic and environmental influences, do not differ systematically in most studies (Lasker & Mascie-Taylor, 1996; Rona *et al.*, 1978; Tambs *et al.*, 1992; Wich, 1983), but some have reported that mother–daughter correlations are the highest (Chrzastek-Spruch, 1977; Tanner & Israelsohn, 1963; Welon & Bielicki, 1971). The data of Welon & Bielicki (1971) are for the same individuals from 8 to 18 years. They show increasing differences in the correlations between mother–daughter and other pairings after 11 years.

Infants of low weight mothers, matched for weight at birth with infants of heavier mothers, are about 3 kg lighter at 7 years (Garn & LaVelle, 1984). Significant effects of maternal and paternal weight–stature indices on the weight–stature indices of their offspring have been reported, but these studies adjusted for few confounding variables (Duran-Tauleria *et al.*, 1995; Hashimoto *et al.*, 1995; Strauss & Knight, 1999; Zive *et al.*, 1992).

In overview, the effects of family influences on growth are small except for the effects of maternal ages older than 30 years on birth weight (Table 3.14). Nevertheless, these influences can be important when they act jointly and the size of the child would have been near the margin of the normal range in the absence of these influences.

Table 3.14. *Approximate adjusted central tendencies for independent effects of family influences on growth*[a]

Effects on birth weight
Maternal age: 2 g per year except
  <20 years = −25 g per year
  >30 years = +15 g per year
Parity: +40 g per parity order except adolescents:
  −76 g per parity order
Employment: small effects except in one study
Pre-pregnancy weight: +10 g per kg from mean
Weight gain during pregnancy: +20 g per kg from mean
Maternal stature: 15 g per cm from mean

Effects on childhood stature
Maternal age: small effects
Parity: −0.5 cm per parity order
Maternal stature: 0.4 cm per cm from mean
Paternal stature: 0.3 cm per cm from mean

[a] The estimated effects on growth during and after infancy are few and uncertain.

Maturation

There are small effects of SES on skeletal maturation rates and the timing of PHV (Deschamps & Benchemsi, 1974; Eiben, 1989; Farkas, 1980; Roche *et al.*, 1975*b*, 1978). In the United States, parental education and income adjusted for family size are significantly related to age at menarche, but not to the timing of stages of sexual maturation (Harlan *et al.*, 1980). In Turkey, however, secondary sexual maturation tends to be slower in low SES groups categorized by income, family size, paternal occupation, and educational level (Onat, 1997).

Older studies in The Netherlands, the United Kingdom, and the United States found associations between low SES and delayed age at menarche (de Wijn, 1966; Douglas & Simpson, 1964; MacMahon, 1973), but SES effects on age at menarche are now absent or small in some countries including Finland, Germany, Sweden, and the United Kingdom (Billewicz *et al.*, 1981*b*; Georgiadis *et al.*, 1997; Laitinen *et al.*, 2001; Lindgren, 1976; Richter, 1989). In other European countries, menarche occurs earlier when the fathers have non-manual occupations (Bielicki *et al.*, 1986; Eiben, 1994; Martuzzi Veronesi & Gueresi, 1994; Pasquet & Ducros, 1980). There may be an opposite relationship for French-Canadians and Norwegians in whom menarche occurs earlier in lower SES groups in association with larger weights (Brundtland *et al.*, 1980; Brundtland & Walløe, 1973; Jeniček & Demirjian, 1974). As the number of siblings increases, menarche tends to occur later, but this association may be present only for low SES groups (Billewicz *et al.*, 1981*b*; Bodzsár & Pápai, 1994; Richter, 1980; Roberts *et al.*, 1971).

URBAN–RURAL DIFFERENCES

In some countries, there are differences in growth and maturation between urban and rural groups; these differences are usually attributed to variations in housing, sanitation, family size, and access to healthcare (Wnuk-Lipinski, 1990). Ethnicity may also be involved. Urban children tend to be larger, but the differences vary markedly among countries (Booth *et al.*, 1999; Hamill *et al.*, 1972; Reading *et al.*, 1993; Shephard *et al.*, 1984). Urban–rural differences are absent in The Netherlands (van Wieringen, 1972) and they have decreased in Italy and Poland (Bielicki & Hulanicka, 1998; Bielicki *et al.*, 2000; Cresta *et al.*, 1982; Ferro-Luzzi & Sofia, 1967; Sforza *et al.*, 1999). In Poland, the statures of conscripts born and raised in urban areas are larger when both parents migrated to the urban environment from a rural area than when both parents were born and raised in urban environments (Kolodziej *et al.*, 2001). Urban children reach menarche earlier than rural children in France, Italy, and Poland (Bielicki & Welon, 1982; Boetsch & Bley, 1980; Ducros & Pasquet, 1978; Martuzzi Veronesi & Gueresi, 1994).

SUBSTANCE ABUSE DURING PREGNANCY

Analyses of the effects of maternal use of tobacco, alcohol, caffeine, marijuana, cocaine, heroin, and methadone during pregnancy are based on self-reports, but data for near-current use can be obtained from chemical testing. Almost all the studies considered here compared mother–infant pairs in which the mothers used particular substances during pregnancy and pairs in which the mothers did not use these substances. In the absence of random assignment to experimental and control groups, adjustments have been made for other maternal characteristics and demographic variables that may influence length of gestation, size at birth, and postnatal growth. These maternal and demographic influences, including substance abuse, jointly explain only about 8% of the variance in gestational age at birth and 40% of the variance in size at birth (Hingson *et al.*, 1982). Ideally, estimates of the effects of a chemical substance would take into account the possible maternal use of other chemicals, which may have additive or interactive effects, but accurate data are rarely available. The estimates in the following tables that relate to size at birth are from analyses made after the exclusion of twins and infants with congenital abnormalities.

When the dependent variable is continuous, such as birth weight, estimates of the effects of specific influences are typically made from multiple regressions in which the independent variables are separated into categories. For example, maternal ages may be grouped to <25 years, 25 to 35 years, and

>35 years. Large samples are required to remove the effects of confounding variables including maternal age, ethnicity, parity, pre-pregnancy weight, weight gain in pregnancy, and educational level. There are also confounding family variables including occupations, income, housing, and ethnicity. Logistic regressions are used when the dependent variable is dichotomous, for example, the presence or absence of LBW. These regressions provide odds ratios for the presence of the selected condition, in association with substance abuse. These ratios have been interpreted as significant when the 95% confidence limits do not include 1.0, although non-significant findings may indicate real trends.

In theory, observed sizes at birth or during infancy could be adjusted for substance abuse during pregnancy, maternal characteristics, and other influences. Each of these adjustments has an associated error and such adjustments can be recommended only when there are large consistent adjusted effects among studies. It is more appropriate to use the estimated effects when interpreting the observed data. Thus infants whose weights at birth are at the 10th percentile would be regarded as less likely to have pathological conditions and have less need for laboratory investigations when influences are operating that have significant associations with deficits in birth weight. These influences may include substance abuse during pregnancy.

Tobacco

It is useful to analyze the effects of maternal tobacco use in terms of number of cigarettes per day because these data can be applied more easily in clinical and public health settings than circulating levels of cotinine or thiocyanate (Prue et al., 1980; Windsor et al., 1989). The results from these approaches are not in full agreement. The results of cotidine and thiocyanate assays explain no more than 50% of the variance in self-reports of the number of cigarettes smoked per day (Kandel & Udry, 1999; Klebanoff et al., 1998a; Pierce et al., 1987; Wagenknecht et al., 1990). The assays reflect recent smoking and there are variations among types of cigarettes and in-

dividuals in the formation and excretion of cotidine and thiocyanate and the reported number of cigarettes smoked may be inaccurate. Cotinine and thiocyanate measures help to validate reports of current tobacco use (Fortmann et al., 1984; Windsor et al., 1989).

With maternal tobacco use, the increases in carboxyhemoglobin in fetal cord blood are larger than those in maternal blood (Cole et al., 1972). The increased carboxyhemoglobin may impair oxygen transport and lead to fetal tissue hypoxia (Longo, 1977; Permutt & Farhi, 1969; Sagone et al., 1973; Younoszai et al., 1968). Furthermore, maternal tobacco use increases fetal carbon monoxide, which reduces the delivery of oxygen to tissues. These changes are important because the fetus cannot adapt by an increased hematocrit or a decrease in the oxygen tension required to saturate hemoglobin (Bureau et al., 1982, 1983). The collagen content of placental villi is increased and the number of fetal capillaries is decreased in women who smoke >10 cigarettes per day during pregnancy (Asmussen, 1980; Mochizuki et al., 1984). These changes, which suggest a reduction in placental blood flow, may be direct effects of nicotine and may reduce fetal growth (Philipp et al., 1984; Socol et al., 1982). Maternal tobacco use may retard fetal growth due to increases in the circulating levels of cyanide, cadmium, and lead (Kuhnert et al., 1987a, b; Siegers et al., 1983) and it may interfere with the transport of amino acids directly or by vasoconstriction (Pastrakuljic et al., 1999).

The effects of maternal tobacco use during pregnancy on the fetus may be mediated through the nutritional status of the mother (Rush, 1974). Mothers who use tobacco during pregnancy have lower circulating levels of amino acids, carotene, and ascorbic acid and increased needs for folic acid, zinc, and vitamins B_{12} and C (Crosby et al., 1977; Institute of Medicine, 1990; Schorah et al., 1978). In most studies, weight gain during pregnancy does not differ with maternal tobacco use and the associations with size at birth remain after adjusting for maternal skinfold thickness (Bosley et al., 1981; d'Souza et al., 1981; Harrison et al., 1983; Keppel & Taffel, 1987).

The intakes of energy are not reduced in mothers who use tobacco during pregnancy, but the intakes of some minerals and vitamins are lower, particularly late in pregnancy (Ellard *et al.*, 1996; Haste *et al.*, 1990; Haworth *et al.*, 1980*a*, *b*).

The apparent effects of maternal tobacco use during pregnancy are not likely to be due to unrecognized inherent differences between mothers who use tobacco and those who do not. Wainwright (1983) analyzed data from successive pregnancies categorized as non-smoking (N; <1 cigarette per day) or smoking (S; >5 cigarettes per day). The changes in birth weight from one pregnancy to the next, compared with N–N sequences, were −67 g for N–S, +75 g for S–N and −34 for S–S sequences. Nordström & Cnattingius (1994) have reported similar data. Furthermore, as discussed later, mothers who stop tobacco use before or early in pregnancy have smaller deficits in the mean birth weight of their infants.

Summarizing statements that the median deficits at birth due to maternal tobacco use during pregnancy are 200 g for weight, 0.6 cm for length, and 0.4 cm for head circumference are approximately accurate, but they do not take into account variations in the number of cigarettes smoked per day. Comparisons among studies are difficult because of variations in the ways in which subjects were grouped, e.g., 1–4 cigarettes per day or 1–9 cigarettes per day. The reported estimated effects vary for the same level of tobacco use (Table 3.15). Thus, in two studies, the infants of mothers who smoked >15 cigarettes per day had estimated deficits in adjusted birth weight of 160 g and 246 g. Perhaps these studies differed in the distributions of the number of cigarettes smoked.

Maternal tobacco use may be associated with deficits in birth weight only at gestational ages older than 34 weeks (Amini *et al.*, 1994; Butler & Alberman, 1969*b*; Lieberman *et al.*, 1994) although reductions in weight and biparietal diameter before 30 weeks gestation have been reported (Chamberlain, 1975; Persson *et al.*, 1978). Maternal tobacco use increases the prevalence of LBW without altering the mean length of gestation (Abel *et al.*,

1991; Day *et al.*, 1992; Fox *et al.*, 1994; McDonald *et al.*, 1992), except in older mothers for whom the length of gestation may be slightly increased (Ahluwalia *et al.*, 1997; Kistin *et al.*, 1996).

The deficit in birth weight due to tobacco use during pregnancy is significant for the infants of normal weight mothers, but not for the infants of overweight mothers (Mitchell & Lerner, 1989). These deficits may increase with the age of the mother from 117 g for those younger than 16 years to 376 g for those 40 years or older (Fox *et al.*, 1994), but Hediger *et al.* (1989) found larger deficits in birth weight with tobacco use during pregnancy in the infants of adolescent mothers than in the infants of older mothers. With tobacco use during pregnancy, skinfold thicknesses at birth are not reduced, but there is a reduction in FFM (Spady *et al.*, 1986).

Some have used multiple linear regressions to calculate the adjusted effects of the number of cigarettes smoked during pregnancy on size at birth (Table 3.16). There is considerable variation among these reports, which may be spurious because the relationships between the number of cigarettes smoked and birth weight deficits appear to be non-linear with smaller additional deficits per cigarette after the use exceeds 20 cigarettes per day (Butler *et al.*, 1972; Cliver *et al.*, 1995; Fried & O'Connell, 1987).

McDonald *et al.* (1992) found significantly increased odds ratios for LBW with tobacco use during pregnancy of 1.6 for <10 cigarettes per day, 2.4 for 10–19 cigarettes per day, and 2.8 for ≥20 cigarettes per day. In mothers who smoked ≥20 cigarettes per day, but stopped tobacco use during the first trimester, the odds ratio for LBW was not significantly increased. These findings are in agreement with those from other studies (Alameda County Low Birth Weight Study Group, 1990; Lieberman *et al.*, 1994; Miller & Jekel, 1987*a*; Seidman *et al.*, 1988). The risk of SGA births is also significantly increased by maternal tobacco use during pregnancy (Heinonen *et al.*, 1999; Hoff *et al.*, 1985; Scott *et al.*, 1981; Sprauve *et al.*, 1999).

The effect of maternal tobacco use during pregnancy on size at birth and the prevalences of LBW

Table 3.15. *Adjusted deficits in size at birth associated with tobacco use during pregnancy*

Author	Number of cigarettes per day	Number of smokers	Size deficit at birth			Trimesters when tobacco used
			Weight (g)	Length (cm)	Head circumference (cm)	
Butler *et al.* (1972)	1–4		94	–	–	2, 3
	5–9		180	–	–	2, 3
	10–19	4660[a]	176	–	–	2, 3
	20–30		208	–	–	2, 3
Cliver *et al.* (1995)	1–19	384	161	0.9	0.3	1, 2, 3
	≥20	183	221	0.9	0.2	1, 2, 3
Cornelius *et al.* (1995)	20	150	202	–	+0.6	1
	20	180	–	0.7	–	2
Cortinovis *et al.* (1994)	1–9	1204/1072[b]	107/88[b]	–	–	1, 2, 3
	≥10	228/245[b]	147/168[b]	–	–	1, 2, 3
Day *et al.* (1992)	≤20	141	204	0.1	0.4	3 or earlier
Dougherty & Jones (1982)	1–15	162	119	–	–	1, 2, 3
	>15	63	160	–	–	1, 2, 3
Ellard *et al.* (1996)	1–12	533	131	–	–	1, 2, 3
	>12	405	101	–	–	1, 2, 3
Fried & O'Connell (1987)	20	667	134	–	–	1
	20		181	–	–	1, 2, 3
Haste *et al.* (1991)	<15	210	–	0.4	0.2	1, 2, 3
	≥15	133	–	0.8	0.1	1, 2, 3
Hebel *et al.* (1988)	1–9	306	100	–	–	1, 2, 3
	10–19	218	40	–	–	1, 2, 3
	20–29	48	99	–	–	1, 2, 3
	≥30	23	29	–	–	1, 2, 3
Hoff *et al.* (1986)	<10	324	80	0.3	–	1 or later
	≥10	144	88	0.5	–	1 or later
Larroque *et al.* (1993)	1–14	133	136	–	–	1 or later
	≥15	55	246	–	–	1 or later
Martin & Bracken (1986)	≥1	1206	137	–	–	1 or later
Zarén *et al.* (1996)	<9	200	147	0.6	0.1	1, 2, 3
	≥10	326	212	0.8	0.3	1, 2, 3

[a] Total sample (smokers + non-smokers).

[b] Males/females.

Table 3.16. *Deficits in weight at birth due to maternal tobacco use during pregnancy estimated from multiple linear regressions*

Author	Number of smokers	Weight deficit	Trimesters when tobacco used
Anderson *et al.* (1984)	671	12.5 g per cig per day	1, 2, 3
Garn & Rosenberg (1986)	11 456[a]	5 g per cig per day European-American males	1 or later
	10 727[a]	6 g per cig per day European-American females	1 or later
	23 339[a]	2 g per cig per day African-American males and females	1 or later
Marbury *et al.* (1984)	1 596	8.8 g per cig per day	3 or earlier

[a] Total sample (smokers + non-smokers).

Table 3.17. *Adjusted deficits in birth weight in relation to the timing of the cessation of maternal tobacco use during pregnancy*

Author	Timing within pregnancy	Number	Mean deficit vs. non-smokers (g)	Mean increase vs. persistent smokers (g)
Butler *et al.* (1972)	4 mo	305	32	208
Cliver *et al.* (1995)	before 30 wk	143	55	134
MacArthur & Knox (1988)	before 6 wk	85	+12	217
	6–16 wk	119	+8	213
	after 16 wk	56	75	120
Rantakallio (1978)	before 6 mo	471	95	–
Rush & Cassano (1983)	1 mo	117	+98	253
	2–6 mo	527	+43	198
	7–8 mo	149	+36	191
	9 mo	50	90	65
Schell *et al.* (1987)	first trimester	4803	–	270

and SGA infants are reduced or absent for mothers who cease smoking during pregnancy (Andrews & McGarry, 1972; Borlee *et al.*, 1978; Lieberman *et al.*, 1994). The benefits of cessation may exceed those noted in Table 3.17 because self-reports of cessation are commonly inaccurate (Kozlowski *et al.*, 1980). In one study, a reduction from mean levels of 12.8 cigarettes per day to 6.4 cigarettes per day during pregnancy was associated with an unadjusted increase of 92 g in weight, 0.6 cm in length, and 0.8 cm in head circumference at birth without a change in the length of gestation (Sexton & Hebel, 1984). Similar data were reported by Ershoff *et al.* (1990). The earlier in pregnancy tobacco use ceases, the larger the reduction in the deficit (MacArthur & Knox, 1988; Rush & Cassano, 1983). Maternal tobacco use during pregnancy is also associated with dose-dependent deficits at birth in adjusted length and head circumference, and there may be a deficit in the ponderal index, but these are smaller than the

Table 3.18. *Adjusted deficits in postnatal growth with maternal tobacco use during pregnancy*

Author	Number of cigarettes per day	Number of smokers	Age (yr)	Deficit in postnatal growth Weight (g)	Length/ stature (cm)	Head circumference (cm)
Butler & Goldstein	1–9	1729	11.0	–	1.0	–
(1973)	≥10	1316	11.0	–	1.5	–
Conter *et al.* (1995)	≥10	407	0.25	97/55	–	–
		407	0.6	17/55	–	–
Day *et al.* (1992)	>20	126	0.7	188	1.0	1.1
Dunn *et al.* (1976)	<10 to >20	120	0.25	260	0.1	–
		120	0.5	140	0.7	–
		120	1.0	140[a]	0.8	–
		120	1.5	360	0.8	–
		120	2.5	450	0.9	–
		120	4.0	260	2.3	–
		120	6.5	1280	2.0	–
Fogelman (1980)	1–10	3023[b]	16.0	–	0.6	–
	>10		16.0	–	0.9	–
Fogelman & Manor	1–9	703/756	23.0	–	0.4/0.0[c]	–
(1988)	10–19	539/506	23.0	–	0.3/0.8[c]	–
	≥20	75/79	23.0	–	0.8/1.6[c]	–
Goldstein (1971)	1–10	5538	7.0	–	0.1	–
	>10		7.0	–	0.3	–
Hardy & Mellits	≥10	88	1.0	250	1.1	0.4
(1972)		88	4.0	190	0.7	0.4
		88	7.0	350	0.0	0.2
Naeye (1981)	1–19	140	7.0	1400	0.7	0.2
Rantakallio (1983)	<10	1474	14.0	–	0.2	–
	≥10	289	14.0	–	0.3	–
Wingerd & Schoen	1–14	562	5.0	–	0.4	–
(1974)	≥15	741	5.0	–	0.9	–

[a] Larger for smokers than non-smokers.
[b] Total sample (smokers + non-smokers).
[c] Males/females.

deficits in weight (Fried *et al.*, 1999; Lindley *et al.*, 2000; Lindsay *et al.*, 1997). When mothers continue tobacco use throughout pregnancy, the deficits at birth may be larger for European Americans than for African Americans (Cliver *et al.*, 1995; Garn & Rosenberg, 1986; Li *et al.*, 1993).

The infants of mothers who used tobacco during pregnancy may have two patterns of growth. The common pattern is a persistence of growth deficits in those who are short and light at birth (Miller *et al.*, 1976; Nilsen *et al.*, 1984). The other pattern is a gradual reduction in the deficits for those born with normal lengths, but low weights. Reported deficits in postnatal growth with maternal tobacco use during pregnancy are summarized in Table 3.18. There may be complete catch-up after infancy. The adjusted differences in weight or stature at three years between the offspring of mothers who ceased tobacco use before 30 weeks of gestation and those of mothers who continued to use tobacco throughout pregnancy are not significant (Fox *et al.*, 1990) and the adjusted deficits in adult stature associated with maternal tobacco use that was continued after the fourth month of pregnancy are not significant (Alberman *et al.*, 1991). In low SES groups, Dunn *et al.* (1976), however, found deficits in weight and stature at 6.5 years for the children of mothers who used tobacco during pregnancy, in comparison with the children of non-smoking mothers. In a similar study Fogelman (1980) found significant deficits in stature at 16 years in the children of mothers who used tobacco during pregnancy.

The levels of maternal urinary cotinine and fetal serum thiocyanate and cotinine are significantly increased for mothers who do not use tobacco, but are exposed to tobacco smoke in their homes. The levels are, however, considerably lower than those for women who use tobacco (Kandel & Udry, 1999; Martinez *et al.*, 1994; Matsukura *et al.*, 1984; Wald *et al.*, 1984). Therefore, one would expect exposure to passive tobacco smoke during pregnancy to have less effect on birth weight than maternal tobacco use during pregnancy (Jensen & Foss, 1981; Pettigrew *et al.*, 1977). In agreement with this expectation, some have reported that passive tobacco smoke is associated with very small adjusted deficits in birth weight that tend to be larger for mothers who use tobacco during pregnancy (Brooke *et al.*, 1989; Eskenazi *et al.*, 1995; Magnus *et al.*, 1984; Martin & Bracken, 1986). The effect of passive tobacco smoke on adjusted size at birth is just significant for weight, but not for length or head circumference (Borlee *et al.*, 1978; MacMahon *et al.*, 1965). There may be a dose-response relationship. These findings may be influenced by demographic variations. Women who do not use tobacco, but whose spouses smoke tobacco, tend to be less educated and are more likely to drink alcohol than those with spouses who do not smoke tobacco (Matanoski *et al.*, 1995). The adjusted deficits in birth weight due to passive paternal

smoking for the offspring of mothers who did not use tobacco during pregnancy have been estimated as 29, 82, and 88 g when the fathers smoked 1 to 10, 11 to 30, or >30 cigarettes per day, respectively (Martinez *et al.*, 1994).

In overview, these reports suggest that exposure to passive tobacco smoke during pregnancy has only a small effect on birth weight. Rubin *et al.* (1986) reported a significant adjusted effect on birth weight when the mothers were exposed to passive tobacco smoke of 6.1 g per cigarette per day; most of the estimates for maternal tobacco use during pregnancy indicate smaller deficits (Table 3.15). Rubin's study has been criticized by Trichopoulos (1986) who considered the adjustment for social class was inadequate and suggested that tobacco use during pregnancy may have been under-reported.

Passive tobacco smoke during pregnancy is not significantly associated with adjusted size at 1 and 5 years and there is a trivial (4 mm) deficit in adjusted stature at 15 to 18 years (Chinn & Rona, 1991; Fried *et al.*, 1999; Rona *et al.*, 1985). Exposure to passive tobacco smoke by the offspring due to post-natal tobacco use by the mother is associated with small deficits in stature at 6 to 11 years (0.4 cm for 1 to 9 cigarettes per day; 0.6 cm for >10 cigarettes per day); the adjusted effects of paternal tobacco use are even smaller (Berkey *et al.*, 1984). The adjusted deficits in postnatal stature from 5 to 11 years are not significantly associated with the number of people in the home who use tobacco (Chinn & Rona, 1991; Rona *et al.*, 1985).

Alcohol

Effects of alcohol intake are estimated from self-reports which tend to be erroneously low (Ernhart *et al.*, 1988; Jacobson *et al.*, 1991; Room, 1977; Verkerk, 1992). Alcohol consumption is expressed in g, oz or number of drinks per day. Conversion factors can transform the data to common units, but uncertainties would remain due to the wide variations among countries in serving sizes and differences among drinks of the same name within and among countries.

The mechanism by which alcohol intake might influence fetal growth is not well established. In experimental animals, placental size and the rate of protein synthesis by the fetus are reduced by alcohol ingestion (Henderson et al., 1980, 1981). Also, the transfer of nutrients from the mother to the fetus may be reduced due to the influence of alcohol on numerous metabolic processes (Isselbacher, 1977). Excessive alcohol intake during pregnancy can result in the fetal alcohol syndrome (Ernhart et al., 1985; Jones & Smith, 1975; Rostand et al., 1990). This condition is not considered in this review which relates to more moderate intakes.

Data relating to the adjusted effects of maternal alcohol intake during pregnancy on size at birth are given in Table 3.19. The effects on weight, length, and head circumference are small and nonsignificant when the intake of alcohol is <1 drink per day, <0.9 oz per day, or <20 g per week (Kline et al., 1987; Moiraghi-Ruggenini et al., 1992; Walpole et al., 1990). At intakes of ≤2 drinks per day, <2 oz per day, or <49 g per week, the reported deficits in weight exceed 65 g in most studies and with intakes >2 drinks per day or >2 oz per day, the reported deficits are 135 to 259 g (Barr et al., 1984; Larroque et al., 1993; Mills et al., 1984; Passaro et al., 1996). Binge drinking at least once in mid-pregnancy combined with an intake of <7 drinks per week is associated with a birth weight deficit of 40 g that increases to 137 g with 7 to 14 drinks per week (Passaro et al., 1996). Some variation among reports could be due to differences in the choice of drinks. The decreases in birth weight are larger with drinking beer or wine than with drinking spirits (Kuzma & Sokol, 1982; Sulaiman et al., 1988) and there is synergism between the effects of tobacco use and alcohol intake on birth weight (Abel & Hannigan, 1995; Windham et al., 1995).

Olsen et al. (1991) reported very small adjusted effects of maternal alcohol intake during pregnancy on length and head circumference at birth with intakes up to 120 g per week; others have found significant effects on length with intakes >1 oz per day averaged over pregnancy and on head circumference with intakes of 1 drink per day during the first month of pregnancy (Day et al., 1989; Fried & O'Connell, 1987). The deficits with alcohol intake in the third trimester are larger than those due to drinking alcohol earlier in pregnancy (Shu et al., 1995). The effect of large alcohol intakes during pregnancy on birth weight may be reduced for those who stop or reduce their drinking in or before the third trimester (Halmesmaki, 1988; Larsson et al., 1985; Rosett et al., 1980, 1983; Smith et al., 1986).

The prevalence of preterm births is increased when 2 or more drinks per day are consumed (Berkowitz et al., 1982). The odds ratio for LBW is not significantly increased when the mothers consume ≤2 drinks per day of alcohol during pregnancy (Marbury et al., 1983; McDonald et al., 1992; Windham et al., 1995). The odds ratio for the infants being SGA is significantly increased when the intake of alcohol during pregnancy is ≥1 drink per day in some studies (Mills et al., 1984; Sokol et al., 1980) or >3 drinks per day in the study of Windham and colleagues (1995).

The mean deficits in size due to alcohol use decrease after birth, but significant effects on weight and length, but not head circumference, continue to at least 8 months (Barr et al., 1984; Streissguth et al., 1981). Alcohol intake during pregnancy is not significantly related to the mean length of gestation, but there is a significantly increased risk of preterm deliveries (Berkowitz et al., 1982; Marbury et al., 1983; Mills et al., 1984; Shiono et al., 1986b).

Caffeine

Some have recorded caffeine intake on the basis of cups of coffee per day, but others have taken all caffeine sources into account. The latter approach is preferable because pregnant women in the United States who drink coffee obtain about 50% of their caffeine intake from other sources (Bracken et al., 1982; Luke, 1982). A cup of coffee contains about 110 mg of caffeine (Debry, 1990), but this value has not been applied to transform the reported data to metric units because cup size varies and coffee is brewed in many ways. Caffeine consumed by a mother crosses the placenta and accumulates in

Table 3.19. *Adjusted deficits in size at birth associated with maternal alcohol consumption during pregnancy*

Author	Amount per day in g, ml, oz or drinks per day, week or month[a]	Number of drinkers	Size deficit at brith		
			Weight (g)	Length (cm)	Head circumference (cm)
Barr et al. (1984)	0.1–0.9 oz per day	145	60	0.5	0.3
	1.0–2.0 oz per day	73	200	1.1	0.5
	>2.0 oz per day	24	160	1.6	0.6
Kaminski et al. (1978)	>1.6 oz per day vs. <1.6 oz per day	8901	58	–	–
Kline et al. (1987)	1 drink per week	201	30	–	–
	2 drinks per week	20	100	–	–
Kuzma & Sokol (1982)	1–19 drinks per month	1858	5	–	–
	≥20 drinks per month	154	102	–	–
Larroque et al. (1993)[b]	7–20 drinks per week	103	32	–	–
	21–34 drinks per week	43	136	–	–
	≥35 drinks per week	23	202	–	–
Mills et al. (1984)	1–2 drinks per day	743	146	–	–
	3–5 drinks per day	113	259	–	–
	≥6 drinks per day	37	199	–	–
Olsen et al. (1983)	1–4 drinks per week	1571	20	–	–
	5–9 drinks per week	153	65	–	–
	≥10 drinks per week	26	175	–	–
Passaro et al. (1996)	<1 drink per week	4090	13	–	–
	1–2 drinks per week	1347	5[c]	–	–
	1–2 drinks per day	164	91	–	–
	≥3 drinks per day	32	135		
Peacock et al. (1991)	1–19 g per week	23	56		–
	20–49 g per week	19	122		–
Shu et al. (1995)	First trimester				
	<1 drink per week	199	29[c]	–	–
	<2 drinks per week	32	64	–	–
	≥2 drinks per week	18	90	–	–
	Second trimester				
	<1 drink per week	139	84[c]	–	–
	<2 drinks per week	26	107[c]	–	–
	≥2 drinks per week	11	12[c]	–	–
	Third trimester				
	<1 drink per week	81	9	–	–
	<2 drinks per week	16	42[c]	–	–
	≥2 drinks per week	20	142	–	–

[a] One drink contains about 12 ml (10 g) of alcohol.

[b] Early pregnancy.

[c] Larger for drinkers than non-drinkers.

Table 3.20. *Adjusted deficits in weight at birth with maternal caffeine use (mg per day) during pregnancy*

Author	Caffeine intake (mg per day)	Number of subjects	Weight deficit (g)
Beaulac-Baillargeon & Desrosiers (1987)	≥300	42	100
Brooke *et al.* (1989)	1–199	547	0
	201–399	617	55
	>400	308	108
Martin & Bracken (1987)	1–150	1794	6
	151–300	707	31
	>300	306	105
Peacock *et al.* (1991)	200–400	128	56
	>400	206	154
Shu *et al.* (1995)	*First trimester*		
	<50	291	27
	≥50 <200	222	28
	≥200 <300	45	52
	≥300	34	24
	Second trimester		
	<50	211	31
	≥50 <200	207	45
	≥200 ≤300	48	103
	≥300	29	17
	Third trimester		
	<50	189	1
	≥50 <200	187	3
	≥200 <300	44	4
	≥300	21	49

the fetus, which lacks the enzymes to metabolize it (Adler, 1970; Dews, 1982). This leads to a reduction in placental blood flow which could restrict fetal growth (Goldstein & Warren, 1962; Horning *et al.*, 1973; Kirkinen *et al.*, 1983). Maternal caffeine use during pregnancy is not significantly related to the length of gestation (Fenster *et al.*, 1991; Fortier *et al.*, 1993; Linn *et al.*, 1982; McDonald *et al.*, 1992). Nevertheless, when combined with tobacco use, caffeine intakes of >400 mg per day are associated with markedly increased risks of preterm birth (Wisborg *et al.*, 1996).

Deficits in weight at birth due to caffeine and coffee intake by mothers during pregnancy are given in Table 3.20. These reports are based on comparisons between adjusted means for groups. They show only modest effects of intakes up to 300 mg per day, but, with such intakes, there is a significant increase in the odds ratio for SGA infants (Fenster *et al.*, 1991; Fortier *et al.*, 1993). When the intakes are judged on the basis of mg of caffeine per day, the estimated mean deficits in weight at birth are small except for intakes >300 mg per day when the deficits are 100 to 154 g. McDonald *et al.* (1992) reported little or no deficit in birth weight with intakes of >10 cups of coffee per day; cup size may have been small in this study. The estimated deficit in birth weight with an intake >4 cups coffee per day is 143 g (Brooke *et al.*, 1989). The effect may be smaller for caffeine intake during the third trimester (Shu *et al.*, 1995). Maternal caffeine intake during pregnancy is not significantly related to weight, length, and head circumference at 8 months (Barr *et al.*, 1984). The odds ratios for LBW are increased significantly when the maternal intake of caffeine during pregnancy is >301 mg per day or the mothers consume >10 cups of coffee per day (Caan & Goldhaber, 1989; Martin & Bracken, 1987; McDonald *et al.*, 1992).

Marijuana

Almost all studies of the effects of maternal marijuana use are based on self-reported data that may underestimate this use. The concentrations of the active ingredients in marijuana vary, as do the amounts used on each occasion. Additionally, data for the use of other drugs is uncertain. Maternal marijuana could affect the fetus because the major ingredient of marijuana (δ-9-THC) crosses the placenta (Harbison & Mantilliplata, 1972).

Marijuana use during pregnancy does not appear to alter the length of gestation in some studies (Fried *et al.*, 1984; Gibson *et al.*, 1983; Linn *et al.*, 1983; Tennes & Blackard, 1980) or the prevalence of LBW in some studies (Day *et al.*, 1991; Fried & O'Connell, 1987; Linn *et al.*, 1983), but Fried *et al.* (1984) found a reduction in the length of gestation. In one study, marijuana use during pregnancy was associated with an unadjusted significant deficit at 8 months in length, without deficits in weight or head circumference (Barr *et al.*, 1984). Surprisingly, the use of six or more joints of marijuana per week during pregnancy may be associated with significant increases in weight and length at 2 years, and in head circumference at 6 to 12 years (Day *et al.*, 1992).

Cocaine

Chemical tests can identify recent users of cocaine, but analyses related to long-term use must be based on self-reports (Ball, 1967). The mechanism by which maternal cocaine use might affect the fetus is not established, but it may reduce uterine and placental blood flow and fetal oxygenation (Woods *et al.*,

1987). Cocaine use after the first trimester of pregnancy reduces the mean length of gestation by 0.7 to 1.8 weeks and increases the prevalence of preterm births (Chasnoff *et al.*, 1989; Cherukuri *et al.*, 1988; Chouteau *et al.*, 1988; Kaye *et al.*, 1989). The prevalence of LBW and SGA infants is increased with cocaine use (Chasnoff *et al.*, 1989; Handler *et al.*, 1991; Keith *et al.*, 1989; MacGregor *et al.*, 1987; Petitti & Coleman, 1990).

Studies in which the mothers used cocaine, but did not use other illegal drugs during pregnancy, show large deficits in adjusted weight, length, and head circumference at birth (Table 3.21). There is a high level of agreement among these studies. These deficits become evident at about 34 weeks of gestation (Amini *et al.*, 1994) and they are smaller if maternal cocaine use occurs only during the first trimester (Chasnoff *et al.*, 1989). The deficits in size for infants of mothers who used multiple drugs, including cocaine, during pregnancy, are about the same as those for infants whose mothers used cocaine only (Chasnoff *et al.*, 1992; Ryan *et al.*, 1987). Perhaps mothers using multiple drugs have smaller cocaine intakes than those who use cocaine only. The infants of mothers who used cocaine, with or without other street drugs, during pregnancy have only small deficits in weight, length, and head circumference at 6 months, but larger deficits in weight and length at 12 to 24 months (Chasnoff *et al.*, 1992) (Table 3.22). In this study, adjustments were not made for intervening variables, but the groups compared were demographically similar.

Heroin and methadone

The use of heroin throughout pregnancy is associated with a reduction in the length of gestation, but there is no reduction when methadone is used (Kandall *et al.*, 1976; Ramer & Lodge, 1975). There are large deficits in weight, length, and head circumference at birth in the infants of mothers who use heroin during pregnancy (Table 3.23) and increased prevalences of LBW infants and of weight and head circumference values more than 2 Z below the mean (Cherukuri *et al.*, 1988; Chouteau *et al.*, 1988;

Table 3.21. *Adjusted deficits in size at birth associated with maternal cocaine use during pregnancy*

Author	Number of subjects	Size deficit at birth		
		Weight (g)	Length (cm)	Head circumference (cm)
Bingol *et al.* (1987)	50	768	3.8	2.0
Chasnoff *et al.* (1989)				
first trimester only	19	276	1.8	1.2
throughout pregnancy	36	607	3.1	1.9
Cherukuri *et al.* (1988)	55	528	–	1.9
Kaye *et al.* (1989)	382	376	–	–
Keith *et al.* (1989)	63	412	–	–
Little *et al.* (1989)	153	325	–	–
MacGregor *et al.* (1987)				
throughout pregnancy	24	705	–	–

Table 3.22. *Adjusted deficits in postnatal growth associated with maternal cocaine use during pregnancy*

Author	Number of subjects	Age (months)	Deficit in postnatal growth		
			Weight (g)	Length (cm)	Head circumference (cm)
Chasnoff *et al.* (1992)	106	3	131	1.1	1.2
	104	6	15	1.1	0.9
	85	12	249	0.6	1.1
	53	18	194	1.2	1.0
	41	24	454	2.2	1.6

Fulroth *et al.*, 1989). The deficits in size at birth are less with methadone use than with heroin use, but are still considerable and there is an increased prevalence of LBW infants (Connaughton *et al.*, 1977; Rosen & Johnson, 1982; Zelson, 1973). Some of the variations among the reports in Table 3.23 may be related to the timing of methadone treatment in relation to gestation (Ramer & Lodge, 1975). After birth, the deficits in size become smaller. The infants of mothers who used heroin or methadone during pregnancy differ little from controls in size at 18 months to 5 years (Chasnoff *et al.*, 1980; Lifschitz *et al.*, 1983; Rosen & Johnson, 1982; Strauss *et al.*, 1979) (Table 3.24).

BREAST-FEEDING

The information in this section can assist the interpretation of growth data during infancy. This review is restricted to publications after 1970 because formula composition changed in recent decades and overfeeding of infants with formula has become less common (Fomon, 1987; Taitz & Lukmanji, 1981; Whitehead & Paul, 1985). The reports considered are from developed countries in which formula is likely to be stored and diluted appropriately.

The initiation of breast-feeding is negatively related to parity, SES, and the duration of hospitalization of the infant after birth, but it is positively

Table 3.23. *Adjusted deficits in size at birth associated with maternal heroin or methadone use during pregnancy*

Author	Number of subjects	Size deficit at birth		
		Weight (g)	Length (cm)	Head circumference (cm)
Heroin				
Lifschitz *et al.* (1985)	25	530	2.3	1.5
Naeye *et al.* (1973)	59	20%[a]	–	–
Methadone				
Lifschitz *et al.* (1985)	26	379	1.9	1.3
Rosen & Johnson (1982)	62	92	–	–
Ryan *et al.* (1987)	43	280	0.9	0.8

[a] Deficit as % of control group.

Table 3.24. *Adjusted deficits in postnatal growth associated with maternal heroin or methadone use during pregnancy*

Author	Number of subjects	Age (year)	Deficit in postnatal growth		
			Weight (g)	Length (cm)	Head circumference (cm)
Heroin					
Lifschitz *et al.* (1983)	22	3	200[a]	0.3	0.3
Methadone					
Lifschitz *et al.* (1983)	21	3	300	1.6	0.6

[a] Larger for users than for non-users.

related to maternal age and prenatal care (Ford & Labbok, 1990; Kistin *et al.*, 1990; Lyon, 1984; Scott *et al.*, 2001; Starbird, 1991). The duration of breast-feeding is negatively related to maternal employment and educational level (Hilson *et al.*, 1997; Ryan, 1997; Starbird, 1991; Visness & Kennedy, 1997). Therefore, adjustments for these and other confounding variables are needed when growth comparisons are made between breast-fed and formula-fed infants. In some studies unadjusted data from breast-fed infants are compared with reference data for a general population. These studies indicate the differences in levels to be expected when data for breast-fed infants are plotted on growth charts, but these differences reflect the combined effects of breast feeding and confounding variables.

Exclusively breast-fed infants

Infants who are exclusively breast-fed receive only breast milk without other liquids or solids except those containing medicines or supplements of vitamins or minerals (World Health Organization Working Group on Infant Growth, 1994). Comparisons of exclusively breast-fed and formula-fed infants in which adjustments were made for confounding variables show exclusively breast-fed infants have small deficits in weight at 8 to 12 months,

Table 3.25. *Adjusted differences in growth associated with exclusive breast feeding (Z-scores in comparison with reference data)*

Author	Country	Duration of exclusive breast-feeding (months)	Number of infants	Age (months)	Weight (g)	Length (cm)
Auestad *et al.* (1997)	United States	≥3	63	1	150	0.12
				2	160	0.08
				4	−100	0.13
				6	−130	−0.06
				9	−260	−0.14
				12	−360	−0.04
Hediger *et al.* (2000)	United States	≥4	894	4–7	−80	−0.06
				8–11	−210	0.03
				12–23	−20	−0.01
				60–71	−60	0.06
Salmenperä *et al.* (1985)	Finland	≥4	68	1	350	−0.23
				2	440	0.23
				3	440	0.25
				4	400	0.22
				5	350	0.18
				6	240	0.06
				7	−20	−0.06
				8	−180	−0.20
				9	−230	−0.22
				10	−320	−0.20
				11	−420	−0.24
				12	−500	−0.23

but the differences in length are near zero (Auestad *et al.*, 1997; Hediger *et al.*, 2000; Salmenperä *et al.*, 1985) (Table 3.25). It has been estimated that each week of exclusive breast-feeding decreases weight at 12 months by 35 g (Kramer *et al.*, 1985*a*). After adjusting for confounding variables, exclusive breast-feeding reduces the prevalence of overweight (BMI >90th percentile) and obesity (BMI >97th percentile) at 6 to 18 years (Kramer, 1981; von Kries *et al.*, 1999).

Comparisons of growth in exclusively breast-fed infants with reference data for general populations, without adjusting for confounding variables, show weight is normal or slightly increased in exclusively breast-fed infants until about 4 months when deficits develop (Butte *et al.*, 1990; Hitchcock

et al., 1981; Krebs *et al.*, 1994; Whitehead, 1983). The situation is less clear for length which appears to be normal to 6 months, but small deficits may develop from 6 to 9 months (Ahn & MacLean, 1980; Hitchcock *et al.*, 1981; Salmenperä *et al.*, 1985; Stuff & Nichols, 1989).

Predominantly breast-fed infants

Predominantly breast-fed infants receive water-based drinks such as fruit juice, sugar-water or tea in addition to breast milk. Dewey *et al.* (1991) reported data for groups of healthy term AGA infants who were predominantly breast-fed or formula-fed. The breast-fed infants received no more than 120 ml per day of other milk or formula and did not receive

Table 3.26. *Adjusted differences in growth associated with predominant breast-feeding (predominantly breast-fed less formula-fed)*

Author	Country	Age (months)	Number of infants	Weight (g)	Length (cm)
Males					
Dewey *et al.* (1992)	United States	1	34	138	0.6
		2	34	131	0.8
		3	34	106	0.4
		4	28	47	0.1
		5	28	−215	0.2
		6	28	−197	−0.1
		7	25	−326	−0.4
		8	25	−380	−0.8
		9	25	−430	−1.0
		10	24	−389	−0.3
		11	24	−550	−0.7
		12	24	−524	−1.0
		18	23	−415	−0.9
Females					
Dewey *et al.* (1992)	United States	1	36	432	0.0
		2	39	10	−0.3
		3	39	−11	−0.2
		4	34	−126	−0.2
		5	34	−243	−0.4
		6	34	−363	−0.7
		7	25	−383	−0.7
		8	25	−416	−0.5
		9	25	−516	−0.6
		10	23	−618	−0.9
		11	23	−709	−0.6
		12	23	−730	−0.1
		18	21	−185	−0.8

solid foods before 4 months. The mothers of the two groups were matched for many important characteristics. The mean weights of both groups were near the 75th percentiles of United States reference data until 3 months, but, from 7 to 18 months, the means for the formula-fed group were near the 60th percentile while those for the breast-fed group were near the 35th percentile. The breast-fed infants were slightly heavier than the formula-fed infants to about 3 months, after which the breast-fed infants were lighter than the formula-fed infants by amounts that increased with age until the difference was about 600 g at 12 months (Table 3.26). There were only small differences in length and head circumference between these feeding groups to 18 months (Dewey & Heinig 1993; Dewey *et al.*, 1990*b*, 1991; Heinig *et al.*, 1993). Similar findings were reported by Michaelsen *et al.* (1994) who found that

Table 3.27. *Unadjusted differences in growth associated with predominant breast-feeding (Z-scores in comparison with reference data)*

Author	Country	Duration of exclusive breast-feeding (months)	Number of infants	Age (months)	Weight	Length
Males						
Michaelsen *et al.* (1994)	Denmark	≥4	14	1	0.21	0.15
				2	0.28	0.20
				3	0.26	0.20
				6	0.02	−0.20
				9	−0.20	−0.35
				12	−0.40	−0.40

the mean Z scores for unadjusted weight and length measurements of predominantly breast-fed infants are slightly positive from birth to 6 months (Z about 0.4), but slightly negative from 9 to 12 months when the deficits increase to 0.4 Z (Table 3.27).

Weight–length relationships

In unadjusted data, infants exclusively breast-fed to 12 months are similar to formula-fed infants in weight-for-length to 3 months, but the mean percentiles of weight-for-length are significantly higher in formula-fed infants from 6 to 24 months (Dewey, 1998; Dewey *et al.*, 1993). The means of weight-for-length and BMI are slightly higher (about 0.2 Z) than reference data to 9 months in exclusively and predominantly breast-fed infants (Ferris *et al.*, 1980; Michaelsen *et al.*, 1994; Salmenperä *et al.*, 1985; World Health Organization Working Group on Infant Growth, 1994).

HORMONAL INFLUENCES

Some hormones have important effects on growth, maturation, and body composition. The control of their secretion, changes in circulating levels with age, and interactions with other hormones are also described. The hormones considered are estradiol, testosterone, thyroid and adrenal hormones, growth hormone, insulin-like growth factor-1 (IGF-1), insulin, and leptin. The role of

gonadotrophins, luteinizing hormone, and follicle-stimulating hormone is discussed in relation to estradiol and testosterone.

There are some fundamental differences among these hormones in the ways they are formed and released. The production and release of estradiol, testosterone, and thyroid, and adrenal hormones are initiated by hypothalamic activation of the pituitary followed by the release of specific pituitary tropic hormones that stimulate the ovaries, testes, thyroid, and adrenal cortex. Most IGF-1 is formed in the liver at rates that are influenced by nutritional status and growth hormone levels. Insulin is formed in the pancreas at rates controlled by blood glucose levels and the amount of visceral adipose tissue. Leptin is produced in adipose tissue at rates that are related to the amount of this tissue.

Estradiol

The female sex steroid hormones are collectively called estrogens. The most potent of these is estradiol, most of which is produced by the ovary, but some is formed by aromatization of androgens in adipose tissue, muscle, and the brain (Edman & MacDonald, 1978; Longcope *et al.*, 1978; Nimrod & Ryan, 1975). Estradiol circulates in a biologically active free form and also linked to sex hormone-binding globulin. In males, most estradiol is formed by aromatization of testosterone and androstene-dione, but some is produced directly by the testes

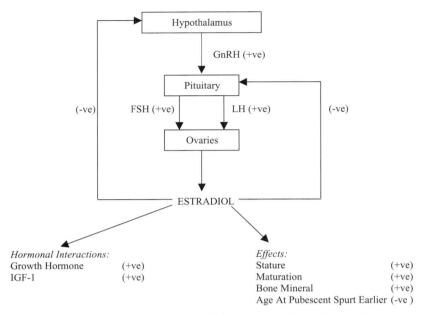

Figure 3.1. A schematic representation of the production of estradiol showing some regulatory influences, hormonal interactions, and effects. GnRH, gonadotropin-releasing hormone; FSH, follicle-stimulating hormone; LH, luteinizing hormone; IGF-1, insulin-like growth factor-1.

(Forney *et al.*, 1981; Simpson *et al.*, 1997; Stanik *et al.*, 1981).

Follicle-stimulating and luteinizing hormones control the formation of estradiol by the ovaries; the latter of these is the more important in the present context because follicle-stimulating hormone is secreted in significant amounts during the prepubescent period, but there are large increases in the circulating levels of luteinizing hormone with the onset of pubescence (Figure 3.1). The release of these tropic substances, which are jointly called gonadotropins, from the pituitary is initiated by peptide secretion cells that are concentrated in the arcuate region of the medial basal part of the hypothalamus (King *et al.*, 1985). These cells, which are limited in activity before pubescence, secrete gonadotropin-releasing hormone (GnRH) in pulses (Bercu *et al.*, 1983; Grumbach & Kaplan, 1990; Knobil, 1990). These pulses become larger and may increase in frequency during pubescence (August *et al.*, 1972; Hassing *et al.*, 1990; Jakacki *et al.*, 1982; Kapen *et al.*, 1975). Nocturnal surges of luteinizing hormone become particularly prominent during

pubescence (Large & Anderson, 1979; Parker *et al.*, 1975). A portal blood system conveys GnRH to receptors on target cells of the anterior pituitary (Gorski, 1990; Hazum & Conn, 1988) (Figure 3.1). Luteinizing hormone stimulates estradiol formation by the granulosa and theca cells within chosen ovarian follicles.

Blood levels of luteinizing hormone are high at birth, but low during childhood due to a relative suppression of GnRH (Jakacki *et al.*, 1982; Wierman & Crowley, 1986). Large increases occur in each sex during pubescence, which can be related to stages of sexual maturation (Albertsson-Wikland *et al.*, 1997; Beitins *et al.*, 1990; Belgorosky *et al.*, 1996; Preece, 1986). Estradiol depresses levels of luteinizing hormone at all ages in a classic negative feedback, but after menarche it is also responsible for the large mid-cycle surges (positive feedback) required for ovulation (Presl *et al.*, 1976; Reiter *et al.*, 1974; Styne, 1995).

Estradiol levels increase from soon after birth to 3 years (Forest, 1990; Winter *et al.*, 1975), but then decrease and remain low before pubescence.

They are higher in females than in males even before pubescence (Kerrigan & Rogol, 1992; Klein *et al.*, 1994, 1996). In females, estradiol levels increase rapidly from 10 to 14 years and are linearly related to stages of breast maturation (Grumbach, 1975; Preece, 1986). In males, the levels change little during childhood and increase only slightly during pubescence (Bourguignon, 1988). Exogenous estradiol obviously accelerates sexual maturation in females (Clark & Rogol, 1996; Conte *et al.*, 1994; Frank, 1995; Morishima, 1996).

Estradiol levels feed back to the hypothalamus and anterior pituitary to modulate ambient levels of the sex steroid (Knobil, 1980; Kulin *et al.*, 1972; Styne, 1995). Sexual maturation is accompanied by reduced sensitivity to negative feedback of gonadal steroids on the hypothalamus or increases in adrenal steroids or baseline hormone production by the hypothalamic–pituitary complex (Ellison, 1998).

Estradiol accelerates skeletal maturation and, if in excess for an extended period during childhood, may reduce adult stature, as occurs in precocious puberty. Estradiol can also increase growth in stature by stimulating the proliferation of osteoblasts, perhaps through increased levels of IGF-1 (Schmid *et al.*, 1989). Estradiol also increases the metabolism of chondrocytes, which may directly increase bone elongation (Rappaport, 1988; Ross *et al.*, 1983). Estradiol levels are reported to be unrelated to statural growth before pubescence, but this may reflect the use of assay methods that have low sensitivity (Belgorosky *et al.*, 1989; Buchanan *et al.*, 1989; Cara *et al.*, 1987; Preece *et al.*, 1980). Some contrary findings have been reported by Caruso-Nicoletti *et al.* (1985).

In each sex, the timing of the pubescent growth spurt is probably determined by estradiol in combination with growth hormone. In males, androgens may stimulate growth after they have been transformed to estrogens. In agonadal females, estrogens stimulate bone elongation without changes in IGF-1 levels, but with possible increases in growth hormone levels (Guyda, 1990). Estrogen deficiency in males, due to defects in the aromatase gene, retards skeletal maturation and leads to adult tallness with a marked delay in epiphyseal fusion and relatively long limbs (Conte *et al.*, 1994; Morishima *et al.*, 1995; Smith *et al.*, 1994).

Estradiol is plentiful in adipose tissue where it may increase the concentration of lipoprotein lipase leading to augmented conversion of triglycerides to free fatty acids that can enter adipocytes (Deslypere *et al.*, 1985; Price *et al.*, 1998). There is a high density of estrogen receptors in subcutaneous adipose tissue (Mizutani *et al.*, 1994; Pedersen *et al.*, 1996). In the obese, estradiol levels are increased and sex hormone-binding globulin levels are decreased; these return to normal with weight loss (Wabitsch *et al.*, 1995; Zumoff & Strain, 1994).

Estrogens are necessary for the achievement of normal peak bone mass (Bachrach & Smith, 1996). Estradiol increases BMD by reducing apoptosis of osteoblasts, and may increase the formation of pre-osteoblasts (Arts *et al.*, 1997; Eriksen *et al.*, 1988; Ernst *et al.*, 1988). Estradiol increases growth hormone secretion by amplifying the pulse amplitude without altering the pulse frequency and it may increase the production of IGF-1 (Bourguignon, 1988; Guyda, 1990; Hindmarsh, 1998; Ross *et al.*, 1983). Consequently, estradiol levels are positively correlated with growth hormone and IGF-1 levels (Faria *et al.*, 1992; Ho *et al.*, 1987; Rosenfield *et al.*, 1983). The growth promoting effects of estrogens are markedly reduced if the levels of growth hormone and IGF-1 are depressed.

Testosterone

The production of testosterone by the testes is initiated by the release of GnRH from the hypothalamus. This hormone travels in the portal system to the anterior pituitary where it stimulates the gonadotrops to produce luteinizing hormone, which, in turn, controls testosterone production (Figure 3.2). Luteinizing hormone is released from the pituitary in pulses that precede increases in testosterone levels by about 30 minutes (Lee *et al.*, 1978; Parker *et al.*, 1975). The androgen binds to receptors on the Leydig cells of the testes leading to the production of male sex steroid. The circulating

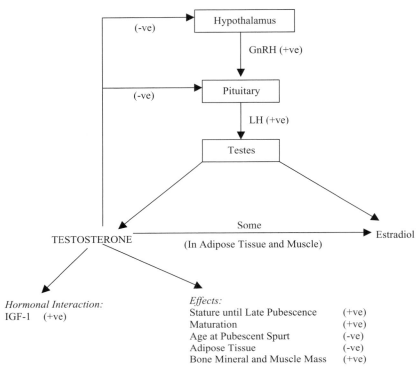

Figure 3.2. A schematic representation of the production of testosterone showing some regulatory influences, hormonal interactions, and effects. GnRH, gonadotropin-releasing hormone; LH, luteinizing hormone; IGF-1, insulin-like growth factor-1.

levels of luteinizing hormone and testosterone are controlled by feedback mechanisms in the hypothalamus and the pituitary (Griggs *et al.*, 1989; Styne, 1995).

A small amount of testosterone circulates in a biologically active free state; 97% is joined to sex hormone-binding globulin that is formed in the liver at a rate that is increased by estradiol, but reduced by testosterone (Anderson, 1974; Horst *et al.*, 1977). The levels of sex hormone-binding globulin decrease with age in males before pubescence in association with increases in free testosterone and growth hormone (Rudd *et al.*, 1986). At its site of action, testosterone is separated from sex hormone-binding protein and diffuses into target cells that have androgen receptors. In these cells testosterone may be converted to dihydrotestosterone or estradiol. Free testosterone is removed from the circulation by the kidneys or converted

to either estrogens or diethyltestosterone in extra-glandular tissues (Hadley, 1996).

Testosterone levels increase markedly during the first day after birth, but decrease during the next several days and then increase to be near adult levels at 2 months (Bidlingmaier *et al.*, 1983; Corbier *et al.*, 1990; Gendrel *et al.*, 1980). The levels then decline gradually until 6 months and remain low to about 11 years (Albertsson-Wikland *et al.*, 1997; August *et al.*, 1972; Kerrigan & Rogol, 1992). Testosterone levels increase 20-fold from 11 to 16 years in males, with positive correlations in timing between the most rapid increase and PHV (Preece, 1986). Testosterone levels are related to genital maturation with especially large increases from stage 4 to stage 5 and large increases near the age of PHV (Grumbach, 1975; Preece, 1986; Zhang *et al.*, 2000). During pubescence, there are large nocturnal pulses of testosterone, but, after pubescence, there is a return to the prepubertal secretion pattern with somewhat higher levels in the morning (Large & Anderson, 1979; Parker et al., 1975). In females, testosterone levels also increase during pubescence, but this change may not be involved in the growth spurt (Apter & Vihko, 1977).

Control of testosterone levels during infancy, childhood, and pubescence seems to be largely through negative feedback mechanisms that may be less important during childhood when the rate of secretion of GnRH is low. Testosterone increases the production of IGF-1 by enhancing the response to growth hormone-releasing hormone and it enlarges the amplitude of growth hormone pulses without affecting their frequency (Eakman et al., 1996; Hindmarsh, 1998; Keenan et al., 1993; Mauras et al., 1994). Despite these effects, testosterone levels are not correlated with those of growth hormone or IGF-1 (Butenandt et al., 1976; Ho et al., 1987; Thompson et al., 1972).

The high levels of testosterone from birth to 2 months do not appear to affect growth. During childhood and pubescence, testosterone accelerates growth in stature and sitting height; this acceleration is much larger when growth hormone is present (Bourguignon, 1988; Buchanan & Preece, 1992). There are significant positive correlations between testosterone levels and stature velocity from 2 years before PHV until PHV, but negative correlations after PHV. There is a similar pattern of change in the correlations between testosterone levels and stature velocity at different stages of genital maturation; these are positive at stages 1 to 3, but negative at stages 4 and 5 (Buchanan & Preece, 1992; Preece, 1986). These findings may reflect stimulation of cartilage growth by testosterone early in pubescence and advancement of epiphyseal fusion late in pubescence.

The timing and duration of pubescence in males is primarily regulated by testosterone, which has direct anabolic effects and acts in association with growth hormone and IGF-1 (Preece et al., 1984). In normal males, testosterone accelerates skeletal and sexual maturation and, at large dosages, can accelerate epiphyseal fusion (Brämswig et al., 1988; Martin et al., 1986; Richman & Kirsch, 1988; Schibler et al., 1974).

Growth in stature is accelerated when testosterone is administered to males with obesity or constitutional delay (Bergadá & Bergadá, 1995; Gregory et al., 1992; Uruena et al., 1992; Zacharin, 2000). This effect may be mediated through IGF-1, which is increased if growth hormone is not deficient (Link et al., 1986; Parker et al., 1984). Alternatively, the increased growth may be due to direct effects on epiphyseal chondrocytes, which have testosterone receptors (Carrascosa et al., 1990). Studies of males with androgen insensitivity show testosterone is not essential for normal pubescent growth, if estrogen levels are normal (Campos & MacGillivray, 1989; Cicognani et al., 1989; Zachmann et al., 1986).

Testosterone tends to reduce total body fat by inhibiting the development of pre-adipocytes and increasing the number of β-adrenoceptors in these cells (Xu et al., 1990). Testosterone also acts directly on adipocytes through androgen receptors (Rebuffé-Scrive et al., 1990; Xu et al., 1990). It increases the density of these receptors and the activity of hormone-sensitive lipase (de Pergola, 2000; de Pergola et al., 1990; Pecquery et al., 1988; Xu et al., 1990). These actions reduce triglyceride uptake and increase lipolysis. Testosterone increases the β-adrenergic activity of catecholamines and thereby increases their antilipolytic actions (Pecquery et al., 1988).

Testosterone increases FFM by enlarging muscle mass (Forbes et al., 1992; Griggs et al., 1989; Mauras et al., 1994; Young et al., 1993). These increases may be due to increased protein synthesis or reduced proteolysis (Alen et al., 1994; Arslanian & Suprasongsin, 1997; Hervey et al., 1981; Mauras et al., 1994). Testosterone levels are positively correlated with BMC, BMD, and calcium retention (Finkelstein et al., 1987; Horowitz et al., 1992; Mauras et al., 1994). These changes may occur through effects on osteoclasts, which have testosterone receptors, and by increased metabolism of chondrocytes (Colvard et al., 1989; Rappaport, 1988; Sasano et al., 1997). When a genetic mutation leads to androgen resistance, bone mineralization is reduced (MacLean *et al.*, 1995; Muñoz-Torres *et al.*, 1995).

Thyroid hormones

The formation of thyroid hormones is initiated by the central nervous system, which activates the hypothalamus to produce thyroid-releasing hormone (Figure 3.3). Thyroid-releasing hormone (TRH) acts on the pituitary to produce thyroid-stimulating hormone (TSH), which acts on the follicular cells in the thyroid gland that are responsible for the synthesis and secretion of thyroid hormones. Thyroglobulin is synthesized within these cells and joined to iodinated tyrosol residues to form thyroxine (T_4), which is the precursor of tri-iodothyronine (Robbins & Rall, 1983). Tri-iodothyronine, which is known as T_3, is the active thyroid hormone. Most T_3 is formed in the liver and kidneys and enters the circulation where it is almost entirely bound to

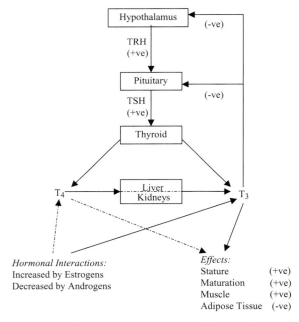

Hormonal Interactions:
Increased by Estrogens
Decreased by Androgens

Effects:
Stature   (+ve)
Maturation   (+ve)
Muscle   (+ve)
Adipose Tissue (-ve)

Figure 3.3. A schematic representation of the production of thyroid hormones showing some regulatory influences, hormonal interactions, and effects. TRH, thyroid-releasing hormone; TSH, thyroid-stimulating hormone; T3, tri-iodothyronine; T4, thyroxine.

proteins. Circulating levels of thyroid hormones are low at birth, but peak during the next few days due to a surge in TRH. The level of TRH is near constant throughout childhood after about 1 month of age, but the levels of T_3 and T_4 decrease (Sack, 1996).

Thyroid hormone levels are influenced by a negative feedback from their circulating levels to TRH receptors in the pituitary. The circulating levels of binding proteins for thyroid hormones are increased by estrogens and reduced by androgens (Sack, 1996). Therefore, estrogens tend to reduce free thyroid hormones and their effects while androgens have opposite actions. Thyroid hormones are necessary for normal growth in early infancy (Letarte, 1985) and they stimulate the maturation of osteoblasts and accelerate ossification through increases in growth hormone and IGF-1 (Katakami *et al.*, 1986; Rappaport, 1988; Vigouroux, 1990). Consequently, skeletal maturation and growth in stature are slow in hypothyroidism and when there is genetic resistance to the actions of thyroid hormones (Fort, 1986; Greenberg *et al.*, 1974; Sack, 1996).

Thyroid hormones increase basal metabolic rate and lipolysis. These hormones degrade and mobilize lipids from adipose tissue and increase the circulating levels of non-esterified fatty acids (Pucci *et al.*, 2000). Thyroid hormone levels are usually normal in the obese, but when the obese lose weight due to dieting, T_3 levels are reduced due to slower deiodination of T_4 (Cavalieri & Rapoport, 1977; Roti *et al.*, 2000). Total body potassium, which is an index of FFM, is significantly correlated ($r = 0.5$) with T_3 levels (Edmonds & Smith, 1981). A considerable amount of T_3 is located within muscle where it stimulates the synthesis and degradation of muscle protein (Brown *et al.*, 1981; Flaim *et al.*, 1978; Nicoloff, 1978; Schwartz, 1983). Total body potassium tends to be reduced in hypothyroidism, but returns to normal with replacement therapy (Bayley *et al.*, 1980).

Thyroid hormones stimulate chondrocyte proliferation and increase bone turnover causing a reduction in BMD (Mora *et al.*, 1999; Saggese *et al.*, 1990). In untreated hyperthyroidism, there are negative correlations ($r = -0.4$) between serum free T_4 levels and BMD (Bayley *et al.*, 1980; Mora *et al.*, 1999; Rosen & Adler, 1992).

Adrenal hormones

The adrenal cortex produces several steroids of which the most important are cortisol, dehydroepiandrosterone (DHEA) and selected mineral corticoids. The median eminence of the hypothalamus secretes corticotropic-releasing hormone (CRH) that is transported to receptors on cells of the anterior pituitary and the brain (Grossman *et al.*, 1982) (Figure 3.4). Adrenocorticotropic hormone (ACTH), which controls the secretion of adrenal cortical hormones, is released by the pituitary in pulses that precede, by about 20 minutes, those in cortisol. The circulating levels of cortisol change little during childhood and pubescence, but they are slightly higher in males than in females (Apter *et al.*, 1979; Grumbach & Styne, 1998; Savage *et al.*, 1975;

Zumoff *et al.*, 1974). About 95% of circulating cortisol is bound to globulin or albumen (Behan *et al.*, 1995). Cortisol is conjugated to glucuronides in the liver, and excreted in the urine.

Insulin is necessary for the effects of cortisol on adipose tissue (Appel & Fried, 1992; Ottosson *et al.*, 1994). When cortisol secretion is increased, sex steroids are reduced, perhaps because of increased leptin which blunts the effect of CRH on gonadotropin-releasing hormone (Huang *et al.*, 1998; Olsen & Ferin, 1987; Richard *et al.*, 2000).

Cortisol enhances growth hormone secretion and reduces the levels of IGF-1 (Rappaport, 1988). Nevertheless, cortisol does not seem important in normal growth and maturation, although excess adrenal glucocorticosteroids are strong growth suppressants. Cortisol acts on adipocytes, and the density of these receptors is inversely related to the circulating levels of glucocorticoids (McDonald & Goldfine, 1988; Rebuffé-Scrive *et al.*, 1985).

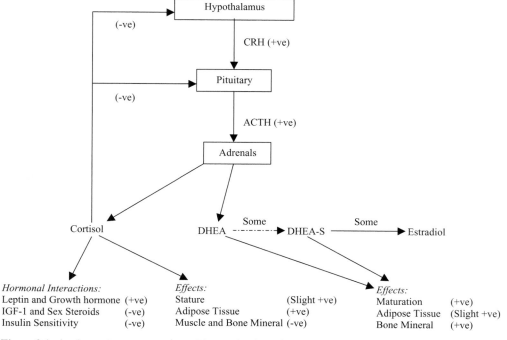

Figure 3.4. A schematic representation of the production of adrenal hormones, showing some regulatory influences, hormonal interactions, and effects. CRH, corticotropic-releasing hormone; ACTH, adrenocorticotropic hormone; DHEA, dehydroepiandrosterone; DHEA-S, dehydroepiandrosterone sulfate.

Cortisol promotes lipid accumulation and reduces lipid mobilization, but these effects may be small (Björntorp & Edén, 1996). In long-term studies, cortisol has a permissive effect on the lipolytic actions of growth hormone (Cigolini & Smith, 1979; Ottosson *et al.*, 1995). It increases the circulating levels of free fatty acids, partly by inhibiting their re-esterification after release by lipolysis (Carter-Su & Okamoto, 1987; Divertie *et al.*, 1991; Guillaume-Gentil *et al.*, 1993). These increased levels of free fatty acids have a negative feedback effect on lipolysis (Rosell & Belfrage, 1979). In obesity, CRH, which reduces energy intake and increases energy expenditure, may be deficient (Arase *et al.*, 1989; Rohner-Jeanrenaud *et al.*, 1989). Cortisol causes muscle wasting by decreasing protein synthesis and converting existing protein to glucose and glycogen. Excess glucocorticoids lead to reductions in bone mineral and decreases in bone formation and maturation rates (Luckert, 1996).

Dehydroepiandrosterone (DHEA) is the major adrenal androgen. It is changed to dehydroepiandrosterone sulfate (DHEA-S) in the zona reticularis of the adrenal, the liver, and the spleen (Falany *et al.*, 1995). Because it has a low clearance rate, DHEA-S accumulates in the plasma and is potentially converted to estradiol in adipose tissue (Labrie *et al.*, 1997; Neville & O'Hare, 1982). The levels of DHEA are high at birth, but decrease early in infancy and remain low until they begin to increase at 7 years. This increase causes mild androgen effects and is a period known as adrenarche. These increases continue throughout the period of growth, but changes are small after 16 years in males and 12 years in females (Apter *et al.*, 1979; Sizonenko & Paunier, 1975; Wierman & Crowley, 1986). DHEA-S levels are positively correlated with skeletal maturity ($p < 0.005$), but have only weak correlations ($r = 0.2$) with body fatness. There are low correlations ($r = 0.3$) between DHEA-S and BMD that reflect the stimulation of osteoblast-like cells to produce osteocalcin and collagen (Chiu *et al.*, 1999; Scheven & Milne, 1998). At the same pubic hair stage, DHEA levels are about 50% higher in females than in males (Apter *et al.*, 1979). Any

effect of DHEA on growth is slight (Palmert *et al.*, 2001). DHEA, however, increases resting metabolic rate, lipid oxidation, and glucose disposal (Apter *et al.*, 1979; Katz *et al.*, 1985; de Pergola, 2000). The circulating levels of DHEA are significantly correlated ($p < 0.001$) with those of cortisol (Apter *et al.*, 1979).

Growth hormone

Growth hormone is produced throughout life in intermittent bursts by the somatotrop cells of the anterior pituitary, but its release is activated by growth hormone-releasing hormone (GHRH) and blocked by somatotropin release-inhibiting hormone (SRIH). The hypophyseal portal system carries GHRH and SRIH from the hypothalamus, where they are formed, to the pituitary (Figure 3.5). Pulse amplitude is mainly controlled by GHRH while SRIH determines the frequency and duration of pulses (Kracier *et al.*, 1988).

The release of growth hormone is also regulated by brain neurotransmitters, including acetylcholine, serotonin, and dopamine, and by blood levels of glucose, IGF-1, estrogens, and testosterone (Rogol, 1989). The release of growth hormone is pulsative from birth with the largest pulses occurring at night during the early hours of sleep (Albertsson-Wikland *et al.*, 1994; Miller et al., 1982; Veldhuis & Johnson, 1992). Growth hormone circulates in a free form and also linked to growth hormone-binding protein (GHBP), which slows clearance and prevents interactions with specific receptors.

The circulating levels of growth hormone decrease during the first few days after birth and become similar in the two sexes until pubescence (Albertsson-Wikland & Rosburg, 1988; Costin *et al.*, 1989; Zadik *et al.*, 1985). Due to effects of estrogens and testosterone, the levels increase during pubescence, particularly in females (Martha *et al.*, 1989; Roemmich *et al.*, 1998; Rose *et al.*, 1991; Zadik *et al.*, 1985) with large increases in pulse size, but no change in pulse frequency (Albertsson-Wikland *et al.*, 1994; Veldhuis & Johnson, 1992). Maximum levels are reached at about the age of PHV

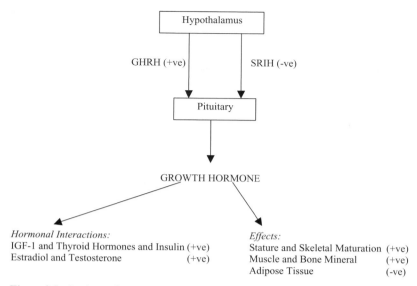

Figure 3.5. A schematic representation of the production of growth hormone showing some regulatory influences, hormonal interactions, and effects. GHRH, growth hormone-releasing hormone; SRIH, somatotropin release-inhibiting hormone; IGF-1, insulin-like growth factor-1.

(Stanhope *et al.*, 1988). The pulses become smaller after pubescence when females have more frequent pulses than males and have higher circulating levels of growth hormone (Hartman *et al.*, 1993).

GHBP reduces the pulse amplitude of growth hormone and increases its half-life (Nilsson *et al.*, 1996). The levels of GHBP are low at birth but increase during childhood and adolescence (Argente *et al.*, 1993; Martha *et al.*, 1991; Massa *et al.*, 1992*a*, *b*; Silbergeld *et al.*, 1989). These levels are higher in females than in males which may reduce the biological activity of growth hormone in females (Juul *et al.*, 2000). In males, there are only small systematic changes in levels of GHBP during pubescence, but they vary considerably and are not correlated with stages of sexual maturation (Juul *et al.*, 2000; Martha *et al.*, 1993). During childhood and adolescence, GHBP levels are correlated ($r = 0.4$) with stature velocity (Gelander *et al.*, 1998). The levels of GHBP are correlated with BMI, %BF, and levels of IGF-1, leptin, and insulin ($r = 0.5$–0.7) (Fernandez-Real *et al.*, 2000; Gelander *et al.*, 1998; Kratzsch *et al.*, 1997; Martha *et al.*, 1992*a*, *b*).

Growth hormone increases the production of IGF-1 (Davoren & Hsueh, 1987; Isaksson *et al.*, 1985; Parker *et al.*, 1984; Tres *et al.*, 1986) which, in turn, inhibits the secretion of growth hormone by increasing the secretion of SRIH (Berelowitz *et al.*, 1981; Martha *et al.*, 1991, 1992*a*). Despite these relationships, growth hormone secretion and circulating levels of IGF-1 are correlated ($r = 0.5$) in pubescent males (Martha *et al.*, 1992*a*).

The production of insulin is increased by growth hormone (Boguszewski *et al.*, 1998; Curry & Bennett, 1973; Zapf *et al.*, 1998), which reduces the synthesis and secretion of growth hormone by a negative feedback mechanism (Chrousos & Gold, 1992; Isaacs *et al.*, 1987). Insulin increases the ability of growth hormone to bind with adipocytes (Gause & Edén, 1985). Growth hormone reduces the effects of insulin on lipolysis by actions distal to receptors (Dietz & Schwartz, 1991; Ottosson *et al.*, 1994).

Growth hormone levels are positively related to the production of sex steroids (Adashi *et al.*, 1985), which increase the amplitude of growth hormone pulses through effects on GHRH, but do not alter their frequency (Mauras *et al.*, 1990; Moll *et al.*,

1986; Ulloa-Augirre *et al.*, 1990; Zeitler *et al.*, 1990). Testosterone increases the secretion of growth hormone; this action may occur after aromatization of testosterone to estradiol (Eakman *et al.*, 1996; Ho *et al.*, 1987; Metzger & Kerrigan, 1994; Thompson *et al.*, 1972). Sex steroids increase the metabolic clearance of growth hormone during pubescence (Rosenbaum & Gertner, 1989). Thyroid hormones increase the secretion of growth hormone; this action is blocked by insulin (Melmed & Slanina, 1985; Yamashita & Melmed, 1986). Growth hormone also stimulates the conversion of T_4 to T_3 (Grunfeld *et al.*, 1988; Jørgensen *et al.*, 1989). Cortisol inhibits the secretion of growth hormone (Chrousos & Gold, 1992).

Growth hormone has a major effect on linear growth, except during early infancy when there are few growth hormone receptors. Growth hormone acts predominantly on prechondrocytes or young differentiating chondrocytes in the germinative layer of the epiphyseal plate (Isaksson *et al.*, 1987; Lowe, 1991; Nilsson *et al.*, 1996). These effects on chondrocytes are partly direct and partly through IGF-1 (Hochberg, 1990; Isaksson *et al.*, 1987). During childhood and pubescence, stature is related more to pulse size than pulse frequency. The effects of growth hormone on statural growth during pubescence may require the presence of insulin and sex steroids (Aynsley-Green *et al.*, 1976; Cheek & Hill, 1974; Copeland *et al.*, 1977). The administration of exogenous growth hormone markedly accelerates growth in children who are deficient in this substance, but it has only small effects on the stature of short normal children; these effects are particularly small in those AGA at birth (Boguszewski *et al.*, 1998; Coutant *et al.*, 1998; Czernichow, 1997; de Zegher & Chatelain, 1998). Maturation is delayed in patients with growth hormone deficiency; this delay is reduced when growth hormone is administered (Bourguignon *et al.*, 1986; Darendeliler *et al.*, 1990; Price *et al.*, 1988; Stanhope *et al.*, 1992).

Growth hormone increases the number of adipocytes (Nilsson *et al.* 1996), but reduces body fat by stimulating lipolysis and increasing basal metabolic rate (di Girolamo *et al.*, 1986; Fagin *et al.*, 1980; Saloman *et al.*, 1989; Vikman *et al.*, 1991). The increased lipolysis may reflect changes in the activity of hormone-sensitive triglyceride lipase and lipoprotein lipase (Dietz & Schwartz, 1991; Ottosson *et al.*, 1995). Some of these actions require the presence of cortisol and insulin (Goodman *et al.*, 1990; Green *et al.*, 1985; Ottosson *et al.*, 1995). Free fatty acids inhibit growth hormone secretion in response to GHRH (Alvarez *et al.*, 1991; Casanueva *et al.*, 1987; Gertner, 1992; Imaki *et al.*, 1985). The increased body fatness seen in growth hormone deficiency is reversed by the administration of growth hormone (Binnerts *et al.*, 1992; Hassan *et al.*, 1996; Kuromaru *et al.*, 1998; Rosen *et al.*, 1993).

In the obese, growth hormone levels are usually low due to smaller and less frequent pulses. These changes may be caused by reduced responses to GHRH (Iranmanesh *et al.*, 1991; Kopelman *et al.*, 1985; Loche *et al.*, 1987; Veldhuis *et al.*, 1991), or increased SRIH (Hartman *et al.*, 1992; Martin-Hernández *et al.*, 1996; Rose *et al.*, 1991; Vanderschueren-Lodeweyckx, 1993). Some investigators have claimed clearance rates of growth hormone are increased in the obese despite increased GHBP which is correlated ($r = 0.5$–0.6) with body fatness and BMI (Fernandez-Real *et al.*, 2000; Juul *et al.*, 2000; Martha *et al.*, 1993; Veldhuis *et al.*, 1991). These changes in clearance rates may be reversed by weight loss (Kelijman & Frohman, 1988; Manglik *et al.*, 1998; Williams *et al.*, 1984).

There is a reduction in FFM with growth hormone deficiency that is reversed by the administration of growth hormone (Bell *et al.*, 1999; Beshyah *et al.*, 1995; Hassan *et al.*, 1996; Saloman *et al.*, 1989). Growth hormone, acting through IGF-1, promotes the differentiation and multiplication of muscle cells (Green *et al.*, 1985; Nilsson *et al.*, 1996; Preece & Holder, 1982). It increases amino acid uptake leading to increased protein synthesis and it reduces proteolysis (Fong *et al.*, 1989; Fryburg *et al.*, 1991; Kostyo, 1968; Underwood & van Wyk, 1985). As a result, muscle mass is increased, but there are only small effects on strength (Bell *et al.*, 1999; Ebert *et al.*, 1988; Saloman *et al.*, 1989; Wieghart *et al.*, 1988).

Measures of BMC and BMD are low in those with growth hormone deficiency, after adjusting for stature, BMI, and skeletal age (Baroncelli et al., 1998; Bing-You et al., 1993; Hyer et al., 1992; Saggese et al., 1996). When growth hormone is administered, BMD increases, perhaps because osteoblasts are stimulated to produce IGF-1, thereby increasing collagen synthesis (Canalis, 1980; Cowan et al., 1999; Johansson et al., 1992; van Wyk, 1984).

Insulin-like growth factor-1

Most insulin-like growth factor (IGF-1) is formed in the liver. This substance has endocrine effects in addition to paracrine and autocrine actions. Almost all circulating IGF-1 is joined to IGF-1 binding protein (IGF-1 BP), which limits its insulin-like actions, modulates interactions with receptors, and extends the half-life (Rechler, 1993; Spencer, 1991). By acting through cell membrane receptors, IGF-1 increases the synthesis of ribonucleic and deoxyribonucleic acids.

From birth through infancy, IGF-1 levels are about 30% of adult levels (Hall & Sara, 1984; Kaplowitz et al., 1982), but increase during childhood to reach adult levels at the beginning of pubescence (Bala et al., 1981; Luna et al., 1983; Rosenfield et al., 1983; Underwood et al., 1980). During pubescence, IGF-1 levels increase to be three to four times the adult levels (Cara et al., 1987; Hasegawa et al., 1996; Juul et al., 1996; Rosenfield et al., 1983). The increases are closely related to stages of sexual maturation during early to mid-pubescence, but not later (Beckett et al., 1998; Luna et al., 1983). The peak level occurs close to the age of PHV after which the circulating levels begin to decrease to adult values (Guyda, 1990; Luna et al., 1983; Preece et al., 1984; Rosenfield et al., 1983). The pubescent increase in IGF-1 levels is partly due to changes in growth hormone levels (Attie et al., 1990). Sex steroids also stimulate IGF-1 production through their effects on growth hormone secretion (Styne, 1995).

Growth hormone stimulates the production of IGF-1 and its expression in epiphyseal cartilage (Bing-You et al., 1993; Schwander et al., 1983; Zapf et al., 1981). Growth hormone acts on cartilage growth through IGF-1, but IGF-1 does not appear to be necessary for the intermediate metabolic actions of growth hormone (Blum & Gluckman, 1996) (Figure 3.6). Negative feedback mechanisms from IGF-1 to the hypothalamus and pituitary affect the secretion of growth hormone (Blum & Gluckman, 1996; Thorner et al., 1992). Insulin secretion is reduced and insulin sensitivity increased by IGF-1 (Beckett et al., 1998; Froesch et al., 1989; Laron, 1999; Mauras et al., 1994). Insulin levels have low ($r = 0.3$) positive correlations with IGF-1 and low negative correlations with IGF-1 BP (Conover et al., 1992; Scheiwiller et al., 1986; Wabitsch et al., 1996).

Nutritional status has a major influence on IGF-1 levels. These levels decrease when calories or protein are deficient; they increase in obesity and decrease with loss of weight (Hochberg et al., 1992; Isley et al., 1983; Vanderschueren-Lodeweyckx, 1993; Wabitsch et al., 1996). The high insulin levels in the obese may increase IGF-1 levels leading to decreases in growth hormone and in IGF-1 BP (Hochberg et al., 1992). The latter decrease could amplify the ability of IGF-1 to inhibit the secretion of growth hormone (Frystyk et al., 1995; Martha et al., 1992b; Vanderschueren-Lodeweyckx, 1993).

The production of IGF-1 in the thyroid is increased by TSH and production in the ovary and testis is increased by gonadotropins. In turn, IGF-1 increases the responses of the ovary and testis to gonadotropins (Bernier et al., 1986; Froesch et al., 1989). The increases in IGF-1 levels during pubescence are correlated with estradiol levels in females; these effects may be mediated by growth hormone (Craft & Underwood, 1984; Rosenfield et al., 1983; Zachmann et al., 1986). Small doses of estradiol or testosterone increase IGF-1 production, but large doses of estradiol reduce IGF-1 production (Blum & Gluckman, 1996; Cuttler et al., 1985; Zapf et al., 1998). The levels of IGF-1 are correlated ($r = 0.6$) with those of testosterone, DHEA-S, and sex hormone-binding globulin (Guyda, 1990; Pfeilschifter et al., 1996).

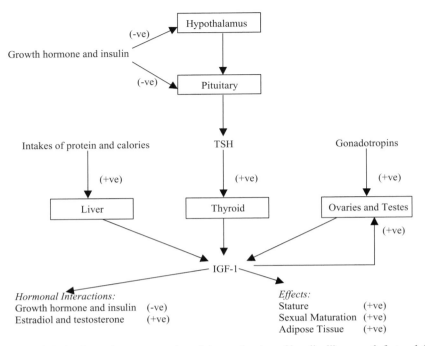

Figure 3.6. A schematic representation of the production of insulin-like growth factor-1 (IGF-1) showing some regulatory influences, hormonal interactions, and effects. TSH, thyroid-stimulating hormone.

The metabolic effects of IGF-1 resemble those of insulin, but its effects on the proliferation and differentiation of many cell types including chondroblasts and osteoblasts are more marked than those of insulin (Blum & Gluckman, 1996; Froesch et al., 1989). Locally produced IGF-1, through a paracrine or autocrine action, accelerates bone elongation by increasing the proliferation of chondrocytes in the epiphyseal plate; this effect is increased by growth hormone (Isaksson et al., 1987; Laron & Klinger, 2000). In early pubescence, but not in late pubescence, stature, weight, and BMI are positively correlated ($r = 0.4$–0.5) with IGF-1 levels (Beckett et al., 1998; Isaksson et al., 1987; Rosenfield et al., 1983). Congenital deficiencies of IGF-1, as in the Laron syndrome, result in shortness from birth that is more marked in the limbs than the trunk and an absence of growth spurts (Laron et al., 1992; Laron, 1999; Walker et al., 1992). Head circumferences are small, but increase with IGF-1 administration up to the age of 10 years (Laron et al., 1992). This suggests that sutural interlocking is delayed in

IGF-1 deficiency, as is skeletal and sexual maturation (Baker et al., 1993; Laron & Klinger, 2000).

Body fat is increased by IGF-1; the age changes in IGF-1 levels resemble those in skinfold thicknesses (Rolland-Cachera et al., 1995). The oxidation and mobilization of lipids and fats from adipose tissue are increased by IGF-1 through cross-reactions with insulin receptors (Dietz & Schwartz, 1991; Poggi et al., 1979; Zapf et al., 1981, 1998). These effects of IGF-1 on body fat are weaker than those of insulin, but both substances stimulate lipid synthesis and compete for receptors (Zapf et al., 1984). Consequently, IGF-1 may limit the stronger lipolytic effect of insulin. The anabolic effects of IGF-1 include increases in FFM, particularly muscle mass. The energy for this anabolic effect comes from increased oxidation of free fatty acids and ketone bodies, and perhaps from increased uptake of glucose by muscle (Guler et al., 1987; Hussain et al., 1993). Protein synthesis is increased and proteolysis is decreased (Douglas et al., 1991; Fryburg, 1994; Jacob et al., 1989; Zapf et al., 1998). The acceleration of

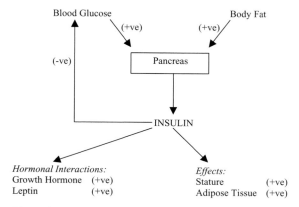

Figure 3.7. A schematic representation of the production of insulin showing some regulatory influences, hormonal interactions, and effects.

bone turnover by growth hormone is mediated by IGF-1 (Bianda *et al.*, 1997).

Insulin

Insulin is produced by β islet cells of the pancreas and is degraded in the liver and kidneys after a half-life of about 5 minutes. It is released by pancreatic cells in response to increases in blood glucose (Figure 3.7). Insulin levels in cord blood are positively correlated with birth weight and ponderal index (Ong *et al.*, 1999*b*). Insulin levels increase more rapidly in females than in males before pubescence (Burke *et al.*, 1986; Jiang *et al.*, 1993; Rönnemaa *et al.*, 1991). The rapid increases during pubescence are due to elevated secretion rates and decreased clearance rates, in association with decreased insulin sensitivity. The circulating levels continue to increase until PHV (Caprio *et al.*, 1989; Smith *et al.*, 1988), but there are only small changes after PHV (Arslanian & Kalhan, 1994; Cook *et al.*, 1993; Potau *et al.*, 1997; Travers *et al.*, 1995). Throughout pubescence, insulin levels are higher in females than males, which may be related to the more rapid maturation of females (Bergström *et al.*, 1996; Burke *et al.*, 1986; Travers *et al.*, 1995) and to sex differences in body fatness (Acerini *et al.*, 2000; de Simone *et al.*, 1995; Zumoff & Strain, 1994). Insulin levels increase from stage 2 to stage 3 of sexual maturity in females, in association with increases in

fatness, but do not increase between these stages in males who become less fat (Travers *et al.*, 1995). Insulin levels are increased in the obese and decrease with weight loss (Brook & Lloyd, 1973; de Simone *et al.*, 1995; Hassink *et al.*, 1996; Manglik *et al.*, 1998) and they are highly correlated ($r = 0.9$) with stature in obese adolescents (de Simone *et al.*, 1995; Loche *et al.*, 1987).

Insulin sensitivity is a measure of the ability of insulin to stimulate whole body disposal of glucose (Amiel *et al.*, 1991). The sensitivity to insulin of some aspects of glucose metabolism decreases during pubescence when there are increases in BMI and body fatness and in circulating levels of growth hormone and IGF-1 (Hindmarsh *et al.*, 1988*a*, *b*). The influence of growth hormone on insulin sensitivity may be independent of IGF-1 (Amiel *et al.*, 1986; Rizza *et al.*, 1982). The decrease in insulin sensitivity during pubescence is independent of estradiol and testosterone (Arslanian & Kalhan, 1994; Caprio *et al.*, 1989, 1994; Cutfield *et al.*, 1990; Travers *et al.*, 1995). As a result of this decrease, insulin-stimulated glucose disposal is reduced and there are increases in lipid oxidation and insulin production by the pancreas. There are inverse relationships between insulin sensitivity and DHEA-S levels (Pang, 1984).

Insulin levels are inversely related to growth hormone secretion by a negative feedback mechanism perhaps related to visceral adipose tissue. The amount of this tissue is related to growth hormone pulse amplitude and insulin levels (Caprio *et al.*, 1995; Manglik *et al.*, 1998; Riedel *et al.*, 1995; Sharp *et al.*, 1987). These effects may be through the pituitary, which has insulin receptors (Yamashita & Melmed, 1986). Insulin levels are correlated ($r = 0.6$) with leptin levels independently of TBF (Butte *et al.*, 1997; Segal *et al.*, 1996). Insulin may inhibit the production of sex hormone-binding globulin and thereby increase the biological activity of testosterone and estradiol. In addition, insulin may stimulate androgen production by the ovary (Barbieri & Hornstein, 1988; Barbieri *et al.*, 1986). Testosterone does not influence insulin sensitivity, except in the obese (Geisthovel *et al.*, 1994;

Wabitsch *et al.*, 1995), but testosterone and growth hormone increase the rate of insulin clearance and growth hormone increases insulin secretion (Acerini *et al.*, 2000; Arslanian & Suprasongsin, 1997; Nielsen, 1982).

Insulin ensures that tissues receive the metabolic fuels needed for growth. It directly stimulates bone and cartilage growth through influences on IGF-1 levels and on receptors for growth hormone and IGF-1 (Milman *et al.*, 1951; Schwartz & Amos, 1968; Silbermann, 1983; van Wyk *et al.*, 1985). Before pubescence, insulin promotes lipid accumulation which results in increased weight, stature, and BMI (Amiel *et al.*, 1991; Björntorp, 1996; Smith *et al.*, 1988). Insulin helps maintain fat stores by increasing the uptake of glucose by adipocytes (Björntorp & Edén, 1996). The glucose is stored as glycogen, which is a source of energy for the formation of fats. Insulin also stimulates the release of free fatty acids that are transported into adipocytes where they combine with glycerol to form triglycerides. Insulin is a strong inhibitor of catecholamine-stimulated lipolysis (Xu *et al.*, 1990) and has an acute inhibitory effect on lipolysis, but may stimulate lipolysis in the long-term presence of cortisol (Björntorp & Smith, 1976). Insulin increases the active transport of glucose and amino

acids into muscle cells where it has protein-sparing effects (Björntorp & Edén, 1996). The high plasma insulin levels of pubescent children tend to increase protein synthesis and the growth of FFM (Amiel *et al.*, 1991; Caprio *et al.*, 1994; Manchester, 1972). The necessary energy comes from glucolysis.

Leptin

Leptin is a polypeptide that is produced by adipocytes and is cleared from the body by the kidneys (Sharma *et al.*, 1997). Leptin is secreted in pulses, with nocturnal increases in adults and perhaps in children (Licinio *et al.*, 1997; Pombo *et al.*, 1997; Sinha *et al.*, 1996; Wolthers *et al.*, 1999); these increases are small in the obese (Saad *et al.*, 1998). Leptin reaches the hypothalamus, where there are leptin receptors (Golden *et al.*, 1997), but the transport of leptin across the blood–brain barrier may be impaired in the obese (Caro *et al.*, 1996; Considine *et al.*, 1996; Schwartz *et al.*, 1996a). The secretion rate of leptin must match current needs because it is not stored by adipocytes. The sympathetic nervous system may provide a negative feedback that regulates leptin synthesis by inhibiting *ob* gene transcription through β_3-adrenoceptors and lipolysis (Trayhurn *et al.*, 1995, 1999) (Figure 3.8). The *ob*

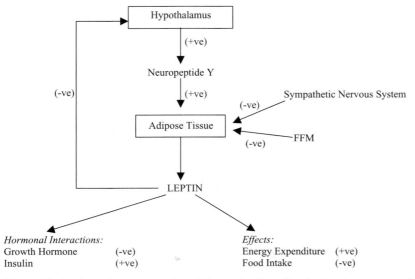

Figure 3.8. A schematic representation of the production of leptin showing some regulatory influences, hormonal interactions, and effects. FFM, fat-free mass.

gene controls the production of leptin; inactivation of this gene causes marked obesity with an early onset (Lönnqvist *et al.*, 1999; Montague *et al.*, 1997). Subcutaneous adipose tissue is the major source of leptin and has more leptin receptors than omental adipose tissue (Hube *et al.*, 1996; Montague *et al.*, 1997).

At birth, leptin levels, adjusted for birth weight, are positively correlated ($r = 0.1$–0.5) with ponderal index and are negatively correlated ($r = -0.3$) with weight gain from birth to 4 months (Lepercq *et al.*, 1999; Mantzoros *et al.*, 1997; Ong *et al.*, 1999*b*). Leptin levels at birth, adjusted for birth weight, are negatively correlated with length ($p < 0.0005$) and head circumference ($p < 0.005$) at birth (Ong *et al.*, 1999*b*). Leptin levels decrease rapidly for 3 days after birth, but increase gradually during childhood (Schubring *et al.*, 1998). These increases are larger in females than in males after adjusting for skinfold thicknesses (Ambrosius *et al.*, 1998; Clayton *et al.*, 1997; Ellis & Nicolson, 1997; Garcia-Mayor *et al.*, 1997). Leptin levels may increase briefly just before pubescence (Ahmed *et al.*, 1999; Mantzoros *et al.*, 1997; Yu *et al.*, 1997). During pubescence, they increase in females, but decrease after stage 3 of sexual maturity in males (Blum *et al.*, 1997; Kiess *et al.*, 1999). The changes in levels with age and the sex differences in leptin levels are partly explained by variations in TBF (Ahmed *et al.*, 1999; Carlsson *et al.*, 1997; Johnson *et al.*, 2001; Kulik-Rechberger *et al.*, 1999), but leptin levels are higher in females than in males independently of body composition (Demerath *et al.*, 1999; Ellis & Nicolson, 1997; Hassink *et al.*, 1996; Rosenbaum *et al.*, 1996). Leptin levels decrease after pubescence in females, which suggests a transient leptin resistance is associated with rapid pubescent growth and maturation.

Leptin stimulates lipolysis and inhibits the maturation of pre-adipocytes (Bai *et al.*, 1996; Frühbeck *et al.*, 1997). In children, leptin levels are positively correlated ($r = 0.5$–0.9) with TBF, BMI, body fat, and skinfold thicknesses (Caprio *et al.*, 1996; Chu *et al.*, 2000; Demerath *et al.*, 1999; Kulik-Rechberger *et al.*, 1999), but are negatively related to FFM (Marshall *et al.*, 2000; Nagy *et al.*, 1997; Reiterer *et al.*, 1999). During pubescence,

regression analyses show leptin levels are negatively related to FFM in males ($b = -0.2$), but there are positive relationships in females ($b = 0.5$) (Ahmed *et al.*, 1999). Nevertheless, leptin levels vary markedly within groups matched for BMI (Lönnqvist *et al.*, 1995; Maffei *et al.*, 1995). The obese tend to have leptin resistance (Caro *et al.*, 1996; Ramsey *et al.*, 1998) and their adipocytes contain more *ob* mRNA (Jéquier & Tappy, 1999). Females with low leptin levels at the beginning of pubescence tend to have large gains in %BF during pubescence ($r = -0.7$) (Ahmed *et al.*, 1999; Ong *et al.*, 1999*a*). Leptin helps regulate body weight by increasing energy expenditure and reducing food intake, perhaps through a decrease in TRH (Blum *et al.*, 1997; Halaas *et al.*, 1995; Schwartz *et al.*, 1996*b*). Food intake is increased in those who lack leptin or have non-functional leptin receptors (Brunner *et al.*, 1997; Clément *et al.*, 1998; Montague *et al.*, 1997; Strobel *et al.*, 1998).

Leptin may be important in sexual maturation through effects on the secretion of GnRH (Aubert *et al.*, 1998), but the relationships are controversial (Mantzoros *et al.*, 1997; Matkovic *et al.*, 1997). Leptin may signal the size of the fat mass to leptin receptors in the hypothalamus and other parts of the brain; these central actions are counteracted by glucocorticoids (Zakrzewska *et al.*, 1997). In a negative feedback mechanism, leptin reduces the production of neuropeptide Y by neurons in the arcuate nucleus of the hypothalamus, which leads to reduced production of leptin (Wang *et al.*, 1997; Wilding *et al.*, 1993; Zarjevski *et al.*, 1993).

The levels of insulin and leptin are positively correlated ($r = 0.4$–0.6), perhaps through their associations with body fat (Kolaczynski *et al.*, 1996; Reiterer *et al.*, 1999; Saad *et al.*, 1998; Widjaja *et al.*, 1997). Infused insulin causes slow increases in leptin levels, apparently due to increases in body fat (Becker *et al.*, 1995; Boden *et al.*, 1996; Considine *et al.*, 1996; Hickey *et al.*, 1996). Increased leptin levels lead to insulin resistance independent of TBF (Chu *et al.*, 2000; Mantzoros *et al.*, 1997; Segal *et al.*, 1996).

In short-term studies, growth hormone decreases leptin levels without detectable changes in body composition (Eilmam *et al.*, 1999); the long-term effects of growth hormone include decreases in leptin levels due to reductions in body fat (Matsuoka *et al.*, 1999*a*). Leptin levels are correlated ($r = 0.5$) with GHBP in normal children, but not in growth hormone deficient children (Bjarnason *et al.*, 1997; Fernandez-Real *et al.*, 2000). There is a low positive correlation ($r = 0.1$) between leptin and IGF-1 levels (Ahmed *et al.*, 1999). The levels of leptin and estradiol are positively correlated ($r = 0.7$), but this relationship is not significant after adjusting for TBF (Ahmed *et al.*, 1999; Elbers *et al.*, 1997). There are low negative correlations between leptin and testosterone levels that may be due to the reduction of TBF by testosterone (Adan *et al.*, 1999; Blum *et al.*, 1997; Ertl *et al.*, 1999; Luukkaa *et al.*, 1998). These findings may reflect a negative feedback of leptin on neuropeptide Y, thereby decreasing the secretion of insulin and corticosteroids (Rohner-Jeanrenaud, 2000).

ETHNIC INFLUENCES

Ethnicity and race are imprecise and ambiguous terms. Imprecision comes from the ancestral heterogeneity within racial and ethnic groups, including dissimilarity within some pairs of spouses that can affect parental reports that are used to assign children to ethnic or racial groups (Hahn, 1999; Keita & Boyce, 2001). Ambiguity results from differences between countries in the geographic origins of those assigned to particular groups. Indians were included with European Americans in some United States studies and Chinese were not included with Asians in some studies made in the United Kingdom (Bhopal *et al.*, 1991). Such variations make it difficult to compare ethnic or racial groups among countries. The term "ethnic group" may be replacing "racial group," because it has less hint of racism and it implies that growth deficits in some groups are amenable to public health programs. Deficits associated with race are commonly considered to be fixed differences of genetic origin. Whichever unsatisfactory term is used, there is a clear need to describe the differences among groups taking into account the influences other than ethnicity or race that can alter growth, maturation, and body composition.

Ethnic classification is typically based on self-report or maternal report. This information can be augmented by data relating to cultural practices and the places of birth of parents and grandparents. Therefore, ethnic classification is largely sociocultural, but there are some genetic differences between ethnic groups. Many consider growth variations within and between populations are determined more by environmental than by genetic influences because there are only small differences in stature and weight between upper SES groups from various populations (Eveleth & Tanner, 1990; Habicht *et al.*, 1974*a*, *b*; Jelliffe & Jelliffe, 1974).

The following account of ethnic influences is restricted to studies from developed countries and to studies in which adjustments were made for confounding variables. The ethnic differences that remain after these adjustments reflect both genetic effects and the effects of environmental influences that were not recognized or measured. Almost all the studies considered were made in the United States where there are large ethnic groups.

African Americans

The estimated adjusted birth weights of African Americans are 86 to 246 g smaller than those of European Americans (Table 3.28). Part of this ethnic difference in birth weight may be due to the increased prevalence of maternal hypertension and hemorrhage during pregnancy in African-American women compared with European-American women and to environmental influences not included in the models (Kempe *et al.*, 1992). These differences in adjusted birth weight are larger for high-risk mothers than for low-risk mothers (Kleinman & Kessel, 1987), perhaps because of genetic differences or uneven distributions of unrecognized environmental influences. Some consider these differences in birth weights between African Americans and European Americans are genetic in origin (Goldenberg *et al.*, 1991; Hulsey *et al.*,

Table 3.28. *Adjusted effects of ethnicity on size at birth in comparison with data for European Americans*

Authors	Number	Weight (g)	Length (cm)	Head circumference (cm)
African Americans				
Cramer (1985)	925	−178	–	–
Friedman *et al.* (1993)	11 075	−186	–	–
Goldenberg *et al.* (1991)	1 205	−148	−0.5	−0.4
Hediger *et al.* (1989)[a]	727	−232	–	–
Horon *et al.* (1983)[b]	576	−108	–	–
Hulsey *et al.* (1991)[b]	6 820	−181	–	–
Kuzma & Sokol (1982)	306	−86	–	–
Shiono *et al.* (1986a)	2 716	−246	–	–
Showstack *et al.* (1984)[b]	3 649	−228	–	–
Singh & Yu (1994)	10 830	−120	–	–
Hispanic Americans				
Cramer (1985)	410	−8	–	–
Kuzma & Sokol (1982)	917	−46	–	–
Shiono *et al.* (1986a)	3 051	−105	–	–
Mexican Americans				
Showstack *et al.* (1984)[b]	1 824	−21	–	–
Puerto Ricans				
Cramer (1985)	72	−258	–	
Hediger *et al.* (1989)[a]	414	−189	–	
Asian Americans				
Shiono *et al.* (1986a)	2 082	−210	–	–
Singh & Yu (1994)				
Chinese	5 757	−115	–	–
Filipino	1 356	−164	–	–
Japanese	531	−235	–	–

[a] Adolescents.

[b] Tobacco use not included in model.

1991; Little & Sing, 1987; Magnus, 1984). In an interesting study, David & Collins (1997) analyzed birth weights among three groups of infants. Compared with European Americans, there were mean deficits of 113 g for infants of African-American mothers who were born in Africa and of 357 g for infants of African-American mothers born in the United States. Nevertheless, the prevalence of LBW is similar for African-American mothers born in Africa and those born in the United States, although the latter have about 25% admixture of genetic material (Adams & Ward, 1973; Cabral *et al.*, 1990; Chakraborty *et al.*, 1991; Fuentes-Afflick *et al.*, 1998).

The adjusted prevalence of LBW infants is higher in African Americans than in European Americans, but it is lower in African Americans of Haitian, West Indian, and Cape Verdean ancestry than in African Americans generally (David & Collins, 1997; Friedman *et al.*, 1993; Miller & Jekel, 1987b). African-American and European-American groups do not differ in the adjusted prevalences of preterm births (Kleinman & Kessel, 1987; Lieberman *et al.*, 1987; Shiono *et al.*, 1986*a*).

Ethnic differences may occur in the effects of confounding influences on birth weight. Maternal educational level is negatively related to the prevalence of VLBW infants for European Americans, but not for African Americans, although the prevalence of moderately LBW infants is related to maternal educational level in both these groups (Kleinman & Kessell, 1987). Differences in educational level may be linked to variations in other influences. In European-American women, half the excess prevalence of LBW infants in those with <12 years education is due to increased tobacco use (Kleinman & Madans, 1985).

Mean head circumference at birth is smaller for African Americans than for European Americans by about 0.5 cm after 30 weeks of gestation (Goldenberg *et al.*, 1991; Wingerd, 1970). The deficits in head circumference and length at birth for African Americans compared with European Americans are small and of brief duration (Goldenberg *et al.*, 1991).

At the same income level, stature tends to be larger in African Americans than in European Americans from 1 to 11 years with larger differences for females than males (Garn *et al.*, 1973, 1974). Stature, adjusted for parental stature, differs only slightly between African Americans and European Americans until 4 years, but African Americans are taller than European Americans by about 1.0 to 2.0 cm from 4 to 9 years (Wingerd *et al.*, 1973). Among females, African Americans have significantly larger unadjusted sums of triceps and subscapular skinfold thicknesses from 10 through 19 years and significantly larger values for %BF from 14 through 19 years, but there is an opposite significant ethnic difference for %BF at 9 through 12 years (Morrison *et al.*, 2001).

Hispanic Americans

Deficits in birth weight among Hispanic Americans are larger for Puerto Ricans, living in the continental United States than for Hispanic Americans in general (Table 3.29). The prevalence of LBW infants does not differ among women grouped by ancestry to Central and South Americans, Cubans, Mexicans, Puerto Ricans, and European Americans, but Hispanic-American women, in combination, have a higher prevalence of moderately LBW infants than European-American women (Fuentes-Afflick *et al.*, 1999). The prevalences of LBW and of preterm births are significantly higher for Mexican-American mothers with long periods of United States residence and marked acculturation than for those who have been in the United States for 5 years or less (Collins & Shay, 1994; Guendelman & English, 1995; Scribner & Dwyer, 1989). Additionally, Hispanic-American mothers born in the United States have a higher prevalence of LBW infants than foreign-born Hispanic-American mothers (Collins & Shay, 1994; Guendelman & Abrams, 1995). These differences may be due to better medical care and nutrition prior to migration. Mexican-American children, in comparison with European Americans at the same poverty index level, have stature deficits throughout childhood that become more marked after 13 years, but they have larger weights and BMI values (Ryan *et al.*, 1990).

Asian Americans

Birth weight is 115 to 235 g smaller in Chinese, Filipino, and Japanese Americans than in European Americans (Shiono *et al.*, 1986*a*; Singh & Yu, 1994; Wang *et al.*, 1994) (Table 3.29). The prevalence of LBW infants is high for mothers from the Philippines, Cambodia, and India and for Korean-American mothers born abroad (Fuentes-Afflick & Hessol, 1997). The prevalence of LBW infants is lower for mothers of Asian, Chinese, and Japanese ancestry born abroad than for those of matching ancestry born in the United States, but this pattern does not occur for mothers of Filipino ancestry (Alexander *et al.*, 1992).

Canadian ethnic groups

Comparisons among groups of mothers with low risk for deficits in birth weight show the mean birth weights of Chinese born in Canada or in China are almost identical, but are about 220 g smaller than the mean birth weight for a comparison group of European Canadians. Nevertheless, the prevalence of LBW infants is increased for Chinese born in Canada, but does not differ between Chinese born in China and European Canadians (Wen *et al.*, 1995). The prevalence of LBW infants is increased for acculturated migrants to Canada from China, Greece, Italy, and southern or southeastern Asia when acculturation is judged by the language spoken in the home (Hyman & Dussault, 1996).

Native Americans

Native Americans living in Canada or the United States have large mean adjusted birth weights. There is an excess prevalence of LGA infants among Cree Indians living near James Bay in northern Canada and for western central Arctic Inuit at Coppermine in British Columbia (Armstrong *et al.*, 1998; Partington & Roberts, 1969; Rodrigues *et al.*, 2000; Schaefer, 1970). These differences may be due to gestational diabetes, which is common in Native Americans, or they may be genetic in origin (Murphy *et al.*, 1993; Pettitt *et al.*, 1985). Native Americans tend to be overweight during childhood (Broussard *et al.*, 1991; Gallaher *et al.*, 1991; Owen *et al.*, 1981; van Duzen *et al.*, 1976).

European ethnic groups

There are some well-established ethnic groups in Europe, but few studies of these groups have adjusted for confounding variables. Within some ethnic groups in England, birth weights for mothers born abroad are smaller than those for mothers born in England; the mean differences by ancestral home are Bangladesh and India 270 g, Caribbean 150 g, and Pakistan 160 g (Alberman *et al.*, 1991).

NUTRITIONAL INFLUENCES

This review of nutritional influences on growth is restricted to intakes of energy, fat, protein, and selected minerals. Vitamins are essential nutrients, but it is unusual for vitamin lack to cause growth deficiencies in developed countries. In general, low intakes lead to reduced body stores and later to the depletion of tissue levels and slow growth. Nutritional influences on growth differ markedly between developed and developing countries. In developed countries, the major nutritional influence on growth is the excessive intake of energy, which may be more marked in low SES groups. In developing countries, intakes of protein and energy are commonly inadequate, particularly in low SES groups.

Precise knowledge of nutritional influences on growth is limited. The necessary studies are complex, especially after the preschool period when it is increasingly difficult to record accurate dietary intakes. In these studies, adjustments should be made for many confounding variables, including SES and the prevalence of diseases. When the influence of one nutrient is to be estimated, the influence of other nutrients must be taken into account, but this has rarely been done, because large samples are needed and many measures must be recorded. Such studies are logistically difficult and expensive.

Energy

Energy intake must be sufficient to meet total energy expenditure and allow for energy deposition in the form of fat and FFM. The energy cost of growth is important during the first few months after birth, but it is only 6% of the total energy requirement at 6 to 12 months (Butte *et al.*, 2000*b*). Total energy requirements increase rapidly from birth to 2 years. These requirements are about 4 MJ per day for each sex at 2 years and increase to 15 MJ per day for males and 11 MJ per day for females at 18 years (Butte *et al.*, 2000*a*).

When energy requirements are not met, there is inadequate weight gain followed by a slowing of growth in stature and of maturation (Lifshitz & Moses, 1986). If data are available from only one examination of an undernourished child, a small value for weight and perhaps for stature is to be expected, but it will not be clear whether these small values are normal and represent constitutional delay

(Boulton & Magarey, 1995; Lauer *et al.*, 2000; Michaelsen, 1997; Simell *et al.*, 2000). Biochemical measures and IGF-1 levels may indicate whether energy intakes are adequate (Cossack, 1988; Underwood, 1996). When low energy intakes are prolonged, deficits in stature develop, particularly when the low intakes occur during a period of rapid growth. With moderately low energy intakes, there are decreases in body fat without changes in FFM or growth in stature. When the energy deficiency is more severe, there may be catabolism of FFM causing a reduction in the stores of protein and zinc together with stunting. When energy intakes are increased for children with low intakes, there are early increases in weight without changes in stature. Later, weights-for-stature return to normal, but deficits in FFM commonly remain that may be corrected by the administration of zinc (Brooke & Wheeler, 1976).

Fat

More than half the calories in breast milk are from fat, which is needed for myelinization of the nervous system (Agostoni *et al.*, 1992). At ages from 7 months to 10 years, low-fat diets, with energy from fat ≤30% of total dietary energy, are not associated with reduced growth in weight, stature, or head circumference (Boulton & Magarey, 1995; Lauer *et al.*, 2000; Michaelsen, 1997; Simell *et al.*, 2000).

Growth deficits may occur when the energy intake from fat is ≤27% of the total energy intake (Shea *et al.*, 1993). The percentage energy intake from fat has only slight relationships with %BF (Atkin & Davies, 2000; Maffeis *et al.*, 1996), but high fat intakes are associated with obesity after controlling for total energy intake and confounding variables (Obarzanek *et al.*, 1994; Ortega *et al.*, 1995; Tucker *et al.*, 1997). When fat intake is low, the diet may be deficient in calcium, iron, and zinc leading to growth retardation (McPherson *et al.*, 1990; Nicklas *et al.*, 1992; Vobecky *et al.*, 1995).

Protein

When a child's diet provides adequate energy, there is usually a sufficient intake of protein. The major dietary sources of protein are meat, fish, cheese, and eggs. Growth from 3 to 6 months is significantly correlated with protein intake (Heinig *et al.*, 1993). Children with a large protein intake at 2 years have earlier rebounds in BMI and increased skinfold thicknesses after controlling for parental BMI and total energy intake at 2 years (Kramer *et al.*, 1985*b*; Rolland-Cachera & Bellisle, 1986; Rolland-Cachera *et al.*, 1996). The intake of protein at 7 to 12 years has small ($r = 0.1$) correlations with overweight and obesity, but it is significantly related to BMC of the distal forearm at ages from 11 to 17 years (Magarey *et al.*, 1999; Rolland-Cachera & Bellisle, 1986).

Minerals

CALCIUM

After infancy the major dietary sources of calcium are dairy products, tofu, and collard greens. Low calcium intakes are common in developed countries resulting in low levels of bone mineral, particularly during mid-pubescence when large increases in bone mineral are normal. Consequently, peak bone mass in young adulthood is reduced and the probability of fractures in middle age and later is increased.

Calcium intakes change little with age from 5 to 17 years, but the percentage absorbed increases in early pubescence (Abrams & Stuff, 1994). During infancy, after adjusting for weight and total energy intake, calcium intake is correlated ($r = 0.3$) with total body BMC (Specker *et al.*, 1999). After adjusting for confounding variables, the intake of calcium and phosphorus combined is positively related to the increments in weight and total body BMC from birth to 6 months, but not from 6 to 12 months (Specker *et al.*, 1997). Dietary supplements of calcium for 1 year or longer during the period of growth increase BMC and BMD by about 1.3% per year (Chan *et al.*, 1995; Lloyd *et al.*, 1996; Rubin *et al.*, 1993; Specker *et al.*, 1997). Calcium intake is closely related to bone mineralization during pubescence. Calcium intake, age, and stage of sexual maturity, in combination, explain 80% of the variance in BMD of lumbar vertebrae for females aged 8 to 18 years (Sentipal *et al.*, 1991).

COPPER

Low intakes of copper may occur in LBW infants and in infants who consume large amounts of cow's milk. As a result, bone mineral is reduced secondary to a defect in lysyl oxidase, which is necessary for the cross-linking of collagen molecules in bone matrix, but growth in length is not affected (Golden, 1988).

IRON

The physiological importance of iron relates to the transport and storage of oxygen and tissue metabolism. Iron deficiency may be due to a low intake of meat, fish, eggs, and fortified cereals or increased gastrointestinal or other losses. A low iron intake does not affect growth until body stores are depleted. Iron deficiency may be associated with retarded pubescent growth, but it is not always clear that this is a causal relationship (Lifshitz & Moses, 1986). Studies of growth, in relation to iron intakes, are inconclusive. Some of these studies have been too short and it is difficult to control for confounding variables, which include general health, illnesses, weight, stature, maturity, and the intakes of other nutrients.

ZINC

Children on vegetarian diets are at risk of zinc deficiency because meat is the major source of zinc. The low zinc content of breast milk may be related to the slow growth of infants who are exclusively breast-fed after 6 months. Zinc is important for the function of intracellular enzymes that influence gene expression, the synthesis of ribonucleic acid, and the growth and differentiation of cells. A small intracellular exchangeable pool of zinc seems necessary for gene expression and chromatin restructuring (Vallee & Falchuk, 1981). Zinc deficiency could impair nucleic acid metabolism, protein synthesis, and the differentiation of myeloblasts through effects on zinc-dependent enzymes and there may be a direct effect on bone growth (Brown et al., 1978; Petrie et al., 1996).

Zinc deficiency may block responses to growth hormone (Cha & Rojhani, 1997; Prasad et al., 1963), decrease the synthesis of IGF-1, and reduce the actions of IGF-1 (Cossack, 1991; Matsui & Yamaguchi, 1995; Yamaguchi & Hashizume, 1994). It may also reduce the secretion of testosterone, insulin, and growth hormone (Coble et al., 1971; Henkin, 1976; Root et al., 1979; Roth & Kirchgessner, 1997), but increase cortisol production (Park et al., 1989). Zinc seems to be essential for the post-receptor binding required for the induction of cell proliferation by IGF-1 and for the actions of nuclear receptors for testosterone, estradiol, and thyroid hormones (Bettger & O'Dell, 1993; Chesters, 1992; Jackson, 1989; MacDonald, 2000). The growth deficits associated with zinc deficiency are not corrected in all children when the levels of growth hormone and IGF-1 return to normal. This suggests that zinc deficiency causes a defect in hormone signaling or that the major effects of zinc deficiency on growth are not hormone-mediated.

Zinc deficiency may be present in some cases of failure-to-thrive. Walravens and colleagues (1989), in a randomized study of such infants aged 8 to 27 months, showed zinc supplementation was associated with a more rapid increase in weight, but not in length. Others have reported an acceleration of statural growth in 25% of short children given zinc supplements (Gibson et al., 1989). Zinc supplementation of short preschool children is associated with increased intakes of energy and protein that may accelerate growth in stature (Krebs et al., 1984; Hambidge, 1986; Walravens et al., 1983). Ghavami-Maibodi and colleagues (1983) found that short children with retarded skeletal ages grow rapidly in stature when given zinc supplements. The response of weight and stature to zinc supplementation of LBW infants may be more marked in females than in males (Castillo-Durán et al., 1995; Friel et al., 1993), but there may be opposite sex differences in the responses of older children (Friis et al., 1997; Ruz et al., 1997; Schlesinger et al., 1992).

HIGH ALTITUDE

The effects of high altitude will be considered separately for size at birth and for growth during infancy, and adolescence. Infants and children who live at high altitudes in developed countries tend to be small (Haas et al., 1980; McClung, 1969; Moore *et al.*, 1982*a*; Yip, 1987), but few studies of such groups have adjusted for other influences on growth. In some high-altitude populations, the deficits in postnatal growth, which are less marked in weight than in length, are largely due to nutritional stress and disease (Frisancho *et al.*, 1975; Leonard *et al.*, 1995). The nutrient intakes of high-altitude populations in the United States are not significantly deficient and median family income in the United States does not differ with altitude (Grahn & Kratchman, 1963; Lichty *et al.*, 1957).

Animal studies show that hypoxia independently retards growth (Hunter & Clegg, 1973; Shaw & Bassett, 1967), but hyperventilation during pregnancy tends to adjust for hypoxia in high-altitude populations (Moore *et al.*, 1982*a*). After birth, there is also some adaptation to hypoxia by increases in lung volume and compliance and pulmonary diffusion (de Meer *et al.*, 1995; Frisancho *et al.*, 1997). Maternal tobacco use during pregnancy probably reduces the supply of oxygen to the fetus. Therefore, it would be expected that tobacco use during pregnancy would have larger adjusted effects on size at birth for infants born at high altitude than for those born at low altitude. This appears to be the case (Moore *et al.*, 1982*b*). There is an increase in natural radiation at high altitudes, but this increase is too small to affect growth (Grahn & Kratchman, 1963; Russell *et al.*, 1960).

Size at birth

Weight, length, and head circumference tend to be small in high-altitude United States populations (Grahn & Kratchman, 1963; Howard *et al.*, 1957; Jensen & Moore, 1997; Unger *et al.*, 1988). These reductions in size are not due to differences in gestational age at birth (Howard *et al.*, 1957; Jensen & Moore, 1997; McCullough *et al.*, 1977; Unger *et al.*, 1988). The effects of high altitude on birth weight become evident at about 30 weeks of gestation and increase from then to term (Grahn & Kratchman, 1963; McCullough *et al.*, 1977; Unger *et al.*, 1988). There is a larger reduction in birth weights for infants born to mothers who migrated as adults to

Table 3.29. *Adjusted deficits in weight at birth for USA populations living at high altitudes*

Author	Altitude (m)	Deficit (g)
Howard *et al.* (1957)	3000 vs. "low altitude"	290
Jensen & Moore (1997)	per 1000 m	102
Lichty *et al.* (1957)	>3000 vs. 1000	380
McCullough *et al.* (1977)	<2130 vs. 2130–2740	54
	<2130 vs. >2740	204
Unger *et al.* (1988)	<1524 vs. 1525–2133	36
	<1524 vs. >2134–2743	115
	<1524 vs. 2743	213
Yip (1987)	<500 vs. 1500–1999	143
	<500 vs. 2000–2499	194
	<500 vs. 2500–3100	351
Yip *et al.* (1988)	<500 vs. 500–999	24
	<500 vs. 1000–1499	68
	<500 vs. 1500–1999	113
	<500 vs. 2000–2500	137

high altitudes than for infants whose mothers lived at high altitudes during childhood, but maternal size is not protective against the effects of high altitude on birth weight (Weinstein & Haas, 1977; Zamudio *et al.*, 1993).

Estimates of the adjusted effects of high altitude on birth weight for United States populations are summarized in Table 3.29. The mean deficits are about 70 g at 1000 m, 100 g at 2000 m, and 350 g at 3000 m. Ideally, the observed and adjusted birth weights of an infant would be plotted on a growth chart for the general population. This procedure could help the management of newly born infants living at altitudes >3000 m. In association with reduced birth weights, some have reported an increased prevalence of LBW infants at altitudes higher than 1000 m to 1500 m (Unger *et al.*, 1988; Yip, 1987), but others have found such increases only at altitudes exceeding 2700 m (Kashiwazaki

Table 3.30. *Differences in size during infancy between infants born and living at 1700 m and those living at 500 m. The differences are expressed as 500 m value less 1700 m value*

	Males			Females		
Age (mo)	Weight (g)	Length (cm)	Head circumference (cm)	Weight (g)	Length (cm)	Head circumference (cm)
Birth	100	0.3	0.3	50	0.2	0.3
1	310	1.3	0.4	270	1.0	0.2
2	210	0.8	0.2	110	0.4	−0.1
6	120	0.4	0.2	0	0.4	0.0
9	80	0.5	0.5	40	0.7	0.3
12	100	0.7	0.5	170	0.3	0.2
18	150	0.0	0.1	180	0.1	0.1
24	170	−0.3	0.1	70	−0.8	0.0

Source: McCammon (1970) and Hamill *et al.* (1977).

et al., 1988; Lichty *et al.*, 1957). These increases are not due to the effects of other risk factors (Jensen & Moore, 1997).

Growth during infancy

There are small deficits in growth from early infancy to 2 years for infants born and living at 1700 m, compared with infants born and living at an altitude of 400 m (Hamill *et al.*, 1977, 1979; McCammon, 1970). These deficits are about 150 g for weight and 0.5 cm for length without a clear age trend or sex difference (Table 3.30). The corresponding deficits in head circumference, which are about 0.3 cm, decrease with age after 12 months.

Growth during childhood and adolescence

Lindsay *et al.* (1994) reported normal growth for United States schoolchildren living at an average altitude of 1211 m, but there is an increased prevalence of lengths or statures ≤5th percentile from birth to 5 years in United States populations living at altitudes above 1500 m (Yip *et al.*, 1988). There are mean deficits of 6 kg for weight and 2.5 cm for stature in "well-off" European-American children aged 10 to 18 years living in Bolivia at an altitude of 3600 m, in comparison with demographically

similar children living at 1500 m (Greksa *et al.*, 1985; Stinson, 1982). When European-American children living in Peru who are aged 3.7 to 15.5 years move from low to high altitudes (3200 m), the rate of growth decreases significantly in males, but this change is less marked in females (Schutte *et al.*, 1983).

MATURITY

Rapidly maturing children tend to have larger than average values for stature and weight beginning in the preschool period. Growth of stature in these children is more rapid than in general groups before and during the pubescent spurt and the ratio sitting height/leg length is increased. Opposite changes occur when pubescence is delayed. The pubescent increase is relatively larger for weight than for stature. Consequently, when pubescence is delayed, the decrease in percentile levels is more marked for weight than for stature. The differences in size associated with the timing of the pubescent spurt increase early in pubescence, but decrease later until they are near zero for young adult statures, but weight remains slightly larger. The literature is considered separately for sexual maturity, age at menarche, skeletal age, and age at PHV.

Table 3.31. *Adjustments (cm) to observed statures for sexual maturity index values[a]*

Sexual maturity index	Age (years)										
	12.0	12.5	13.0	13.5	14.0	14.5	15.0	15.5	16.0	16.5	17.0
Males											
1.0	+4	+5	+7	+10	−	−	−	−	−	−	−
1.5	+2	+3	+5	+8	+13	−	−	−	−	−	−
2.0	0	0	+2	+5	+11	+11	−	−	−	−	−
2.5	−2	−3	0	+3	+8	+8	−	−	−	−	−
3.0	−4	−5	−2	0	+5	+5	+6	−	−	−	−
3.5	−	−8	−5	−3	+3	+3	+4	−	−	−	−
4.0	−	−10	−7	−5	0	0	+2	+4	+6	+7	−
4.5	−	−	−10	−8	−3	−3	0	+2	+3	+4	+3
5.0	−	−	−12	−10	−5	−5	0	0	0	0	0
Females											
1.5	+5	+6	−	−	−	−	−	−	−	−	−
2.0	+4	+5	+6	−	−	−	−	−	−	−	−
2.5	+2	+3	+4	+5	−	−	−	−	−	−	−
3.0	0	+2	+3	+3	+3	−	−	−	−	−	−
3.5	−2	0	+1	+2	+1	+1	+1	+1	−	−	−
4.0	−4	−2	0	0	+1	0	0	0	−	−	−
4.5	−5	−3	−1	−1	0	0	0	0	−	−	−
5.0	−	−5	−3	−2	−1	0	0	0	−	−	−

[a] The sexual maturity index is the average of the numerical values of stages for pubic hair and genitalia in males and of pubic hair and breasts in females.
Source: Wilson, DM *et al.* (1987) *American Journal of Diseases of Children*, 141: 565–570, Copyright (1987) American Medical Association.

Sexual maturity

Within chronological age groups, sexual maturity index (SMI) values, which combine pubic hair stages in each sex with genital stages in males and breast stages in females, are positively related to stature. The values in Table 3.31 from Wilson *et al.* (1987) are intended for use in adjusting recorded statures for unusual levels of sexual maturity for age. Using these data, a male with a value of 4 for SMI at 12.5 years should have 10 cm subtracted from his observed stature before this is compared with reference data. Males aged 13 years with an SMI index of 1.0 are 19 cm shorter than those with an SMI index

of 5.0. These data are not directly applicable to most reference data because Wilson calculated them from the central 80% of the distributions of SMI values in a national United States survey conducted from 1966 to 1970.

Many other reports document similar differences in size associated with variations in sexual maturity. Coy *et al.* (1986) reported data from females aged 13 or 14 years who were placed in mature and immature groups based on the occurrence or absence of both breast stage 5 and menarche. The more mature females were the heavier by about 7 kg at 13 and 14 years, and they were taller by 4 cm at 13 years and 2 cm at 14 years. Douglas & Simpson (1964) grouped males aged 15 years as pre-adolescent, adolescent, or mature using the development of genitalia, pubic and axillary hair, and the pitch of the voice. After adjusting for family size and social class, there were small differences in stature between the groups at 7 years (mature less adolescent = 1.1 cm; adolescent less pre-adolescent = 2.0 cm). The corresponding differences were 2.0 cm and 1.8 cm at 11 years and they were 5.6 cm and 7.1 cm at 15 years. Clinically, it is useful to know the mean total increments in stature after early stages of sexual maturity are reached. Buckler (1995) reported mean growth in stature for males of 28.1 cm after genital stage 2 and of 21.0 cm in females after breast stage 2. Stature at stage 2 of sexual maturation is closely correlated ($r = 0.8$) with adult stature (Buckler & Green, 2001). In males, other sexual maturity data are available from testicular volume, which is related to stature and weight. The mean 6-month increments in stature and weight from 12.5 to 14.0 years are larger in those with testicular volumes ≥ 15 ml than in those with volumes ≤ 5.0 ml. The mean differences at 14 years between these groups are about 16.0 cm for stature and 13.0 kg for weight (Barghini, 1987).

Body mass index and body composition measurements change with sexual maturation. Large increases in %BF occur from stage 1 to stage 2 in each sex, but %BF is consistently larger in females than in males at matching maturity stages and the increase in %BF from stage 4 to stage 5 is much larger in females than in males (Boot et al., 1997a). Males exceed females in FFM at all sexual maturity stages. The increases in FFM from stage 1 to stage 2 are similar in each sex, but the increases after stage 3 are considerably larger in males than in females. Males who are advanced in sexual maturity at 14 years have larger arm muscle areas at ages from 6 months to 4 years (Mills et al., 1986).

Values for total body BMC are larger in males than females by amounts that increase gradually with maturation. In each sex, BMC and BMD of the distal forearm increase linearly with sexual maturity stages except for a decrease in BMD from stage 1 to stage 3 (Magarey et al., 1999). After adjusting for weight, stages of sexual maturation are highly correlated ($r = 0.9$) with lumbar spine BMD (Rubin et al., 1993).

Age at menarche

Age at menarche is inversely related to body size and BMI during childhood (Laitinen et al., 2001). Barghini & Previtera (1975) compared Italians with menarche at 10 years with those who reached menarche at 14 years. Those with an early menarche were taller by 5.1 cm at 6 years, and by 15 cm at 11 years, but by only 2.0 cm at 15 years. In another study, comparisons between those with menarche at 10 years and those with menarche at 16 years show the early menarche group was taller by 5.2 cm at 10 years and by 13.3 cm at 11 years but there were only small differences in stature at 16 and 17 years. In this latter study, weight was larger in the early menarche group by 3.1 kg at 6 years, 16.3 kg at 11 years, and 3.6 kg at 17 years (Barghini, 1987). Similar findings have been reported for English and Japanese children (Boothby et al., 1952; Douglas & Simpson, 1964; Miller et al., 1972; Yoneyama et al., 1988).

Age at menarche is positively correlated ($r = 0.4$) with stature at menarche, but it is negatively correlated ($r = -0.6$) with the interval from menarche to the cessation of growth in stature (Buckler, 1995; Roche, 1989; Sumiya et al., 1999). The balance between these associations limits the effects of age at menarche on adult stature. Wellens et al. (1992) placed young European-American women in four groups depending on age at menarche. Compared with those who reached menarche before 12 years, those with menarche at 12.0 to 12.99 years, 13.0 to 13.99 years, and >13.99 years were taller by 0.4 cm, 1.6 cm, and 2.7 cm, respectively. Some reports that age at menarche has only slight correlations ($r = 0.0$–0.2) with adult stature may be influenced by the lack of data for definitive adult statures (Bielicki & Hauspie, 1994; Szemik, 1980; Tanner et al., 1976).

The median total increase in stature after menarche is about 7.5 cm, but is 10 cm for those with menarche at 10 years and 4 cm for those with menarche at 16 years (Roche & Davila, 1972; Roche, 1989; Singleton *et al.*, 1975). Smaller estimates have been reported by Sumiya *et al.* (1999) who used predicted adult statures, and by Buckler (1995) who used statures at 17 years as the adult values. During childhood and adulthood, BMC and BMD of the distal forearm are negatively related to age at menarche; these measures increase markedly from 1 year before to 1 year after menarche (Fox *et al.*, 1993; Magarey *et al.*, 1999).

Skeletal maturity

Skeletal ages are related to current body size (Low *et al.*, 1964) and to future growth in stature (Bayley & Pinneau, 1952; Garn *et al.*, 1961; Roche *et al.*, 1975b; Tanner *et al.*, 1983a, b). The correlations between skeletal age and either weight or stature increase from $r = 0.4$ at 7 years to $r = 0.7$ at 11 years, but decrease to $r = 0.2$ at 15 years (Beunen *et al.*, 1997a). The differences in stature associated with a change of 2 Z (1.6 to 2.0 years) in skeletal age increase from 4.0 cm at 4 years to 11.0 cm at 15 years in males and to 8.0 cm at 11 years in females (Himes, 1999b). The corresponding differences in weight are 1.0 kg at 2 years increasing to 12 kg at 15 years in males and 10 kg at 12 years in females (Figure 3.9).

Age at peak height velocity

There are negative correlations ($r = -0.3$) between age at PHV and stature during the prepubescent period, but positive correlations with stature at PHV (Bielicki, 1976; Largo *et al.*, 1978; Lindgren, 1978; Tanaka *et al.*, 1988). Age at PHV is negatively correlated ($r = -0.5$ to -0.7) with the rate of growth in stature at PHV and with the total gain in stature during the pubescent spurt (Bourguignon, 1991; Largo *et al.*, 1978; Lindgren, 1978; Tanner *et al.*, 1966b). In Swedish children grouped by whether age at PHV differs from the mean by more than 2 years, males with late PHV are about 2 cm shorter to 9 years than those with an early PHV. These differences increase to 15 years after which they decrease. Females with late PHV are taller than the early PHV group of females by less than 2.0 cm to 4 years after which the difference increases to 8.0 cm at 10 years and 13.1 cm at 12 years (Hägg & Taranger, 1991, 1992). The mean increment in stature from PHV to 18 years is 15.7 cm in males and 14.0 cm from PHV to 17 years in females (Buckler, 1995). Roche & Davila (1972), however, found median increments from PHV to 28 years of 17.8 cm for males and 15.8 cm for females. Age at PHV is positively correlated ($r = 0.5$) with the total growth in stature after 18 years (Hulanicka & Kotlarz, 1983).

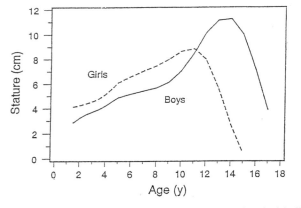 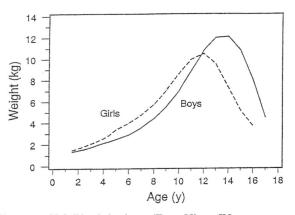

Figure 3.9. Variations in stature and weight associated with differences of 2.0 Z in skeletal age. (From Himes JH, Maturation-related deviations and misclassification of stature and weight in adolescence, *American Journal of Human Biology*, 11:499–504, 1999, with permission of Wiley-Liss, Inc., a subsidiary of John Wiley & Sons, Inc.)

The growth charts of Tanner & Davies (1985), which are derived from cross-sectional data, display large estimated differences in stature for groups with an early PHV (-2 Z) and groups with a late PHV ($+2$ Z). The median estimated statures of the rapidly maturing males approximately match the 75th percentile of the total population to 10 years and then briefly exceed the 95th percentile before returning to the median level. The median statures of the slowly maturing males are at the 25th percentile level to 11 years, and become smaller than the 5th percentile for a brief period before they increase to match the median. There are corresponding differences for females. These patterns of change are similar to those described by Tanaka et al. (1988) and are in agreement with reports that age at PHV does not have important relationships with adult stature (Bielicki & Hauspie, 1994; Georgiadis et al., 1997; Qin et al., 1996; Sumiya et al., 1999). Some reports may be flawed because the last recorded statures may not have been adult values. This problem is exemplified by the work of Hägg & Taranger (1991) who found males with late PHV are 3 cm shorter at 17 years than those with near-average ages at PHV, but are 1.4 cm taller at 18 years and 4 cm taller at 21 years.

4 · Secular changes in growth and maturity

The term *secular changes* refers to increases or decreases in populations over time that are associated with dates of birth. An increase in median weight for 10-year-old females in a particular country from 1950 to 1990 is a secular change. Secular increases are said to be positive, whereas secular decreases are said to be negative. The terms *increase* and *decrease* describe changes in numerical values and not the possible consequences. A positive secular change in weight may have negative health and social consequences. The data needed to analyze secular changes must come from repeated surveys of a national, ethnic, or regional population spanning at least 10 years. These surveys must use the same sampling procedures and techniques of measurement. Migration to or from the study population between the surveys may invalidate the conclusions. Additionally, differences between surveys may be affected by selective changes in response rates, which could be associated with social pressures on those with unusual body size.

Data from repeated surveys of a defined population can be compared in several ways. Some report secular changes in means for measures of size during childhood (Bodzsár, 1998; Jaeger, 1998). This approach may be dictated by the nature of the earlier reports, but the utility of the information provided is limited. Usually, secular changes are much larger in the upper parts of the distributions than in the means; this may have health consequences in the case of weight, BMI, and skinfold thicknesses. Others have reported Z scores relative to a common set of reference values (Vignerová & Bláha, 1998). This is appropriate for normally distributed measures, but not for weight and skinfold thicknesses that are positively skewed.

Large BMI values in children have current and later health consequences. This has led to analyses of secular changes in BMI based on differences between surveys in the prevalence of values larger than stipulated cut-off levels (Troiano *et al.*, 1995). To analyze secular changes in BMI, Flegal & Troiano (2000) compared the total distributions from two surveys using mean difference plots (Cleveland, 1979). These plots compare many selected percentile levels between surveys and thereby show age- and sex-specific differences in general levels and in parts of the distributions. A similar method of analysis was used by Moreno *et al.* (2000) who plotted Z scores, derived from independent reference data, for one survey against corresponding Z scores for a later survey. Many reports of secular changes in age at menarche are restricted to the medians or means. Comparisons of the total distributions of age at menarche by cumulative frequency plots, as made by Bodzsár (1998), are more informative.

In clinical growth assessment, the use of reference data from surveys made before recent secular changes would lead to only small errors during infancy in most developed countries. At older ages, there would be many errors in the selection of those with unusual values (Flegal & Troiano, 2000). The use of old reference data can have marked effects on the results of epidemiological studies. Commonly, it would tend to decrease the reported prevalence of short statures and increase the reported prevalence of large weights.

Reviews of secular changes typically list numerous studies, many of which do not meet the criteria necessary to provide accurate estimates. Despite the limitations of the reported data, none doubt that large positive secular changes have occurred in the growth and maturation of children within many countries during the past century. Chapter 4 describes the secular changes during the past 40 years in the growth and maturation of children in developed countries, and considers the determinants of these changes, and possible effects of secular changes on the relevance of long-term serial growth studies.

CHANGES IN SIZE

Marked secular increases in size during childhood occurred in Belgium from 1960 to 1980–2 (Hauspie et al., 1997; Vercauteren et al., 1998). Stature increased by about 0.5 cm per decade, and weight increased by about 0.1 kg per decade after the ages of 8 years in males and 6 years in females, except for females older than 15 years for whom the increases in weight were about 0.2 kg per decade. In Brussels, stature increased from 1960 to 1980 by 0.7 cm per decade to 16 years in each sex, and by 1.1 cm per decade at ages older than 19 years in males and 16 years in females (Twiesselmann, 1969; Vercauteren & Susanne, 1985). The 95th percentiles for BMI in Flemish children aged 12 to 18 years increased markedly from 1969 to 1993 (Hulens et al., 2001). The increases were 0.9 to 1.9 units in males at 13 and 14 years and 2.7 units in females at 15 years.

Secular changes in the Czech Republic are documented from large representative national surveys made in 1971, 1981, 1991, and 1995–6 (Vignerová & Bláha, 1998). Stature increased in each sex from 1961 to 1991 by about 0.8 cm per decade at 1 to 6 years, 1.0 cm per decade at 7 to 9 years, and 1.5 cm per decade at older ages, except for a decrease of 0.6 cm per decade in females at 18 years (Tables 4.1–4.4). In each sex, the secular increases in stature became larger during pubescence and smaller after pubescence, which may reflect changes in the timing of the pubescent spurt. From 1961 to 1991, the means for weight increased by up to 0.5 kg per decade until 9 years in each sex, and they exceeded 1.0 kg per decade in males from 11 to 18 years and in females from 11 to 14 years (Vignerová & Bláha, 1998).

Data from annual examinations of Danish males aged 7 to 13 years show only small secular changes in BMI from 1950 to 1975, but the prevalence of values >95th percentiles of data recorded from 1930 to 1934 increased by about 10% in data recorded from 1975 to 1985 (Thomsen et al., 1999). Comparisons between two longitudinal studies of Parisian children, one of children born in 1953 and the other of children born in 1984, allow some conclusions about secular changes (Deheeger et al., 1994; Sempé et al., 1979). The secular changes in median statures increase with age from 0.5 cm per decade at 5 years in each sex to 0.9 cm per decade in males and 1.5 cm per decade in females at 8 years. The secular changes in weight are small. The sample born in 1984 has an excess of BMI values >97th percentile of the 1953 sample. Although median skinfold thicknesses do not differ between these samples, median limb circumferences are larger for the 1984 cohort than for the 1953 cohort, suggesting a secular increase in FFM, but not in TBF. These conclusions are tentative because the samples were relatively small and unrepresentative, as in almost all longitudinal studies.

Repeated surveys in Jena (Germany) from 1964 to 1995 show birth weight decreased slightly from 1978 to 1996, but weight for males increased by 0.4 to 0.7 kg per decade at 7 and 8 years and by more than 1.0 kg per decade from 9 to 14 years. In females, the increases in weight exceeded 1.0 kg per decade at almost all ages from 7 to 14 years. Secular changes in skinfold thicknesses were reported as changes in percentages of the means. From 1975 to 1985, thicknesses at the triceps, subscapular, and suprailiac sites decreased by 9% to 25% in males and by 11% to 22% in females. This decade of decreases was succeeded by one of increases that ranged from 12% to 40% in males and 11% to 38% in females. Stature increased by about 1.0 cm per decade in each sex at 7 to 9 years, and by about 2.0 cm per decade after 11 years. From 1975 to 1995, sitting height increased in each sex by more than 0.5 cm per decade

Table 4.1. *Changes per recent decade in mean length and stature (cm) in males*

Age (years)	Czech Republic	Germany (Jena)	Hungary (Budapest)	Japan	The Netherlands	Poland		United Kingdom		United States
						Poznán	Kraków	England	Scotland	
Birth	–	–0.2	–	–0.1	–	–	–	–	–	–
1	+0.7	–	–	+0.8	–0.1	–	–	–	–	–0.1
2	+0.6	–	–	+0.9	0.0	–	–	–	–	0.0
3	+0.7	–	+0.6	+1.2	+0.2	–	–	–	–	0.3
4	+0.8	–	–0.9	+1.3	+0.2	–	–	–	–	–0.6
5	+0.9	–	–0.1	+1.4	+0.5	–	–	+0.4	+1.4	–0.1
6	+0.9	–	–0.9	+1.7	+0.7	–	–	+0.4	+1.4	–0.5
7	+1.1	+0.7	–0.5	+1.7	+0.9	+1.7	–0.4	+0.5	+0.8	+0.2
8	+0.9	+1.0	+0.5	+2.1	+1.0	+1.4	0.0	+0.6	+1.7	+0.3
9	+1.1	+1.2	+0.6	+2.1	+1.1	+0.5	+0.6	+1.6	+1.5	+0.6
10	+1.0	+1.2	+0.8	+2.4	+1.1	+1.0	+1.0	+0.7	+1.6	+0.7
11	+1.2	+2.0	+1.5	+2.7	+1.1	+1.4	0.0	–	–	+0.7
12	+1.5	+1.4	+0.7	+3.2	+1.4	+1.1	+0.9	–	–	+1.1
13	+1.4	+3.2	–0.2	+3.6	+1.8	+2.1	+0.7	–	–	+0.8
14	+1.8	+2.0	+1.5	+3.1	+2.2	+2.7	+0.7	–	–	+0.6
15	+1.7	–	+1.8	+2.2	+1.8	+4.9	0.0	–	–	+0.5
16	+1.7	–	+1.2	+2.0	+1.7	+1.9	+0.8	–	–	+0.7
17	+1.9	–	+0.8	+1.8	+1.6	+1.1	+0.7	–	–	+0.8
18	+1.5	–	+0.3	–	+1.6	+1.3	+0.4	–	–	–0.4
19	–	–	–	–	+1.7	–	–	–	–	–0.6
Survey years	1961–91	1964–95	1984–1994–5	1960–90	1965–97	1972–91	1971–83	1972–94	1972–94	1978–91

Sources: Czech Republic, Vignerová & Bláha, 1998; Germany, Jaeger, 1998; Hungary, Németh & Eiben, 1997; Japan, Takaishi, 1995; The Netherlands, Fredriks *et al.*, 2000*a*; van Wieringen, 1986; Poland, Bielicki & Hulanicka, 1998; England and Scotland, Hughes *et al.*, 1997; United States, Kuczmarski *et al.*, 2000, 2002; Najjar & Rowland, 1987.

Table 4.2. *Changes per recent decade in mean length and stature (cm) in females*

Age (years)	Czech Republic	Germany (Jena)	Hungary (Budapest)	Japan	The Netherlands	Poland		United Kingdom		United States
						Poznán	Kraków	England	Scotland	
Birth	–	–0.2	–	–0.3	–	–	–	–	–	–
1	+0.6	–	–	+0.6	+0.1	–	–	–	–	+0.2
2	+0.5	–	–	+0.9	–0.1	–	–	–	–	–0.1
3	+0.8	–	–	+1.1	+0.1	–	–	–	–	+0.6
4	+0.8	–	–	+1.5	+0.1	–	–	–	–	+0.2
5	+1.0	–	–	+1.5	+0.3	–	–	0.0	+0.9	+0.2
6	+1.1	–	–	+1.8	+0.4	–	–	+0.6	+1.3	–1.4
7	+1.1	+1.3	+0.3	+2.0	+0.6	+1.4	+0.2	+0.8	+1.0	–0.3
8	+1.0	+0.9	+1.0	+2.1	+0.8	+1.1	0.0	+0.6	+1.2	+0.2
9	+1.3	+1.5	+1.7	+2.3	+1.0	+0.9	+0.5	+0.9	+1.6	+0.7
10	+1.3	+1.1	+1.1	+2.5	+1.3	+2.5	+0.3	+0.9	+1.4	+0.4
11	+1.4	+2.0	+1.7	+2.7	+1.6	+0.7	+0.1	–	–	+1.4
12	+1.6	+2.2	+1.3	+2.5	+1.3	+1.9	+0.7	–	–	+0.3
13	+1.6	+1.7	+0.3	+2.2	+1.1	+1.4	+0.7	–	–	+0.2
14	+1.6	+0.3	+0.4	+1.9	+1.1	+2.2	+1.4	–	–	+0.3
15	+1.5	–	+0.5	+1.5	+1.0	+2.8	+0.6	–	–	–0.1
16	+1.3	–	+1.9	+1.4	+1.0	+1.6	+0.1	–	–	–0.2
17	+2.2	–	+2.4	+1.4	+1.0	+1.3	+0.1	–	–	+0.1
18	–0.6	–	+2.1	–	+1.1	+1.3	0.0	–	–	+0.4
19	–	–	–	–	+1.3	–	–	–	–	+0.4
Survey years	1961–91	1964–95	1984–1994/5	1960–90	1965–97	1972–91	1971–83	1972–94	1972–94	1978–91

Sources: Czech Republic, Vignerová & Bláha, 1998; Germany, Jaeger, 1998; Hungary, Németh & Eiben, 1997; Japan, Takaishi, 1995; The Netherlands, Fredriks *et al.*, 2000*a*; van Wieringen, 1986; Poland, Bielicki & Hulanicka, 1998; England and Scotland, Hughes *et al.*, 1997; United States, Kuczmarski *et al.*, 2000, 2002; Najjar & Rowland, 1987.

Table 4.3. *Changes per recent decade in mean weight (kg) in males*

| Age (years) | Czech Republic | Germany (Jena) | Hungary (Budapest) | United Kingdom | | United States |
				England	Scotland	
Birth	–	–	–	–	–	
1	+0.1	–	–	–	–	−0.1
2	+0.3	–	–	–	–	0.0
3	+0.5	–	+0.3	–	–	0.0
4	+0.2	–	+0.2	–	–	0.0
5	+0.3	–	+0.8	+.01	+0.5	+0.1
6	+0.3	–	−0.1	+0.1	+0.6	+0.1
7	+0.4	+0.7	+0.2	+0.3	+0.5	+0.5
8	+0.5	+0.4	+0.4	+0.4	+1.0	+0.9
9	+0.5	+1.1	+1.0	+1.1	+1.3	+1.4
10	+0.6	+1.1	+1.8	+0.9	+1.5	+0.4
11	+0.9	+1.7	+2.5	–	–	+1.0
12	+1.2	+1.4	+2.2	–	–	+2.1
13	+1.2	+3.7	+0.3	–	–	+1.8
14	+1.6	+1.4	+1.0	–	–	+3.6
15	+1.3	–	+0.1	–	–	+2.1
16	+1.3	–	+2.0	–	–	+0.7
17	+1.7	–	+0.7	–	–	+2.7
18	+1.7	–	+1.3	–	–	0.0
19	–	–	–	–	–	–
Survey years	1961–91	1964–95	1984–1994–5	1972–94	1972–94	1978–91

Sources: Czech Republic, Vignerová & Bláha, 1998; Germany, Jaeger, 1998; Hungary, Németh & Eiben, 1997; England and Scotland, Hughes *et al.*, 1997; United States, Kuczmarski *et al.*, 2000, 2002; Najjar & Rowland, 1987.

at ages older than 8 years and by more than 1.0 cm per decade at most ages after 11 years (Jaeger, 1998). The increases in the ratio sitting height/stature from 1975 to 1995 were small in this study, indicating that secular changes were similar in trunk length and leg length (Kromeyer-Hauschild & Jaeger, 2000). Age-matched data for German parents measured during childhood and their like-sex offspring, with a mean difference in birth dates of 33 years, do not show secular changes in head circumference from infancy to 13 years (Brandt, 1989).

Németh & Eiben (1997) reported comparisons between surveys made in 1985 and 1995 of children in Budapest. Both surveys included children aged

3 to 18 years and "as far as possible the same educational institutions." Stature increased between surveys by <1.0 cm except at 14 to 16 years in males and at 8 to 12 years and 16 to 17 years in females. Weight increased by >2.0 kg at 11 to 12 years and at 16 years in males and at 10, 11, 15, and 18 years in females, but weight-for-stature changed only slightly. These secular changes were influenced by migration into Budapest from rural areas where statures are considerably smaller (Bodzsár, 1998).

Some reports of secular changes in Japan use data from annual surveys that each included about 600 000 children. Comparisons between the 1960 and 1990 surveys show length at birth did not

Table 4.4. *Changes per recent decade in mean weight (kg) in females*

Age (years)	Czech Republic	Germany (Jena)	Hungary (Budapest)	United Kingdom		United States
				England	Scotland	
1	0.0	–	–	–	–	0.0
2	+0.1	–	–	–	–	+0.1
3	+0.2	–	−0.8	–	–	0.0
4	+0.2	–	0.0	–	–	+0.4
5	+0.3	–	+0.7	+0.0	+0.4	+0.3
6	+0.4	–	0.0	+0.2	+0.7	+0.2
7	+0.4	+1.2	0.0	+0.5	+0.8	+0.7
8	+0.5	+0.7	+0.9	+0.4	+1.1	+0.9
9	+0.5	+1.2	+1.7	+0.8	+1.5	+1.1
10	+0.5	+0.7	+2.1	+0.9	+1.6	+0.8
11	+0.6	+1.5	+2.2	–	–	+1.0
12	+0.6	+1.1	+0.3	–	–	+1.1
13	+0.9	+1.1	+1.0	–	–	+2.1
14	+0.6	+1.3	+1.8	–	–	+0.7
15	+0.2	–	+2.5	–	–	+0.4
16	+0.0	–	+1.3	–	–	+1.4
17	+0.1	–	+1.7	–	–	+1.2
18	+0.1	–	+2.3	–	–	+1.0
19	–	–	–	–	–	–
Survey years	1961–91	1964–95	1984–1994–5	1972–94	1972–94	1978–91

Sources: Czech Republic, Vignerová & Bláha, 1998; Germany, Jaeger, 1998; Hungary, Németh & Eiben, 1997; England and Scotland, Hughes *et al.*, 1997; United States, Kuczmarski *et al.*, 2000, 2002; Najjar & Rowland, 1987.

change in males, but it decreased by 0.3 cm per decade in females (Takaishi, 1995). Stature increased in each sex by more than 1.0 cm per decade after 3 years and by more than 2.0 cm per decade after 8 years in males and 7 years in females. These increases are mainly due to changes in leg length (Leung *et al.*, 1993; Tanner *et al.*, 1982). Similar findings relating to the large contribution of changes in leg length to secular changes in stature have been reported from Belgium (Vercauteren *et al.*, 1998) and Italy (Cresta *et al.*, 1982; Sanna & Soro, 2000), but secular increases in stature in Hungary and Poland have been due to similar changes in leg and trunk lengths or slightly smaller increases in leg length than trunk length (Bochénska 1978; Bodzsár

& Pápai, 1994; Charzewski & Bielicki, 1978; Górny, 1977; Kromeyer-Hauschild & Jaeger, 2000). Surveys made in 1979 and 1989 in the Chiba Prefecture (Japan) show the percentage prevalence of relative weights >120% increased by 4.4% in males and 2.4% in females at ages 6 to 11 years, but the increases were smaller at 12 to 16 years (Shirai *et al.*, 1990).

Particularly interesting data come from national surveys made in The Netherlands with stratified sampling in 1965, 1979, and 1997 (Fredriks *et al.*, 2000a; Roede & van Wieringen, 1985; van Wieringen *et al.*, 1971). From the first to the third of these surveys, length (stature) increased by only 0.1 cm per decade from 1 to 2 years in males and

Table 4.5. *Changes per recent decade for weight (kg) in males at selected percentile levels*

Age (years)	The Netherlands			United States		
	10th percentile	50th percentile	90th percentile	10th percentile	50th percentile	90th percentile
1	−0.2	−0.2	−0.2	−0.1	−0.1	0.0
2	+0.0	−0.1	−0.3	+0.1	0.0	−0.1
3	−0.1	−0.2	−0.2	−0.1	0.0	0.0
4	−0.1	−0.1	−0.1	0.0	−0.1	−0.1
5	0.0	0.0	0.0	−0.1	+0.2	0.0
6	0.0	+0.2	0.0	−0.4	0.0	+0.6
7	+0.1	+0.2	0.0	+0.1	+0.3	+0.9
8	+0.3	+0.2	0.0	+0.2	+0.2	+3.0
9	+0.4	+0.3	+0.3	+0.3	+0.8	+2.9
10	+0.6	+0.6	+0.7	−0.1	+0.6	+1.4
11	+0.6	+0.7	+1.2	+1.3	+1.0	+0.5
12	+0.4	+0.5	+1.3	+1.6	+2.4	+1.9
13	+0.4	+0.2	+0.5	+0.8	+1.7	+2.3
14	+0.8	+0.7	+1.5	+1.3	+1.9	+2.3
15	+1.2	+1.6	+1.2	+0.3	+1.5	+5.5
16	+1.0	+1.4	+1.1	0.1	+0.6	+2.8
17	+0.7	+1.1	+1.1	+1.7	+1.9	+4.9
18	+1.1	+1.2	+1.4	0.0	−1.2	+3.4
19	+1.6	+1.4	−0.4	–	–	–
Survey years		1979–97			1978–91	

Sources: The Netherlands, Fredriks *et al.*, 2000a; van Wieringen, 1986; United States, Kuczmarski *et al.*, 2000, 2002; Najjar & Rowland, 1987.

from 1 to 4 years in females. The increases became larger with age and exceeded 1.0 cm per decade after 7 years in males and 8 years in females. The increases per decade in stature from the first to the second survey were larger than those from the second to the third survey indicating that the secular increase in stature is slowing in The Netherlands. Weight increased from 1980 to 1997 by <0.5 kg per decade until 9 years in each sex, but by >1.0 kg per decade in males older than 14 years. From 1979 to 1997, changes in the 10th, 50th, and 90th percentile levels for weight were similar to 10 years in males and 8 years in females; at older ages the secular changes in weight were largest at the 90th percentile level and smallest at the 10th percentile level (Tables 4.5 and 4.6).

In Norway, birth weights for married mothers did not change from 1960 to 1980, but the mean for unmarried mothers increased about 180 g. This may reflect increased social support for unmarried mothers. The secular increase in stature from 1960 to 1975 was about 0.8 cm per decade at ages from 8 years to 18 years (Brundtland *et al.*, 1980).

Data from Poland describe secular changes in Poznán from 1972 to 1991 and in Kraków from 1971 to 1983 (Bielicki & Hulanicka, 1998). In Poznán, stature for each sex increased by at least 1.0 cm per decade at most ages from 7 to 18 years, but in Kraków most of the increases were 0.5 cm per decade or less. The stature of male conscripts in Poland increased from 1965 to 1995 by 1.8 cm per decade (Bielicki *et al.*, 2000). This increase was

Table 4.6. *Changes per recent decade for weight (kg) in females at selected percentile levels*

Age (yr)	The Netherlands			United States		
	10th percentile	50th percentile	90th percentile	10th percentile	50th percentile	90th percentile
1	−0.2	−0.3	−0.3	0.0	0.0	−0.1
2	−0.1	−0.1	−0.2	0.0	+0.1	+0.2
3	0.0	0.0	0.0	+0.3	+0.2	+0.1
4	0.0	0.0	+0.1	+0.1	+0.3	+0.2
5	+0.1	0.0	+0.1	0.0	+0.3	0.0
6	0.0	+0.1	+0.3	0.0	−0.1	+0.3
7	+0.1	+0.1	+0.5	+0.1	+0.1	+0.8
8	+0.4	+0.3	+0.8	+0.4	+0.4	+2.0
9	+0.6	+0.3	+1.0	0.0	+0.8	+1.8
10	+0.6	+0.5	+1.3	+0.2	+0.7	+2.7
11	+0.6	+0.9	+1.6	+0.8	+1.3	+0.2
12	+0.6	+1.3	+1.4	+0.4	+0.5	+1.4
13	+0.7	+1.1	+1.0	+0.9	+2.1	+4.6
14	+0.7	+0.9	+0.8	+1.3	+0.4	+2.8
15	+0.2	+0.7	+0.8	+0.6	+0.9	+1.8
16	+0.5	+0.5	+0.5	+0.8	+0.2	+2.8
17	+0.6	+0.3	+0.2	+0.4	+1.1	+3.3
18	+0.3	+0.1	+0.3	+0.1	+1.6	+6.1
19	0.0	−0.5	−0.6	−	−	−
Survey years		1965–80			1978–91	

Sources: The Netherlands, Fredriks *et al.*, 2000*a*; van Wieringen, 1986; United States, Kuczmarski *et al.*, 2000, 2002; Najjar & Rowland, 1987.

larger in those from rural areas which is consistent with reports that secular increases in the statures of children have been larger in Polish villages than in cities and towns (Hulanicka *et al.*, 1990; Waliszko *et al.*, 1980). During the past few decades, the statures of conscripts became more variable with an increased prevalence of statures that were larger or smaller than the mean by more than 12 cm. BMI also became more variable with an increased prevalence of conscripts with BMI >25 kg/m^2 or <18 kg/m^2 (Bielicki *et al.*, 2000).

Moreno and colleagues (2000) analyzed data obtained at annual examinations made from 1985 to 1995 of children aged 6 to 14 years in Aragon (Spain). The prevalence of overweight, defined as a BMI value >95th percentile of Hamill and colleagues (1977), increased by 14% in males and 17% in females at 6 to 7 years, but the increase at 13 to 14 years was only 4.0% in males and near zero in females.

Surveys of Stockholm public school children in 1960 and 1970 show small secular changes (Cernerud & Lindgren, 1991). During this decade, the statures of males increased by 0.2 cm at 7 years and by 0.9 cm at 13 years. The corresponding increases for females were 0.5 and 0.8 cm. A later survey of children aged 7 years showed very small changes in stature and BMI in each sex from 1960 to 1990 (Cernerud, 1993). Other Swedish surveys show stature at 8 to 15 years increased by about 1.0 cm from 1976 to 1986, but the increase had almost ceased by 1991 (Lindgren, 1976; Lindgren *et al.*, 1995; Lindgren & Strandell, 1986). The means for BMI in Swedish children increased from 1976 to 1986 by about 0.5 units at 10 to 13 years and by about 1.0 units at 14 and 15 years (Lindgren *et al.*, 1995). In Swedish conscripts there were increases from 1976 to 1986 of 1.3 cm in stature, 5.1 kg in weight, and 1.3 units in BMI (Lindgren, 1998; Lindgren *et al.*, 1995; Taranger, 1984). The secular increases in Swedish conscripts may have ceased about 1984 for stature, but continued for weight and BMI (Lindgren, 1998).

Marked secular changes occurred from 1972 to 1994 in the stature and weight of children of European ancestry living in the United Kingdom. From 5 to 7 years, the secular increases for English children were about 0.5 cm per decade and those for Scottish children were about 1.0 cm per decade. The increase in stature for children aged 8 through 10 years was about 0.8 cm per decade for English children and 1.5 cm per decade for Scottish children. Weight increased by about 0.3 kg per decade for English children and by 0.6 kg per decade for Scottish children at 5 through 8 years. From 8 through 10 years, the secular increases in weight were about 1.0 kg per decade in each group. Weight-for-stature changed little in English and Scottish children aged 5 through 8 years, but there were increases in those aged 9 and 10 years that were larger for Scottish children than for English children (Hughes *et al.*, 1997). These positive secular changes in size after infancy are in contrast to suggestive evidence of negative secular changes in skinfold thicknesses for English infants (Hutchinson-Smith, 1973; Tanner & Whitehouse, 1975; Whitehead & Paul, 1984; Whitelaw, 1977).

Surveys of inner-city children in England during 1983 and 1994 show stature increased by 1.5 to 1.7 cm per decade for Afro-Caribbean, Urdū, Punjabi and Gujarati-speaking Indian males, but by 0.8 cm per decade in other Indian groups and by 1.2 cm per decade in European-English children. For females, the largest increases occurred in "other Indian groups" (3.0 cm per decade) and those for European-English children (1.6 cm per decade)

exceeded those for Afro-Caribbean children (1.1 cm per decade). The increases for weight in males and females were largest in the Indian groups. Among males, they were smallest for the European-English children, but among females, they were smallest for the Afro-Caribbean children (Chinn *et al.*, 1998).

Freeman *et al.* (1995) constructed growth charts for the United Kingdom using data collected from 1978 to 1990 in three national samples and two regional samples. The differences between their percentiles and those of Tanner *et al.* (1966*a*, *b*), who used data collected from 1954 to 1965 in or near London, cannot be expressed per decade because the years of data collection varied within both sets of data. The midpoints of the timing of data collection are about 1960 for the Tanner data and 1984 for the Freeman data. Differences between these sets of data should not be interpreted as measures of secular changes although they may be suggestive. They are important because they indicate the need to use current reference data when there is evidence of secular changes. Comparisons between the Freeman and Tanner charts suggest that the medians for stature increased from 1960 to 1984 by 1.0 to 1.5 cm at 2 to 11 years, but by 0.8 cm at 16 years in each sex. The smaller increase at 16 years, compared with 11 years, suggests that the pubescent spurt may have occurred earlier in the Freeman samples. Somewhat surprisingly, the increases in stature were larger at the 3rd percentile level than at the 97th percentile level. The medians for weight changed only slightly at almost all ages. The increases in weight were larger at the 3rd percentile than the 97th percentile for males to 11 years, but not at older ages. In females aged 11 years, the 3rd percentile level increased by 1.3 kg, but the 97th percentile decreased by 2.1 kg. The pattern of change was different for values at 18 years for which the 3rd percentile decreased by 1.3 kg, but the 97th percentile increased by 1.3 kg. Variations in the nature of the samples that were combined by each of these authors and errors in smoothing could have contributed to the differences between the Tanner and Freeman percentiles.

In the United States, median birth weights did not change from 1979 to 1996 (Ventura *et al.*, 1998). This may be due, in part, to an increased prevalence of multiple births (Paneth, 1995; Ventura *et al.*, 1998). The prevalence of LBW infants increased slightly for European Americans, but not for African Americans during this period (Ventura *et al.*, 1998). Migration could be partly responsible. National United States surveys show the prevalence of overweight, defined as >95th percentile of weight-for-length (stature) from Hamill and colleagues (1977), increased from 1971 to 1994 for females aged 4 or 5 years, but not at younger ages. The prevalences for males aged 4 or 5 years increased for African Americans and Mexican Americans (Ogden *et al.*, 1997). National United States surveys with midyears of data collection in 1978 and 1991 show stature increased by less than 1.0 cm per decade at almost all ages (Kuczmarski *et al.*, 2000, 2002; Najjar & Rowland, 1987) (Tables 4.1 and 4.2). The small size of this increase may be attributed, at least in part, to effects of migration. During the same period, weight increased only slightly in males to 10 years, but increased by 0.5 kg per decade from 6 through 10 years in females (Tables 4.3 and 4.4). After 10 years, the increase was about 1.4 kg per decade for males and 0.8 kg per decade for females. The increases in the 10th percentile levels for weight were less than 0.5 kg per decade before 9 years in males and 8 years in females; at most older ages, they were 0.6 to 0.8 kg per decade. The 90th percentile levels increased by <0.3 kg per decade to 9 years in males and 7 years in females, but by >1.0 kg per decade after 11 years in males and from 9 to 13 years in females (Tables 4.5 and 4.6).

The prevalence of overweight and obesity, determined from national United States surveys, increased from 1963–70 to 1976–80 and from then to 1988–91 (Troiano *et al.*, 1995). In this analysis, overweight was defined as a BMI value larger than the 85th percentile from the 1963–70 survey, and obesity was defined as a BMI larger than the 95th percentile from that survey. The prevalence of overweight, using the midyears of surveys, increased from 1966 to 1990 by about 40% for overweight and

by 100% for obesity at 6 to 17 years in each sex. The increases in the prevalence of overweight and obesity were much larger for African Americans than for European Americans.

In a further analysis, Flegal & Troiano (2000) compared BMI data from surveys made from 1963 to 1970 with data from a survey made from 1988 to 1994. The medians increased by 0.3 to 0.5 units per decade in each sex at 6 to 10 years and by slightly larger amounts at 11 to 17 years. The increases were generally larger in males than in females. Using mean-difference plots, they demonstrated that the changes were small except in the parts of the distributions above the 90th percentiles. The only exception occurred in females at 15 to 16 years for which the increases at levels from the 50th to the 90th percentile were larger than those for levels above the 90th percentile. The findings of Flegal & Troiano are consistent with a model in which the distribution of BMI in young adults represents a combination of a *normal* population component and an *overweight* population component (Moll *et al.*, 1991; Price *et al.*, 1990, 1991; Sørensen *et al.*, 1989). The overweight component could be a group with an accentuated response to the environment that may be mediated by genotype, or genetic–environment interactions, or this group may be exposed to a different environment.

Gortmaker *et al.* (1987) considered triceps skinfold thicknesses >85th percentile of the 1963–5 United States survey indicated obesity and values >95th percentile indicated super-obesity. Analyzing data from national United States surveys, they found the prevalence of obesity increased from 1963–5 to 1976–80 by 61% in males and 46% in females at 6 to 11 years and by 18% in males and 58% in females at 12 to 17 years. The prevalence of super-obesity increased by 122% in males and 70% in females at 6 to 11 years and by 41% in males and 87% in females at 12 to 17 years.

Findings from national United States surveys are in general agreement with those from regional studies. The mean ponderal index for children aged 10 years in the Bogalusa (Louisiana) Heart Study increased by 0.8 units per decade between

1973 and 1988 (Nicklas *et al.*, 1993). In a later analysis from this study, Freedman *et al.* (2000) found stature increased from 1973 to 1992 by 0.7 cm per decade in those aged 15 to 17 years. These increases, which were similar in each sex, did not differ between African Americans and European Americans. Surveys in 1956–65, 1977, and 1994–5 of African Americans aged 11 to 15 years in low socioeconomic areas of Philadelphia show marked secular increases in the prevalence of values for BMI and triceps and subscapular skinfold thicknesses >85th and >95th national percentiles. These increases were larger for females than males (Gordon-Larsen *et al.*, 1997; Krogman, 1970). In Navajo Indians living in Arizona, stature at 6 to 13 years increased from 1955 to 1997 by about 1.8 cm per decade in males and 2.5 cm per decade in females. The increases in weight during the same period were 1.0 kg per decade at 7 years and 3.5 kg per decade at 12 years in males; the increases in females were 2.0 kg per decade at 7 years and 3.0 kg per decade at 12 years. The increases in BMI were 0.5 to 1.0 units per decade in each sex (Eisenmann *et al.*, 2000).

CHANGES IN MATURITY

Reports of secular changes in the timing of PHV are scarce because definition of this age requires a considerable amount of serial data for each individual. In Belgian females, PHV occurred slightly earlier in 1980 than in 1960 (Susanne & Vercauteren, 1997). In females attending an upper middle class school in Japan, age at PHV and the rate of growth in stature at PHV did not change from 1960 to 1970, but the stature attained at PHV increased by 1.2 cm (Takaishi & Kikuta, 1989). Some analyses of national Japanese data use the mean ages at which the largest increments between annual means of stature and weight occurred as indices of ages at PHV and PWV. These ages decreased by about 0.2 years per decade in males and 0.1 years per decade in females from 1960 to 1980 (Ali *et al.*, 2000; Ali & Ohtsuki, 2000; Ji *et al.*, 1995; Matsumoto, 1982). In the Fels Longitudinal Study, age at PHV does not differ

within parent–offspring pairs of like sex (Bock & Sykes, 1989).

National surveys show age at menarche continues to decrease slowly in Hungary, The Netherlands, and Poland (Bodzsár, 1998; Fredriks *et al.*, 2000a; Hulanicka & Waliszko, 1991; Laska-Mierzejewska & Luczak, 1993), but decreases have ceased in Belgium, the Czech Republic, Italy (except Sardinia), and Norway (Brundtland *et al.*, 1980; Floris & Sanna, 1998; Vercauteren *et al.*, 1998; Vignerová & Bláha, 1998). Regional studies in Germany, Spain, and Sweden show an absence of recent changes (Gonzáles Apraiz & Rebato, 1995; Lindgren, 1998; Richter, 1981), while other regional studies in England and Germany show increases in age at menarche (Barth *et al.*, 1984; Dann & Roberts, 1984; Richter, 1989). A study of urban children in Upper Silesia (Poland) shows increases in age at menarche for those with fathers who are professionals, engineers, small businessmen, or unskilled workers, but decreases for those with fathers who are managers, police, or machine-tool operators (Hulanicka *et al.*, 1994). Apparently socioeconomic conditions changed differentially among these groups. Information about secular changes in the distributions of ages at menarche is limited. Variability of ages at menarche has decreased in France, in an industrial area of Poland, and in the Basque region of Spain (Bodzsár, 1998; Ducros & Pasquet, 1978; Gonzáles Apraiz & Rebato, 1995). In Belgium, the 10th and 50th percentiles for age at menarche did not change from 1950 to 1970, but the 90th percentile became earlier by about 1 year (Vercauteren & Susanne, 1985). In Germany also, there has been a reduction in the prevalence of late ages at menarche (Richter & Kern, 1980; Winter, 1962).

In The Netherlands, percentile levels for age at menarche changed little from 1965 to 1997, but most stages of secondary sexual maturity were reached 0.1 or 0.2 years per decade earlier in 1997 than in 1965 (Fredriks *et al.*, 2000a; van Wieringen *et al.*, 1971) (Table 4.7). The 90th percentile for the timing of genital stage 5, however, became later by 0.5 years per decade, while the 10th and 50th percentile ages

Table 4.7. *Changes (year per decade) in timing of stages of sexual maturation in The Netherlands (1965–97)*

	10th percentile	50th percentile	90th percentile
Males			
Pubic hair 2	–	0.0	−0.2
Pubic hair 3	−0.1	−0.2	−0.2
Pubic hair 4	−0.2	−0.2	−0.2
Pubic hair 5	−0.3	−0.3	−0.1
Genitalia 2	–	+0.1	−0.3
Genitalia 3	0.0	−0.1	−0.1
Genitalia 4	−0.1	−0.1	−0.1
Genitalia 5	−0.2	−0.2	+0.5
Females			
Pubic hair 2	0.0	−0.1	−0.2
Pubic hair 3	+0.1	−0.1	−0.2
Pubic hair 4	0.0	−0.2	−0.3
Pubic hair 5	−0.3	−0.4	0.0
Genitalia 2	0.0	−0.1	−0.1
Genitalia 3	0.0	−0.1	−0.2
Genitalia 4	+0.1	−0.2	−0.3
Genitalia 5	−0.1	−0.3	–

Source: Fredriks *et al.* (2000*a*).

became earlier by 0.2 years per decade. This implies a large change in the variance of maturational timing for genital stage 5 that could be partly due to differences between surveys in methods of assessment or variations in sampling at about 15 years when stage 5 occurs. A testicular volume of 4 ml was reached 0.5 years earlier in the 1965 Dutch survey than in the 1997 survey, but the median ages at which volumes of 8 ml and 12 ml were reached do not differ between these surveys (Fredriks *et al.*, 2000*a*). In Sweden, secondary sexual maturation has become slightly more rapid in recent decades with the exception of genital stage 5 (Karlberg *et al.*, 1976; Lindgren, 1996). Secular changes have not occurred recently in rates of skeletal maturation for Tokyo children (Matsuoka *et al.*, 1999*a*).

RECENT SLOWING OF SECULAR CHANGES

Secular increases in stature and decreases in age at menarche may cease because the populations studied have reached their genetic potentials or the environments are no longer changing in ways that lead to these secular changes. Secular increases in stature have slowed, but not stopped, in Belgium, the Czech Republic, Hungary, Japan, The Netherlands, Sweden, and the United States (Cernerud, 1993; Fredriks *et al.*, 2000*a*; Kuczmarski *et al.*, 2000, 2002; Németh & Eiben, 1997; Takaishi, 1995; Thomsen *et al.*, 1999; Vercauteren & Susanne, 1985; Vignerová & Bláha, 1998). The decrease in age at menarche has slowed or ceased in Belgium, the Czech Republic, Germany (Jena), Hungary, Norway, Sweden, The Netherlands, and Poland (Bodzsár & Susanne, 1998; Fredriks *et al.*, 2000*a*; Hulanicka & Waliszko, 1991; Jaeger, 1998; Laska-Mierzejewska & Luczak, 1993; Malina, 1990). The age at PHV may have stopped decreasing in Japan about 1975 (Murata, 1993).

The situation is different for body fatness. Weight, weight-for-stature, and BMI continue to increase in Belgium, Germany, The Netherlands, Spain, the United Kingdom, and the United States. Weight-for-stature has decreased in Sweden and is near stable in the Czech Republic and Germany (Lindgren, 1998; Vignerová & Bláha, 1998). It must be concluded that the genetic potential for these indices of body fatness has not been reached, and that the environments of most developed countries continue to change in ways that favor increases in body fatness.

DETERMINANTS OF SECULAR CHANGES

The secular changes in size and maturational timing of children during the past four decades reflect alterations in environmental influences relating to nutrition, physical activity, socioeconomic status, and health. It is difficult to separate the effects of these influences because they are commonly interrelated.

The almost complete absence of secular changes in weight from birth to 2 years is unexplained. Increases in the prevalence of preterm and first-born births may balance any effects of increased pre-pregnancy weight on size during infancy and migration may have affected the results of some studies (Alberman et al., 1991; Kramer et al., 1998; Rosenberg, 1988). Secular changes in weight become evident at about 2 years and increase from then until after pubescence when they decrease. This pattern suggests that alterations in the determinants of secular changes, or the responses to them, are age-dependent and may be influenced by alterations in the timing of the pubescent spurt.

The nutritional influences on secular changes concern dietary intakes and reserves and, in regard to body fatness, the emphasis is on energy balance. Weight and BMI have increased in Japan during the past four decades, although the energy intakes of children have decreased or remained the same (Murata, 2000). The secular increases in the growth of Japanese children may be related to increases in the intakes of meat and milk and a reduced intake of rice (Takahashi, 1984). Protein intake at 2 years, as a percentage of energy intake, is positively correlated ($r = 0.2$) with BMI and subscapular skinfold thickness at 8 years, after adjusting for parental BMI, and it is negatively associated with age at rebound in BMI suggesting that it may be a risk factor for overweight in adulthood (Rolland-Cachera et al., 1995). Animal protein intakes have increased for children in Japan, but have decreased or remained the same in the United States (Albertson et al., 1992; Murata, 2000; Nicklas et al., 1993; Stephen & Wald, 1990).

A large part of the secular changes in weight and BMI may be attributable to reductions in physical activity. Regular childhood physical activity has decreased in the United States and Japan (Committee on the Surveillance Project of the Condition of Children's Health, 1998; Ross & Pate, 1987). The mean levels of physical activity decrease during adolescence in each sex (Bradley et al., 2000; Janz et al., 1992; Telama & Yang, 2000; van Mechelen et al., 2000). These decreases may now occur at younger ages than in the past due to the acceleration of maturation.

Epidemiological studies of physical activity are based on reported data, which may make the observed relationships weaker than the actual ones. In cross-sectional data for Spanish children, levels of physical activity are not related to BMI, but they are related to skinfold thicknesses in females (Sarria et al., 1987). National data for United States children show weak associations of physical activity with skinfold thicknesses and that increased physical activity is associated with larger BMI values in males, but not females, perhaps due to self-selection (Andersen et al., 1998). A controlled study in which children were evaluated before and after a lengthy program of physical activity showed increases in muscle mass, but only small changes in weight, BMI, and skinfold thicknesses (Beunen et al., 1992). Others have found that high-intensity physical activity three to five times each week reduces the risk of overweight in males, but not females (McMurray et al., 2000).

Increased television viewing by children may be a determinant of the secular increase in body fatness. Television viewing could have such an effect by reductions in physical activity and metabolic rates and by increases in the intakes of high-calorie snacks while children watch television (Klesges et al., 1993; Murray & Kippax, 1978; Ross & Pate, 1987; Tucker, 1986; Williams & Handford 1986). The number of hours of television viewing per day by children has near zero correlations ($r = 0.01$–0.04) with weight, BMI, and skinfold thicknesses (Andersen et al., 1998; Gortmaker et al., 1996; Guillaume et al., 1997). Nevertheless, data from a national United States survey made in 1988–94 of children aged 6 to 16 years show the risk of having BMI >85th percentile is higher for those watching television 4 or more hours per day than for those watching for 1 hour or less (Epstein et al., 2000).

There have been a few serial studies. Du Rant and colleagues (1994) enrolled children aged 3 to 4 years and studied them for 3 years. The time spent viewing television, based on observations, was not correlated with BMI or the sum of skinfolds. In females aged 12 years at baseline and examined 7, 12, and 24 months later, hours of television viewing on weekdays, which is 60% of the total weekly viewing (Shannon *et al.*, 1991), is not significantly correlated with changes in BMI or triceps skinfold thickness or with changes in the level of physical fitness (Robinson *et al.*, 1993). Any effects of television viewing on body fatness may be affected by socioeconomic differences. Shannon and colleagues (1991) found the time spent viewing television is correlated with skinfold thicknesses only in "less affluent districts," perhaps because these children eat more while they watch television.

The construct socioeconomic status is commonly indexed by combinations of family income, that may be adjusted for the number in the family, and the occupations and educational levels of the parents (Padez & Johnston, 1999; Pařízková & Berdychová, 1977). Socioeconomic status is correlated positively with stature, but negatively with weight and BMI in developed countries (Rolland-Cachera & Bellisle, 1986; Schaefer *et al.*, 1998). Therefore, one would expect improvements in SES for a nation would be associated with increased stature, but decreased weight and BMI. This is not commonly observed, perhaps because improved standards of living for a country are associated with decreases in the prevalence of marked poverty and changes in the cut–off levels for low SES.

Secular changes are generally greater for groups with large deficits in size or maturational timing than for those with small deficits. This pattern is reflected in the secular decreases in urban–rural differences in Austria, the Czech Republic, Poland, Portugal, and Spain (Bielicki & Waliszko, 1991; Bláha & Vignerová, 1999; Fernandez *et al.*, 1994; Padez & Johnston, 1999; Prado, 1984; Weber *et al.*, 1995). These differential secular changes may reflect larger environmental changes in rural than in urban areas, but those in stature may occur because urban populations are near their genetic potential.

Alterations in SES, with associated differences in diet, energy expenditure, and levels of health care, may have influenced secular changes (Bielicki

& Waliszko, 1991; Bielicki *et al.*, 2000; Facchini & Russo, 1982; Takamura *et al.*, 1988). Some have estimated the adjusted effects of the components of socioeconomic status. In Dutch data, Fredriks *et al.* (2000*a*) found differences in mean statures dependent on region (north vs. south, +0.4 SD), family size (1 vs. ≥4, +0.2 SD), children's educational level (higher secondary vs. special, +0.2 SD), and parental educational level (higher secondary vs. primary, +0.2 SD). These authors suggested that the influence of parental education may be associated with differences in the use of alcohol and tobacco and the consumption of fruit and vegetables. The effects of SES on stature have become smaller in recent Dutch surveys, which would be expected for a population that is approaching its genetic potential for growth in stature (Fredriks *et al.*, 2000*a*; Roede & van Wieringen, 1985). Similarly the effect of paternal educational level on the statures of offspring has decreased in Belgium (Bodzsár & Susanne, 1998; Vercauteren, 1993). In young Polish men, the effect of paternal education on BMI has become non-significant, but remains significant for stature (Bielicki *et al.*, 2000). Secular changes in stature have been larger in lower socioeconomic groups, therefore the differences between high and low socioeconomic groups have decreased in Belgium, England, Poland, and Sweden (Bielicki & Waliszko, 1991; Hauspie *et al.*, 1997; Hulanicka & Waliszko, 1991; Lindgren, 1994; Rona & Chinn, 1986; Vercauteren, 1993; Weber *et al.*, 1995). Child health has improved due to more widespread access to clean drinking water, vaccination, and preventive health care, and the more efficient disposal of waste with a reduction in the prevalence of chronic and acute illnesses. These changes may have reduced the prevalence of small weights and statures (Burgemeijer *et al.*, 1998). Family income is significantly related to length (stature) in Polish children. These effects increase from 1 to 6 years and then change little as older childhood ages are considered (Hauspie *et al.*, 1996). Per capita family income and paternal occupation are highly correlated ($r = 0.8$) with stature of offspring in France (Chamla, 1983), but not in Sweden (Lindgren, 1976).

Secular increases in body fatness during childhood may be influenced by changes in some family characteristics. Secular trends toward higher education, smaller family sizes and maternal employment could contribute to the secular increases in body fatness. Fredriks *et al.* (2000*b*) found adjusted effects on childhood BMI for parental educational level (higher secondary vs. primary, +0.3 SD), family size (1 vs. ≥4, +0.2 SD), and maternal employment (yes vs. no, +0.05 SD). Rates of maternal employment have increased; this may be associated with *increased* physical activity and *decreased* television viewing by children (Guillaume *et al.*, 1997). The cause–effect relationships are, however, uncertain.

The determinants of the secular changes in age at menarche are poorly understood. In Hungary, the secular decreases in age at menarche have been more marked in large cities than in medium-sized and small cities, which is the reverse of what has occurred in body size (Bodzsár, 1998). In a Polish study, age at menarche increased from 1981 to 1991 for daughters of managers and police, but decreased for daughters of small businessmen, professionals, unskilled workers, and coal miners (Bielicki & Hulanicka, 1998). Effects of family size on age at menarche are small in Hungary (Bodzsár, 1998), but age at menarche in France is 0.5 years younger for only children than for those in families with six or more siblings (Ducros & Pasquet, 1978). These findings are relevant to secular changes in the timing of menarche because families with one child have become more common.

SECULAR CHANGES AND LONG-TERM SERIAL GROWTH STUDIES

Secular changes could affect the usefulness of long-term serial growth studies. Consequently, any such changes must be documented within long-term studies, if the designs allow, and in the populations to which the findings will be applied. The effects of secular changes on the applicability of findings from long-term serial studies may not be serious for the types of analyses that should be made of long-term serial data. Furthermore, the slowing or cessation of secular changes in stature within developed countries may reduce this source of bias for future studies, but potential biases remain in regard to weight, BMI, and skinfold thicknesses.

Some serial growth studies include many examinations during childhood that continue during adulthood. These studies allow descriptions of long-term growth patterns and of relationships between parameters of functions fitted to serial data for individual children, or values at defined points on growth curves, to size in adulthood (Guo *et al.*, 1994, 1997*b*; Kouchi *et al.*, 1985*a*, *b*; Nieto *et al.*, 1992; Rolland-Cachera *et al.*, 1989). Other studies have applied follow-up designs that relate status data for size in childhood to health in adulthood. For example, Must and colleagues (1992) selected growth study participants who had BMI values >75th percentile from a national United States survey (Cronk & Roche, 1982) at two or more annual examinations during pubescence and a group with small BMI values during the pubescent age range. The morbidity and mortality of these groups were compared at a mean age of 73 years. Because the data were from individuals born about 1920, secular changes may have influenced the applicability of the results. Present-day children with small or large BMI values may not be subject to the health consequences indicated by these data.

Most long-term serial studies enrolled a single cohort. Notable examples are the European studies that began in 1954 and were coordinated by the Centre International de l'Enfance (Falkner *et al.*, 1980). Because sampling was not random, these studies cannot provide cross-sectional reference data from representative groups. The utility of reference data developed from such studies may be reduced further by secular changes. This statement is not a criticism of these studies because they were not designed to provide reference data. The aim was to investigate patterns of change in size and relationships among variables measured at various ages. They remain an important, and in some cases, a unique source of data for such analyses.

A few long-term serial growth studies have enrolled participants over many years. The best example is the Fels Longitudinal Study (Kettering, Ohio) in which about 15 infants were enrolled at birth each year from 1929 to the present (Roche, 1992). The applicability of data from the Fels Study could be reduced by secular changes within the study or in general populations to which the findings may be applied. Secular changes within the Fels Study must be considered before data from annual groups are pooled for analyses. These changes can be estimated by regressions on year of birth and (year of birth)^2 and by comparisons among groups defined by decade of birth. Some analyses of Fels data have shown only small secular changes (Byard & Roche, 1984; Kouchi *et al.*, 1985*a*, *b*; Roche *et al.*, 1986; Siervogel *et al.*, 1991), but there have been significant recent increases in weight and BMI that are predominantly due to increases in FFM rather than TBF, except for females older than 14 years in whom the changes are mainly due to increases in TBF (Maynard *et al.*, 2001).

Secular changes in growth patterns have been small in the Fels Study, except for weight after 10 years in each sex and BMI after 8 years in females (Byard & Roche, 1984; S. Guo, personal communication; Kouchi *et al.*, 1985*a*, *b*). Secular changes have not occurred in age-to-age correlations (tracking) for size, or in sibling correlations for size at an age in other studies (Susanne, 1980; Thomson, 1955; Wright *et al.*, 1994). Methods still work well that were developed decades ago in long-term serial studies to predict adult stature from childhood variables (Roche *et al.*, 1975*b*; Tanner *et al.*, 1983*a*) and to describe serial data using mathematical models (Bayley & Pinneau, 1952; Jenss & Bayley, 1937; Preece & Baines, 1978). Secular changes in the general population could reduce the applicability of

reference data for increments or patterns of change that were derived from data for groups enrolled many years ago. Data for secular changes in increments and patterns of change are scarce, but the need to continually update reference data for them is shown by the well-established secular changes in status values.

A study may be designed to reduce or eliminate secular trends within the study. Such a design was applied in the Nymegan (Netherlands) Growth Study by Prahl-Andersen and colleagues (1979) who included six cohorts that differed in birth dates, ages at enrollment, and ages at examinations. Because the groups in the Nymegan study were born no more than 6 years apart, inter-cohort differences are more likely to be due to sampling variations than to secular changes. An alternative design, called "shingling" or "laddering," may help avoid secular changes. In this approach, participants are enrolled in the same year at, for example, birth, 4, 8, 12, and 16 years and measured serially for 4 years. A complete description of growth from birth to 20 years might result if the groups were consistent with regard to sampling, enrollment, and examination response rates, and if the parts of the total curve join smoothly on final analysis. Such a study will not allow analyses of tracking or the accurate estimation of risk for a later outcome during intervals longer than the range of ages for which each group was measured.

Long-term serial growth studies should be supported because they are the only source of some particular types of important information. They should not be used as sources of cross-sectional reference data because the samples are necessarily unrepresentative and such data must be current if secular changes are present or suspected.

5 · Significance of human growth

The significance of human growth and maturity is derived from their relationships to the future size, proportions, and composition of the body, and from associations of growth measures with current and future risk factors for serious diseases.

EARLY GROWTH AND LATER GROWTH

Growth status and growth rates at young ages are related to later growth. Tracking of status values, which is usually estimated from age-to-age correlations, is described in Chapter 3. The relationships between the timing of spurts, particularly the pubescent spurt, and body size before and after these spurts are considered together with other effects of maturity in Chapter 3.

Size at birth

In overview, birth weight has only slight relationships to weight, length, BMI, and skinfold thickness during infancy and childhood, but it is negatively correlated with the rates of growth in weight and length during early infancy. Small positive effects of length at birth on stature may persist until adulthood. The effects of birth weight on size during infancy decrease with age. The differences in weight, calculated as Z scores, among groups of infants with different birth weights, decrease markedly to 18 months, but change little from 18 months to 5 years (Binkin *et al.*, 1988). There are low to moderate correlations ($r = 0.2$–0.5) of birth weight with weight in infancy (Boryslawski, 1988; Cole, 1995; Wright *et al.*, 1994) and with BMI and skinfold thicknesses at 2 years and 5 years (Rolland-Cachera *et al.*, 1996). Studies that adjusted for some

confounding variables have estimated that increases in birth weight of 1.0 kg have effects of 0.8 to 1.2 kg on weight at 6 months and 0.3 kg on weight at 12 months (Kramer *et al.*, 1985*b*; Roche *et al.*, 1993).

The effects of birth weight on size during childhood and young adulthood tend to increase with age for weight and BMI, but not for stature (Allison *et al.*, 1995; Muramatsu *et al.*, 1990; Phillips & Young, 2000; Zive *et al.*, 1992). Birth weight is not significantly related to adult weight after adjustments are made for parity and maternal BMI (Stettler *et al.*, 2000), but extreme birth weight groups of males differ by 10 kg in weight at 18 to 26 years when confounding variables are not taken into account (Tuvemo *et al.*, 1999*a*). Birth weight has a low correlation ($r = 0.2$) with adult stature (Weyer *et al.*, 2000). The adjusted effect of birth weight on adult stature is also small, being about 3 cm per kg (Table 5.1). A moderate correlation ($r = 0.5$) of birth weight with stature during pubescence was found in one survey, but others found much lower correlations or insignificant effects of birth weight on stature before and during pubescence (Bacallao *et al.*, 1996; Kromeyer *et al.*, 1997; Malina *et al.*, 1999).

Birth weight has negative correlations with weight and length increments during the first few months after birth, but is positively correlated with weight increments later in infancy (Bergman & Bergman, 1986; Boulton, 1981; Hoffmans *et al.*, 1988; Persson, 1985). Each 1.0 kg increase in birth weight is associated with an adjusted decrease of 260 g in the weight gain from birth to 3 months (Fergusson *et al.*, 1980).

There is a lack of agreement among reports concerning the effects of birth weight on BMI at

Table 5.1. *Adjusted effects of birth weight on stature*

Author	Number of children	Age (years)	Effects[a]
Alberman *et al.* (1991)	7 710	>18	3.1 cm per kg
Goldstein (1971)	5 538	7	2.1 cm per kg
Kuh & Wadsworth (1989)	1 667	36	3.2 cm per kg
Seidman *et al.* (1993)[b]	30 083	17	3.3 cm per kg; males
			2.8 cm per kg; females

[a] Expressed in relation to deviation of birth weight from the mean.
[b] Incomplete adjustments for intervening variables.

older ages. The correlations between adjusted birth weights and BMI during infancy and childhood are low ($r = 0.3$ at 4 years; $r = 0.1$ at 7 to 12 years), but the odds ratio for the presence of high BMI values at 4 to 17 years increases significantly with birth weight until it is >2.0 in those with birth weights >4500 g vs. those with birth weights 3000 to 3499 g (Curhan *et al.*, 1996*a*, *b*; Maffeis *et al.*, 1994; Seidman *et al.*, 1991*b*). These relationships may be related to the rate of sexual maturation, which is positively correlated ($r = 0.2$) with birth weight (Bacallao *et al.*, 1996). Adults with birth weights >4500 g have significantly greater risks of BMI values >29.2 kg/m^2 then those with normal birth weights (Curhan *et al.*, 1996*a*, *b*; Phillips & Young, 2000). Birth weight is positively related to weight, stature, and BMI in young adulthood, but its effect on BMI during pubescence is smaller than that of parental BMI (Frisancho, 2000; Tuvemo *et al.*, 1999*a*). Despite these findings, groups of young men with BMI >30 kg/m^2 or <24.9 kg/m^2 differ by only 126 g in mean adjusted birth weights (Sørensen *et al.*, 1997).

Variations in the timing of slow fetal growth may be responsible for the differences between some reports of relationships between birth weight and BMI or relative weight in adulthood (Allison *et al.*, 1995; Charney *et al.*, 1976). Studies of men born during the Dutch famine winter of 1944–5 show an increased prevalence of overweight (>120% relative weight) in those whose mothers were undernour-

ished early in pregnancy, but a reduced prevalence in those whose mothers were undernourished late in pregnancy (Ravelli *et al.*, 1976). Birth weight has a low correlation ($r = 0.2$) with FFM in adulthood and is not correlated with TBF in adulthood (Weyer *et al.*, 2000). Adjusted birth weights have low positive correlations ($r = 0.02$–0.2) with the sums of skinfold thicknesses at 1 to 4 years (Kramer *et al.*, 1985*b*; Zive *et al.*, 1992), and almost zero correlations ($p = -0.02$) with subscapular skinfold thickness at 14 to 16 years (Barker *et al.*, 1997).

Length at birth has positive, but modest, relationships with length and stature at older ages, but its effects are considerable for those in extreme length groups. There are correlations of +0.1 to +0.5 between length at birth and length or stature at 3 months to 5 years (Boryslawski, 1988; Buschang *et al.*, 1985; Dine *et al.*, 1979); each 1.0 cm difference in length at birth is associated with an increase of 0.9 cm in length at 3 months (Fergusson *et al.*, 1980). Those with lengths at birth <5th percentile have mean statures at 7 years near the 20th percentile for males and the 10th percentile for females (Garn *et al.*, 1977). Similarly, young men grouped by length at birth (<47 cm, >55 cm) differ in mean stature by 10 cm (Tuvemo *et al.*, 1999*a*).

Length at birth is negatively correlated ($r = -0.6$) with the rate of growth in length during the first few months after birth (Bergman & Bergman, 1986; Fergusson *et al.*, 1980). Each 1.0 cm increase in length at birth is associated with an adjusted decrease of 0.9 cm in length increments from birth to 3 months, but is not related to the increments from 1 to 3 years (Eliot & Deniel, 1977; Fergusson et al., 1980; Holmes et al., 1977; Pomerance & Krall, 1981). Infants with head circumferences at birth <5th percentile have mean head circumferences at the 20th percentile at 7 years (Garn et al., 1977).

Infants born SGA at term mature at normal rates, but they are small during childhood and have mean deficits in adjusted adult values of 5.0 cm for stature and 1.5 kg for weight (Leger et al., 1998; Paz et al., 1993; Westwood et al., 1983). In these infants, head circumference increases slowly to 6 months, after which the rate of growth is normal (Colle et al., 1976; Fitzhardinge & Inwood, 1989; Tenovuo et al., 1987). Term SGA infants, who are also short at birth, have rapid increments in length to 12 months, after which the rate of growth is near the mean. Such infants, who are stunted during childhood (stature $< -2 Z$), have deficits in adult stature of 8 cm for males and 10 cm for females (Chaussain et al., 1994).

Size in infancy

Weight, stature, and BMI in infancy are moderately correlated with the corresponding measurements in childhood and adulthood. Weights at 3 and 6 months are correlated ($r = 0.3–0.6$) with weights at 5 years and in adulthood (Dine et al., 1979; Molinari et al., 1995). Length at 1 month and the increment in length from 1 month to 2 years have significant correlations ($r = 0.3$) with 18-year values for stature in each sex and for weight in females, but not for head circumference or BMI in either sex (Kouchi et al., 1985b). There are low correlations ($r = 0.3$) between BMI or relative weight during infancy and the corresponding measurements in childhood and adolescence (Dine et al., 1979; Muramatsu et al., 1990; Rolland-Cachera et al., 1989; Siervogel et al., 1991). Weights at 12 and 18 months are correlated ($r = 0.2–0.4$) with total body BMC and BMC of the lumbar spine and femoral neck at 5, 10, and 21 years, after adjusting for current weight (Cooper et al., 1995).

Size during childhood and adolescence

The correlations of weight, stature, and BMI during childhood and adolescence with matching measures at 16 to 18 years increase slowly with age except for a decrease during pubescence (Braddon et al., 1986; Furusho, 1968; Molinari et al., 1995; Rolland-Cachera et al., 1989). These correlations are closer for stature than for weight or BMI. Therefore, it is not surprising that most overweight children do not become overweight adults (Power et al., 1997; Rolland-Cachera et al., 1987; Whitaker et al., 1997).

The relationships between BMI at 7 to 15 years and BMI in adulthood are significant (Guo et al., 1994; Rolland-Cachera et al., 1989; Sørensen & Sonne-Holm, 1988; Valdez et al., 1996). As an example of such studies, Braddon and colleagues (1986) found correlations between BMI in childhood and at 36 years of 0.2 to 0.4 at 7 years and 0.5 to 0.6 at 11 and 14 years. The probabilities of large BMI values at 35 years, in relation to percentile values for BMI during childhood and adolescence, increase with childhood percentile level. These probabilities differ little with age from 2 to 7 years, but increase considerably at older childhood ages and do not decrease during pubescence (Guo et al., 1994, 2000b). There are significant negative correlations between the age at rebound in BMI and the maximum BMI value in young adulthood (Rolland-Cachera et al., 1989).

MATURITY

Relationships between maturity and function are particularly evident when term and preterm infants are compared at birth. By definition, these groups differ in gestational age, which is closely related to maturity. Ethnic differences must be considered when assessing the normality of sexual maturity status for age in the United States. As noted in Chapter 3, African Americans are advanced in age at menarche and in sexual maturation in comparison with European Americans and Hispanic Americans (S. Guo, unpublished data; Herman-Giddens et al., 2001). The importance of maturity in the prediction of adult stature is considered in Chapter 2 and its relationships to size and body composition are described in Chapter 3.

Until pubescence, skeletal maturity is positively correlated with body size, %BF, TBF, and FFM and with bone widths and cortical thicknesses, which are related to bone mineral (Low *et al.*, 1964; Roche, 1980) and it is negatively related to the timing of the rebound in BMI (Rolland-Cachera *et al.*, 1984). Age at menarche is negatively related to adult values for weight, BMI, and skinfold thicknesses, but not FFM (Kirchengast *et al.*, 1998; Laitinen *et al.*, 2001; Ness, 1991; Wellens *et al.*, 1992) and it has a slight positive relationship with adult stature (Garn & Clark, 1975). Age at menarche is positively related to BMD in adulthood, but is not related to fracture risk (Armamento-Villereal *et al.*, 1992; Finkelstein *et al.*, 1992; Mallmin *et al.*, 1994; Rauch *et al.*, 1999). Stages of sexual maturation, independently of age and stature, have significant positive relationships with systolic blood pressure at 11 to 16 years, but not with diastolic pressure (Cho *et al.*, 2001).

Skeletal and sexual maturity have significant positive correlations with maximum oxygen consumption, physical-working capacity, and static strength in children aged 6 to 16 years. These correlations, which are higher in males than females, reflect associations with muscle mass and they are independent of weight and stature in some groups (Beunen *et al.*, 1981; Jones *et al.*, 2000; Katzmarzyk *et al.*, 1997; Malina, 1994). Because the level of maturity differs within chronological age groups, some children in the pubescent age range are assisted in physical performance by their maturity levels, but others are hindered. The associations between rates of maturation and physical performance are age-dependent. Until the end of the pubescent period, those with rapid maturation perform better on tests of physical performance than those with slow maturation; at older ages the differences in performance between these groups are reversed (Lefevre *et al.*, 1990). These and similar findings led Mafulli (1996) to suggest that physical performance standards during pubescence be set relative to maturity instead of chronological age.

GROWTH AND DISEASE

Size at birth

Fetal growth retardation before the third trimester has similar effects on weight and length, but retardation in the third trimester causes a larger deficit in weight than in length. Therefore, weight-for-length and the ponderal index at birth may indicate the timing of growth retardation *in utero*. The ponderal index and the ratio arm circumference/head circumference may have stronger relationships to morbidity and mortality than those of birth weight (Danielian *et al.*, 1992; Deter *et al.*, 1990; Patterson & Pouliot, 1987; Villar *et al.*, 1990). Small birth weights are related to increased morbidity and mortality during infancy, especially in the perinatal period (McIntyre *et al.*, 1999; Samuelsen *et al.*, 1998; Tzoumaka-Bakoula, 1993). Large birth weights (>4000 g) are associated with increased rates of caesarian section and perinatal mortality (Taffel, 1994).

There has been considerable interest in the possible relationship of small weights and lengths at birth to hypertension and coronary heart disease in adulthood. Additionally, small weights-for-length at birth may be related to later insulin resistance, and small head circumferences may affect cognitive performance and increase the risk of Alzheimer's disease. Effects of size at birth on later blood pressure could be due to undernutrition late in pregnancy, maternal hypertension, genetic mechanisms, or variations in IGF-1 concentrations. Birth weight, after adjustments for sex, gestational age, and length, has weak relationships with systolic blood pressure at 1 week (10.8 mm Hg per kg birth weight), 3 months (−4.3 mm Hg per kg birth weight), and 4 years (−19.7 mm Hg per kg birth weight) as reported by Launer and colleagues (1993). A weak inverse relationship may persist between birth weight and blood pressure throughout childhood and adulthood, but the literature is not in agreement (Curhan *et al.*, 1996*a*, *b*; Hulman *et al.*, 1998; Taittonen *et al.*, 1996). The reports of relationships between birth weight and blood pressure in adolescence are particularly inconsistent (Laor et al., 1997; Law & Shiell, 1996; Matthes et al., 1994; Seidman et al., 1991a). Some have found weak relationships between birth weight and blood pressure that are independent of gestational age at birth and the mother's use of tobacco, alcohol, and oral contraceptives during pregnancy and are strengthened by adjusting for current weight, stature, and BMI (Pharoah et al., 1998; Uiterwaal et al., 1997; Vestbo et al., 1996). Birth weight explains only 2% of the variance in blood pressure during adulthood, after adjusting for age and current BMI (Vestbo et al., 1996). There is no association between birth weight and blood pressure at 8 years in those born before 34 weeks gestation, which suggests that the weak inverse relationships with birth weight at later gestational ages may result from poor fetal nutrition late in pregnancy (Morley et al., 1994). Adult blood pressures also tend to be higher in those with small values at birth for length, head circumference, or the ponderal index (Barker et al., 1992b; Law et al., 1991; Martyn et al., 1995a).

The relationship between birth weight and later blood pressure is not explained by more rapid growth during infancy in those with small birth weights (Law et al., 1993), but small birth weights are followed by high IGF-1 levels during childhood that may directly affect the growth of blood vessels leading to increases in blood pressure (Fall et al., 1995b; King et al., 1985; Nakao-Hayashi et al., 1992). A small birth weight is commonly associated with increased maternal blood pressure during pregnancy, but adjustments for maternal blood pressure may not alter the relationships between birth weight and blood pressure during childhood (Churchill et al., 1997; Taittonen et al., 1996; Taylor et al., 1998; Whincup et al., 1992). In men with a family history of hypertension, the ponderal index at birth is significantly negatively correlated ($r = -0.4$) with systolic blood pressure and fasting blood glucose (Melander et al., 1999). Walker and colleagues (1998) concluded that one-quarter of the effect of birth weight on blood pressure is of genetic origin.

The prevalence of coronary heart disease is inversely related to weight, length, and abdominal circumference at birth (Barker et al., 1989a, b, 1992b, 1993b; Koupilová et al., 1999; Osmond et al., 1993; Rich-Edwards et al., 1995a). This relationship may be non-linear; there is also an increased prevalence of coronary heart disease in those with large birth weights (Barker, 1994). The inverse relationship of birth weight to coronary heart disease is changed only slightly by adjustments for tobacco use, BMI, and systolic blood pressure in adulthood or maternal blood pressure during pregnancy (Barker, 1997; Barker et al., 1989a) but it is strengthened when birth weight is adjusted for gestational age (Leon et al., 1998). In partial conflict with these findings, all-cause mortality from 6 to 90 years is not increased in twins despite their small birth weights (Christensen et al., 1995).

Birth weight and the ponderal index at birth are inversely related to levels of low-density lipoprotein cholesterol in childhood and adulthood (Boulton et al., 1999; Valdez et al., 1994). The prevalence of the metabolic syndrome in adults is inversely related to birth weight (Curhan et al., 1996a, b; Phillips et al., 1996; Valdez et al., 1994; Yarbrough et al., 1998). This syndrome is a combination of increased blood pressure, dyslipidemia, and glucose intolerance that can lead to cardiovascular disease and insulin-dependent diabetes mellitus (IDDM). Birth weight may be negatively related to concentrations of Factor VII in each sex and to plasma fibrinogen in men, but not women (Barker et al., 1992a; Martyn et al., 1995b). These changes in Factor VII and fibrinogen, which may reflect slow growth of the liver in utero due to blood being shunted to the brain to compensate for undernutrition, could predispose to atheroma and thrombosis. Males with small birth weights have significantly increased activity of serum glutamic oxaloacetic transaminase and serum glutamic pyruvic transaminase indicating changes in liver function (Koziel et al., 2001). In old age, after adjusting for gestational age at birth and current cardiovascular risk factors, those with small birth weights have reduced compliance of the carotid and brachial arteries, but not the arteries of the lower limbs (Leeson et al., 1997; Martyn et al., 1998). This decreased compliance may be due to reduced synthesis of scleroprotein elastin before birth and during infancy (Martyn & Greenwald, 1997).

Length at birth may be inversely related to the prevalence of fatal coronary heart disease in men (Martyn *et al.*, 1995*a*). Forsén and colleagues (1997) did not find such a relationship, but noted a significant increase in the prevalence of fatal coronary heart disease for those with small weights-for-length at birth. A small ponderal index at birth, combined with a small head circumference, is more strongly related to an increased prevalence of fatal coronary heart disease than birth weight (Barker *et al.*, 1993*a*; Eriksson *et al.*, 1999).

Despite some conflicting reports, there is considerable evidence that small birth weights and small ponderal indices at birth are related to the later prevalence of non-insulin-dependent diabetes mellitus (NIDDM). Small birth weights are related to increased insulin resistance and high insulin levels in childhood and adulthood that may persist after adjustments for current body size (Dabelea *et al.*, 1999; Hofman *et al.*, 1997; Phillips *et al.*, 1994*a, b*; Whincup *et al.*, 1997). Reports of the relationships between birth weight and glucose tolerance in adults are not in agreement (Lindsay *et al.*, 2000; McCance *et al.*, 1994; Phillips *et al.*, 1994*a, b*; Valdez *et al.*, 1994). The prevalence of NIDDM in European-American adults is inversely related to birth weight after adjusting for adult BMI (Curhan *et al.*, 1996*a, b*; Hales *et al.*, 1991; Rich-Edwards *et al.*, 1999), but this relationship is U-shaped in Pima Indians living in Arizona (McCance *et al.*, 1994). Length at birth is not related to the prevalence of IDDM during childhood (Podar *et al.*, 1999), but a small ponderal index at birth, indicating undernutrition late in gestation, is related to increased insulin resistance and reduced glucose tolerance in children and adults, and an increased prevalence of NIDDM in adults (Barker *et al.*, 1993*a*; Lithell *et al.*, 1996; Phillips *et al.*, 1994*a, b*; Whincup *et al.*, 1997).

The prevalence of asthma in childhood and adulthood is inversely related to weight, length, and head circumference at birth, but has a U-shaped relationship to the ponderal index (Fergusson *et al.*, 1997; Shaheen *et al.*, 1999; Xu *et al.*, 2000). A large head circumference at birth is related to increased levels of IgE in cord blood and during childhood independently of maternal IgE and tobacco use (Fergusson *et al.*, 1997; Godfrey *et al.*, 1994; Gregory *et al.*, 1999; Oryszczyn *et al.*, 1999). Small birth weights are associated with decreased pulmonary function and an increased risk of fatal chronic obstructive airway disease and lung cancer in adults (Barker *et al.*, 1991). In those with large birth weights the prevalence of prostate cancer is increased (Chyou *et al.*, 1994; Tibblin *et al.*, 1995).

Head circumference at birth is related closely to brain size and reflects the development of neurons, dendrites, synapses, and myelin (Brandt, 1981; Bray *et al.*, 1969; Cooke *et al.*, 1977). Therefore, small head circumferences during early infancy can have permanent functional consequences. Head circumference in childhood and adulthood is positively related to cognitive performance (Graves *et al.*, 1996) and it is inversely related to the age of onset, prevalence, and rate of progression of Alzheimer's disease (Graves *et al.*, 1996; Schofield *et al.*, 1995, 1997).

Size in infancy

Growth during infancy and childhood is a guide to health status, but normal growth does not guarantee normal health. Abnormal growth, although commonly associated with disease, may occur when disease is absent. Despite these limitations, a slow rate of weight gain during infancy is a sensitive index of current disease (Kristiansson & Fällström, 1981). In developing countries, mortality rates are closely related to weight-for-age and stature-for-age in infants and preschool children (Pelletier *et al.*, 1993; Victora *et al.*, 1990), but these relationships are not evident in developed countries (Bairagi *et al.*, 1985; Chen *et al.*, 1980).

Slow weight gains during infancy, not necessarily sufficient to meet the criteria for failure-to-thrive, are related to the prevalence of left ventricular hypertrophy and coronary heart disease in adulthood (Osmond *et al.*, 1993; Vijayakumar *et al.*, 1995;

Zureik et al., 1996), perhaps because of increases in Factor VII and plasma fibrinogen (Martyn et al., 1995b). Nevertheless, large weights during infancy are associated with increased left ventricular mass in adulthood, which is related to hypertension and increased mortality rates (Osmond et al., 1993; Levy et al., 1990; Vijayakumar et al., 1995; Zureik et al., 1996). In men, but not women, the prevalence of fatal coronary heart disease is increased in those with small birth weights and may be increased further if weight at 1 year is small (Barker et al., 1989b; Fall et al., 1995b; Osmond et al., 1993). These puzzling reports of different effects in males and females require replication.

Increased insulin resistance precedes impaired glucose tolerance, which may be followed by NIDDM. Associations between these conditions and small birth weight could be due to fetal malnutrition that impairs general fetal growth or the growth of pancreatic β cells by reducing pancreatic blood flow while cerebral blood flow is maintained. Blood shunting to the brain may also permanently reduce insulin-responsive glucose transport systems in skeletal muscle (Simmons et al., 1985, 1993). The association of reduced glucose tolerance with small birth weight remains after adjusting for adult BMI (Dabelea et al., 1999; Lithell et al., 1996; Phillips et al., 1994a, b).

Size in childhood and adolescence

During childhood and adolescence, tracking of weight and BMI is important in relation to the risk of overweight and obesity in adulthood when these conditions are related to the prevalence of cardiovascular diseases and NIDDM. Children with large BMI values at ages older than 6 years should be monitored carefully and other risk factors for cardiovascular disease should be investigated (Gillum, 1999; Guo et al., 1994, 2000b; Himes & Dietz, 1994; Laitinen et al., 2001). Although it is desirable to limit weight gains in children and adolescents, effective programs are labor-intensive and expensive, compliance is commonly poor, and these programs may lead to eating disorders in some children.

The gain in weight from childhood to adolescence is more closely related to all-causes mortality than is weight status at birth or during infancy (Rhoads & Kagan, 1983). Children with large values for weight or BMI have significant increases in blood pressure, all-causes mortality, fatal coronary heart disease, atherosclerosis, colon cancer, stroke, and gout during adulthood, after adjusting for childhood stature and BMI in adulthood (di Pietro et al., 1994; Eriksson et al., 1999; Gunnell et al., 1998; Nieto et al., 1992). There is, however, a negative association between BMI at 10 to 14 years and the prevalence of pre-menopausal breast cancer (Le Marchand et al., 1988).

Any relationship between small birth weights and fatal coronary heart disease may be mediated, in part, by the development of obesity during childhood or adulthood (Eriksson et al., 1999; Frankel et al., 1996; Leon et al., 1996). The prevalence of fatal coronary heart disease is increased in men with small ponderal indices at birth and large BMI values at 7 years with a hazard ratio for extreme quartiles of 5.36 (Eriksson et al., 1999). Cardiovascular and all-causes mortalities are significantly increased after 12 years for those with childhood BMI values >75th percentile compared with those with BMI values from the 25th to 49th percentile (Gunnell et al., 1998).

The mechanisms underlying some of these associations may relate to circulating levels of lipids and blood pressure levels. Sixty percent of children with BMI values >95th percentile have large values of one or more risk factors for cardiovascular disease including increases in total cholesterol, low-density lipoprotein cholesterol, triglycerides, insulin and blood pressure, and 20% have large values of two or more risk factors (Freedman et al., 1999). There are similar relationships between large values of weight-for-stature and skinfold thicknesses with risk factors (Steinberger et al., 1995). Increments in the ponderal index during childhood are positively correlated ($r = 0.2$) with changes in low-density lipoprotein cholesterol levels (Freedman et al., 1985; Gidding et al., 1995). At 9 through 13 years, the correlations of BMI and skinfold thicknesses with high-density lipoprotein cholesterol are −0.2 and those with triglycerides and apolipoprotein B are 0.3 (Morrison *et al.*, 1996; Teixeira *et al.*, 2001). In children, large %BF values (>25%, males; >30%, females) are associated with increases in blood pressure and low-density lipoprotein cholesterol and decreases in high-density lipoprotein cholesterol (Dwyer & Blizzard, 1996; Williams *et al.*, 1992). Values of %BF at 7 to 11 years are positively related to an atherogenic index based on serum levels that compare the sum of low-density lipoprotein cholesterol and apolipoprotein B with the product of apolipoprotein A-1 and high- density lipoprotein cholesterol (Gutin *et al.*, 1994). Increases in weight during childhood, relative to stature, are positively correlated with changes in blood pressure during childhood and with blood pressure in adulthood ($r = 0.2$–0.4) (Guo *et al.*, 1998*b*); these relationships are closer in those with small birth weights (Jiang *et al.*, 1993). The risk of developing the metabolic syndrome in adulthood is strongly related to BMI at 7 years and the changes in BMI and the ponderal index from birth to 7 years (Vanhala *et al.*, 1999). Childhood weight and BMI are related to current insulin resistance and to insulin levels and the prevalence of NIDDM in adulthood (Hyppönen *et al.*, 1999; Pettitt *et al.*, 1985; Whincup *et al.*, 1997). Childhood stature, independently of BMI and sexual maturity stage, has significant positive relationships with systolic and diastolic blood pressure in each sex at 11 to 16 years (Cho *et al.*, 2001).

ADULT STATURE

The tracking of stature from childhood to adulthood leads to expectations that the relationships between childhood stature and disease in adulthood would be similar to those between adult stature and disease. Stature in young adults is negatively related to overall morbidity and to the risk of coronary heart disease (Blumchen & Jette, 1992; Rich-Edwards *et al.*, 1995*b*; Rimm *et al.*, 1995; Tuvemo *et al.*, 1999*a*). This relationship may be mediated through associations with blood pressure, high-density lipoprotein cholesterol, and forced expiratory volume (Walker

et al., 1989). Short adult statures are associated with reduced work capacity and, in women, with obstetric complications and small birth weights of offspring (Butler & Alberman, 1969*a*; Camilleri, 1981; Spurr, 1988; Tuvemo *et al.*, 1999*a*). Adults with short statures have an increased prevalence of Alzheimer's disease (Abbott *et al.*, 1998). Tallness is positively related to the prevalence of breast and colon cancer and nuclear and subcapsular cataracts (Albanes *et al.*, 1988; Caulfield *et al.*, 1999; Swanson *et al.*, 1988; van den Brandt *et al.*, 1997).

ASSESSMENT OF GROWTH

This section presents an overview, with some extensions, of material presented in Chapters 1 and 2. The growth of children is typically evaluated as part of a regular physical examination. Status values for growth are plotted on charts for the general population or for a special group to which the child belongs, e.g., preterm very low birth weight, Turner syndrome. Further steps are indicated, if the status values are unusual. These may include scheduling further examinations at which serial data will be plotted on status charts. Other useful information will be obtained by the calculation of increments and their comparison with reference data. Some growth evaluations are made because the parents or the children are concerned about unusual growth and whether a disease is present that is affecting growth. In the evaluation of a child growing at an unusual rate, the possible influences of parental stature, rate of maturation, and maternal substance abuse should be considered, and predictions of adult stature may be useful.

Growth charts are the primary tools for the recognition of unusual growth. These charts should be used by nurses and clinicians who have been trained to measure accurately, and to plot data correctly. These plots will show the extent to which children are small or large for each measure. Nurses and clinicians should acknowledge and understand the importance of growth data in the clinical management of infants and children. Clinicians should consider the relationships between the measured variables and other available data before completing their interpretations of growth status and progress.

Growth data from serial examinations allow judgements about growth rates. Serial data can be plotted on a status chart and the successive points for each variable joined by straight lines. These lines will show whether the values remain large or small and whether the levels are increasing or decreasing relative to reference percentiles. With this procedure it is not possible, however, to determine the extent to which the rates of growth indicated by changes in status levels are unusual. Such interpretations require the calculation and plotting of increments. A growth assessment may lead to the recognition of failure-to-thrive (FTT) during infancy. This condition is present when the weight is less than the 3rd or 5th percentile and the rate of weight gain is less than the 3rd percentile (Fomon, 1993). Consequently, to evaluate the possibility that FTT may be present, serial weight data showing an infant is underweight should be used to calculate increments that can be compared with reference data (Guo et al., 1991; Roche et al., 1989a, b; Roche & Himes, 1980). Of course, if the percentile level for weight is very low, intervention may be necessary before the criteria for FTT are present. It must be understood that FTT is not a disease, but a descriptive term that indicates the presence of a problem for which the cause should be sought.

Catch-up growth may be recognized. Recognition of catch-up growth, which refers particularly to length or stature, requires a decrease in relative levels for status that is followed by an increase. Catch-up growth is complete if the percentile levels are the same before the decrease and after the increase, but is incomplete if the increase does not eliminate the deficit. The extent to which catch-up growth is complete is a measure of the success of intervention.

Growth assessments may lead to the identification of decanalization. The term decanalization refers to the crossing of two or more percentile curves by lines connecting serial data points for a child (Li et al., 1998; Park et al., 1997). Decanalization can occur in any variable and may be directed toward or from the median. Additionally, transitions can occur from within the normal range (5th to 95th, 3rd to 97th percentiles or ±2 Z) to levels outside the normal range. Such transitions, which may not meet the criteria for decanalization, are relatively common, but many reflect small changes in relative levels. Decanalization in weight and head circumference during 6-month intervals is common from birth to 6 months and decanalization in weight from below the 90th percentile to above the 95th percentile is common throughout infancy. Infants with large birth weights tend to have decanalizations during infancy that are decreases in relative levels for weight. Infants with tall parents tend to have decanalizations in length that are increases in percentile levels, whereas infants with short parents tend to have decanalizations in length that are decreases. Decanalization is less common after infancy, even when calculated for 2-year intervals, but it is relatively common for decreases in weight from above the 10th percentile to below the 5th percentile (Li et al., 1998). Decanalization of stature during pubescence is significantly related to skeletal age. Slowly maturing children tend to have decanalizations in stature that are decreases early in pubescence and decanalizations that are increases late in pubescence. Opposite tendencies occur in rapidly maturing children.

The pubescent growth spurt is clearly evident in serial data for individuals, but it is obscured in data from cross-sectional surveys, which are averages from children maturing at various rates. Therefore, growth data from normal children who mature rapidly tend to increase in percentile levels early in pubescence, but they approximate their prepubescent percentile levels late in pubescence. Normal children who mature slowly tend to decrease in percentile levels early in the pubescent period, but later return to about their prepubertal levels.

The detection of abnormal growth at one or a series of examinations does not constitute a diagnosis. Abnormal growth does not necessarily indicate the presence of disease and normal growth does not necessarily indicate the absence of disease. Because the prevalence of disease is increased in those with abnormal growth, it is recommended that unusual status values at one examination be considered carefully, together with physical examination data. It might be decided to schedule a follow-up examination or some immediate laboratory investigations may be made. If the child is overweight or obese, as judged by BMI or skinfold thicknesses, the clinician should determine whether the levels of blood pressure and selected lipids are high because these constitute risk factors for disease that are commonly present in overweight and obese children (Freedman *et al.*, 1999; Himes & Dietz, 1994). A slow weight gain may lead to the assessment of social influences involving the family, such as substance abuse, income level, criminal behavior, and whether the child is under the care of a social service agency. Slow weight gains are significantly correlated ($r = 0.4$) with psychosocial scores derived from these social influences (Kristiansson & Fällström, 1981).

The statures of children are correlated with the statures of their parents. Therefore, parental stature should be taken into account when children's statures are evaluated. The average stature of the parents (mid-parent stature) is used for adjustment purposes because the paternal and maternal effects on the statures of children are approximately equal. Adjustment factors specific for sex, mid-parent stature and the current observed stature of the child should be added to the observed statures of children with short parents and subtracted from the observed statures of children with tall parents (Himes *et al.*, 1981; Prokopec, 1977; Sorva *et al.*, 1989; Tanner *et al.*, 1970). Both the observed and the adjusted data should be plotted on growth charts. These adjustment factors allow the stature of the child to be compared with the stature distributions of children matched for parental stature. Removing the effects of parental stature makes the effects of other influences more apparent. Parent–child correlations for statures are similar in many developed countries (Furusho, 1968; Himes *et al.*, 1981; Mueller, 1976; Otsuki, 1956) suggesting that adjustments such as those developed by Himes are widely applicable.

Percentiles for the ages at which stages of sexual maturity are observed have been added to some growth charts. These data allow a clinician to determine the normality of the rate of sexual maturation of a child. Rapidly maturing children tend to be tall and slowly maturing children tend to be short until about 17 years in males and 15 years in females. Consequently, observed statures should be adjusted for the rates of maturation using the age- and sex-specific differences in stature associated with different stages of sexual maturation. Wilson and colleagues (1987) used the average of the stages for pairs of organs in each sex (pubic hair and genitalia in males, pubic hair and breasts in females), which is problematic because the stages are discordant in 40% of children. These estimated differences allow the statures of rapidly or slowly maturing children to be judged after removing estimated maturational effects. The differences in stature associated with parental stature and rate of maturation are not independent. Therefore, the sum of these adjustments should not be applied in the assessment of a child's growth.

There are growth charts for LBW preterm infants from birth to 36 months gestation-adjusted age (Guo *et al.*, 1996, 1997*a*, 1998*b*; Roche *et al.*, 1997). When these charts are used, the chronological age must be corrected for the degree of prematurity. Thus, for the assessment of an infant born at a gestational age of 34 weeks (6 weeks preterm), the chronological (postnatal) age would be reduced by 6 weeks to obtain the gestation-adjusted age (GAA). These growth charts present reference data for status and for 2- and 3-week increments for weight, length, and head circumference and percentiles of weight-for-length and head circumference-for-length. These sex-specific data are for two birth weight groups (VLBW <1500 g; moderately LBW 1501–2500 g). The charts for LBW preterm infants were developed from a sample that included SGA and AGA infants. The differences in growth between SGA infants and the total sample were estimated for each quartile of the distributions (Guo *et al.*, 1998*a*). These differences, which change with age, should be used to adjust the observed

Table 5.2. *Selected disease-specific growth charts for common conditions*

Condition	Author	Country	Number of children	Age (years)[a]	Variables[a]
Down syndrome	Cronk *et al.* (1988)	United States	730	0.1–18	S,W
Down syndrome	Palmer *et al.* (1992)	United States	421	B–3	HC
Turner syndrome	Milani *et al.* (1994)	Italy	772	1–20	S
Turner syndrome	Rongen-Westerlaken *et al.* (1997)	Sweden, The Netherlands	598	B–18	S,W, BMI[b]

[a] B, birth; HC, head circumference; S, stature; W, weight.
[b] Also annual increments for stature.

data for SGA infants before comparisons are made with the charts. Throughout infancy, weight and length are significantly larger in moderately LBW preterm infants than in VLBW preterm infants, but the increments are generally larger for the moderate LBW group. In both VLBW and moderately LBW preterm infants, weight-for-length tends to be larger than general reference data at lengths from 56 to 64 cm, but not at larger lengths. After the age of 36 months, the growth of preterm LBW infants should be compared with reference data from general populations, though moderate deficits are common.

The postnatal growth of twins born preterm can be judged by comparison with reference data for singleton infants born at similar gestational ages and with similar birth weights (Guo *et al.*, 1996, 1997*a*, 1998*b*), or by the use of growth charts specific for twins (Bossi *et al.*, 1993). Disease-specific growth charts allow the growth of a child to be compared with that of untreated children with the same disease. If such comparisons show the child's growth differs markedly from what is expected in those with the disease, a cause other than the recognized disease should be sought. This approach is recommended when the disease is relatively homogeneous and the child being evaluated has not received specific therapy for the disease. Table 5.2 lists selected disease-specific growth charts for relatively homogeneous common conditions. Their accuracy is uncertain for

populations other than those from which they were derived.

The use of incremental reference data is recommended when status values are outside the normal range. They allow more rapid detection and improved evaluation of unusual growth rates than is possible when serial data are plotted on growth charts for status (Roche *et al.*, 1989*a*). Reference data for increments are available for 1-, 2-, and 3-month intervals during infancy and for 6-month intervals from birth to 18 years (Baumgartner *et al.*, 1986; Guo *et al.*, 1988, 1991; Roche *et al.*, 1989*a*, *b*; Roche & Himes, 1980). In the recent United States charts for increments, the values are plotted vertically above the ages at the ends of the intervals. This simplifies their application because it is not necessary to calculate the ages at the midpoints of the intervals, as is required with some other charts. Reference data for increments during annual intervals are less useful clinically than those for shorter intervals because long delays are required before the rates of growth can be established. Annual data are unaffected by possible seasonal variations in growth, which may be real in Canada and England for school-age children (Marshall, 1971; Mirwald & Bailey, 1997).

The evaluation of increments for individuals requires that the measuring techniques be appropriate and efforts must be made to ensure the same observer measures the child at the beginning and end

Table 5.3. *Approximate deficits in size at birth with maternal substance abuse during pregnancy*

Substance	Amount per day	Size deficit at birth		
		Weight (g)	Length (cm)	Head circumference (cm)
Tobacco	9 cigarettes	120	0.7	0.1
	10–19 cigarettes	108	0.9	0.3
	≥20 cigarettes	190	0.6	0.3
Alcohol	≤2 drinks	65	–	–
	>2 drinks	150	–	–
Caffeine	>4 cups coffee	220	–	–
Cocaine	–	500	2.0	2.0
Heroin	–	600	2.4	1.7
Methadone	–	350	1.5	1.5

of each interval. Errors of measurement can lead to false positives in the detection of abnormal growth rates; this problem is reduced markedly if the final judgement of growth rate is delayed until two successive increments are available.

The intervals can begin at any age. Thus a 6-month interval may be from 10.2 to 10.8 years. In clinical situations, it is unlikely that children will be measured at exactly the same intervals as those in the reference data. Linear mathematical adjustments (interpolations) can convert the observed increments to approximate what they would have been if the examinations had matched the reference intervals. The easiest method is to convert the actual interval into months and decimals of months, divide the observed increment by the actual number of months in the interval, and multiply the result by the number of months in the reference interval. The length of the intervals between examinations that allows reasonably accurate comparisons of the observed increments with reference increments is a function of several factors. Adjustments for larger variations in interval lengths are acceptable when the rate of change in the variable is relatively constant and the errors of measurement are small. Recalculating an increment observed over a short time, e.g., 4 months, to estimate a 6-month increment increases the variance and the opposite occurs when an increment observed during a longer interval than

the reference is adjusted to estimate what the increment would have been during an interval matching those of the reference data. Consequently, there should be less concern about adjusted values near the margins of the distributions for increments derived from intervals shorter than the reference intervals, but more concern about those derived from intervals longer than the reference intervals.

It is important to consider the possible effects of maternal substance abuse during pregnancy on size at birth and on postnatal growth. The accuracy of information about substance abuse is uncertain because it is based on maternal reports. Commonly dosages are unknown and multiple substances may be taken. The reported effects of substance abuse on size at birth, after adjustments for other determinants of growth, are largest for weight, intermediate for length, and smallest for head circumference (Table 5.3). The most reliable estimates of the effects of substance abuse on size at birth are those for maternal tobacco and alcohol use, which appear to be dose-dependent. The effects of cocaine and heroin use on size at birth are particularly large. While the assessment of size at birth should take substance abuse into account, mathematical adjustments are not recommended.

Maternal substance abuse during pregnancy can cause deficits in postnatal growth (Table 5.4). Tobacco use is likely to persist after delivery resulting

Table 5.4. *Approximate deficits in postnatal growth with maternal substance abuse during pregnancy*

| Substance | Age (years) | Deficit in postnatal growth | |
		Weight (g)	Length (cm)
Tobacco >10 cigarettes	0.5–2.0	200	0.9
per day	2.1–6.0	300	0.9
	6.1–12.0	870	0.8
	12.1–18.0	1280	1.0
Cocaine	0.25	90	0.7
	0.5	+50	0.4
	1.0	10	+0.2
	1.5	90	0.1
	2.0	540	1.8
Heroin	3.0	+200	0.3
Methadone	3.0	300	1.6

in exposure of the infant to passive smoke. This might explain why the deficits in weight with tobacco use increase after birth. The postnatal deficits in length and head circumference during infancy with maternal substance abuse are small, except for cocaine, heroin, and methadone use. Adjustments to observed size data for infants and children to correct for maternal substance abuse are not recommended.

ASSESSMENT OF NUTRITIONAL STATUS

In an illuminating review, Dwyer (1991) defines nutritional status as "the condition of the body resulting from the intake, absorption, and utilization of food as well as from factors of pathological significance. Its full assessment requires anthropometric, dietary and biochemical data in addition to clinical observations." The development of an abnormal nutritional status begins with cellular changes; later there are alterations in body functions, clinical signs, and body measurements (Devlin & Horton, 1988; Waterlow, 1986). Anthropometric data are potentially informative about stores of energy (fat),

protein (muscle), and calcium (bone mineral). Information about other body stores is not available from anthropometric data.

Weight is sensitive to acute diseases, such as diarrhea, but only chronic diseases cause changes in length or stature. Therefore, measures of weight and stature, in combination, can suggest whether a state of malnutrition is acute or chronic. Weight is an approximate guide to fat mass, but it is misleading when total body water or muscle mass differs from normal (Brans *et al.*, 1974; Enzi *et al.*, 1981; Petersen *et al.*, 1988).

Weight is commonly adjusted for body length by calculating BMI or by deriving weight-for-length, both of which are highly correlated with TBF (Norgan & Ferro-Luzzi, 1982). In children, BMI is correlated ($r = 0.5$–0.8) with %BF and skinfold thicknesses and with TBF ($r = 0.7$–0.9) (Malina & Katzmarzyk, 1999; Pietrobelli *et al.*, 1998; Widhalm & Schönegger, 1999). These correlations have led to the inclusion of BMI and skinfold thicknesses in some equations that predict %BF or TBF from anthropometric data (Boileau *et al.*, 1985; Dezenberg *et al.*, 1999; Goran *et al.*, 1996; Slaughter *et al.*, 1988). One principal component explains most of the variation among skinfold sites (Jackson & Pollock, 1976; Mueller & Reid, 1979) showing that measurements at only a few skinfold sites are needed to grade body fatness. The triceps and subscapular sites are the ones measured most commonly and are included in numerous predictive equations. For the recognition of large %BF values in children, BMI has high specificity (few false positives) but low sensitivity (misses many true positives) in normal children and in those with diseases (Himes & Bouchard, 1989; Lazarus et al., 1996; Reilly et al., 2000; Warner et al., 1997); skinfold thicknesses may be more effective than BMI for this purpose in adolescent males (Himes & Bouchard, 1989). Age-independent cut-off values of BMI for categorization of adults as overweight (>25 kg/m^2) or obese (>30 kg/m^2) have been established, but there is a lack of consensus about corresponding values for children (Cole et al., 2000; National Heart, Lung and Blood Institute, United States, 1998; World Health Organization, 1998).

Cross-sectional areas of adipose tissue and "muscle plus bone" in the arm and calf, calculated from skinfold thicknesses and circumferences at the same levels, are inaccurate due to non-circularity and variations in the thickness of subcutaneous adipose tissue around the arm (de Koning et al., 1986; Forbes et al., 1988). Adipose tissue areas estimated in this way are more highly correlated with TBF than are skinfold thicknesses at the same levels, but are not more highly correlated with %BF (Himes et al., 1980). These areas may assist the monitoring of children in whom skinfold thicknesses decrease while adipose tissue areas increase, as can occur when there are large increases in cross-sectional areas of "muscle plus bone." Revised equations have been developed that estimate these areas more accurately for adults (Heymsfield et al., 1982; Martine et al., 1997). The relationships between estimates of tissue areas from these revised equations and TBF or %BF have not been established.

One major aspect of nutritional status is the body store of protein, which is closely related to muscle and FFM. The independent variables in equations that estimate FFM from anthropometric data

in children include weight, stature, and limb circumferences with positive coefficients, and skinfold thicknesses with negative coefficients (Guo et al., 1989; Lohman et al., 1975; Slaughter et al., 1978). Differences from reference values for these measures are guides to whether FFM is large or small for age and sex. Total body BMC is correlated with weight ($r = 0.8$) in preterm infants and with a combination of weight and stature ($r = 0.9$) in females aged 9 to 21 years (Chan, 1992; Katzman et al., 1991). These findings are notable in relation to nutritional status because there are very high correlations ($r > 0.95$) between total body BMC and total body calcium (Heymsfield et al., 1989; Mazess & Barden, 1988).

The preceding paragraphs relate to nutritional status at one examination. Increments in anthropometric data can be compared with reference data and interpreted as guides to changes in nutritional status. The validity of conclusions about changes in nutritional status from changes in measurements of skinfold thicknesses and body composition is uncertain unless the changes are large (Ballor, 1996).

GROWTH SCREENING

Comparisons between growth measurements and reference data can identify infants and children with status values that are large or small and children with rapid or slow growth can be recognized if these measurements are repeated after an interval. These procedures are applied frequently in clinical practice where repeated measurements are common and may be referred to as growth monitoring. The term growth screening is used when they are applied in epidemiological surveys.

To be effective, public health screening programs related to child growth must meet the following criteria:

- The health problem sought is important and has an asymptomatic phase.
- Changes in growth or maturation occur at an early stage of the disease when treatment is more effective.

- Suitable diagnostic tests and effective treatments are available.
- The cost of the survey and the treatment of those identified are justifiable relative to the total public cost of healthcare (Hakama, 1991; Wilson & Jungner, 1968).

Screening is applied universally at birth to detect LBW infants and those with unusual head circumferences. This is clearly justified because the recognition of unusual values can lead to early changes in clinical management. Screening for short stature during childhood may meet the recommended criteria in relation to growth hormone deficiency and Turner syndrome, but may not be justified for Down syndrome and many congenital abnormalities because specific treatment is not available. Treatment is not needed for familial short stature and constitutional growth delay, but parents and children can be reassured by knowing a disease is not present and those with constitutional growth delay can be helped by knowing their growth deficits will decrease after a delayed pubescence. If the aim of the program is to identify overweight children, the health problem is less important before 6 years than at older ages because tracking is limited at young ages. It would be difficult to justify screening of preschool children because effective treatment at any age is labor-intensive and expensive and many of those with large BMI values will not have large values at older ages. Lengthy intervention by a multi–disciplinary team can be effective, but is not widely available (Epstein, 1993a, b; Epstein & Goldfield, 1999) and should be restricted to overweight children who have other risk factors for chronic diseases.

Cut-off levels must be set in any screening program. In growth screening, these levels are usually the 3rd and 97th percentiles or the 5th and 95th percentiles on selected charts. The choice of percentiles further from the median leads to the identification of fewer divergent children and therefore a reduced burden on families and on clinics where detailed physical examinations and laboratory tests will be performed. The percentage prevalence of confirmed diseases among those identified as unusual in size during the screening will be higher, however, if more extreme cut-off levels are applied. For example, Lacey & Parkin (1974a) found 82% of those with statures <3rd percentile had organic diseases. The prevalence of newly diagnosed organic disease, among those found to have unusual statures in a screening program, will be increased if referrals are made only after the observed statures have been adjusted for mid-parent stature (Himes et al., 1981). This will exclude referral examinations for those with familial short stature or familial tall stature.

Children with growth values that are unusual, but within the cut-off levels, may also require careful attention. Another examination a few months later during infancy or 6 to 12 months later for older children will provide additional status data and information about the rate of growth that can assist a decision about the need for referral. When growth charts are used in screening, and when they are used clinically, knowledge of the age groupings used in chart construction is necessary for accurate interpretation. In the United States growth charts, the values plotted above a particular age apply to that age. They were derived from subjects with a 1-month age range. For example, the values shown for 8-year-old males represent children aged 8 years ± 0.5 months (Flegal, 2000). In the growth charts of Lhotská et al. (1993), however, the data for 8-year-old children are derived from those aged 8.00 to 8.99 years with a mean age close to 8.5 years.

The selection of appropriate anthropometric instruments and their calibration is essential for accurate screening. In one screening program, the instruments for measuring stature were compared against a meter rod, which provided readings varying from 98.6 to 102.8 cm (Voss et al., 1990). The reference data used must match the expected distributions in the study population to allow a valid selection of children with unusual statures. This has not always been the case (Ahmed et al., 1993; Voss et al., 1992).

Serial data from clinical examinations or repeated surveys, when plotted on growth charts for status, can lead to the recognition of unusual growth rates, but unusual rates of growth can be detected earlier and assessed more accurately by the use of incremental growth charts (Guo *et al.*, 1989, 1992; Roche & Himes, 1980; Roche *et al.*, 1989*a*; Sempé *et al.*, 1979). The errors of measurements must be small for calculated increments to be useful (Hall, 2000; Roche & Himes, 1980; Voss *et al.*, 1991). Small errors can be easily achieved for weight, if time of weighing is consistent in relation to food intake. It is difficult, but not impossible, to obtain accurate incremental data for length during infancy (Guo *et al.*, 1989; van den Broeck *et al.*, 2000). Errors in the measurement of length and stature can be minimized with attention to detail and continuous quality control, ideally by having every measurement repeated by another technician. It is difficult to achieve small measurement errors in public health screening which typically includes thousands of children measured at multiple sites. Increments derived from some screening programs may not assist the identification of unusual growth (Voss *et al.*, 1991).

The effectiveness of public health screening for abnormal child growth is indicated by reports from four studies with remarkably similar findings. In a study of 20 388 English children aged 3.0 to 4.5 years, Ahmed and colleagues (1993) identified 260 children with statures <3rd percentile. Seventy-six of these were lost to follow-up at 4.5 years, but, as a result of the survey, 67 new diagnoses were made in the remaining 184 children. Most of the diagnoses were familial short stature and constitutional growth delay. In addition, they found cases of growth hormone deficiency (2), Noonan syndrome (4), psychosocial dwarfism (1), and Turner syndrome (2). Voss *et al.* (1992) reported findings from 14 346 English children aged 5 to 8 years. They identified 180 children with statures <3rd percentile. After excluding minority children and those with known organic diseases, there remained 152 short children. The treatable diseases found as a result of the survey were one case each of celiac disease, growth hormone deficiency, hypothyroidism, and lead poisoning.

Among 2256 English children born in 1960, Lacey & Parkin (1974*b*) found 111 with statures <3rd percentile at 11 years. They examined 98 of

these short children and found 16 had organic diseases of which only three were treatable. In other cohorts, born in 1961 and 1962, they found 12 children with statures <−3 *Z* all of whom were diagnosed as idiopathic short stature. The authors concluded that a short child requires investigation to exclude organic disease only when there are other signs such as anemia or pyorrhea, or the stature is <−3 *Z*, or the stature level is falling further below reference values over the course of a year. Lindsay and colleagues (1994) obtained annual stature increments for 79 495 United States children aged 5 to 11 years to estimate the prevalence of growth hormone deficiency. Of these children, 578 had statures <−2 *Z* *and* a growth rate <5.0 cm per year. As a result of the survey, they identified 16 children with undiagnosed growth hormone deficiency. In addition, 53 children had medical conditions, most of which had been diagnosed prior to the survey. These included six cases of Turner syndrome and single cases of celiac disease, congenital heart disease, genetic disorders, and hypothyroidism. The remaining 509 children with short stature and slow growth rates had familial short stature, constitutional delay, or idiopathic short stature.

The studies reviewed were designed to estimate the utility of surveys that aim to detect undiagnosed abnormal growth in stature that is due to treatable conditions. They did not identify enough children to justify such surveys for public health purposes. Children with markedly deviant values for weight and stature are likely to be investigated independently of a survey. Screening for BMI may be considered justified because there is marked tracking from childhood to adulthood. Treatment of those with large values is difficult, but is indicated when there are concurrent risk factors for chronic disease, as is common in overweight or obese children and adolescents. Some benefit is derived from a survey when children with unusual growth are investigated and a disease is not found. The reassurance that follows, combined with a prediction of adult stature, can be useful to the child and the family.

There are few reports of the outcomes of growth assessments in clinical practice. Green &

MacFarlane (1983), in data from a growth clinic, found that among children with statures <3rd percentile, the percentage prevalences of particular conditions among those diagnosed were: constitutional delay, 15%; hypothyroidism, 5%; hypopituitarism, 15%; short normal, 33%; and Turner syndrome, 5%. Most tall children (stature >97th percentile) were diagnosed as congenital adrenal hyperplasia, familial tall stature, Klinefelter syndrome, Marfan syndrome or thyrotoxicosis.

References

Abbott, R. D., White, L. R., Ross, G. W., Petrovitch, H., Masaki, K. H., Snowdon, D. A. and Curb, J. D. (1998). Height as a marker of childhood development and late-life cognitive function: The Honolulu-Asia Aging Study. *Pediatrics*, **102**:602–9.

Abel, E. L. and Hannigan, J. H. (1995). 'J-shaped' relationship between drinking during pregnancy and birth weight: Reanalysis of prospective epidemiological data. *Alcohol and Alcoholism*, **30**:345–55.

Abel, T. D., Baker, L. C. and Ramsey, C. N., Jr. (1991). The effects of maternal smoking on infant birth weight. *Family Medicine*, **23**:103–7.

Abrams, B. F. and Laros, R. K. (1986). Prepregnancy weight, weight gain and birth weight. *American Journal of Obstetrics and Gynecology*, **154**:503–9.

Abrams, B. F. and Selvin, S. (1995). Maternal weight gain pattern and birth weight. *Obstetrics and Gynecology*, **86**:163–9.

Abrams, S. A. and Stuff, J. E. (1994). Calcium metabolism in girls: Current dietary intakes lead to low rates of calcium absorption and retention during puberty. *American Journal of Clinical Nutrition*, **60**:739–43.

Acerini, C. L., Cheetham, T. D., Edge, J. A. and Dunger, D. B. (2000). Both insulin sensitivity and insulin clearance in children and young adults with Type I (insulin-dependent) diabetes vary with growth hormone concentrations and with age. *Diabetologia*, **43**:61–8.

Ackland, F. M., Stanhope, R., Eyer, C., Hamill, G., Jones, J. and Preece, M. A. (1988). Physiological growth hormone secretion in children with short stature and intra-uterine growth retardation. *Hormone Research*, **30**:241–5.

Adair, L. S. (1999). Filipino children exhibit catch-up growth from age 2 to 12 years. *Journal of Nutrition*, **129**:1140–8.

Adams, J. and Ward, R. H. (1973). Admixture studies and the detection of selection. *Science*, **180**:1137–43.

Adan, L., Bussières, L., Trivin, C., Souberbielle, J. C. and Brauner, R. (1999). Effect of short-term testosterone treatment on leptin concentrations in boys with pubertal delay. *Hormone Research*, **52**:109–12.

Adashi, E. Y., Resnick, C. E., d'Ercole, A. J., Svobeda, M. E. and van Wyk, J. J. (1985). Insulin-like growth factors as intraovarian regulators of granulosa cell growth and function. *Endocrine Reviews*, **6**:400–20.

Adler, I. D. (1970). The problem of caffeine mutagenicity. In *Chemical Mutagenesis in Mammals and Man*, eds. F. Vogel and B. Rohr, pp. 383–403. New York: Springer-Verlag.

Agostoni, C., Salari, P. and Riva, E. (1992). Metabolic needs, utilization and dietary sources of fatty acids in childhood. *Progress in Food and Nutrition Science*, **16**:1–49.

Ahluwalia, I. B., Grummer-Strawn, L. and Scanlon, K. S. (1997). Exposure to environmental tobacco smoke and birth outcome: Increased effects on pregnant women aged 30 years or older. *American Journal of Epidemiology*, **146**:42–7.

Ahmed, M. L., Allen, A. D., Sharma, A., MacFarlane, J. A. and Dunger, D. B. (1993). Evaluation of a district growth screening programme: The Oxford Growth Study. *Archives of Disease in Childhood*, 69:361–5.

Ahmed, M. L., Ong, K. K. L., Morrell, D. J., Cox, L., Drayer, N., Perry, L., Preece, M. A. and Dunger, D. B. (1999). Longitudinal study of leptin concentrations during puberty: Sex differences and relationship to changes in body composition. *Journal of Clinical Endocrinology and Metabolism*, 84: 899–905.

Ahn, C. H. and MacLean, W. C. (1980). Growth of the exclusively breast-fed infant. *American Journal of Clinical Nutrition*, 33:183–92.

Aicardi, G., Vignolo, M., Milani, S., Naselli, A., Magliano, P. and Garzia, P. (2000). Assessment of skeletal maturity of the hand–wrist and knee: A comparison among methods. *American Journal of Human Biology*, 12:610–15.

Alameda County Low Birth Weight Study Group (1990). Cigarette smoking and the risk of low birth weight: A comparison in black and white women. *Epidemiology*, 1:201–5.

Albanes, D., Jones, D. Y., Schatzkin, A. G., Micozzi, M. S. and Taylor, P. R. (1988). Adult stature and risk of cancer. *Cancer Research*, 48:1658–62.

Alberman, E., Emanuel, I., Filakti, H. and Evans, S. J. W. (1992). The contrasting effects of parental birthweight and gestational age on the birthweight of offspring. *Paediatric and Perinatal Epidemiology*, 6:134–44.

Alberman, E., Filakti, H., Williams, S., Evans, S. J. W. and Emanuel, I. (1991). Early influences on the secular change in adult height between the parents and children of the 1958 birth cohort. *Annals of Human Biology*, 18:127–36.

Albertson, A. M., Tobelmann, R. C., Engstrom, A. and Asp, E. H. (1992). Nutrient intakes of 2- to 10-year-old American children: 10 year trends. *Journal of the American Dietetic Association*, 92:1492–6.

Albertsson-Wikland, K., Boguszewski, M. and Karlberg, J. (1998). Children born small-for-gestational age: Postnatal growth and hormonal status. *Hormone Research, Supplement*, 49:7–13.

Albertsson-Wikland, K. and Karlberg, J. (1997). Postnatal growth of children born small for gestational age. *Acta Paediatrica, Supplement*, 423:193–5.

Albertsson-Wikland, K. and Rosberg, S. (1988). Analyses of 24-hour growth hormone profiles in children: Relation to growth. *Journal of Clinical Endocrinology and Metabolism*, 67:493–500.

Albertsson-Wikland, K., Rosberg, S., Karlberg, J. and Groth, T. (1994). Analysis of 24-hour growth hormone profiles in healthy boys and girls of normal stature: Relation to puberty. *Journal of Clinical Endocrinology and Metabolism*, 78:1195–1201.

Albertsson-Wikland, K., Rosberg, S., Lannering, B., Dunkel, L., Selstom, G. and Norjavarra, E. (1997). Twenty-four-hour profiles of luteinizing hormone, follicle-stimulating hormone, testosterone, and estradiol levels: A semilongitudinal study throughout puberty in healthy boys. *Journal of Clinical Endocrinology and Metabolism*, 82:541–9.

Albertsson-Wikland, K., Wennergren, G., Wennergren, M., Vilbergsson, G. and Rosberg, S. (1993). Longitudinal follow-up of growth in children born small for gestational age. *Acta Paediatrica*, 82: 438–43.

Alen, M., Hakkinen, A. and Komi, P. (1994). Changes in neuromuscular performance and muscle fiber characteristics of elite power athletes self-administering androgen and anabolic steroids. *Acta Physiologica Scandinavica*, 122:535–44.

Alexander, G. R., Baruffi, G., Mor, J. and Kieffer, E. (1992). Maternal nativity status and pregnancy outcome among U.S.-born Filipinos. *Social Biology*, 39:278–04.

Alexander, G. R., Hulsey, T. C. and Smeriglio, V. L. (1990). Factors influencing the relationship between a newborn assessment of gestational maturity and the gestational age interval. *Paediatric and Perinatal Epidemiology*, 4:135–48.

Alfieri, A. and Gatti, I. (1976). Accrescimento staturo-ponderale in gemelli nel primo anno di vita (Growth in weight and stature during the first year of life in twins). *Acta Medica Auxologica*, 8:239–44.

Alfieri, A., Gatti, I. and Alfieri, A. C. (1987). Weight and height growth in twins and children born in the last decade. *Acta Geneticae Medicae et Gemellologiae*, 36:209–11.

Ali, A., Lestrel, P. E. and Ohtsuki, F. (2000). Secular trends for takeoff and maximum adolescent growth for eight decades of Japanese cohort data. *American Journal of Human Biology*, **12**:702–12.

Ali, M. A. and Ohtsuki, F. (2000). Estimation of maximum increment age in height and weight during adolescence and the effect of World War II. *American Journal of Human Biology*, **12**:363–70.

Allison, D. B., Heshka, S., Neale, M. C. and Heymsfield, S. B. (1994). Race effects in the genetics of adolescents' body mass index. *International Journal of Obesity*, **18**:363–8.

Allison, D. B., Paultre, F., Heymsfield, S. B. and Pi Sunyer, F. X. (1995). Is the intra-uterine period really a critical period for the development of adiposity? *International Journal of Obesity*, **19**:397–402.

Alp Günöz, H., Darendeliler, F., Tunali, D., Bundak, R., Saka, N. and Neyzi, O. (1989). Growth velocity in different types of short stature. In *Perspectives in the Science of Growth and Development*, ed. J.M.Tanner, pp. 307–11. London: Smith-Gordon.

Altemeier, W. A., O'Connor, S. M., Sherrod, K. B. and Vietze, P. M. (1985). Prospective study of antecedents for nonorganic failure to thrive. *Journal of Pediatrics*, **106**:360–5.

Alvarez, C. V., Mallo, F., Burguera, B., Cacicedo, L., Diéguez, C. and Casanueva, F. F. (1991). Evidence for a direct pituitary inhibition by free fatty acids of *in vivo* growth hormone responses to growth hormone-releasing hormone in the rat. *Neuroendocrinology*, **53**:185–9.

Ambrosius, W. T., Compton, J. A., Bowsher, R. R. and Pratt, J. H. (1998). Relation of race, age, and sex hormone differences to serum leptin concentrations in children and adolescents. *Hormone Research*, **49**:240–6.

Amiel, S. A., Caprio, S., Sherwin, R. S., Plewe, G., Haymond, M. W. and Tamborlane, W. V. (1991). Insulin resistance of puberty: A defect restricted to peripheral glucose metabolism. *Journal of Clinical Endocrinology and Metabolism*, **72**:277–82.

Amiel, S. A., Sherwin, R. S., Simonsson, D. C., Lauritano, A. and Tamborlane, W. (1986). Impaired insulin action in puberty: A contributing factor to poor glycemic control in adolescents with diabetes. *New England Journal of Medicine*, **315**:215–19.

Amini, S. B., Catalano, P. M., Hirsch, V. and Mann, L. I. (1994). An analysis of birth weight by gestational age using a computerized perinatal data base, 1975–1992. *Obstetrics and Gynecology*, **83**:342–52.

Amit, Y., Jabbour, S. and Arad, I. D. (1993). Standards of skinfold thickness and anthropometric indices in term Israeli newborn infants. *Israel Journal of Medical Sciences*, **29**:632–5.

Ancel, P. Y., Saurel-Cubizolles, M.-J., di Renzo, G. C., Papiernik, E. and Bréart, G. (1999). Very and moderate preterm births: Are the risk factors different? *British Journal of Obstetrics and Gynaecology*, **106**:1162–70.

Andersen, R. E., Crespo, C. J., Bartlett, S. J., Cheskin, L. J. and Pratt, M. (1998). Relationship of physical activity and television watching with body weight and level of fatness among children: Results from the Third National Health and Nutrition Examination Survey. *Journal of the American Medical Association*, **279**:938–42.

Anderson, D. C. (1974). Sex-hormone-binding globulin. *Clinical Endocrinology*, **3**:69–95.

Anderson, G. D., Bildner, I. N., McClemont, S. and Sinclair, J. C. (1984). Determinants of size at birth in a Canadian population. *American Journal of Obstetrics and Gynecology*, **150**:236–44.

Andrews, J. and McGarry, J. M. (1972). A community study of smoking in pregnancy. *Journal of Obstetrics and Gynaecology of the British Commonwealth*, **79**:1057–73.

Angehrn, V., Zachmann, M. and Prader, A. (1979). Silver–Russell syndrome: Observations in 20 patients. *Helvetica Paediatrica Acta*, **34**:297–308.

Appel, B. and Fried, S. (1992). Effects of insulin and dexamethasone on lipoprotein lipase in human adipose tissue. *American Journal of Physiology*, **262**:695–9.

Apter, D., Pakarinen, A., Hammond, G. L. and Vihko, R. (1979). Adrenocortical function in puberty. *Acta Paediatrica Scandinavica*, **68**:599–604.

Apter, D. and Vihko, R. (1977). Serum pregnenolone, progesterone, 17-hydroxyprogesterone, testosterone and 5-alpha-dihydrotestosterone during female puberty. *Journal of Clinical Endocrinology and Metabolism*, **45**:1039–48.

Arase, K., Shargill, N. S. and Bray, G. A. (1989). Effects of corticotropin releasing factor on genetically obese (fatty) rats. *Physiology and Behavior*, **45**:565–70.

Arbuckle, T. E. and Sherman, G. J. (1989). An analysis of birth weight by gestational age in Canada. *Canadian Medical Association Journal*, **140**:157–65.

Arden, N. K., Baker, J., Hogg, C., Baan, K. and Spector, T. D. (1996). The heritability of bone mineral density, ultrasound of the calcaneus and hip axis length: A study of postmenopausal twins. *Journal of Bone and Mineral Research*, **11**:530–4.

Arden, N. K. and Spector, T. D. (1997). Genetic influences on muscle strength, lean body mass, and bone mineral density: A twin study. *Journal of Bone and Mineral Research*, **12**:2076–81.

Argente, J., Barrios, V., Pozo, J., Muñoz, M. F., Hervas, F., Stene, M. A. and Hernandez, M. (1993). Normative data for insulin-like growth factors (IGFs), IGF-binding proteins and growth hormone-binding protein in a healthy Spanish pediatric population: Age- and sex-related changes. *Journal of Clinical Endocrinology and Metabolism*, **77**:1522–8.

Armamento-Villereal, R., Villereal, D. T., Avioli, L. and Civitelli, R. (1992). Estrogen status and heredity are major determinants of premenopausal bone mass. *Journal of Clinical Investigation*, **90**:2464–71.

Armstrong, I. E., Robinson, E. J. and Gray-Donald, K. (1998). Prevalence of low and high birthweight among the James Bay Cree of Northern Quebec. *Canadian Journal of Public Health*, **89**:419–20.

Arslanian, S. A. and Kalhan, S. C. (1994). Correlations between fatty acid and glucose metabolism: Potential explanation of insulin resistance of puberty. *Diabetes*, **43**:908–14.

Arslanian, S. A. and Suprasongsin, C. (1997). Testosterone treatment in adolescents with delayed puberty: Changes in body composition, protein, fat, and glucose metabolism. *Journal of Clinical Endocrinology and Metabolism*, **82**:3213–20.

Arts, J., Kuipfer, G. G. J. M., Janssen, J. M. M. F., Gustafsson, J., Lowik, C. W. G. M., Pols, H. A. P. and Van Leeuwen, J. P. T. M. (1997). Differential expression of estrogen receptors A and β mRNA during differentiation of the human osteoblast SV-HFO cells. *Endocrinology*, **138**:5067–70.

Asaka, A., Imaizumi, Y. and Inouye, E. (1980). Analysis of multiple births in Japan. *Japanese Journal of Human Genetics*, **25**:213–18.

Ashizawa, K. (1994). TW2 skeletal maturation, growth and age at menarche in Tokyo girls. *Humanbiologica Budapestiensis*, **25**:261–5.

Ashworth, A. (1969). Growth rates in children recovering from protein-calorie malnutrition. *British Journal of Nutrition*, **23**:835–45.

Asmussen, I. (1980). Ultrastructure of the villi and fetal capillaries in placentas from smoking and nonsmoking mothers. *British Journal of Obstetrics and Gynaecology*, **87**:239–45.

Atkin, L.-M. and Davies, P. S. W. (2000). Diet composition and body composition in preschool children. *American Journal of Clinical Nutrition*, **72**:15–21.

Attie, M. K., Ramirez, N. R., Conte, F. A., Kaplan, S. L. and Grumbach, M. M. (1990). The pubertal growth spurt in eight patients with true precocious puberty and growth hormone deficiency: Evidence for a direct role of sex steroids. *Journal of Clinical Endocrinology and Metabolism*, **71**:975–83.

Aubert, M. L., Pierroz, D. D., Gruaz, N. M., d'Allèves, V., Vuagnat, B. A. M., Pralong, F. P., Blum, W. F. and Sizonenko, P. C. (1998). Metabolic control of sexual function and growth: Role of neuropeptide Y and leptin. *Molecular and Cellular Endocrinology*, **140**:107–13.

Aubert, R., Betoulle, D., Herbeth, B., Siest, G. and Fumeron, F. (2000). 5-HT$_{2A}$ receptor gene polymorphism is associated with food and alcohol intake in obese people. *International Journal of Obesity*, **24**:920–4.

Auestad, N., Montalto, M. B., Hall, R. T., Fitzgerald, K. M., Wheeler, R. E., Connor, W. E., Neuringer, M., Connor, S. L., Taylor, J. A. and Hartmann, E. E. (1997). Visual acuity, erythrocyte fatty acid composition, and growth in term infants fed formulas with long chain polyunsaturated fatty acids for one year. *Pediatric Research*, **41**:1–10.

August, G. P., Grumbach, M. M. and Kaplan, S. L. (1972). Hormonal changes in puberty. III. Correlation of plasma testosterone, LH, FSH, testicular size, and bone age with male pubertal development.

Journal of Clinical Endocrinology and Metabolism, **34**: 319–26.

Aynsley-Green, A., Zachmann, M. and Prader, A. (1976). Interrelation of the therapeutic effects of growth hormone and testosterone on growth in hypopituitarism. *Journal of Pediatrics*, **89**:992–9.

Bacallao, J., Amador, M. and Hermelo, M. (1996). The relationship of birthweight with height at 14 and with the growing process. *Nutrition*, **12**:250–4.

Bachrach, B. E. and Smith, E. P. (1996). The role of sex steroids in bone growth and development: Evolving new concepts. *Endocrinologist*, **6**:362–8.

Bachrach, L. K., Hastie, T., Wang, M.-C., Narasimhan, B. and Marcus, R. (1999). Bone mineral acquisition in healthy Asian, Hispanic, Black, and Caucasian youth: A longitudinal study. *Journal of Clinical Endocrinology and Metabolism*, **84**:4702–12.

Bai, Y., Zhan, S., Kim, K.-S., Lee, J.-K. and Kim, K.-H. (1996). Obese gene expression alters the ability of 30A5 preadipocytes to respond to lipogenic hormones. *Journal of Biological Chemistry*, **271**:13939–42.

Bailey, D. A. (1997). The Saskatchewan Pediatric Bone Mineral Accrual Study: Bone mineral acquisition during the growing years. *International Journal of Sports Medicine, Supplement*, **18**:191–4.

Bairagi, R., Chowdhury, M. K., Kim, Y. J. and Curlin, G. T. (1985). Alternative anthropometric indicators of mortality. *American Journal of Clinical Nutrition*, **42**:296–306.

Baker, J., Liu, J.-P., Robertson, E. J. and Efstatiadis, A. (1993). Role of insulin-like growth factors in embryonic and postnatal growth. *Cell*, **75**:73–82.

Baker, L. A., Reynolds, C. and Phelps, E. (1992). Biometrical analysis of individual growth curves. *Behavioral Genetics*, **22**:253–64.

Bala, R. M., Lopatka, J., Leung, A., McCoy, E. and McArthur, R. G. (1981). Serum immunoreactive somatomedin levels in normal adults, pregnant women at term, children at various ages and children with constitutionally delayed growth. *Journal of Clinical Endocrinology and Metabolism*, **52**:508–12.

Balcazar, H. and Haas, J. (1990). Classification schemes of small-for-gestational age and type of intrauterine growth retardation and its implications to early

neonatal mortality. *Early Human Development*, **24**:219–30.

Ball, J. C. (1967). The reliability and validity of interview data obtained from 59 narcotic drug addicts. *American Journal of Sociology*, **72**:650–4.

Ballard, J. L., Novak, K. K. and Driver, M. (1979). A simplified score for assessment of fetal maturation of newly born infants. *Journal of Pediatrics*, **95**:769–74.

Ballor, D. L. (1996). Exercise training and body composition changes. In *Human Body Composition*, eds. A. F. Roche, S. B. Heymsfield and T. Lohman, pp. 287–304. Champaign, IL: Human Kinetics.

Bandini, L. G., Schoeller, D. A., Cyr, H. N. and Dietz, W. H. (1990). Validity of reported energy intake in obese and nonobese adolescents. *American Journal of Clinical Nutrition*, **52**: 421–5.

Bar, A., Linder, B., Sobel, E. H., Saenger, P. and di Martino-Nardi, J. (1995). Bayley–Pinneau method of height prediction in girls with central precocious puberty: Correlation with adult height. *Journal of Pediatrics*, **126**:955–8.

Barbaglia, M., Ardizzi, A., Mostert, M., Moreni, G., Guzzaloni, G., Grugni, G. and Morabito, F. (1989). Reliability of height prediction methods in childhood adiposity. *Acta Medica Auxologica*, **21**:5–12.

Barbero, G. J. and Shaheen, E. (1967). Environmental failure to thrive: A clinical view. *Journal of Pediatrics*, **71**:639–44.

Barbieri, R. L. and Hornstein, M. D. (1988). Hyperinsulinaemia and ovarian hyperandrogenism: Cause and effect. *Endocrinology and Metabolism Clinics of North America*, **17**:685–703.

Barbieri, R. L., Makris, A. and Randall, R. W. (1986). Insulin stimulates androgen accumulation in incubations of ovarian stroma obtained from women with hyperandrogenism. *Journal of Clinical Endocrinology and Metabolism*, **62**:904–10.

Barera, G., Mora, S., Brambilla, P., Ricotti, A., Menni, L., Beccio, S. and Bianchi, C. (2000). Body composition in children with celiac disease and the effects of a gluten-free diet: A prospective case-control study. *American Journal of Clinical Nutrition*, **72**:71–5.

Barghini, G. (1987). La dinamica dello sviluppo fisico in relazione alla maturazione sessuale nei due sessi (Changes in physical development in relation to sexual maturation in the two sexes). *Acta Medica Auxologica*, 19:133–41.

Barghini, G. and Previtera, A. (1975). Sviluppo staturale ed età del menarca (Growth in stature and age at menarche). *Acta Medica Auxologica*, 7:217–20.

Barker, D. J. P. (1994). Outcome of low birthweight. *Hormone Research*, 42:223–30.

Barker, D. J. P. (1997). Maternal nutrition, fetal nutrition, and disease in later life. *Nutrition*, 13:807–13.

Barker, D. J. P., Gluckman, P. D., Godfrey, K. M., Harding, J. E., Owens, J. A. and Robinson, J. S. (1993*a*). Fetal nutrition and cardiovascular disease in adult life. *Lancet*, 341:938–41.

Barker, D. J. P., Godfrey, K. M., Fall, C., Osmund, C., Winter, P. D. and Shaheen, S. O. (1991). Relation of birth weight and childhood respiratory infection to adult lung function and death from chronic obstructive airways disease. *British Medical Journal*, 303:671–5.

Barker, D. J. P., Godfrey, K. M., Osmond C. and Bull, A. (1992*a*). The relation of fetal length, ponderal index and head circumference to blood pressure and the risk of hypertension in adult life. *Paediatric and Perinatal Epidemiology*, 6:35–44.

Barker, D. J. P., Hales, C. N., Fall, C. H. D., Osmond, C., Phipps, K. and Clark, P. M. S. (1993*b*). Type 2 (non-insulin-dependent) diabetes mellitus, hypertension and hyperlipidaemia (syndrome X): Relation to reduced fetal growth. *Diabetologia*, 36:62–7.

Barker, D. J. P., Meade, T. W., Fall, C. H. D., Lee, A., Osmond, C., Phipps, K. and Stirling, Y. (1992*b*). Relation of fetal and infant growth to plasma fibrinogen and factor VII concentrations in adult life. *British Medical Journal*, 304:148–52.

Barker, D. J. P., Osmond, C., Golding, J., Kuh, D. and Wadsworth, M. E. J. (1989*a*). Growth in utero, blood pressure in childhood and adult life, and mortality from cardiovascular disease. *British Medical Journal*, 298:564–7.

Barker, D. J. P., Winter, P. D., Osmond, C., Margetts, B. and Simmonds, S. J. (1989*b*). Weight in infancy and death from ischaemic heart disease. *Lancet*, ii:577–80.

Barker, M., Robinson, S., Osmond, C. and Barker, D. J. (1997). Birth weight and body fat distribution in adolescent girls. *Archives of Disease in Childhood*, 77:381–3.

Barkhouse, L. B., Fahey, J., Gillespie, C. T. and Cole, D. E. C. (1989). Quantitating the effect of cystic fibrosis on linear growth by mathematical modelling of longitudinal growth curves. *Growth, Development and Aging*, 53:185–90.

Baroncelli, G. I., Bertelloni, S., Ceccarelli, C. and Saggese, G. (1998). Measurement of volumetric bone mineral density accurately determines degree of lumbar undermineralization in children with growth hormone deficiency. *Journal of Clinical Endocrinology and Metabolism*, 83:3150–4.

Barr, D. G., Shmerling, D. H. and Prader, A. (1972). Catch-up growth in malnutrition, studied in celiac disease after institution of gluten-free diet. *Pediatric Research*, 6:521–7.

Barr, H. M., Streissguth, A. P., Martin, D. C. and Herman, C. S. (1984). Infant size at 8 months of age: Relationship to maternal use of alcohol, nicotine, and caffeine during pregnancy. *Pediatrics*, 74:336–41.

Barth, H., Döbler, T. and Amon, K. (1984). Zur Stagnation der Menarche (Stabilization of the age at menarche). *Ärztliche Jugendkunde*, 75:303–7.

Basso, O., Olsen, J., Johansen, A. M. T. and Christensen, K. (1997). Change in social status and risk of low birth weight in Denmark: Population-based cohort study. *British Medical Journal*, 315:1498–1502.

Bate, T. W. P., Price, D. H., Holme, C. A. and McGucken, R. B. (1984). Short stature caused by obstructive apnoea during sleep. *Archives of Disease in Childhood*, 59:78–80.

Baumgartner, R. N., Roche, A. F. and Himes, J. H. (1986). Incremental growth tables: Supplementary to previously published charts. *American Journal of Clinical Nutrition*, 43:711–22.

Bayley, N. and Davis, F. C. (1935). Growth changes in bodily size and proportions during the first three years: A developmental study of sixty-one children by repeated measurements. *Biometrika*, 27:26–87.

Bayley, N. and Pinneau, S. R. (1952). Tables for predicting adult height from skeletal age: Revised for use with the Greulich–Pyle hand standards. *Journal of Pediatrics*, 40:423–41.

Bayley, T. A., Harrison, J. E., McNeill, K. G. and Mernagh, J. R. (1980). Effect of thyrotoxicosis and its treatment on bone mineral and muscle mass. *Journal of Clinical Endocrinology and Metabolism*, **50**: 916–20.

Beaulac-Baillargeon, L. and Desrosiers, C. (1987). Caffeine–cigarette interaction on fetal growth. *American Journal of Obstetrics and Gynecology*, **157**:1236–40.

Becker, D. J., Ongemba, L. M., Brichard, V., Henquin, J.-C. and Britchard, S. M. (1995). Diet- and diabetes-induced changes of *ob* gene expression in rat adipose tissue. *Letters: Federation of European Biochemical Societies*, **371**:324–8.

Beckett, P. R., Wong, W. W. and Copeland, K. C. (1998). Developmental changes in the relationship between IGF-I and body composition during puberty. *Growth Hormone and IGF Research*, **8**:283–8.

Behan, D. P., Desouza, E. B., Lowry, P. J., Potter, E., Sawchenko, P. and Vale, W. W. (1995). Corticotropin releasing factor (CRF) binding protein: a novel regulator of CRF and related peptides. *Frontiers in Neuroendocrinology*, **16**:362–82.

Beitins, I. Z., Padmanabhan, V., Kasa-Vubu, J., Kleter, G. B. and Sizonenko, P. C. (1990). Serum bioactive follicle-stimulating hormone concentrations from prepuberty to adulthood: a cross-sectional study. *Journal of Clinical Endocrinology and Metabolism*, **71**:1022–7.

Belgorosky, A., Chahin, S., Chaler, E., Maceiras, M. and Rivarola, M. A. (1996). Serum concentrations of follicle stimulating hormone and luteinizing hormone in normal girls and boys during prepuberty and at early puberty. *Journal of Endocrinological Investigation*, **19**:88–91.

Belgorosky, A., Martinez, A., Heinrich, J. J. and Rivarola, M. A. (1989). Lack of correlation of serum estradiol with growth velocity during male pubertal growth. *Acta Endocrinologica*, **116**:579–83.

Bell, W., Davies, J. S., Evans, W. D. and Scanlon, M. F. (1999). Strength and its relationship to changes in fat-free mass, total body potassium, total body water and IGF-1 in adults with growth hormone deficiency: Effect of treatment with growth hormone. *Annals of Human Biology*, **26**:63–78.

Benso, L., Aicardi, G., Fabris, C. and Milani, S. (1999). What longitudinal studies can tell us about fetal growth. In *Human Growth in Context*, vol. 2, eds. F. E. Johnston, B. Zemel and P. B. Eveleth, pp. 41–50. London: Smith-Gordon.

Bercu, B. B., Lee, B. C., Spiliotis, B. E., Pineda, J. L., Denman, D. W., Hoffman, H. J. and Bioron, T. J. (1983). Male sexual development in the monkey. II. Cross-sectional analysis of pulsatile hypothalamic–pituitary secretion in castrated males. *Journal of Clinical Endocrinology and Metabolism*, **6**:1227–35.

Berelowitz, M., Szabo, M., Frohman, L. A., Firestone, S., Chu, L. and Hintz, R. L. (1981). Somatomedin mediates growth hormone negative feedback by effects on both the hypothalamus and the pituitary. *Science*, **212**:1279–81.

Bergadá, I. and Bergadá, C. (1995). Long term treatment with low dose testosterone in constitutional delay of growth and puberty: Effect on bone age, maturation and pubertal progression. *Journal of Pediatric Endocrinology and Metabolism*, **8**:117–22.

Bergman, P. and Goracy, M. (1984). The timing of adolescent growth spurts of ten body dimensions in boys and girls of the Wrocław Longitudinal Twin Study. *Journal of Human Evolution*, **13**:339–47.

Bergman, R. L. and Bergman, K. E. (1986). Nutrition and growth in infancy. In *Human Growth: A Comprehensive Treatise*, 2nd edn, vol. 3, Methodology, eds. F. Falkner and J. M. Tanner, pp. 389–413. New York: Plenum Press.

Bergsten-Brucefors, A. (1976). A note on the accuracy of recalled age at menarche. *Annals of Human Biology*, **3**:71–3.

Bergström, E., Hernell, O., Persson, L. Å. and Vessby, B. (1996). Insulin resistance syndrome in adolescents. *Metabolism: Clinical and Experimental*, **45**: 908–14.

Berkey, C. S., Dockery, D. W., Wang, X., Wypij, D. and Ferris, B., Jr. (1993). Longitudinal height velocity standards for US adolescents. *Statistics in Medicine*, **12**:403–14.

Berkey, C. S., Reed, R. B. and Valadian, I. (1983). Longitudinal growth standards for preschool children. *Annals of Human Biology*, **10**:57–67.

Berkey, C. S., Ware, J. H., Speizer, F. E. and Ferris, B. G. (1984). Passive smoking and height growth of preadolescent children. *International Journal of Epidemiology*, **13**:454–8.

Berkowitz, C. D. and Sklaren, B. C. (1984). Environmental failure to thrive: The need for intervention. *American Family Physician*, **29**:191–9.

Berkowitz, G. S., Holford, T. R. and Berkowitz, R. L. (1982). Effects of cigarette smoking, alcohol, coffee, and tea consumption on preterm delivery. *Early Human Development*, **7**:239–50.

Berkowitz, G. S., Kelsey, J. L., Holford, T. R. and Berkowitz, R. L. (1983). Physical activity and the risk of spontaneous preterm delivery. *Journal of Reproductive Medicine*, **28**:581–8.

Bernier, M., Chatelain, P. C., Mather, J. P. and Saez, J. M. (1986). Regulation of gonadotrophin receptors, gonadotrophin responsiveness, and cell multiplication by somatomedin-C and insulin in cultured pig Leydig cells. *Journal of Cellular Physiology*, **129**:257–63.

Beshyah S. A., Freemantle, C., Thomas, E., Rutherford, O., Page, B., Murphy, M. and Johnston, D. G. (1995). Abnormal body composition and reduced bone mass in growth hormone deficient hypopituitary adults. *Clinical Endocrinology*, **42**:179–89.

Bettger, L. and O'Dell, B. L. (1993). Physiological roles of zinc in the plasma membrane of mammalian cells. *Journal of Nutritional Biochemistry*, **4**:194–207.

Bettinelli, A. (1991). Growth in children with chronic renal failure undergoing conservative treatment. *Acta Medica Auxologica*, **23**:61–7.

Beunen, G., Lefevre, J., Ostyn, M., Renson, R., Simons, J. and van Gerven, D. (1990). Skeletal maturity in Belgian youths assessed by the Tanner–Whitehouse method (TW2). *Annals of Human Biology*, **17**:355–76.

Beunen, G. and Malina, R. (1988). Growth and physical performance relative to the timing of the adolescent spurt. *Exercise and Sport Sciences Reviews*, **16**:503–40.

Beunen, G., Thomis, M., Maes, H. H., Loos, R., Malina, R. M., Claessens, A. L. and Vlietinck, R. (2000). Genetic variance of adolescent growth in stature. *Annals of Human Biology*, **27**:173–86.

Beunen, G. P., Malina, R. M., Lefevre, J., Classens, A. L., Renson, R. and Simons J. (1997*a*). Prediction of adult stature and noninvasive assessment of biological maturation. *Medicine and Science in Sports and Exercise*, **29**:225–30.

Beunen, G. P., Malina, R. M., Lefevre, J., Claessens, A. L., Renson, R., van den Eynde, B., van Reusel, B. and Simons, J. (1997*b*). Skeletal maturation, somatic growth and physical fitness in girls 6–16 years of age. *International Journal of Sports Medicine*, **18**:413–19.

Beunen, G. P., Malina, R. M., Renson, R., Simons, J., Ostyn, M. and Lefevre, J. (1992). Physical activity and growth, maturation and performance: A longitudinal study. *Medicine and Science in Sports and Exercise*, **24**:576–85.

Beunen, G. P., Malina, R. M., Van't Hof, M. A., Simons, J., Ostyn, M., Renson, R. and Van Gerven, D. (1988). *Adolescent Growth and Motor Performance: A Longitudinal Study of Belgian Boys*. HKP Sport Science Monograph Series. Champaign, IL: Human Kinetics.

Beunen, G. P., Ostyn, M., Simons, J., Renson, R. and van Gerven, D. (1981). Chronological age and biological age as related to physical fitness in boys 12 to 19 years. *Annals of Human Biology*, **8**:321–31.

Beynen, A. C., Katan, M. B. and van Zutphen, L. F. (1987). Hypo and hyperresponders: Individual differences in the response of serum cholesterol concentration to changes in diet. *Advances in Lipid Research*, **22**:115–71.

Bhopal, R. S., Phillimore, P. and Kohli, H. S. (1991). Inappropriate use of the term "Asian": An obstacle to ethnicity and health research. *Journal of Public Health Medicine*, **13**:244–6.

Bianda, T., Hussain, M. A., Glatz, Y., Bopuillon, R., Froesch, E. R. and Schmid, C. (1997). Effects of short-term insulin-like growth factor-1 or growth hormone treatment on bone turnover, renal phosphate re-absorption and 1,25 dihydroxyvitamin D_3 production in healthy man. *Journal of Internal Medicine*, **241**:143–50.

Bidlingmaier, F., Dorr, H. G., Eisenmenger, W., Kuhnle, U. and Knoor, D. (1983). Testosterone and androstenedione concentrations in human testis and epididymis during the first two years of life. *Journal of Clinical Endocrinology and Metabolism*, **57**:311–15.

Bielicki, T. (1975). Interrelationships between various measures of maturation rate in girls during adolescence. *Studies in Physical Anthropology*, **1**:51–64.

Bielicki, T. (1976). On the relationships between maturation rate and maximum velocity of growth during adolescence. *Studies in Physical Anthropology*, **3**:79–84.

Bielicki, T. and Hauspie, R. C. (1994). On the independence of adult stature from the timing of the adolescent growth spurt. *American Journal of Human Biology*, **6**:245–7.

Bielicki, T. and Hulanicka, B. (1998). Secular trend in stature and age at menarche in Poland. In *Secular Growth Changes in Europe*, eds. É. B. Bodzsár and C. Susanne, pp. 263–79. Budapest: Eötvös University Press.

Bielicki, T., Koniarek, J. and Malina, R. M. (1984). Interrelationships among certain measures of growth and maturation rate in boys during adolescence. *Annals of Human Biology*, **11**:201–10.

Bielicki, T., Szklarska, A., Wełon, Z. and Malina, R. M. (2000). Variation in the body mass index among young adult Polish males between 1965 and 1995. *International Journal of Obesity*, **24**:658–62.

Bielicki, T., Waliszko, A., Hulanicka, B. and Kotlarz, K. (1986). Social-class gradients in menarcheal age in Poland. *Annals of Human Biology*, **13**:1–11.

Bielicki, T. and Waliszko, H. (1991). Urbanization-dependent gradients in stature among Polish conscripts in 1976 and 1986. *American Journal of Human Biology*, **3**:419–24.

Bielicki, T. and Wełon, Z. (1982). Growth data as indicators of social inequalities: The case of Poland. *Yearbook of Physical Anthropology*, **25**:153–67.

Billewicz, W. Z., Fellowes, H. M. and Thomson, A. M. (1981*a*). Pubertal changes in boys and girls in Newcastle-upon-Tyne. *Annals of Human Biology*, **8**:211–20.

Billewicz, W. Z., Fellowes, H. M. and Thomson, A. M. (1981*b*). Menarche in Newcastle-upon-Tyne girls. *Annals of Human Biology*, **8**:313–20.

Bingol, N., Fuchs, M., Diaz, V., Stone, R. D. and Gromisch, D. S. (1987). Teratogenicity of cocaine in humans. *Journal of Pediatrics*, **110**:93–6.

Bing-You, R. G., Denis, M.-C. and Rosen, C. J. (1993). Low bone mineral density in adults with previous hypothalamic-pituitary tumors: Correlation with serum growth hormone responses to GH-releasing hormone, insulin like growth factor I and IGF binding protein 3. *Calcified Tissue International*, **52**:183–7.

Binkin, N. J., Rust, K. R. and Williams, R. L. (1988). Racial differences in neonatal mortality: What causes of death explain the gap? *American Journal of Diseases of Children*, **142**:434–40.

Binnerts, A., Deurenberg, P., Swart, G. R., Wilson, J. H. P. and Lamberts, S. W. J. (1992). Body composition in growth hormone deficient adults. *American Journal of Clinical Nutrition*, **55**:918–23.

Bithoney, W. G., Dubowitz, H. and Egan, H. (1992). Failure to thrive/growth deficiency. *Pediatrics in Review*, **13**:453–9.

Bithoney, W. G., McJunkin, J., Michalek, J., Egan, H., Snyder, J. and Munier, A. (1989). Prospective evaluation of weight gain in both nonorganic and organic failure-to-thrive children: An outpatient trial of a multidisciplinary team intervention strategy. *Developmental and Behavioral Pediatrics*, **10**:27–31.

Bithoney, W. G. and Rathbun, J. M. (1983). Failure to thrive. In *Developmental–Behavioral Pediatrics*, eds. M. D. Levine, W. B. Carey, A. C. Crocker and R. T. Gross, pp. 557–72. Philadelphia, PA: WB Saunders.

Bjarnason, R., Boguszewski, M., Dahlgren, J., Gelander, L., Kriström, B., Rosberg, S., Carlsson, B., Albertsson-Wikland, K. and Carlsson, L. M. S. (1997). Leptin levels are strongly correlated with those of GH-binding protein in prepubertal children. *European Journal of Endocrinology*, **137**:68–73.

Bjerkedal, T. and Skjærven, R. (1980). Percentiler for fødselsvekt og isse-hællengde i forhold til svangerskapsvarighet for levendefødte enkeltfødte (Percentiles for birth weight and crown heel length in relation to duration of pregnancy for live born singletons). *Tidsskrift for Den Norske Lægeforening*, **16**:1088–91.

Björntorp, P. (1996). The regulation of adipose tissue distribution in humans. *International Journal of Obesity*, **20**:291–302.

Björntorp, P. and Edén, S. (1996). Hormonal influences on human body composition. In *Human Body Composition*, eds. A. F. Roche, S. B. Heymsfield and T. G. Lohman, pp. 329–44. Champaign, IL: Human Kinetics.

Björntorp, P. and Smith, U. (1976). The effect of fat cell size on subcutaneous adipose tissue metabolism. *Frontiers of Matrix Biology*, **2**:37–47.

Black, M. M., Dubowitz, H., Hutcheson, J., Berenson-Howard, J. and Starr, R. H., Jr. (1995).

A randomized clinical trial of home intervention for children with failure to thrive. *Pediatrics*, **95**: 807–14.

Bláha, P. and Vignerová, J. (1999). *Vývoj tĕlesných parametrŭ Èeských dĕtí a mládeže se zamĭoením na rozmĭry hlavy (0–16 let) (Development of Somatic Parameters of Czech Children and Adolescents Focused on Cephalic Parameters (0–16 years)*. Prague: National Institute of Public Health, Czech Republic.

Bleker, O. P., Breur, W. and Huidekoper, B. L. (1979). A study of birth weight, placental weight, and mortality of twins as compared to singletons. *British Journal of Obstetrics and Gynaecology*, **86**:111–18.

Bleker, O. P., Oosting, J. and Hemrika, D. J. (1988). On the cause of the retardation of fetal growth in multiple gestations. *Acta Geneticae Medicae et Gemellologiae*, **37**:41–6.

Blom, L., Persson, L. A. and Dahlquist, G. (1992). A high linear growth is associated with an increased risk of childhood diabetes mellitus. *Diabetologia*, **35**:528–33.

Blum, W. F., Englaro, P., Hanitsch, S., Juul, A., Hertel, N. T., Müller, J., Skakkebaek, N. E., Heiman, M. L., Birkett, M., Attanasio, A. M., Kiess, W. and Rascher, W. (1997). Plasma leptin levels in healthy children and adolescents: Dependence on body mass index, body fat mass, gender, pubertal stage, and testosterone. *Journal of Clinical Endocrinology and Metabolism*, **82**:2904–10.

Blum, W. F. and Gluckman, P. D. (1996). Insulin-like growth factors. In *Pediatrics and Perinatology*, eds. P. D. Gluckman and M. A. Heymann, pp. 314–23. London: Edward Arnold.

Blumchen, G. and Jette, M. (1992). Relationship between stature and coronary heart disease in a German male population. *International Journal of Cardiology*, **36**:35–45.

Bochénska, Z. (1978). *Zmiany w rozwoju osobniczym czlowieka w œwietle trendów sekularnych i ró¿nic spo³ecznych (Changes in the Processes of Human Growth and Physical Development Considered in the Light of Secular Trends and Social Differentiation)*. Kraków, Poland: Academy of Physical Education.

Bock, R. D. and Sykes, R. C. (1989). Evidence for continuing secular increase in height within families in the United States. *American Journal of Human Biology*, **1**:143–8.

Bock, R. D. and Thissen, D. M. (1976). Fitting multi-component models for growth in stature. *Proceedings of the 9th International Biometric Conference*, **1**:431–43.

Bock, R. D. and Thissen, D. M. (1980). Statistical problems of fitting individual growth curves. In *Human Physical Growth and Maturation Methodologies and Factors*, eds. F. E. Johnston, A. F. Roche and C. Susanne, pp. 265–90. New York: Plenum Press.

Boden, G., Chen, X., Mozzoli, M. and Ryan, I. (1996). Effect of fasting on serum leptin in normal human subjects. *Journal of Clinical Endocrinology and Metabolism*, **81**:3419–23.

Bodzsár, É. B. (1998). Secular growth changes in Hungary. In *Secular Growth Changes in Europe*, eds. É. B. Bodzsár and C. Susanne, pp. 175–205. Budapest: Eötvös University Press.

Bodzsár, É. B. and Pápai, J. (1994). Secular trend in body proportions and composition. In *Children and Youth at the End of the 20th Century, Human Biology*, ed. O. G. Eiben, pp. 245–54. Budapest: Humanbiologica Budapestiensis.

Bodzsár, É. B. and Susanne, C. (eds.) (1998). *Secular Growth Changes in Europe*. Budapest: Eötvös University Press.

Boetsch, G. and Bley, D. (1980). Age des premières règles dans une population de filles scolarisées Parisiennes (Age at menarche in a group of Parisian schoolchildren). *Bulletins et Mémoires de la Société d'Anthropologie de Paris*, **7**:3–6.

Bogin, B. A. (1978). Seasonal pattern in the rate of growth in height of children living in Guatemala. *American Journal of Physical Anthropology*, **49**:205–10.

Boguszewski, M., Albertsson-Wikland, K., Aronsson, S., Gustafsson, J., Hagenäs, L., Westgren, U., Westphal, O., Lipsanen-Nyman, M., Sipilä, I., Gellert, P., Müller, J. and Madsen, B. (1998). Growth hormone treatment of short children born small-for-gestational-age: The Nordic Multicentre Trial. *Acta Paediatrica*, **87**:257–63.

Boileau, R. A., Lohman, T. G. and Slaughter, M. H. (1985). Exercise and body composition of children and growth. *Scandinavian Journal of Sports*, **7**:17–27.

Bonjour, J. P., Theintz, G., Buchs, B., Slosman, D. and Rizzoli, R. (1991). Critical years and stages of puberty for spinal and femoral bone mass accumulation during adolescence. *Journal of Clinical Endocrinology and Metabolism*, **73**:555–63.

Böös, N., Wallin, A., Gbedegbegnon, T., Aebi, M. and Boesch, C. (1993). Quantitative MR imaging of lumbar intervertebral disks and vertebral bodies: Influence of diurnal water content variations. *Radiology*, **188**:351–4.

Boot, A. M., Bouquet, J., de Ridder, M. A. J., Krenning, E. P. and de Muinck Keizer-Schrama, S. M. P. F. (1997a). Determinants of body composition measured by dual-energy X-ray absorptiometry in Dutch children and adolescents. *American Journal of Clinical Nutrition*, **66**:232–8.

Boot, A. M., de Ridder, M. A. J., Pols, H. A. P., Krenning, E. P. and de Muinck Keizer-Schrama, S. M. P. F. (1997b). Bone mineral density in children and adolescents: Relation to puberty, calcium intake, and physical activity. *Journal of Clinical Endocrinology and Metabolism*, **82**:57–62.

Booth, M. L., Macaskill, P., Lazarus, R. and Baur, L. A. (1999). Sociodemographic distribution of measures of body fatness among children and adolescents in New South Wales, Australia. *International Journal of Obesity*, **23**:456–62.

Boothby, E. J., Guy, M. A. and Davies, T. A. L. (1952). The growth of adolescents. *Great Britain Ministry of Health, Monthly Bulletin*, **11**:208–23.

Borecki, I. B., Bonney, G. E., Rice, T., Bouchard, C. and Rao, D. C. (1993). Influence of genotype-dependent effects of covariates on the outcome of segregation analysis of the body mass index. *American Journal of Human Genetics*, **53**:676–87.

Borlee, I., Bouckaert, M. F., Lechat, M. F. and Misson, C. B. (1978). Smoking patterns during and before pregnancy: Weight, length and head circumference of progeny. *European Journal of Obstetrics, Gynecology, and Reproductive Biology*, **8**:171–7.

Boryslawski, K. (1988). Structure of monthly increments of length, weight and head circumference in the first year: A pure longitudinal study of 200 Wrocław infants. *American Journal of Human Biology*, **15**:205–12.

Boscherini, B., Cappa, M., Colabucci, F., Galasso, C., Giannotti, A. and Pasquino, A. M. (1993). Growth and pubertal spurt in Williams syndrome. *Acta Medica Auxologica*, **25**:185–9.

Bosley, A. R. J., Sibert, J. R. and Newcombe, R. G. (1981). Effects of maternal smoking on fetal growth and nutrition. *Archives of Disease in Childhood*, **56**:727–9.

Bossi, A., Cortinovis, I., Milani, S., Gallo, L., Kusterman, A. and Morotti, R. (1993). Twins: Neonatal standards for weight, length and head-size. *Acta Medica Auxologica*, **25**:81–96.

Bouchalová, M., Horáckova, E., Bystry, J. and Omelka, B. F. (1978). Vztah mezi výškami rodièù a jejich dìti ve viku 6,5–12 let (Relation between the heights of parents and their children). *Ceskoslovenská Pediatrie*, **33**:422–9.

Bouchard, C. and Pérusse, L. (1988). Heredity and body fat. *Annual Review of Nutrition*, **8**:259–77.

Bouchard, C., Pérusse, L., Leblanc, C., Tremblay, A. and Thériault, G. (1988). Inheritance of the amount and distribution of human body fat. *International Journal of Obesity*, **12**:205–15.

Bouchard, C., Tremblay, A., Després, J. P., Nadeau, A., Lupien, P. J., Thériault, G., Dussault, J., Moorjani, S., Pinault, S. and Fournier, G. (1990). The response to long-term overfeeding in identical twins. *New England Journal of Medicine*, **322**:1477–82.

Bouchard, C., Tremblay, A., Després, J. P., Thériault, G., Nadeau, A., Lupien, P. J., Moorjani, S., Prudhomme, D. and Fournier, G. (1994). The response to exercise with constant energy intake in identical twins. *Obesity Research*, **2**:400–10.

Boulton, J. (1981). Nutrition in childhood and its relationships to early somatic growth, body fat, blood pressure, and physical fitness. *Acta Paediatricia Scandinavica, Supplement*, **284**:1–85.

Boulton, T. J., Garnett, S. P., Cowell, C. T., Baur, L. A., Magarey, A. M. and Landers, M. C. G. (1999). Nutrition in early life: Somatic growth and serum lipids. *Annals of Medicine, Supplement*, **31**:7–12.

Boulton, T. J. C. and Magarey, A. M. (1995). Effects of differences in dietary fat on growth, energy and nutrient intake from infancy to eight years of age. *Acta Paediatrica*, **84**:146–50.

Bourguignon, J.-P. (1988). Linear growth as a function of age at onset of puberty and sex steroid dosage: Therapeutic implications. *Endocrine Reviews*, 9:467–488.

Bourguignon, J.-P. (1991). Growth and timing of puberty: Reciprocal effects.*Hormone Research*, 36:131–5.

Bourguignon, J.-P., Vandeweghe, M., Vanderschueren-Lodeweyckx, M., Malvaux, P., Wolter, R., Du Caju, M. and Ernould, C. (1986). Pubertal growth and final height in hypopituitary boys: A minor role of bone age at onset of puberty. *Journal of Clinical Endocrinology and Metabolism*, 63:376–82.

Bracken, M. B., Bryce-Buchanan, C., Silten, R. and Srisuphan, W. (1982). Coffee consumption during pregnancy. *New England Journal of Medicine*, 306:1548–9.

Braddon, F. E. M., Rodgers, B., Wadsworth, M. E. J. and Davies, J. M. C. (1986). Onset of obesity in a 36-year birth cohort study. *British Medical Journal*, 293:299–303.

Bradley, C. B., McMurray, R. G., Harrell, J. S. and Deng, S. (2000). Changes in common activities of 3rd through 10th graders: The CHIC Study. *Medicine and Science in Sports and Exercise*, 32:2071–8.

Brämswig, J. H., Fasse, M., Holthoff, M. L., von Lengerke, H. J., von Petrykowski, W. and Schellong, G. (1990). Adult height in boys and girls with untreated short stature and constitutional delay of growth and puberty: Accuracy of five different methods of height prediction. *Journal of Pediatrics*, 117:886–91.

Brämswig, J. H., Lengerke, H. J. von, Schmidt, H. and Schellong, G. (1988). The results of short-term (6 months) high-dose testosterone treatment on bone age and adult height in boys of excessively tall stature. *European Journal of Pediatrics*, 148:104–6.

Brandt, I. (1981). Brain growth, fetal malnutrition, and clinical consequences. *Journal of Perinatal Medicine*, 9:3–26.

Brandt, I. (1986). Growth dynamics of low-birth-weight infants with emphasis on the perinatal period. In *Human Growth: A Comprehensive Treatise*, 2nd edn, vol. 2, *Postnatal Growth*, eds. F. Falkner and J. M. Tanner, pp. 415–75. New York: Plenum Press.

Brandt, I. (1989). Growth of head circumference: Relationship to brain development, parent–child correlations and secular trends. In *Perspectives in the Science of Growth and Development*, ed. J. M. Tanner, pp. 221–6. London: Smith-Gordon.

Brandt, I. and Hansmann, M. (1977). Ultrasound diagnosis of intrauterine growth retardation and postnatal catch-up growth in head circumference: A combined longitudinal study. In *Anthropology of Maternity*, eds. A. Dolezad and J. Gutvirth, pp. 227–36. Prague: Charles University.

Brandt, I. and Reinken, L. (1988). Die Wachstumsgeschwindigkeit gesunder Kinder in den ersten 16 Lebensjahren: Longitudinale Entwicklungsstudie Bonn–Dortmund (The growth velocity of healthy children in the first 16 years of life: The Longitudinal Study of Development in Bonn, Dortmund). *Klinische Pediatrie*, 200:451–6.

Brans, Y. W., Sumners, J. E., Dweck, H. S. and Cassady, G. (1974). A noninvasive approach to body composition in the neonate: Dynamic skinfold measurements. *Pediatric Research*, 8:215–22.

Bray, P. F., Shields, D. W. and Wolcott, G. J. (1969). Occipitofrontal head circumference: An accurate measure of intracranial volume. *Journal of Pediatrics*, 75:303–5.

Brook, C. G. D., Huntley, R. M. C. and Slack, J. (1975). Influence of heredity and environment in determination of skinfold thickness in children. *British Medical Journal*, ii:719–21.

Brook, C. G. D. and Lloyd, J. K. (1973). Adipose cell size and glucose tolerance in obese children. *Archives of Disease in Childhood*, 48:301–4.

Brook, C. G. D., Murset, G., Zachmann, M. and Prader, A. (1974). Growth in children with 45,XO Turner's syndrome. *Archives of Disease in Childhood*, 49: 789–95.

Brooke, O. and Wheeler, E. F. (1976). High energy feeding in protein-energy malnutrition. *Archives of Disease in Childhood*, 51:968–77.

Brooke, O. G., Anderson, H. R., Bland, J. M., Peacock, J. L. and Stewart, C. M. (1989). Effects on birth weight of smoking, alcohol, caffeine, socioeconomic factors, and psychosocial stress. *British Medical Journal*, 298:795–801.

Brooke, O. G., Butters, F., Wood, C., Bailey, P. and Tukmachi, F. (1981). Size at birth from 37–41 weeks gestation: Ethnic standards for British infants of both sexes. *Journal of Human Nutrition*, **35**:415–30.

Brooks, A. A., Johnson, M. R., Steer, P. J., Pawson, M. E. and Abdalla, H. I. (1995). Birthweight: Nature or nurture? *Early Human Development*, **42**:29–35.

Brooks-Gunn, J., Warren, M. P., Rosso, J. and Gargiulo, J. (1987). Validity of measures of girls' pubertal status. *Child Development*, **58**:829–41.

Broussard, B. A., Johnson, A., Himes, J. H., Story, M., Fichtner, R., Hauck, F., Bachman-Carter, K., Hayes, J., Frohlich, K., Gray, N., Valway, S. and Gohdes, D. (1991). Prevalence of obesity in American Indians and Alaska Natives. *American Journal of Clinical Nutrition*, **53**:S1535–42.

Brown, E. D., Chan, W. and Smith, J. C. (1978). Bone mineralization during a developing zinc deficiency. *Proceedings of the Society for Experimental Biology and Medicine*, **157**:211–14.

Brown, J. E., Jacobson, H. N., Askue, L. H. and Peick, M. G. (1980). Influence of pregnancy weight gain on the size of infants born to underweight women. *Obstetrics and Gynecology*, **57**:13–17.

Brown, J. G., Bates, P. C., Holliday, M. A. and Millward, D. J. (1981). Thyroid hormones and muscle protein turnover: The effect of thyroid-hormone deficiency and replacement in thyroidectomized and hypophysectomized rats. *Biochemical Journal*, **194**:771–82.

Brown, M., Ahmed, M. L., Clayton, K. L. and Dunger, D. B. (1994). Growth during childhood and final height in type 1 diabetes. *Diabetic Medicine*, **11**:182–7.

Brundtland, G. H., Liestøl, K. and Walløe, L. (1980). Height, weight and menarcheal age of Oslo school children during the last 60 years. *American Journal of Human Biology*, **7**:307–22.

Brundtland, G. H. and Walløe, L. (1973). Menarcheal age in Norway: Halt in the trend toward earlier maturation. *Nature*, **241**:478–9.

Brunner, E., Shipley, M. J., Blane, D., Smith, G. D. and Marmot, M. G. (1999). When does cardiovascular risk start? Past and present socioeconomic circumstances and risk factors in adulthood. *Journal of Epidemiology and Community Health*, **53**:757–64.

Brunner, L., Nick, H.-P., Cumin, F., Chiesi, M., Baum, H.-P., Whitebread, S., Stricker-Krongrad, A. and Levens, N. (1997). Leptin is a physiologically important regulator of food intake. *International Journal of Obesity*, **21**:1152–60.

Buchanan, C. R., Cox, L. A., Dunger, D. B., Baines-Preece, J., Morrell, D. J. and Preece, M. A. (1989). A longitudinal study of serum insulin-like growth factor I, growth and pubertal development. *Hormone Research, Supplement*, **31**:51.

Buchanan, C. R. and Preece, M. A. (1992). Hormonal control of bone growth. In *Bone*, vol. 6, ed. B. K. Hall, pp. 53–89. Boca Raton, FL: CRC Press.

Buckler, J. M. H. (1978). Variations in height throughout the day. *Archives of Disease in Childhood*, **53**:762.

Buckler, J. M. H. (1990). *A Longitudinal Study of Adolescent Growth*. London: Springer-Verlag.

Buckler, J. M. H. (1995). The problems of interpreting growth data at puberty. In *Essays on Auxology*, eds. R. Hauspie, G. Lindgren and F. Falkner, pp. 178–93. Welwyn Garden City, U.K.: Castlemead Publications.

Buckler, J. M. H. and Green, M. (1994). Birth weight and head circumference standards for English twins. *Archives of Disease in Childhood*, **71**:516–21.

Buckler, J. M. H. and Green, M. (2001). Observations on growth and adult height in boys and girls delayed in puberty. *Acta Medica Auxologica*, **33**:1–12.

Bulmer, M. G. (1970). *The Biology of Twinning in Man*. Oxford, U.K.: Clarendon Press.

Bureau, M. A., Monette, J., Shapcott, D., Pare, C., Mathieu, J. L., Lippé, J., Blovin, D., Berthiaume, Y. and Begin, R. (1982). Carboxyhemoglobin concentration in fetal cord blood and in blood of mothers who smoked during labor. *Pediatrics*, **69**:371–3.

Bureau, M. A., Shapcott, D., Berthiaume, Y., Monette, J., Blouin, D., Blanchard, P. and Begin, R. (1983). Maternal cigarette smoking and fetal oxygen transport: A study of P50, 2, 3-diphosphoglycerate, total hemoglobin, hematocrit, and type F hemoglobin in fetal blood. *Pediatrics*, **72**:22–6.

Burgemeijer, R. J. F., Fredriks, A. M., van Buuren, S., Verloove-Vanhorick, S. P. and Wit, J. M. (1998). *Groeidiagrammen: Handleiding bij het meten en wegen* van kinderen en het invullen van groeidiagram. *(Growth Charts: Instructions to Measure Height and Weight in Children and How to Use Growth Charts)*. Houten, The Netherlands: Bohn Stafleu von Loghum.

Burke, G. L., Webber, L. S., Srinivasan, S. R., Radhakrishnamurthy, B., Freedman, D. S. and Berenson, G. S. (1986). Fasting plasma glucose and insulin levels and their relationship to cardiovascular risk factors in children: Bogalusa Heart Study. *Metabolism: Clinical and Experimental*, **35**:441–6.

Burns, T. L., Moll, P. P., Rost, C. A. and Lauer, R. M. (1987). Mothers remember birthweights of adolescent children: The Muscatine Ponderosity Family Study. *International Journal of Epidemiology*, **16**:550–5.

Burstein, S. and Rosenfeld, R. L. (1987). Constitutional delay in growth and development. In *Growth Abnormalities*, eds. R. L. Hintz and R. L. Rosenfield, pp. 167–85. New York: Churchill Livingstone.

Buschan, V. and Paneth, N. (1991). The reliability of neonatal head circumference measurement. *Journal of Clinical Epidemiology*, **44**:1027–35.

Buschang, P. H., Tanguay, A. and Demirjian, A. (1985). Growth instability of French-Canadian children during the first three years of life. *Canadian Journal of Public Health*, **76**:191–4.

Butenandt, O., Eder, R., Wohlfarth, K., Bidlingmaier, F. and Knorr, D. (1976). Mean 24 hr growth hormone and testosterone concentrations in relation to pubertal growth spurt in boys with normal or delayed puberty. *European Journal of Pediatrics*, **122**:85–92.

Butler, G. E., McKie, M. and Ratcliffe, S. G. (1990). The cyclical nature of prepubertal growth. *American Journal of Human Biology*, **17**:177–98.

Butler, M. G. and Meaney, F. J. (1987). An anthropometric study of 38 individuals with Prader–Labhart–Willi syndrome. *American Journal of Medical Genetics*, **26**:445–55.

Butler, N. R. and Alberman, E. D. (1969a). Maternal factors affecting duration of pregnancy, birthweight and fetal growth. In *Perinatal Problems: The Second Report of the 1958 British Perinatal Mortality Survey under the Auspices of the National Birthday Trust Fund*, eds. N. R. Bulter and E. D. Alberman, pp. 47–55. London: Churchill Livingstone.

Butler, N. R. and Alberman, E. D. (1969b). The effects of smoking in pregnancy. In *Perinatal Problems: The*

Second Report of the 1958 British Mortality Survey under the Auspices of the National Birthday Trust Fund, eds. N. R. Bulter and E. D. Alberman, pp. 72–84. London: Churchill Livingstone.

Butler, N. R. and Goldstein, H. (1973). Smoking in pregnancy and subsequent child development. *British Medical Journal*, **iv**:573–5.

Butler, N. R., Goldstein, H. and Ross, E. M. (1972). Cigarette smoking in pregnancy: Its influence on birth weight and perinatal mortality. *British Medical Journal*, **ii**:127–30.

Butte, N. F., Heinz, C., Hopkinson, J., Wong, W., Shypailo, R. and Ellis, K. (1999). Fat mass in infants and toddlers: Comparability of total body water, total body potassium, total body electrical conductivity, and dual-energy X-ray absorptiometry. *Journal of Pediatric Gastroenterology and Nutrition*, **29**:184–9.

Butte, N. F., Hopkinson, J. M. and Nicolson, M. A. (1997). Leptin in human reproduction: Serum leptin levels in pregnant and lactating women. *Journal of Clinical Endocrinology and Metabolism*, **82**:585–9.

Butte, N. F., Hopkinson, J. M., Wong, W. W., O'Brian Smith, E. and Ellis, K. J. (2000a). Body composition during the first 2 years of life: An updated reference. *Pediatric Research*, **47**:578–84.

Butte, N. F., Wong, W. W., Ferlic, L., O'Brian Smith, E., Klein, P. D. and Garza, C. (1990). Energy expenditure and deposition of breast-fed and formula-fed infants during early infancy. *Pediatric Research*, **28**:631–40.

Butte, N. F., Wong, W. W., Hopkinson, J. M., Heinz, C. J., Mehta, N. R. and O'Brian Smith, E. (2000b). Energy requirements derived from total energy expenditure and energy deposition during the first 2 years of life. *American Journal of Clinical Nutrition*, **72**:1558–69.

Byard, P. and Roche, A. F. (1984). Secular trend for recumbent length and stature in the Fels Longitudinal Growth Study. In *Human Growth and Development*, eds. J. Borms, R. Hauspie, A. Sand, C. Susanne and M. Hebbelinck, pp. 209–14. New York: Plenum Press.

Caan, B. J. and Goldhaber, M. K. (1989). Caffeinated beverages and low birth weight: A case-control study. *American Journal of Public Health*, **79**:1299–1300.

Cabral, H., Fried, L. E., Levenson, S., Amaro, H. and Zuckerman, B. (1990). Foreign-born and U.S.-born black women: Differences in health behaviors and birth outcomes. *American Journal of Public Health*, **80**:70–2.

Cameron, N. (1980). Conditional standards for growth in height of British children from 5.0 to 15.99 years of age. *Annals of Human Biology*, **7**:331–7.

Cameron, N., Tanner, J. M. and Whitehouse, R. H. (1982). A longitudinal analysis of the growth of limb segments in adolescence. *Annals of Human Biology*, **9**:211–20.

Camilleri, A. P. (1981). The obstetric significance of short stature. *European Journal of Obstetrics, Gynaecology, and Reproductive Biology*, **12**:347–56.

Campbell, P. T., Katzmarzyk, P. T., Malina, R. M., Rao, D. C., Pérusse, L. and Bouchard, C. (2001). Stability of adiposity phenotypes from childhood and adolescence into young adulthood with contribution of parental measures. *Obesity Research*, **9**:394–400.

Campos, S. P. and MacGillivray, M. H. (1989). Sex steroids do not influence somatic growth in childhood. *American Journal of Diseases of Childhood*, **143**:942–3.

Canalis, E. (1980). Effect of insulin like growth factor I on DNA and protein synthesis in cultured rat calvaria. *Journal of Clinical Investigation*, **66**:709–16.

Capitanio, M. A. and Kirkpatrick, J. A. (1969). Widening of the cranial sutures: A roentgen observation during periods of accelerated growth in patients treated for deprivation dwarfism. *Radiology*, **92**:53–9.

Caprio, S., Cline, G., Boulware, S. D., Permanente, C., Shulman, G. I., Sherwin, R. S. and Tamborlane, W. V. (1994). Effects of puberty and diabetes on metabolism of insulin-sensitive fuels. *American Journal of Physiology*, **266**:E885–91.

Caprio, S., Hyman, L. D., Limb, C., McCarthy, S., Lange, R., Sherwin, R. S., Shulman, G. and Tamborlane, W. V. (1995). Central adiposity and its metabolic correlates in obese adolescent girls. *American Journal of Physiology*, **269**:E118–26.

Caprio, S., Plewe, G., Diamond, M. P., Simonson, D. C., Boulware, S. D., Sherwin, R. S. and Tamborlane, W. V. (1989). Increased insulin secretion in puberty: A compensatory response to reductions in insulin sensitivity. *Journal of Pediatrics*, **114**:963–7.

Caprio, S., Tamborlane, W. V., Silver, D., Robinson, C., Leibel, R., McCarthy, S., Grozman, A., Belous, A., Maggs, D. and Sherwin, R. S. (1996).

Hyperleptinemia: An early sign of juvenile obesity. Relations to body fat deposits and insulin concentrations. *American Journal of Physiology, Endocrinology and Metabolism*, **34**:E626–30.

Cara, J. F., Rosenfield, R. L. and Furlanetto, R. W. (1987). A longitudinal study of the relationship of plasma somatomedin-C concentration to the pubertal growth spurt. *American Journal of Diseases of Children*, **141**:562–4.

Cardon, L. R. (1996). Height, weight, and obesity. In *Pediatric Endocrinology*, 3rd edn, ed. F. Lifshitz, pp. 165–72. New York: Marcel Dekker.

Carlsson, B., Ankarberg, C., Rosberg, S., Norjavaara, E., Albertsson-Wikland, K. and Carlsson, L. M. S. (1997). Serum leptin concentrations in relation to pubertal development. *Archives of Disease in Childhood*, **77**:396–400.

Carlucci, P., Rusconi, R., Appiani, A. C., Ardissino, G., Daccò, V. and Testa, S. (2000). Pubertal growth and its role in chronic renal failure. *Acta Medica Auxologica*, **32**:147–51.

Caro, J. F., Kolaczynski, J. W., Nyce, M. R., Ohannesian, J. P., Opentanova, I., Goldman, W. H., Lynn, R. B., Zhang, P. L., Sinha, M. K. and Considine, R. V. (1996). Decreased cerebrospinal-fluid/serum leptin ratio in obesity: A possible mechanism for leptin resistance. *Lancet*, **348**:159–61.

Carrascosa, A., Audi, L., Ferrandez, M. A. and Ballabriga, A. (1990). Biological effects of androgens and identification of specific dihydrotestosterone-binding sites in cultured human fetal epiphyseal chondrocytes. *Journal of Clinical Endocrinology and Metabolism*, **70**:134–40.

Carter-Su, C. and Okamoto, K. (1987). Effect of insulin and glucocorticoids on glucose transporters in rat adipocytes. *American Journal of Physiology*, **252**:E441–53.

Cartwright, A. and Smith, C. (1979). Some comparisons of data from medical records and from interviews with women who had recently had a live birth or stillbirth. *Journal of Biosocial Science*, **11**:49–64.

Caruso-Nicoletti, M., Cassorla, F. G., Skerda, M. C., Ross, J. L., Loriaux, D. L. and Cutler, G. B., Jr. (1985). Short-term low dose estradiol accelerates ulnar growth in boys. *Journal of Clinical Endocrinology and Metabolism*, **61**:896–8.

Casanueva, F. F., Villanueva, L., Dieguéz, C., Diaz, Y., Cabranes, J. A., Szoke, B., Scanlon, M. F., Schally, A. V. and Ferández-Cruz, A. (1987). Free fatty acids block growth hormone hormone-stimulated GH secretion in man directly at the pituitary. *Journal of Clinical Endocrinology and Metabolism*, 65:634–42.

Casey, P. H. (1983). Failure to thrive: A reconceptualization. *Developmental and Behavioral Pediatrics*, 4:63–6.

Casey, P. H. and Arnold, W. C. (1983). Growth rebound in infants with severe failure to thrive. *Pediatric Research*, 17:94.

Casey, P. H. and Arnold, W. C. (1985). Compensatory growth in infants with severe failure to thrive. *Southern Medical Journal*, 78:1057–60.

Casey, P. H., Kelleher, K. J., Bradley, R. H., Kellogg, K. W., Kirby, R. S. and Whiteside, L. (1994). A multifaceted intervention for infants with failure to thrive. *Archives of Pediatrics and Adolescent Medicine*, 148:1071–7.

Casey, P. H., Kelleher, K. J., Bradley, R. H., Kellogg, K. W., Kirby, R. S. and Whiteside, L. (1995). A multifaceted intervention for infants with failure to thrive. *Archives of Pediatrics and Adolescent Medicine*, 149:1039–41.

Casey, P. H., Kraemer, H. C., Bernbaum, J., Yogman, M. W. and Sells, J. C. (1991). Growth status and growth rates of a varied sample of low birth weight, preterm infants: A longitudinal cohort from birth to three years of age. *Journal of Pediatrics*, 119: 599–605.

Casey, V. A., Dwyer, J. T., Coleman, K. A. and Valadian, I. (1992). Body mass index from childhood to middle age: a 50-y follow-up. *American Journal of Clinical Nutrition*, 56:14–18.

Castillo-Durán, C., Rodriguez, A., Venegas, G., Avarez, P. and Icaza, G. (1995). Zinc supplementation and growth of infants born small-for-gestational age. *Journal of Pediatrics*, 127:206–11.

Caulfield, L. E., West, S. K., Barron, Y. and Cid-Ruzafa, J. (1999). Anthropometric status and cataract: The Salisbury Eye Evaluation Project. *American Journal of Clinical Nutrition*, 69:237–42.

Cavalieri, R. R. and Rapoport, B. (1977). Impaired peripheral conversion of thyroxine to triiodothyronine. *Annual Review of Medicine*, 28:57–65.

Cawley, R. H., McKeown, T. and Record, R. G. (1954). Parental stature and birth weight. *Human Genetics*, 6:448–55.

Cernerud, L. (1993). Height and body mass index of 7-year-old Stockholm school children from 1940 to 1990. *Acta Paediatrica*, 82:304–5.

Cernerud, L. (1994). Are there still inequalities in height and body mass index of Stockholm children? *Scandanavian Journal of Social Medicine*, 22:161–5.

Cernerud, L. and Lindgren, G. W. (1991). Secular changes in height and weight of Stockholm schoolchildren born in 1933, 1943, 1953 and 1963. *Annals of Human Biology*, 18:497–505.

Cha, M. C. and Rojhani, A. (1997). Zinc deficiency inhibits the direct growth effect of growth hormone on the tibia of hypophysectomized rats. *Biological Trace Element Research*, 59:99–111.

Chagnon, Y. C., Wilmore, J. H., Borecki, I. B., Gagnon, J., Pérusse, L., Chagnon, M., Collier, G. R., Leon, A. S., Skinner, J. S., Rao, D. C. and Bouchard, C. (2000). Associations between the leptin receptor gene and adiposity in middle-aged Caucasian males from the HERITAGE Family Study. *Journal of Clinical Endocrinology and Metabolism*, 85:29–34.

Chakraborty, R., Kamboh, I. and Ferrell, R. E. (1991). "Unique" alleles in admixed populations: A strategy for determining "hereditary" population differences of disease frequencies. *Ethnicity and Disease*, 1: 245–56.

Chamberlain, R. (1975). Birthweight and length of gestation. In *British Births 1970*, ed. R. Chamberlain, pp. 48–88. London: Heinemann Medical Books.

Chamla, M. C. (1983). L'évolution récente de la stature en Europe occidentale (Période 1960–1980) (The recent change in stature in Western Europe, Period 1960–1980). *Bulletins et Mémoires de la Société d'Anthropologie de Paris*, 10:195–224.

Chan, G., Hoffman, K. and McMurry, M. (1995). Effects of dairy products on bone and body composition in pubertal girls. *Journal of Pediatrics*, 126:551–6.

Chan, G. and Mileu, L. (1985). Posthospitalization growth and bone mineral status of normal preterm infants. *American Journal of Diseases of Children*, 139:896–8.

Chan, G. M. (1992). Performance of dual-energy X-ray absorptiometry in evaluating bone, lean body mass, and fat in pediatric subjects. *Bone and Mineral Research*, 7:369–74.

Chang, T. C., Robson, S. C. and Spencer, J. A. D. (1993). Neonatal morphometric indices of fetal growth: Analysis of observer variability. *Early Human Development*, 35:37–43.

Chaning-Pearce, S. M. and Solomon, L. (1986). A longitudinal study of height and weight in black and white Johannesburg children. *South African Medical Journal*, 70:743–6.

Charney, E., Goodman, H. C., McBride, M., Lyon, B. and Pratt, R. (1976). Childhood antecedents of adult obesity: Do chubby infants become obese adults? *New England Journal of Medicine*, 295: 6–9.

Charzewski, J. and Bielicki, T. (1978). Is the secular trend in stature associated with relative elongation on the limbs? *Homo*, 29:176–81.

Chase, P. and Martin, H. (1970). Undernutrition and child development. *New England Journal of Medicine*, 282:933–9.

Chasnoff, I., Griffith, D., Freier, C. and Murray, J. (1992). Cocaine/polydrug use in pregnancy: Two-year follow-up. *Pediatrics*, 89:284–9.

Chasnoff, I., Hatcher, R. and Burns, W. (1980). Early growth patterns of methadone-addicted infants. *American Journal of Diseases of Children*, 134: 1049–51.

Chasnoff, I. J., Lewis, D. E., Griffith, D. R. and Willey, S. (1989). Cocaine and pregnancy: Clinical and toxicological implications for the neonate. *Clinical Chemistry*, 35:1276–8.

Chatelain, P. and Nicolino, M. (1994). Intrauterine growth retardation and Silver–Russell syndrome: Demography, auxology and response to growth hormone treatment in the Kabi International Growth Study. In *Progress in Growth Hormone Therapy: 5 Years of KIGS*, eds. M. B. Ranke and R. Gunnarsson, pp. 230–9. Mannheim, Germany: J. & J. Verlag.

Chaussain, J. L., Colle, M. and Ducret, J. P. (1994). Adult height in children with prepubertal short

stature secondary to intrauterine growth retardation. *Acta Paediatrica, Supplement*, 399:72–3.

Cheek, D. and Hill, D. (1974). Effect of growth hormone on cell and somatic growth. In *Handbook of Physiology*, eds. R. Greep and E. Astwood, pp. 159–85. Washington, D. C.: American Physiological Society.

Chen, L. C., Chowdhury, A. K. M. A. and Huffman, S. L. (1980). Anthropometric assessment of energy-protein malnutrition and subsequent risk of mortality among preschool aged children. *American Journal of Clinical Nutrition*, 33:1836–45.

Chen, S., Vohr, B. and Oh, W. (1993). Effects of birth order, gender, and intrauterine growth retardation on the outcome of very low birth weight in twins. *Journal of Pediatrics*, 123:132–6.

Cherukuri, R., Minkoff, H., Feldman, J., Parekh, A. and Glass, L. (1988). A cohort study of alkaloidal cocaine ("crack") in pregnancy. *Obstetrics and Gynecology*, 72:147–51.

Chesters, J. K. (1992). Trace element–gene interactions. *Nutrition Reviews*, 50:217–23.

Chiarelli, F., Tumini, S., Verrotti, A., Verini, M. and Morgese, G. (1988). Growth in pediatric and adolescent diabetes mellitus. *Acta Medica Auxologica*, 20:115–24.

Chinn, S., Hughes, J. M. and Rona, R. J. (1998). Trends in growth and obesity in ethnic groups in Britain. *Archives of Disease in Childhood*, 78:513–17.

Chinn, S. and Rona, R. J. (1991). Quantifying health aspects of passive smoking in British children aged 5 to 11 years. *Journal of Epidemiology and Community Health*, 45:188–94.

Chinn, S., Rona, R. J., Gulliford, M. C. and Hammond, J. (1992). Weight for height in children aged 4 to 12 years: A new index compared to the normalized body mass index. *European Journal of Clinical Nutrition*, 46:489–500.

Chiu, K. M., Keller, E. T., Crenshaw, T. D. and Gravenstein, S. (1999). Carnitine and dehydroepiandrosterone sulfate induce protein synthesis in porcine primary osteoblast-like cells. *Calcified Tissue International*, 64:527–33.

Cho, S. D., Mueller, W. H., Meininger, J. C., Liehr, P. and Chan, W. (2001). Blood pressure and sexual maturity in adolescents: The Heartfelt Study. *American Journal of Human Biology*, **13**:227–34.

Chouteau, M., Namerou, P. B. and Leppert, P. (1988). The effects of cocaine abuse on birthweight and gestational age. *Obstetrics and Gynecology*, **72**:351–4.

Christensen, K., Vaupel, J. W., Holm, N. V. and Yashin, A. I. (1995). Mortality among twins after age 6: Fetal origins hypothesis versus twin method. *British Medical Journal*, **310**:432–6.

Chrousos, G. and Gold, P. (1992). The concept of stress and stress system disorders. *Journal of the American Medical Association*, **267**:1244–52.

Chrzastek-Spruch, H. M. (1977). Some genetic problems in physical growth and development. *Acta Geneticae Medicae et Gemellologiae*, **26**:205–19.

Chrzastek-Spruch, H. M. (1979). Genetic control of some somatic traits in children, as assessed by longitudinal studies. *Studies in Human Ecology*, **3**:27–51.

Chrzastek-Spruch, H. M., Susanne, C., Hauspie, R. C. and Kozlowska, M. A. (1989). Individual growth patterns and standards for height and height velocity based on the Lublin Longitudinal Growth Study. In *Perspectives in the Science of Growth and Development*, ed. J. M. Tanner, pp. 161–6. London: Smith-Gordon.

Chrzastek-Spruch, H. M. and Wolánski, N. (1969). Body length and weight in newborns, and infant growth connected with parents stature and age. *Genetica Polonica*, **10**:257–62.

Chu, N.-F., Wang, D.-J., Shieh, S.-M. and Rimm, E. B. (2000). Plasma leptin concentrations and obesity in relation to insulin resistance syndrome components among school children in Taiwan: The Taipei Children Heart Study. *International Journal of Obesity*, **24**:1265–71.

Churchill, D., Perry, I. J. and Beevers, D. G. (1997). Ambulatory blood pressure in pregnancy and fetal growth. *Lancet*, **349**:7–10.

Chyou, P.-H., Nomura, A. M. Y. and Stemmermann, G. N. (1994). A prospective study of weight, body mass index and other anthropometric measurements in relation to site-specific cancers. *International Journal of Cancer*, **57**:313–17.

Cicognani, A., Cacciari, E., Tacconi, M., Pascucci, M. G., Tonioli, S., Pirazzoli, P. and Balsamo, A. (1989). Effects of gonadectomy on growth hormone, IGF-I, and sex steroids in children with complete and incomplete androgen insensitivity. *Acta Endocrinologica*, **121**:777–83.

Cigolini, M. and Smith, U. (1979). Human adipose tissue in culture. VIII. Studies on the insulin-antagonistic effect of glucocorticoids. *Metabolism: Clinical and Experimental*, **28**:502–10.

Clapp J. F. III and Capeless, E. L. (1990). Neonatal morphometrics after endurance exercise during pregnancy. *American Journal of Obstetrics and Gynecology*, **163**:1805–11.

Clapp J. F. III and Dickstein, S. (1984). Endurance exercise and pregnancy outcome. *Medicine and Science in Sports and Exercise*, **16**:556–62.

Clark, P. A. and Rogol, A. D. (1996). Growth hormones and sex steroid interactions at puberty. *Endocrinology and Metabolism Clinics of North America*, **25**:665–81.

Clark, P. J. (1956). The hereditability of certain anthropometric characters as ascertained from measurements of twins. *American Journal of Human Genetics*, **8**:49–54.

Clausson, B., Lichtenstein, P. and Cnattingius, S. (2000). Genetic influence on birthweight and gestational length determined by studies in offspring of twins. *British Journal of Genetics*, **107**:357–81.

Clayton, P. E., Gill, M. S., Hall, C. M., Tillmann, V., Whatmore, A. J. and Price, D. A. (1997). Serum leptin through childhood and adolescence. *Clinical Endocrinology*, **46**:727–33.

Clément, K., Dina, C., Basdevant, A., Chastang, N., Pelloux, V., Lahlou, N., Berlan, M., Langin, D., Guy-Grand, B. and Frougel, P. (1999). A sib-pair analysis study of 15 candidate genes in French families with morbid obesity. *Diabetes*, **48**:398–402.

Clément, K., Ruiz, J., Cassard-Doulcier, A. M., Bouillaud, R., Ricquier, D., Basdevant, A., Guy-Grand, B. and Froguel, P. (1996). Additive effect of A→G (-3826) variant of the uncoupling protein gene and the Trp64Arg mutation of the beta3-adrenergic gene on weight gain in morbid obesity. *International Journal of Obesity*, **12**:1062–6.

Clément, K., Vaisse, C., Lahlou, N., Cabrol, S., Pelloux, V., Cassuto, D., Gourmelen, M., Dina, C., Chambaz, J., Lacorte, J. M., Basdevant, A., Bougnères, P.,

Lebouc, Y., Froguel, P. and Guy-Grand, B. (1998). A mutation in the human leptin receptor gene causes obesity and pituitary dysfunction. *Nature*, **392**:398–401.

Cleveland, W. S. (1979). Robust locally weighted regression and smoothing scatterplots. *Journal of the American Statistical Association*, **74**:829–36.

Cliver, S. P., Goldenberg, R. L., Cutter, G. R., Hoffman, H. J., Davis, R. O. and Nelson, K. G. (1995). The effect of cigarette smoking on neonatal anthropometric measurements. *Obstetrics and Gynecology*, **85**:625–30.

Cnattingius, S., Axelsson, O., Eklund, G. and Lindmark, G. (1985). Smoking, maternal age, and fetal growth. *Obstetrics and Gynecology*, **66**: 449–52.

Cnattingius, S., Bergstrom, R., Lipworth, L. and Kramer, M. S. (1998). Prepregnancy weight and the risk of adverse pregnancy outcomes. *New England Journal of Medicine*, **338**:147–52.

Coble, Y. D., Jr., Bardin, C. W., Ross, G. T. and Darby, W. T. (1971). Studies of endocrine function in boys with retarded growth, delayed sexual maturation and zinc deficiency. *Journal of Clinical Endocrinology and Metabolism*, **32**:361–7.

Cogswell, M. A., Serdula, M. K., Hungerford, D. W. and Yip, R. (1995). Gestational weight gain among average-weight and overweight women: What is excessive? *American Journal of Obstetrics and Gynecology*, **172**:705–12.

Cohen, S. B., Dulitzky, M., Lipitz, S., Mashiach, S. and Schiff, E. (1997). New birth weight nomograms for twin gestation on the basis of accurate gestational age. *American Journal of Obstetrics and Gynecology*, **177**:1101–4.

Cole, P. V., Hawkins, L. H. and Roberts, D. (1972). Smoking during pregnancy and its effects on the fetus. *Journal of Obstetrics and Gynaecology of the British Commonwealth*, **97**:782–7.

Cole, T. J. (1993). Seasonal effects on physical growth and development. *Society for the Study of Human Biology, Symposium* 35:89–106.

Cole, T. J. (1995). Conditional reference charts to assess weight gain in British infants. *Archives of Disease in Childhood*, **73**:8–16.

Cole, T. J. (1998). Seasonality of growth. In *The Cambridge Encyclopedia of Human Growth and Development*, eds. S. J. Ulijaszek, F. E. Johnston and M. A. Preece, p. 223. Cambridge, U.K.: Cambridge University Press.

Cole, T. J., Bellizzi, M. C., Flegal, K. M. and Dietz, W. H. (2000). Establishing a standard definition for child overweight and obesity worldwide: International survey. *British Medical Journal*, **320**:1–6.

Cole, T. J. and Green, P. J. (1992). Smoothing reference centile curves: The LMS method and penalized likelihood. *Statistics in Medicine*, **11**:1305–19.

Cole, T. J., Paul, A. A., Eccles, M. and Whitehead, R. G. (1989). The use of a multiple growth standard to highlight the effect of diet and infection on growth. In *Perspectives in the Science of Growth and Development*, ed. J. M. Tanner, pp. 91–100. London: Smith-Gordon.

Cole, T. J. and Roede, M. J. (1999). Centiles of body mass index for Dutch children aged 0–20 years in 1980: A baseline to assess recent trends in obesity. *Annals of Human Biology*, **26**:303–8.

Colle, E., Schiff, D., Andrew, G., Bauer, C. B. and Fitzhardinge, M. D. (1976). Insulin responses during catch-up growth of infants who were small for gestational age. *Pediatrics*, **57**:363–71.

Collell, M., Pavia, C., Sanz, M. C. and Eighian, B. (1993). Growth and pubertal development in diabetic children: A longitudinal study. *Acta Medica Auxologica* 25:43–9.

Collins, C. E., MacDonald-Wicks, L., Rowe, S., O'Loughlin, E. V. and Henry, R. L. (1999). Normal growth in cystic fibrosis associated with a specialized centre. *Archives of Disease in Childhood*, **81**:241–6.

Collins, J. W. and Shay, D. K. (1994). Prevalence of low birth weight among Hispanic infants with United States-born and foreign-born mothers: The effect of urban poverty. *American Journal of Epidemiology*, **139**:184–92.

Colvard, D. S., Eriksen, E. F., Keeting, P. E., Wilson, E. M., Lubahn, D. B., French, F. S., Riggs, B. L. and Spelsberg, T. C. (1989). Identification of androgen receptors in normal human osteoblast-like cells. *Proceedings of the National Academy of Sciences (U.S.A.)*, **86**:854–7.

Committee on the Surveillance Project on the Condition of Children's Health (1998). Report for 1997. Tokyo: Japanese Society of School Health.

Comuzzie, A. G., Blangero, J., Mahaney, M. C., Mitchell, B. D., Hixson, J. E., Samollow, P. B., Stern, M. P. and MacCluer, J. W. (1995). Major gene with sex-specific effects influences fat mass in Mexican Americans. *Genetic Epidemiology*, **12**:475–88.

Comuzzie, A. G., Blangero, J., Mitchell, B. D., Stern, M. P. and MacCluer, J. W. (1993). Segregation analysis of fat mass and fat free mass. *Genetic Epidemiology*, **10**:340–1.

Connaughton, J., Reeser, D., Schut, J. and Finnegan, L. (1977). Perinatal addiction: Outcome and management. *American Journal of Obstetrics and Gynecology*, **129**:679–86.

Conover, C. A., Lee, P. D. K., Kanaley, J. A., Clarkson, J. T. and Jensen, M. D. (1992). Insulin regulation of insulin-like growth factor binding protein-I in obese and nonobese humans. *Journal of Clinical Endocrinology and Metabolism*, **74**:1355–60.

Considine, R. V., Sinha, M. K., Heiman, M. L., Kriauciunas, A., Stephens, T. W., Nyce, M. R., Ohannesian, J. P., Marco, C. C., McKee, L. J., Bauer, T. L. and Caro, J. F. (1996). Serum immunoreactive leptin concentrations in normal-weight and obese humans. *New England Journal of Medicine*, **334**: 292–5.

Constantine, N. A., Kraemer, H. C., Kendall-Tackett, K. A., Bennett, F. C., Tyson, J. E. and Gross, R. T. (1987). Use of physical and neurologic observations in assessment of gestational age in low birth weight infants. *Journal of Pediatrics*, **110**:921–8.

Constantini, D., Rozzoni, C. S. and Arban, D. (1991). Growth evaluation in cystic fibrosis. *Acta Medica Auxologica* **23**:85–90.

Conte, F. A., Grumbach, M. M., Ito, Y., Fisher, C. R. and Simpson, E. R. (1994). A syndrome of female pseudohermaphrodism, hypergonadotropic hypogonadism, and multicystic ovaries associated with missense mutations in the gene encoding aromatase (P450arom). *Journal of Clinical Endocrinology and Metabolism*, **78**:1287–92.

Conter, V., Cortinovis, I., Rogari, P. and Riva, L. (1995). Weight growth in infants born to mothers who smoked during pregnancy. *British Medical Journal*, **310**:768–71.

Cook, J., Altman, D., Moore, D., Topp, S., Holland, W. and Elliott, A. (1973). A survey of the nutritional status of schoolchildren: Relation between nutrient intake and socio-economic factors. *British Journal of Preventive and Social Medicine*, **27**:91–9.

Cook, J. S., Hoffman, R. P., Stene, M. A. and Hansen, J. R. (1993). Effects of maturational stage on insulin sensitivity during puberty. *Journal of Clinical Endocrinology and Metabolism*, **77**:725–30.

Cooke, R. W. I., Lucas, A., Yudkin, P. L. N. and Pryse-Davies, J. (1977). Head circumference as an index of brain weight in the fetus and newborn. *Early Human Development*, **1**:145–9.

Cooper, C., Cawley, M., Bhalla, A., Egger, P., Ring, F., Morton, L. and Barker, D. (1995). Childhood growth, physical activity, and peak bone mass in women. *Journal of Bone and Mineral Research*, **10**:940–7.

Copeland, K. C., Paunier, L. and Sizonenko, P. C. (1977). The secretion of adrenal androgens and growth patterns of patients with hypogonadotropic hypogonadism and idiopathic delayed puberty. *Journal of Pediatrics*, **91**:985–90.

Corbier, P., Dehennin, L., Castanier, M., Mebazaa, A., Edwards, D. A. and Roffi, J. (1990). Sex differences in serum luteinizing hormone and testosterone in the human neonate during the first few hours after birth. *Journal of Clinical Endocrinology and Metabolism*, **71**:1344–8.

Corey, L. A., Eaves, L. J., Mellen, B. G. and Nance, W. E. (1986). Testing for developmental changes in gene expression on resemblance for quantitative traits in kinships of twins: Application to height, weight, and blood pressure. *Genetic Epidemiology*, **3**:73–88.

Corey, L. A., Nance, W. E., Kang, K. W. and Christian, J. C. (1979). Effects of type of placentation on birthweight and its variability in monozygotic and dizygotic twins. *Acta Geneticae Medicae et Gemellologiae*, **28**:41–50.

Cornelius, M. D., Taylor, P. M., Geva, D. and Day, N. L. (1995). Prenatal tobacco and marijuana use among adolescents: Effects on offspring gestational age, growth, and morphology. *Pediatrics*, **95**: 738–45.

Corney, G., Thompson, B., Campbell, D. M., MacGillivray, I., Seedburgh, D. and Timlin, D. (1979). The effect of zygosity on the birthweight of twins in Aberdeen and Northeast Scotland. *Acta Geneticae Medicae et Gemellologiae*, **28**:353–60.

Cortinovis, I., Bossi, A. and Milani, S. (1993). Longitudinal growth charts for weight, length and head-circumference of Italian children up to three years. *Acta Medica Auxologica*, **25**:13–29.

Cortinovis, I., Conter, V., Rogari, P. and Riva, L. (1994). Mother's smoking habit in pregnancy and weight growth up to six months in Italian infants. *Humanbiologica Budapestiensis*, **25**:167–74.

Cossack, Z. T. (1988). Effect of zinc level in the refeeding diet in previously starved rats on plasma somatomedin C levels. *Journal of Pediatric Gastroenterology and Nutrition*, **7**:441–5.

Cossack, Z. T. (1991). Decline in somatomedin-C, insulin-like growth factor-1, with experimentally induced zinc deficiency in human subjects. *Clinical Nutrition*, **10**:284–91.

Costin, G., Kaufman, F. R. and Brasel, J. A. (1989). Growth hormone secretory dynamics in subjects with normal stature. *Journal of Pediatrics*, **115**:537–44.

Coutant, R., Carel, J.-C., Letrait, M., Bouvattier, C., Chatelain, P., Coste, J. and Chaussain, J.-L. (1998). Short stature associated with intrauterine growth retardation: Final height of untreated and growth hormone-treated children. *Journal of Clinical Endocrinology and Metabolism*, **83**:1070–4.

Coutinho, R., David, R. J. and Collins, J. W., Jr. (1997). Relation of parental birth weights to infant birth weight among African Americans and Whites in Illinois: A transgenerational study. *American Journal of Epidemiology*, **146**:804–9.

Cowan, F. J., Evans, W. D. and Gregory, J. W. (1999). Metabolic effects of discontinuing growth hormone treatment. *Archives of Disease in Childhood*, **80**: 517–23.

Cowell, C. T. (1994). Growth hormone therapy in idiopathic short stature in the Kabi International Growth Study. In *Progress in Growth Hormone Therapy: 5 Years of KIGS*, eds. M. B. Ranke and R. Gunnarsson, pp. 216–29. Mannheim, Germany: J. & J. Verlag.

Coy, J. F., Lowry, R. K. and Ratkowsky, D. A. (1986). Longitudinal growth study of Tasmanian children, 1967–1983. *Medical Journal of Australia*, **144**:677–9.

Craft, W. H. and Underwood, L. E. (1984). Effect of androgens on plasma somatomedin-C/insulin-like growth factor. I. Responses to growth hormone. *Clinical Endocrinology*, **20**:549–54.

Cramer, J. C. (1985). Racial and ethnic differences in birthweight: The role of income and financial assistance. *Demography*, **32**:231–47.

Cremers, M. J. G. (1993). The analysis of growth in Dutch children with Down syndrome. In *Down Syndrome and Sports*, ed. M. J. G. Cremers, pp. 71–91. Utrecht, The Netherlands: Drukkerji Elinkwijk.

Cresta, M., Barberini, G., Calandra, P. L., Cialfa, E., de Majo, A. M., Gradoni, L., Gramiccia, M., Mariani, M., Passarello, P., Pozio, E., Pulicant, A. R., Ricci, M. and Vecchi, F. (1982). *Studio biologico su 26 anni di storia di una comunità montana dell'Italia Meridionale (Prov. di Salerno) (Biological Study of 26 Years of History of a Mountain Community in Southern Italy; Province of Salerno)*. Quaderni della Nutrizione Monograph Series no. 2. Bologna, Italy: National Institute of Nutrition.

Cronk, C., Crocker, A. C., Pueschel, S. M., Shea, A. M., Zackai, E., Pickens, G. and Reed, R. B. (1988). Growth charts for children with Down syndrome: 1 month to 18 years of age. *Pediatrics*, **81**:102–10.

Cronk, C. E. and Roche, A. F. (1982). Race and sex-specific reference data for triceps and subscapular skinfolds and weight/stature. *American Journal of Clinical Nutrition*, **35**:347–54.

Crosby, W., Metcoff, J., Costiloe, J., Mameesh, M., Sandstead, H., Jacob, R., McClain, P., Jacobson, G., Reid, W. and Burns, G. (1977). Fetal malnutrition: An appraisal of correlated factors. *American Journal of Obstetrics and Gynecology*, **128**:22–31.

Crowne, E. C., Shalet, S. M., Wallace, W. H. B., Eminson, D. M. and Price, D. A. (1990) Final height in boys with untreated constitutional delay in growth and puberty. *Archives of Disease in Childhood*, **65**:1109–12.

Crowne, E. C., Shalet, S. M., Wallace, W. H. B., Eminson, D. M. and Price, D. A. (1991). Final height in girls with untreated constitutional delay in growth and puberty. *European Journal of Pediatrics*, **150**:708–12.

Curhan, G. C., Chertow, G. M., Willett, W. C., Spiegelman, D., Colditz, G. A., Manson, J. E., Speizer, F. E. and Stampfer, M. J. (1996*a*). Birth weight and adult hypertension and obesity in women. *Circulation*, **94**:1310–15.

Curhan, G. C., Willett, W. C., Rimm, E. B., Spiegelman, D., Ascherio, A. L. and Stampfer, M. J. (1996*b*). Birth weight and adult hypertension, diabetes mellitus, and obesity in U.S. men. *Circulation*, **94**:3246–50.

Curry, D. L. and Bennett, L. I. (1973). Dynamics of insulin release by perfused rat pancreases: Effect of hypophysectomy, growth hormone, adrenocorticotrophic hormone and hydrocortisone. *Endocrinology*, **93**:602–9.

Curtis, J. A., Cormode, E., Laski, B., Toole, J. and Howard, N. (1982). Endocrine complications of topical and intra-lesional corticosteroid therapy. *Archives of Disease in Childhood*, **57**:204–7.

Cutfield, W. S., Bergman, R. N., Menon, R. K. and Sperling, M. A. (1990). The modified minimal model: Application to measurement of insulin sensitivity in children. *Journal of Clinical Endocrinology and Metabolism*, **70**:1644–50.

Cuttler, L., van Vliet, G., Conte, F. A., Kaplan, S. L. and Grumbach, M. M. (1985). Somatomedin-C levels in children and adolescents with gonadal dysgenesis: Differences from age-matched normal females and effect of chronic estrogen replacement therapy. *Journal of Clinical Endocrinology and Metabolism*, **60**:1087–91.

Czernichow, P. (1997). Growth hormone treatment of short children born small for gestational age. *Acta Paediatrica, Supplement*, **423**:213–15.

Dabelea, D., Pettitt, D. J., Hanson, R. L., Imperatore, G., Bennett, P. H. and Knowler, W. C. (1999). Birth weight, type 2 diabetes, and insulin resistance in Pima Indian children and young adults. *Diabetes Care*, **22**:944–50.

Damon, A. (1965). Notes on anthropometric technique. II. Skinfolds – right and left sides; held by one or two hands. *American Journal of Physical Anthropology*, **23**:305–6.

Damon, A., Damon, S. T., Reed, R. B. and Valadian, I.

(1969). Age at menarche of mothers and daughters with a note on accuracy of recall. *Human Biology*, **41**:161–75.

Daniel, W. A., Feinstein, R. A. and Howard-Pebbles, P. (1982). Testicular volume of adolescents. *Journal of Pediatrics*, **101**:1010–12.

Danielian, P. J., Allman, A. C. and Steer, P. J. (1992). Is obstetric and neonatal outcome worse in fetuses who fail to reach their own growth potential? *British Journal of Obstetrics and Gynaecology*, **99**:452–4.

Danker-Hopfe, H. (1986). Menarcheal age in Europe. *Yearbook of Physical Anthropology*, **29**:81–112.

Dann, T. C. and Roberts, D. F. (1984). Menarcheal age in University of Warwick students. *Journal of Biosocial Science*, **16**:511–19.

Darendeliler, F., Hindmarsh, P. C., Preece, M. A., Cox, L. and Brook, C. G. D. (1990). Growth hormone increases rate of pubertal maturation. *Acta Endocrinologica*, **122**:414–16.

David, R. J. and Collins, J. W. (1997). Differing birth weight among infants of U.S.-born blacks, African-born blacks, and U.S.-born whites. *New England Journal of Medicine*, **337**:1209–14.

Davie, R. (1972). Influences on physical growth. In *From Birth to Seven: The Second Report of the National Child Development Study (1958 Cohort); with Full Statistical Appendix*, eds. R. Davie, N. Butler, H. Goldstein, E. Alberman, E. Ross and P. Wedge, pp. 81–6. Atlantic Highlands, NJ: Humanities Press.

Davies, P. S. W., Valley, R. and Preece, M. A. (1988). Adolescent growth and pubertal progression in the Silver–Russell syndrome. *Archives of Disease in Childhood*, **63**:130–5.

Davoren, J. B. and Hsueh, J. W. (1987). Growth hormone increases ovarian levels of immunoreactive somatomedin C/insulin-like growth factor I *in vivo*. *Endocrinology*, **118**:888–90.

Day, N., Cornelius, M., Goldschmidt, L., Richardson, G., Robles, N. and Taylor, P. (1992). The effects of prenatal tobacco and marijuana use on offspring growth from birth through 3 years of age. *Neurotoxicology and Teratology*, **14**:407–14.

Day, N., Sambamoorthi, U., Taylor, P., Richardson, G., Robles, N., Jhon, Y., Scher, M., Stoffer, D., Cornelius, M. and Jasperse, D. (1991). Prenatal marijuana use and neonatal outcome. *Neurotoxicology and Teratology*, **13**:329–34.

Day, N. L., Jasperse, D., Richardson, G., Robles, N., Sambamoorthi, U., Taylor, P., Scher, M., Stoffer, D. and Cornelius, M. (1989). Prenatal exposure to alcohol: effect on infant growth and morphologic characteristics. *Pediatrics*, **84**:536–41.

de Bruin, N. C., van Velthoven, K. A. M., Stijnen, T., Juttmann, R. E., Degenhart, H. J. and Visser, H. K. A. (1995). Body fat and fat-free mass in infants: New and classic anthropometric indexes and prediction equations compared with total-body electrical conductivity. *American Journal of Clinical Nutrition*, **61**:1195–1205.

de Koning, F. L., Binkhorst, R. A., Kauer, J. M. G. and Thijssen, H. O. M. (1986). Accuracy of an anthropometric estimate of the muscle and bone area in a transversal cross-section of the arm. *International Journal of Sports Medicine*, **7**:246–9.

de Meer, K., Heymans, H. S. A. and Zijlstra, W. G. (1995). Physical adaptation of children to life at high altitude. *European Journal of Pediatrics*, **154**:263–72.

de Pergola, G. (2000). The adipose tissue metabolism: Role of testosterone and dehydroepiandrosterone. *International Journal of Obesity*, **24**:S59–S63.

de Pergola, G., Xu, X., Yang, S., Giorgino, R. and Björntorp, P. (1990). Up-regulation of androgen receptor binding in male rat fat pad adipose precursor cells exposed to testosterone: Study in a whole cell assay system. *Journal of Steroid Biochemistry and Molecular Biology*, **4**:553–8.

de Sanctis, V., Atti, G., Banin, P., Orzincolo, C., Cavallini, A. R., Patti, D. and Vullo, C. (1991). Growth in thalassaemia major. *Acta Medica Auxologica*, **23**:29–36.

de Simone, M., Farello, G., Palumbo, M., Gentile, T., Ciuffreda, M., Olioso, P., Cinque, M. and de Matteis, F. (1995). Growth charts, growth velocity and bone development in childhood obesity. *International Journal of Obesity*, **19**:851–7.

de Spiegelaere, M., Dramaix, M. and Hennart, P. (1998). The influence of socioeconomic status on the incidence and evolution of obesity during early adolescence. *International Journal of Obesity*, **22**:268–74.

de Wijn, J. F. (1966). Estimation of age at menarche in a population. In *Somatic Growth of the Child*, eds. J. J. van der Werff ten Bosch and A. Haak, pp. 16–24. Leiden, The Netherlands: Stenfert-Kroese.

de Zegher, F. and Chatelain, P. G. (1998). Growth hormone treatment of short children born small for gestational age: Epianalysis of controlled studies and clinical experience. In *Progress in Growth Hormone Therapy: 10 years of KIGS*, eds. M. Ranke and P. Wilton, pp. 305–19. Mannheim, Germany: J. & J. Verlag.

Debry, G. (1990). *Le café: Sa composition, sa consommation, ses incidences sur la santé (Coffee: Its Composition, its Consumption, its Effects on Health)*. Nancy, France: Centre de Nutrition.

Deheeger, M., Rolland-Cachera, M.-F., Labadie, M.-D. and Rossignol, C. (1994). Étude longitudinale de la croissance et de l'alimentation d'enfants examinés de l'âge de 10 mois à 8 ans. *Cahiers de Nutrition et Diététique*, **29**:1–8.

Demerath, E. W., Towne, B., Wisemandle, W. A., Blangero, J., Chumlea, W. C. and Siervogel, R. M. (1999). Serum leptin concentration, body composition, and gonadal hormones during puberty. *International Journal of Obesity*, **23**:678–85.

Deming, J. (1957). Application of the Gompertz curve to the observed pattern of growth in length of 48 individual boys and girls during the adolescent cycle of growth. *Human Biology*, **29**:83–122.

Demirjian, A. and Brault-Dubuc, M. (1985). *Croissance et développement de l'enfant Québécois de la naissance à six ans (Growth and Development of Québec Children from Birth to Six Years)*. Montreal, Quebec, Canada: Les Presses de l'Université.

Demirjian, A., Buschang, P. H., Tanguay, R. and Patterson, D. K. (1985). Interrelationships among measures of somatic, skeletal, dental, and sexual maturity. *American Journal of Orthodontics*, **88**:433–8.

Demoulin, F. (1998). Secular trend in France. In *Secular Growth Changes in Europe*, eds. B. E. Bodzsár and C. Susanne, pp. 109–34. Budapest: Eötvös University Press.

Deschamps, J.-P. and Benchemsi, N. (1974). Maturation osseuse et conditions socio-économiques et culturelles (Étude transversale sur 4526 enfants et adolescents)

(Skeletal maturation and socio-economic and cultural conditions; cross-sectional study of 4526 children and adolescents). *Revue de Pédiatrie*, **10**:143–51.

Deslypere, J. P., Verdonck, L. and Vermeulen, A. (1985). Fat tissue: A steroid reservoir and site of steroid metabolism. *Journal of Clinical Endocrinology and Metabolism*, **61**:564–70.

Deter, R. L., Harris, R. B. and Hill, R. M. (1990). Neonatal growth assessment score: A new approach to the detection of intrauterine growth retardation in the new-born. *American Journal of Obstetrics and Gynecology*, **162**:1030–6.

Devlin, J. T. and Horton, E. S. (1988). Hormone and nutrient interactions. In *Modern Nutrition in Health and Disease*, eds. M. E. Shils and V. R. Young, pp. 570–85. Philadelphia, PA: Lea & Febiger.

Dewey, K. G. (1998). Growth characteristics of breast-fed compared to formula-fed infants. *Biology of the Neonate*, **74**:94–105.

Dewey, K.G. and Heinig, M.J. (1993). Are new growth charts needed for breastfed infants? *Breastfeeding Abstracts* **12**:35–6.

Dewey, K. G., Heinig, M. J., Nommsen, L. A. and Lönnerdäl, B. (1990*a*). Growth patterns of breast-fed infants during the first year of life: The DARLING Study. In *Breastfeeding, Nutrition, Infection and Infant Growth in Developed and Emerging Countries*, eds. S. A. Atkinson, L. A. Hanson and R. K. Chandra, pp. 269–82. St. John's, Newfoundland, Canada: ARTS Biomedical Publishers.

Dewey, K. G., Heinig, M. J., Nommsen, L. A. and Lönnerdäl, B. (1990*b*). Low energy intake and growth velocities of breast-fed infants: Are there functional consequences? In *Activity, Energy Expenditure and Energy Requirements of Infants and Children*, eds. B. Schurch and N. Scrimshaw, pp. 35–43. Cambridge, MA: Nestlé Foundation.

Dewey, K. G., Heinig, M. J., Nommsen, L. A. and Lönnerdäl, B. (1991). Adequacy of energy intake among breast-fed infants in The DARLING Study: Relationships to growth velocity, morbidity, and activity levels. *Journal of Pediatrics*, **119**:538–47.

Dewey, K. G., Heinig, M. J., Nommsen L. A., Peerson, J. M. and Lönnerdäl, B. (1993). Breast-fed infants are leaner than formula-fed infants at 1 yr of age: The DARLING Study. *American Journal of Clinical Nutrition*, **57**:140–5.

Dewey, K. G., Heinig, M. J., Nommsen, L. A., Peerson, J. M. and Lönnerdäl, B. (1992). Growth of breast-fed and formula-fed infants from 0 to 18 months: The DARLING Study. *Pediatrics*, **89**:1035–41.

Dews, P. B. (1982). Caffeine. *Annual Review of Nutrition*, **2**:323–41.

Dezenberg, C. V., Nagy, T. R., Gower, B. A., Johnson, R. and Goran, M. I. (1999). Predicting body composition from anthropometry in pre-adolescent children. *International Journal of Obesity*, **23**:253–9.

di Girolamo, M., Edén, S., Enberg, G., Isaksson, O., Lonnroth, P., Hall, K. and Smith, U. (1986). Specific binding of human growth hormone but not insulin-like growth factors by human adipocytes. *Letters: Federation of European Biochemical Societies*, **205**:15–19.

di Pietro, L., Mossberg, H.-O. and Stunkard, A. J. (1994). A 40-year history of overweight children in Stockholm: Life-time overweight, morbidity, and mortality. *International Journal of Obesity*, **18**:585–90.

Dickerman, Z., Loewinger, J. and Laron, Z. (1984). The pattern of growth in children with constitutional tall stature from birth to age 9 years. *Acta Paediatrica Scandinavica*, **73**:530–6.

Dietz, J. and Schwartz, J. (1991). Growth hormone alters lipolysis and hormone-sensitive lipase activity in 3T3-F442A adipocytes. *Metabolism: Clinical and Experimental*, **40**:800–6.

Dine, M. S., Gartside, P. S., Glueck, C. J., Rheines, L., Greene, G. and Khoury, P. (1979). Where do the heaviest children come from? A prospective study of white children from birth to 5 years of age. *Pediatrics*, **63**:1–7.

Divertie, G., Jensen, M. and Miles, J. (1991). Stimulation of lipolysis in humans by physiological hypercortisolemia. *Diabetes*, **40**:1228–32.

Domargård, A., Särnblad, S., Kroon, M., Karlsson, I., Skeppner, G. and Åman, J. (1999). Increased prevalence of overweight in adolescent girls with type 1 diabetes mellitus. *Acta Paediatrica*, **88**:1223–8.

Dombrowski, M. P., Wolfe, H. M., Brans, Y. W., Saleh, A. A. and Sokol, R. J. (1992). Neonatal morphometry, relation to obstetric, pediatric, and menstrual estimates of gestational age. *American Journal of Diseases of Children*, **146**:852–6.

d'Orazio, C., Pederzini, F. and Mastella, G. (1991). Relationship between auxological variables and respiratory disease in cystic fibrosis. *Acta Medica Auxologica*, **23**:79–84.

Dougherty, C. R. S. and Jones, A. D. (1982). The determinants of birth weight. *American Journal of Obstetrics and Gynecology*, **144**:190–200.

Douglas, J. W. B. and Simpson, H. R. (1964). Height in relation to puberty, family size, and social class: A longitudinal study. *Minerva Pediatrica*, **42**:20–35.

Douglas, R. G., Gluckman, P. D., Ball, K., Breier, B. and Shaw, J. H. (1991). The effects of infusion of insulinlike growth factor (IGF) I, IGF II, and insulin on glucose and protein metabolism in fasted lambs. *Journal of Clinical Ultrasound*, **88**:614–22.

Drillien, C. M. (1958). A longitudinal study of the growth and development of prematurely and maturely born children. II. Physical development. *Archives of Disease in Childhood*, **33**:423–32.

d'Souza, S. W., Black, P. and Richards, B. (1981). Smoking in pregnancy: Associations with skinfold thickness, maternal weight gain, and fetal size at birth. *British Medical Journal*, **282**:1661–3.

d'Souza, S. W., Vale, J., Sims, D. G. and Chiswick, M. L. (1985). Feeding, growth, and biochemical studies in very low birthweight infants. *Archives of Disease in Childhood*, **60**:215–18.

du Rant, R. H., Baranowski, T., Johnson, M. and Thompson, W. O. (1994). The relationship among television watching, physical activity, and body composition of young children. *Pediatrics*, **94**:449–55.

Ducros, A. and Pasquet, P. (1978). Évolution de l'âge d'apparition des premières règles (ménarche) en France (Change in age at menarche in France). *Biométrie Humaine (Paris)*, **13**:35–43.

Duke, P. M., Litt, I. F. and Gross, R. T. (1980). Adolescents' self-assessment of sexual maturation. *Pediatrics*, **66**:918–20.

Dunn, H. G., McBurney, A. K., Ingram, S. and Hunter, C. M. (1976). Maternal cigarette smoking during pregnancy and the child's subsequent development. I. Physical growth to the age of 6.5 years. *Canadian Journal of Public Health*, **67**:499–505.

Dupae, E., Defrise-Gussenhoven, E. and Susanne, C. (1982). Genetic and environmental influences on body measurements on Belgian twins. *Acta Geneticae Medicae et Gemellologiae*, **31**:139–44.

Duran-Tauleria, E., Rona, R. J. and Chinn, S. (1995). Factors associated with weight for height and skinfold thickness in British children. *Journal of Epidemiology and Community Health*, **49**:466–73.

Dwyer, J. T. (1991). Concept of nutritional status and its measurement. In *Anthropometric Assessment of Nutritional Status*, ed. J. H. Himes, pp. 5–28. New York: Wiley-Liss.

Dwyer, T. and Blizzard, C. L. (1996). Defining obesity in children by biological endpoint rather than population distribution. *International Journal of Obesity*, **20**:472–80.

Eakman, G. D., Dallas, J. S., Ponder, S. W. and Keenan, B. S. (1996). The effects of testosterone and dihydrotestosterone on hypothalamic regulation of growth hormone secretion. *Journal of Clinical Endocrinology and Metabolism*, **81**:1217–23.

Ebert, K., Low, M., Overstrom, E., Buonom, F. C., Baile, C. A., Roberts, T. M., Lee, A., Mandel, G. and Goodman, R. H. (1988). A Moloney MLV-rat somatotropin fusion gene produces biologically active somatotropin in a transgenic pig. *Molecular Endocrinology*, **2**:277–83.

Edman, C. D. and MacDonald, P. C. (1978). Effect of obesity on conversion of plasma androstenedione to estrone in ovulatory and anovulatory young women. *American Journal of Obstetrics and Gynecology*, **130**:456–61.

Edmonds, C. J. and Smith, T. (1981). Total body potassium in relation to thyroid hormones and hyperthyroidism. *Clinical Science*, **60**:311–18.

Edwards, A. G. K., Halse, P. C. and Parkin, J. M. (1990). Recognising failure to thrive in early childhood. *Archives of Disease in Childhood*, **65**:1263–5.

Edwards, L. E., Alton, I. R., Barrada, M. I. and Hakanson, E. Y. (1979). Pregnancy in the underweight woman: Course, outcome, and growth patterns of the infant. *American Journal of Obstetrics and Gynecology*, **135**:297–302.

Edwards, L. E., Dickes, W. F., Alton, I. R. and Hakanson, E. Y. (1978). Pregnancy in the massively obese: Course, outcome and prognosis of the infant. *American Journal of Obstetrics and Gynecology*, **131**:479–83.

Ehrenkranz, R. A., Younes, N., Lemons, J. A., Fanarof, A. A., Donovan, E. F., Wright, L. L., Katsikiotis, V., Tyson, J. E., Oh, W., Shankaran, S., Bauer, C. R., Korones, S. B., Stoll, B. J., Stevenson, D. K. and Papile, L.-A. (1999). Longitudinal growth study of hospitalized very low birth weight infants. *Pediatrics*, **104**:280–9.

Eiben, O. G. (1989). Educational level of parents as a factor influencing growth and maturation. In *Perspectives in the Science of Growth and Development*, ed. J. M. Tanner, pp. 227–34. London: Smith-Gordon.

Eiben, O. G. (1994). The Kormend Growth Study: Data to secular growth changes in Hungary. *Humanbiologica Budapestiensis*, **25**:205–19.

Eiben, O. G., Barabás, A. and Pantó, E. (1991). The Hungarian National Growth Study. *Humanbiologica Budapestiensis*, **21**:1–121.

Eiben, O. G. and Pantó, E. (1988). Some data on growth of Hungarian youth in function of socio–economic factors. *Anthropologie*, **26**:19–23.

Eid, E. E. (1971). A follow-up study of physical growth following failure to thrive with special references to a critical period in the first year of life. *Acta Paediatrica Scandinavica*, **60**:39–48.

Eiholzer, U., Bodmer, P., Bühler, M., Döhmann, U., Meyer, G., Reinhard, P., Schimert, G., Varga, G., Wälli, R., Largo, R. and Molinari, L. (1998). Longitudinal monthly body measurements from 1 to 12 months of age: A study by practitioners for practitioners. *European Journal of Pediatrics*, **157**:547–52.

Eiholzer, U., Boltshauser, E., Frey, D., Molinari, L. and Zachmann, M. (1988). Short stature: A common feature in Duchenne muscular dystrophy. *European Journal of Pediatrics*, **147**:602–5.

Eilmam, A., Lindgren, A. C., Norgren, S., Kamel, A., Skwirut, C., Bang, P. and Marcus, C. (1999). Growth hormone treatment downregulates serum leptin levels in children independent of changes in body mass index. *Hormone Research*, **52**:66–72.

Eisenmann, J. C., Katzmarzyk, P. T., Arnall, D. A., Kanuho, V., Interpreter, C. and Malina, R. M. (2000). Growth and overweight of Navajo youth: Secular changes from 1955 to 1997. *International Journal of Obesity*, **24**:211–18.

Ekvall, S. W. (1993). Growth grids for special conditions. In *Pediatric Nutrition in Chronic Diseases and Developmental Disorders: Prevention, Assessment, and Treatment*, ed. S. W. Ekvall, pp. 435–9. New York: Oxford University Press.

Elbers, J. M. H., Asscheman, H., Seidell, J. C., Frölich, M., Meinders, A. E. and Gooren, L. J. G. (1997). Reversal of the sex difference in serum leptin levels upon cross-sex hormone administration in transexuals. *Journal of Clinical Endocrinology and Metabolism*, **82**:3267–70.

Eliot, N. and Deniel, M. (1977). Étude de la croissance de 0 à 18 mois (Study of growth from birth to 18 months). *Archives Françaises de Pédiatrie*, **34**:23–6.

Ellard, G. A., Johnstone, F. D., Prescott, R. J., Wang, J. X. and Mao, J. H. (1996). Smoking during pregnancy: The dose dependence of birthweight deficits. *British Journal of Obstetrics and Gynaecology*, **103**:806–13.

Ellestein, N. S. and Ostrov, B. E. (1985). Growth patterns in children hospitalized because of caloric-deprivation failure to thrive. *American Journal of Diseases of Children*, **139**:164–6.

Elliman, A. M., Bryan, E. M., Elliman, A. D. and Harvey, D. R. (1992). Gestational age correction for height in preterm children to seven years of age. *Acta Paediatrica*, **81**:836–9.

Ellis, K. J. and Nicolson, M. (1997). Leptin levels and body fatness in children: Effects of gender, ethnicity, and sexual development. *Pediatric Research*, **42**:484–8.

Ellison, P. T. (1982). Skeletal growth, fatness and menarcheal age: A comparison of two hypotheses. *Human Biology*, **54**:269–82.

Ellison, P. T. (1998). Sexual maturation. In *The Cambridge Encyclopedia of Human Growth and Development*, eds. S. J. Ulijaszek, F. E. Johnston and M. A. Preece, pp. 227–9. Cambridge, U.K.: Cambridge University Press.

Elster, A. D., Bleyl, J. L. and Craven, T. E. (1991). Birth weight standards for triplets under modern obstetric care in the United States, 1984-1989. *Obstetrics and Gynecology*, **77**:387–93.

Elwood, P. C., Sweetnam, P. M., Gray, O. P., Davies, D. P. and Wood, P. D. P. (1987). Growth of children from 0–5 years: With special reference to mother's smoking in pregnancy. *Annals of Human Biology*, 14:543–57.

Emanuel, I., Filakti, H., Alberman, E. and Evans, S. J. W. (1992). Intergenerational studies of human birth weight from the 1958 birth cohort. I. Evidence for a multigenerational effect. *British Journal of Obstetrics and Gynaecology*, 99:67–74.

Emmerson, A. J. B. and Savage, D. C. L. (1988). Height at diagnosis in diabetes. *European Journal of Pediatrics*, 147:319–20.

Endo, K., Yanagi, H., Hirano, C., Hamaguchi, H., Tsuchiya, S. and Tomura, S. (2000). Association of Trp64Arg polymorphism of the β_3-adrenergic receptor gene and no association of Gln223Arg polymorphism of the leptin receptor gene in Japanese school children with obesity. *International Journal of Obesity*, 24:443–9.

English, P. C. (1978). Failure to thrive without organic reason. *Pediatric Annals*, 7:774–81.

Engstrom, J. L., Kavanaugh, K., Meier, P. P., Boles, E., Hernandez, J., Wheeler, D. and Chuffo, R. (1995). Reliability of in-bed weighing procedures for critically ill infants. *Neonatal Network*, 4:27–33.

Enzi, G., Inelman, E. M., Rubaltelli, F. F., Zanardo, V. and Favoretto, L. (1982). Postnatal development of adipose tissue in normal children on strictly controlled calorie intake. *Metabolism: Clinical and Experimental*, 31:1029–34.

Enzi, G., Zanardo, V., Caretta, F., Inelmen, E. M. and Rubaltelli, F. (1981). Intrauterine growth and adipose tissue development. *American Journal of Clinical Nutrition*, 34:1785–90.

Epstein, L. H. (1993*a*). Methodological issues and ten-year outcomes for obese children. *Annals of the New York Academy of Sciences*, 699:237–49.

Epstein, L. H. (1993*b*). New developments in childhood obesity. In *Obesity: Theory and Therapy*, 2nd edn, eds. A. J. Stunkard and T. A. Wadden, pp. 301–12. New York: Raven Press.

Epstein, L. H. and Goldfield, G. S. (1999). Physical activity in the treatment of childhood overweight and obesity: Current evidence and research issues. *Medicine and Science in Sports and Exercise*, 31:S553–9.

Epstein, L. H., Wu, Y.-W. B., Paluch, R. A., Cerny, F. J. and Dorn, J. P. (2000). Asthma and maternal body mass index are related to pediatric body mass index and obesity: Results from the Third National Health and Nutrition Examination Survey. *Obesity Research*, 8:575–81.

Eriksen, E. F., Colvard, D. S., Berg, C. J., Graham, M. L., Mann, K. G., Spelsberg, T. C. and Riggs, B. L. (1988). Evidence of estrogen receptors in normal human osteoblast-like cells. *Science*, 241: 84–6.

Eriksson, J. G., Forsén, T., Tuomilehto, J., Winter, P. D., Osmond, C. and Barker, D. J. P. (1999). Catch-up growth in childhood and death from coronary heart disease: Longitudinal study. *British Medical Journal*, 318:427–31.

Ernhart, C. B., Morrow-Tlucak, M., Sokol, R. J. and Martier, S. (1988). Under-reporting of alcohol use in pregnancy. *Alcoholism, Clinical and Experimental Research*, 12:506–11.

Ernhart, C. B., Wolf, A. W., Linn, P. L., Sokol, R. J., Kennard, M. J. and Filipovich, H. F. (1985). Alcohol related birth defects: Syndromal anomalies, intrauterine growth retardation, and neonatal behavioral assessment. *Alcoholism, Clinical and Experimental Research*, 9:447–53.

Ernst, M., Schmid, C. and Froesch, E. R. (1988). Enhanced osteoblast proliferation and collagen gene expression by estradiol. *Proceedings of the National Academy of Sciences (U.S.A.)*, 85:2307–10.

Ershoff, D., Quinn, V. P., Mullen, P. D. and Lairson, D. R. (1990). Pregnancy and medical cost outcomes of a self-help prenatal smoking cessation program in a health maintenance organization. *Public Health Reports*, 105:340–7.

Ertl, T., Funke, S., Sarkany, I., Szabo, I., Rascher, W., Blum, W. F. and Sulyok, E. (1999). Postnatal changes of leptin levels in full-term and preterm neonates: Their relation to intrauterine growth, gender and testosterone. *Biology of the Neonate*, 75: 167–76.

Eskenazi, B., Prehn, A. W. and Christianson, R. E. (1995). Passive and active maternal smoking as measured by serum cotinine: The effect on birthweight. *American Journal of Public Health*, 85:395–8.

Eveleth, P. B. and Tanner, J. M. (1990). *Worldwide Variation in Human Growth*, 2nd edn, Cambridge, U.K.: Cambridge University Press.

Evers, S. E. and Hooper, M. D. (1995). Dietary intake and anthropometric status of 7 to 9 year old children in economically disadvantaged communities in Ontario. *Journal of the American College of Nutrition*, **14**:595–603.

Evers, S. E. and Hooper, M. D. (1996). Anthropometric status and diet of 4 to 5 year old low income children. *Nutrition Research*, **16**:1847–59.

Fabiani, E., Rossini, M., Ratsch, I. M., Catassi, C. and Giorgi, P. L. (1991). Height velocity in coeliac children during the first years of gluten-free diet: A retrospective study on 29 patients. *Acta Medica Scandinavica*, **23**:99–101.

Facchini, F. and Russo, G. (1982). Secular anthropometric changes in a sample of Italian adults. *Journal of Human Evolution*, **11**:703–14.

Fagin, K., Lackey, S., Reagan, C. and di Girolamo, M. (1980). Specific binding of growth hormone by rat adipocytes. *Endocrinology*, **107**:608–15.

Falany, C. N., Comer, K. A., Dooley, T. P. and Glatt, H. (1995). Human dehydroepiandrosterone sulfotransferase: Purification, molecular cloning, and characterization. *Annals of the New York Academy of Sciences*, **774**:59–72.

Falkner, F., Pernot-Roy, M.P., Habich, H., Sénécal, J. and Massé, G. (1980). Some international comparisons of physical growth in the first two years of life. In *Growth and Development of the Child: 25 Years of Internationally Coordinated Activities*, ed. F. Falkner, pp. 18–27. Paris: Centre Internationale de l'Enfance.

Fall, C. H. D., Osmond, C., Barker, D. J. P., Clark, P. M. S., Hales, C. N., Stirling, Y. and Meade, T. W. (1995*a*). Fetal and infant growth and cardiovascular risk factors in women. *British Medical Journal*, **310**:428–32.

Fall, C. H. D., Vijayakumar, M., Barker, D. J. P., Osmond, C. and Duggleby, S. (1995*b*). Weight in infancy and prevalence of coronary heart disease in adult life. *British Medical Journal*, **310**:17–19.

Fang, J., Madhavan, S. and Alderman, M. H. (1999). Low birth weight: Race and maternal nativity – impact of community income. *Pediatrics*, **103**:E51–6.

Faria, A. C. S., Bekenstein, L. W., Booth, R. A., Jr., Vaccaro, V. A., Asplin, C. M., Veldhuis, J. D., Thorner, M. O. and Evans, W. S. (1992). Pulsatile growth hormone release in normal women during the menstrual cycle. *Journal of Clinical Endocrinology and Metabolism*, **36**:591–6.

Farkas, G. (1980). Veranderungen des Menarche-Medianwertes nach dem Beruf der Mutte (Variation in the median age at menarche based on mother's occupation). *Ärztliche Jugendkunde*, **71**:62–7.

Faulkner, R. A., Bailey, D. A., Drinkwater, D. T., Wilkinson, A. A., Houston, C. S. and McKay, H. A. (1993). Regional and total body bone mineral content, bone mineral density, and total body tissue composition in children 8–16 years of age. *Calcified Tissue International*, **53**:7–12.

Faust, M. S. (1977). Somatic development of adolescent girls. *Monographs of the Society for Research in Child Development*, **42**:1–90.

Feinleib, M., Garrison, R. J., Fabsitz, R., Christian, J. C., Hrubec, Z., Borhani, N. O., Kannel, W. B., Rosenman, R., Schwartz, J. T. and Wagner, J. D. (1977). The NHLBI twin study of cardiovascular disease risk factors: Methodology and summary of results. *American Journal of Epidemiology*, **106**:284–95.

Fenster, L. and Coye, M. J. (1990). Birthweight of infants born to Hispanic women employed in agriculture. *Archives of Environmental Health*, **445**:46–52.

Fenster, L., Eskenazi, B., Windham, G. C. and Swan, S. H. (1991). Caffeine consumption during pregnancy and fetal growth. *American Journal of Public Health*, **81**:458–61.

Fenton, T., McMillan, D. D. and Sauve, R. S. (1990). Nutrition and growth analysis of very low birth weight infants. *Pediatrics*, **86**:378–83.

Fergusson, D. M., Crane, J., Beasley, R. and Horwood, L. J. (1997). Perinatal factors and atopic disease in childhood. *Clinical and Experimental Allergy*, **27**:1394–1401.

Fergusson, D. M., Horwood, L. J. and Shannon, F. T. (1980). Length and weight gain in the first three months of life. *Human Biology*, **52**:169–80.

Fernandez, G., Marrodán, M. D. and Andrés, S. (1994). Sociogeografia y variabilidad estacional de la menarquia en la población Madriléa (Social geography and variability in menarcheal status in the population of Madrid). In *Biologia de las poblaciones humanas: Problemas metodológicos e interpretación ecológica*, eds. C. Bernis, C. Varea, F. Robles and A. Gonzalez, pp. 807–12. Madrid: Universidad Autónoma de Madrid.

Fernandez-Real, J.-M., Granada, M. L., Ruzafa, A., Casamitjana, R. and Ricart, W. (2000). Insulin sensitivity and secretion influence the relationship between growth hormone-binding-protein and leptin. *Clinical Endocrinology*, **52**:159–64.

Ferrari, S., Rizzoli, R., Slosman, D. and Bonjour, J.-P. (1998). Familial resemblance for bone mineral mass is expressed before puberty. *Journal of Clinical Endocrinology and Metabolism*, **83**:358–61.

Ferris, A. G., Laus, M. J., Hosmer, D. W. and Beal, V. A. (1980). The effect of diet on weight gain in infancy. *American Journal of Clinical Nutrition*, **33**:2635–42.

Ferro-Luzzi, G. and Sofia, F. (1967). Rapporto su altezze, pesi e altri indici di stato di nutrizione dei bambini Italiani (età 6-11 anni) (Relationship of stature, weight and other indices to nutritional status in Italian children; age 6-11 years). *Quaderni della Nutrizione (Bologna)*, **27**:269–92.

Finkelstein, J. S., Klibanski, A., Neer, R. M., Greenspan, S. L., Rosenthal, D. I. and Crowley, W. F., Jr. (1987). Osteoporosis in men with idiopathic hypogonadotropic hypogonadism. *Annals of Internal Medicine*, **106**:354–61.

Finkelstein, J. S., Neer, R. M., Biller, B. M. K., Crawford, J. D. and Klibanski, A. (1992). Osteopenia in men with a history of delayed puberty. *New England Journal of Medicine*, **326**:600–4.

Fischbein, S. and Nordqvist, T. (1978). Profile comparisons of physical growth for monozygotic and dizygotic twin pairs. *Annals of Human Biology*, **5**:321–8.

Fitzhardinge, P. M. and Inwood, S. (1989). Long-term growth in small-for-date children. *Acta Paediatrica Scandinavica*, **349**:27–33.

Flaim, K. E., Li, J. B. and Jefferson, L. S. (1978). Effect of thyroxine on protein turnover in rat skeletal muscle. *American Journal of Physiology*, **235**:E231–6.

Flegal, K. M. (2000). The effects of age categorization on estimates of overweight prevalence for children. *International Journal of Obesity*, **24**:1636–41.

Flegal, K. M. and Troiano, R. P. (2000). Changes in the distribution of body mass index of adults and children in the US population. *International Journal of Obesity*, **24**:807–8.

Fleisher, T. A., White, R. M., Broder, S., Nissley, S. P., Blaese, R. M., Mulvihill, J. J., Olive, G. and Waldmann, T. A. (1980). X-linked hypogammoglobulinemia and isolated growth hormone deficiency. *New England Journal of Medicine*, **302**:1429–34.

Floris, G. and Sanna, E. (1998). Some aspects of the secular trends in Italy. In *Secular Growth Changes in Europe*, eds. É. B. Bodzsár and C. Susanne, pp. 207–32. Budapest: Eötvös University Press.

Fogelman, K. (1980). Smoking in pregnancy and subsequent development of the child. *Child: Care, Health and Development*, **6**:233–49.

Fogelman, K. R. and Manor, O. (1988). Smoking in pregnancy and development into early adulthood. *British Medical Journal*, **297**:1233–6.

Fomon, S. J. (1987). Reflections on infant feeding in the 1970s and 1980s. *American Journal of Clinical Nutrition*, **46**:171–82.

Fomon S. J. (1993). *Nutrition of Normal Infants*. St. Louis, MO: Mosby.

Fomon, S. J., Haschke, F., Ziegler, E. E. and Nelson, S. E. (1982). Body composition of reference children from birth to age 10 years. *American Journal of Clinical Nutrition*, **35**:1169–75.

Fong, Y., Rosenbaum, M., Tracey, K. J., Raman, D., Hesse, D. G., Matthews, D. E., Leibel, R. L., Gertner, J. M., Fischman, D. A. and Lowry, S. F. (1989). Recombinant growth hormone enhances muscle myosin heavy-chain mRNA accumulation and amino acid accrual in humans. *Proceedings of the National Academy of Sciences (U.S.A.)*, **86**:3371–4.

Forbes, G. B., Brown, M. R. and Griffiths, H. J. L. (1988). Arm muscle plus bone area: Anthropometry and CAT scan compared. *American Journal of Clinical Nutrition*, **47**:929–31.

Forbes, G. B., Porta, C. R., Herr, B. E. and Griggs, R. C. (1992). Sequence of changes in body composition induced by testosterone and reversal of changes after drug is stopped. *Journal of the American Medical Association*, **267**:397–9.

Ford, K. and Labbok, M. (1990). Who is breast-feeding? Implications of associated social and biomedical variables for research on the consequences of method of infant feeding. *American Journal of Clinical Nutrition*, 52:451–6.

Forest, M. G. (1990). Pituitary gonadotropin and sex steroid secretion during the first two years of life. In *Control of the Onset of Puberty*, eds. M. Grumbach, P. C. Sizonenko and M. L. Aubert, pp. 451–78. Baltimore, MD: Williams & Wilkins.

Forney, P. J., Milewich, L., Chen, G. T., Garlock, J. L., Schwarz, B. E., Edman, C. D. and MacDonald, P. L. (1981). Aromatization of androstenedione to estrone by human adipose tissue *in vitro*: Correlation with adipose tissue mass, age, and endometrial neoplasia. *Journal of Clinical Endocrinology and Metabolism*, 53:192–7.

Forsén, T., Eriksson, J. G., Tuomilehto, J., Teramo, K., Osmond, C. and Barker, D. J. P. (1997). Mother's weight in pregnancy and coronary heart disease in a cohort of Finnish men: Follow-up study. *British Medical Journal*, 315:837–40.

Fort, P. (1986). Thyroid disorders in infancy. In *Human Growth: A Comprehensive Treatise*, 2nd edn, vol. 2, *Postnatal Growth*, eds. F. Falkner and J. M. Tanner, pp. 437–56. New York: Plenum Press.

Fortier, I., Marcoux, S. and Beaulac-Baillargeon, L. (1993). Relation of caffeine intake during pregnancy to intrauterine growth retardation and preterm birth. *American Journal of Epidemiology*, 137:931–40.

Fortmann, S. P., Rogers, T., Vranizan, K., Haskell, W. L., Solomon, D. S. and Farquhar, J. W. (1984). Indirect measures of cigarette use: Expired-air carbon monoxide versus plasma thiocyanate. *Preventive Medicine*, 13:127–35.

Fox, K. M., Magaziner, J., Sherwin, R., Scott, J. C., Plato, C.C., Nevitt M. and Cummings, S. (1993). Reproductive correlates of bone mass in elderly women. *Journal of Bone and Mineral Research* 8: 901–8.

Fox, N. L., Sexton, M. and Hebel, J. R. (1990). Prenatal exposure to tobacco: Effects on physical growth at age three. *International Journal of Epidemiology*, 19:66–71.

Fox, P. T., Elston, M. D. and Waterlow, J. C. (1981). *Medical Aspects of Food Policy*, Sub-Committee on Nutritional Surveillance, Department of Health and Social Security, 2nd Report. London: Her Majesty's Stationery Office.

Fox, S. H., Koepsell, T. D. and Daling, J. R. (1994). Birth weight and smoking during pregnancy: Effect modification by maternal age. *American Journal of Epidemiology*, 139:1008–15.

François, S., Benmalek, A., Guaydier-Souquières, G., Sabatier, J.-P. and Marcelli, C. (1999). Heritability of bone mineral density. *Revue du Rhumatisme, English Edition*, 66:146–51.

Frank, D. A. (1986). Biologic risk in "nonorganic" failure to thrive: Diagnostic and therapeutic implications. In *New Directions in Failure to Thrive: Implications for Research and Practice*, ed. D. Drotar, pp. 17–26. New York: Plenum Press.

Frank, D. A. and Zeisel, S. H. (1988). Failure to thrive. *Pediatric Clinics of North America*, 35:1187–1206.

Frank, G. R. (1995). The role of estrogen in pubertal skeletal physiology: Epiphyseal maturation and mineralization of the skeleton. *Acta Paediatrica*, 84:627–30.

Frankel, S., Elwood, P. C., Sweetnam, P. M., Yarnell, J. G. W. and Davey Smith, G. (1996). Birth weight, body mass index in middle age, and incident coronary heart disease. *Lancet*, 348:1478–80.

Frasier, S. D. (1986). Tall stature and excessive growth syndromes. In *Human Growth: A Comprehensive Treatise*, 2nd edn, vol. 2, *Postnatal Growth*, ed. F. Falkner and J. M. Tanner, pp. 197–213. New York: Plenum Press.

Fredriks, A. M., van Buuren, S., Burgmeijer, R. J. F., Meulmeester, J. F., Beuker, R. J., Brugman, E., Roede, M. J., Verloove-Vanhorick, S. P. and Witt, J.-M. (2000*a*). Continuing positive secular growth change in The Netherlands 1955–1997. *Pediatric Research*, 47:316–23.

Fredriks, A. M., van Buuren, S., Wit, J. M. and Verloove-Vanhorick, S. P. (2000*b*). Body index measurements in 1996–7 compared with 1980. *Archives of Disease in Childhood*, 82:107–12.

Freedman, D. S., Burke, G. L., Harsha, D. W., Srinivasan, S. R., Cresanta, J. L., Webber, L. S. and Berenson, G. S. (1985). Relationship of changes in obesity to serum lipid and lipoprotein changes in childhood and adolescence. *Journal of the American Medical Association*, 254:515–20.

Freedman, D. S., Dietz, W. H., Srinivasan, S. R. and Berenson, G. S. (1999). The relation of overweight to cardiovascular risk factors among children and adolescents: The Bogalusa Heart Study. *Pediatrics*, **103**:1175–82.

Freedman, D. S., Khan, L. K., Serdula, M. K., Srinivasan, S. R. and Berenson, G. S. (2000). Secular trends in height among children during 2 decades: The Bogalusa Heart Study. *Archives of Pediatrics and Adolescent Medicine*, **154**:155–61.

Freeman, J. V., Cole, T. J., Chinn, S., Jones, P. R. M., White, E. M. and Preece, M. A. (1995). Cross-sectional stature and weight reference curves for the UK 1990. *Archives of Disease in Childhood*, **73**:17–24.

Frerichs, R. R., Harsha, D. W. and Berenson, G. S. (1979). Equations for estimating percentage of body fat in children 10–14 years old. *Pediatric Research*, **13**:170–4.

Fried, P. A. and O'Connell, C. M. (1987). A comparison of the effects of prenatal exposure to tobacco, alcohol, cannabis, and caffeine on birth size and subsequent growth. *Neurotoxicology and Teratology*, **9**: 79–85.

Fried, P. A., Watkinson, B. and Gray, R. (1999). Growth from birth to early adolescence in offspring prenatally exposed to cigarettes and marijuana. *Neurotoxicology and Teratology*, **21**:513–25.

Fried, P. A., Watkinson, B. and Willan, A. (1984). Marijuana use during pregnancy and decreased length of gestation. *American Journal of Obstetrics and Gynaecology*, **150**:23–7.

Friedman, D. J., Cohen, B. B., Mahan, C. M., Lederman, R. I., Vezina, R. J. and Dunn, V. H. (1993). Maternal ethnicity and birthweight among Blacks. *Ethnicity and Disease*, **3**:225–69.

Friel, J. K., Andrews, W. L., Matthew, J., Long, D. R., Cornel, A. M., Cox, M., McKim, E. and Zerbe, G. O. (1993). Zinc supplementation in very low birth weight infants. *Journal of Pediatric Gastroenterology and Nutrition*, **17**:97–104.

Friis, H., Ndhlovu, P., Mduluza, T., Kaondera, K., Sandström, B., Michaelsen, K. F., Vennervald, B. J. and Christensen, N. O. (1997). The impact of zinc supplementation on growth and body composition: A randomized, controlled trial among rural Zimbabwean schoolchildren. *European Journal of Clinical Nutrition*, **51**:38–45.

Frisancho, A. R. (2000). Prenatal compared with parental origins of adolescent fatness. *American Journal of Clinical Nutrition*, **72**:1186–90.

Frisancho, A. R., Borkan, G. A. and Klayman, J. E. (1975). Pattern of growth of lowland and highland Peruvian Quechua of similar genetic composition. *Human Biology*, **47**:233–43.

Frisancho, A. R., Frisancho, H. G., Albalak, R., Villain, M., Vargas, E. and Soria, R. (1997). Developmental, genetic, and environmental components of lung volumes at high altitude. *American Journal of Human Biology*, **9**:191–203.

Frisancho, A. R., Gilding, N. and Tanner, S. (2001). Growth of leg length is reflected in socio-economic differences. *Acta Medica Auxologica*, **33**:47–50.

Froesch, E. R., Guler, H. P., Schmid, C., Ernst, M., Zenobi, P. and Zapf, J. (1989). Growth promotion by insulin-like growth factor. I. Endocrine and autocrine regulation. In *Perspectives in the Science of Growth and Development*, ed. J. M. Tanner, pp. 251–63. London: Smith-Gordon.

Frühbeck, G., Aguado, M. and Martinez, J. A. (1997). In vitro lipolytic effect of leptin on mouse adipocytes: evidence for a possible autocrine/paracrine role of leptin. *Biochemical and Biophysical Research Communications*, **240**:590–4.

Fryburg, D. A. (1994). Insulin-like growth factor I exerts growth hormone- and insulin-like actions on human muscle protein metabolism. *American Journal of Physiology*, **267**:E331–6.

Fryburg, D. A., Gelfand, R. A. and Barrett, E. J. (1991). Growth hormone acutely stimulates forearm muscle protein synthesis in normal humans. *American Journal of Physiology*, **260**:E499–504.

Frystyk, J., Vestbo, E., Skjaerbaek, C., Mogensen, C. E. and Orskov, H. (1995). Free insulin-like growth factors in human obesity. *Metabolism: Clinical and Experimental*, **44**:37–44.

Fuentes-Afflick, E. and Hessol, N. A. (1997). Impact of Asian ethnicity and national origin on infant birth weight. *American Journal of Epidemiology*, **145**: 148–55.

Fuentes-Afflick, E., Hessol, N. A. and Pérez-Stable, E. J. (1998). Maternal birthplace, ethnicity, and low birth weight in California. *Archives of Pediatrics and Adolescent Medicine*, **152**:1105–12.

Fuentes-Afflick, E., Hessol, N. A. and Pérez-Stable, E. J. (1999). Testing the epidemiologic paradox of low birth weight in Latinos. *Archives of Pediatrics and Adolescent Medicine*, **153**:147–53.

Fujimora, M. and Seryu, J.-I. (1977). Velocity of head growth during the perinatal period. *Archives of Disease in Childhood*, **52**:105–12.

Fujisawa, T., Ikegami, H., Kawaguchi, Y. and Ogihara, T. (1998). Meta-analysis of the association of Trp64Arg polymorphism of β_3-adrenergic receptor gene with body mass index. *Journal of Clinical Endocrinology and Metabolism*, **83**:2441–4.

Fulroth, R., Phillips, B. and Durand, D. J. (1989). Perinatal outcome of infants exposed to cocaine and/or heroin *in utero*. *American Journal of Diseases of Children*, **143**:905–10.

Furusho, T. (1968). On the manifestation of genotypes responsible for stature. *Human Biology*, **40**:437–55.

Gallaher, M. M., Hauck, F. R., Yang-Oshida, M. and Serdula, M. K. (1991). Obesity among Mescalero preschool children. *American Journal of Diseases of Children*, **145**:1262–5.

Gampel, B. (1965). The relation of skinfold thickness in the neonate to sex, length of gestation, size at birth and maternal skinfolds. *Human Biology*, **29**:29–37.

Garcia-Mayor, R. V., Andrade, M. A., Rios, M., Lage, M., Dieguez, C. and Casanueva, F. F. (1997). Serum leptin levels in normal children: Relationship to age, gender, body mass index, pituitary–gonadal hormones, and pubertal stage. *Journal of Clinical Endocrinology and Metabolism*, **82**:2849–55.

Garn, S. M. and Clark, D. C. (1975). Nutrition, growth, development, and maturation: Findings from the Ten-State Nutrition Survey of 1968–1970. *Pediatrics*, **56**:306–19.

Garn, S. M., Clark, D. C. and Trowbridge, F. L. (1973). Tendency toward greater stature in American Black children. *American Journal of Diseases of Children*, **126**:164–6.

Garn, S. M. and LaVelle, M. (1984). Interaction between maternal size and birth size and subsequent weight gain. *American Journal of Clinical Nutrition*, **40**:1120–1.

Garn, S. M. and LaVelle, M. (1985). Two-decade follow-up of fatness in early childhood. *American Journal of Diseases of Children*, **139**:181–5.

Garn, S. M., Owen, G. M. and Clark, D. C. (1974). The question of race differences in stature norms. *Ecology of Food and Nutrition*, **3**:319–20.

Garn, S. M. and Pesick, S. D. (1982). Relationship between various maternal body mass measures and size of the newborn. *American Journal of Clinical Nutrition*, **36**:664–8.

Garn, S. M., Pesick, S. D. and Pilkington, J. J. (1984). The interaction between prenatal and socioeconomic effects on growth and development in childhood. In *Human Growth and Development*, eds. J. Borms, R. Hauspie, A. Sand, C. Susanne and M. Hebbelinck, pp. 59–70. New York: Plenum Press.

Garn, S. M., Rohmann, C. G. and Apfelbaum, B. (1961). Complete epiphyseal union of the hand. *American Journal of Physical Anthropology*, **17**: 365–72.

Garn, S. M. and Rosenberg, K. R. (1986). Definitive quantification of the smoking effect on birthweight. *Ecology of Food and Nutrition*, **19**:61–5.

Garn, S. M., Shaw, H. A. and McCabe, K. D. (1977). Birth size and growth appraisal. *Journal of Pediatrics*, **90**:1049–51.

Garn, S. M., Shaw, H. A. and McCabe, K. D. (1978). Effect of maternal smoking on weight and weight gain between pregnancies. *American Journal of Clinical Nutrition*, **31**:1302–3.

Garrow, J. S. and Pike, M. C. (1967). The long-term prognosis of severe infantile malnutrition. *Lancet*, i:1–4.

Gasser, T., Kneip, A., Binding, A., Prader, A. and Molinari, L. (1991). The dynamics of linear growth in distance, velocity and acceleration. *Annals of Human Biology*, **18**:187–205.

Gasser, T., Köhler, W., Müller, H. G., Kneip, A., Largo, R., Molinari, L. and Prader, A. (1984). Velocity and acceleration of height growth using kernel estimation. *Annals of Human Biology*, **11**:397–411.

Gasser, T., Müller, H. G., Köhler, W., Prader, A., Largo, R. and Molinari, L. (1985). An analysis of the mid-growth and adolescent spurts of height based on acceleration. *Annals of Human Biology*, **12**:129–48.

Gasser, T., Ziegler, P., Kneip, A., Prader, A., Molinari, L. and Largo, R. H. (1993). The dynamics of growth of weight, circumferences and skinfolds in distance, velocity and acceleration. *Annals of Human Biology*, **20**:239–59.

Gause, I. and Edén, S. (1985). Hormonal regulation of growth hormone binding and responsiveness in adipose tissue and adipocytes of hypophysectomized rats. *Journal of Endocrinology*, **105**:331–7.

Gayle, H. D., Yip, R., Frank, M. J., Nieburg, P. and Binkin, N. J. (1988). Validation of maternally reported birth weights among 46 637 Tennessee WIC program participants. *Public Health Reports*, **103**:143–7.

Geisthovel, F., Olbrich, M., Frorath, B., Thiemann, M. and Weitzell, R. (1994). Obesity and hypertestosteronaemia are independently and synergistically associated with elevated insulin concentrations and dyslipidaemia in pre-menopausal women. *Human Reproduction*, **9**:610–16.

Geithner, C. A., Satake, T., Woynarowska, B. and Malina, R. M. (1999). Adolescent spurts in body dimensions: Average and model sequences. *American Journal of Human Biology*, **11**:287–95.

Gelander, L., Bjarnason, R., Carlsson, L. M. S. and Albertsson-Wikland, K. (1998). Growth hormone-binding protein levels over one year in healthy prepubertal children: Intraindividual variation and correlation with height velocity. *Pediatric Research*, **43**:256–61.

Gelander, L., Karlberg, J. and Albertsson-Wikland, K. (1994). Seasonality in lower leg length velocity in prepubertal children. *Acta Paediatrica*, **83**:1249–54.

Gendrel, D., Chaussain, J.-L., Roger, M. and Job, J.-C. (1980). Simultaneous postnatal rise of plasma LH and testosterone in male infants. *Journal of Pediatrics*, **97**:600–2.

Georgiadis, E., Mantzoros, C. S., Evangelopoulou, C. and Spentzos, D. (1997). Adult height and menarcheal age of young women in Greece. *Annals of Human Biology*, **24**:55–9.

Gerald, C.F. (1978). *Applied Numerical Analysis*. Reading, MA: Addison-Wesley.

Gertner, J. M. (1992). Growth hormone actions on fat distribution and metabolism. *Hormone Research, Supplement*, **38**:41–3.

Gerver, W. J. M. and de Bruin, R. (2001). *Paediatric Morphometrics: A Reference Manual*, 2nd edn. Utrecht, The Netherlands: Wetenschappelijke Uitgeverij Bunge.

Ghavami-Maibodi, S. Z., Collipp, P. J., Castro-Magana, M., Stewart, C. and Chen, S. Y. (1983). Effect of oral zinc supplements on growth, hormonal levels and zinc in healthy short children. *Annals of Nutrition and Metabolism*, **27**:214–19.

Giani, U., Filosa, A. and Causa, P. (1996). A non-linear model of growth in the first year of life. *Acta Paediatrica*, **85**:7–13.

Gibson, A. T. and Wales, J. K. H. (1994). Episodic growth in premature infants: Abstracts of the 7th International Congress of Auxology. *Humanbiologica Budapestiensis*, **24**:34.

Gibson, G. T., Bayhurst, P. A. and Colley, D. P. (1983). Maternal alcohol, tobacco and cannabis consumption and the outcome of pregnancy. *Australian and New Zealand Journal of Obstetrics and Gynaecology*, **23**:5–19.

Gibson, R. S., Vanderkooy, P. D. S., MacDonald, A. C., Goldman, A., Ryan, B. A. and Berry, M. (1989). A growth-limiting, mild zinc-deficiency syndrome in some southern Ontario boys with low height percentiles. *American Journal of Clinical Nutrition*, **49**:1266–73.

Gidding, S. S., Bao, W. H., Srinivasan, S. R. and Berenson, G. S. (1995). Effects of secular trends in obesity on coronary risk factors in children: The Bogalusa Heart Study. *Journal of Pediatrics*, **127**:868–74.

Gillum, R. F. (1999). Distribution of waist-to-hip ratio, other indices of body fat distribution and obesity and associations with HDL cholesterol in children and young adults aged 4–19 years: The Third National Health and Nutrition Examination Survey. *International Journal of Obesity*, **23**:556–63.

Gilsanz, V., Roe, T. F., Mora, S., Costin, G. and Goodman, W. G. (1991). Changes in vertebral bone density in black girls and white girls during childhood and puberty. *New England Journal of Medicine*, **325**:1597–1600.

Ginsburg, E., Škarić-Jurić, T., Kobyliansky, E., Malkin, I. and Rudan, P. (2001). Evidence on major gene control of cortical index in pedigree data from Middle Dalmatia, Croatia. *American Journal of Human Biology*, **13**:398–408.

Glaser, H. H., Heagarty, M. C., Bullard, D. M. and Pivchik, E. C. (1968). Physical and psychological development of children with early failure to thrive. *Journal of Pediatrics*, **73**:690–8.

Gloria-Bottini, F., Lucarini, N., La Torre, M., Lucarelli, P. and Bottini, E. (2001). Birth weight and parental PBM1 alleles. *American Journal of Human Biology*, **13**:417–20.

Godfrey, K. M., Barker, D. J. P. and Osmond, C. (1994). Disproportionate fetal growth and raised IgE concentration in adult life. *Clinical and Experimental Allergy*, **24**:641–8.

Gofin, J. (1979). The effect on birthweight of employment during pregnancy. *Journal of Biosocial Science*, **11**:259–67.

Gohlke, B. C., Khadilkar, V. V., Skuse, D. and Stanhope, R. (1998). Recognition of children with psychosocial short stature: A spectrum of presentation. *Journal of Pediatric Endocrinology and Metabolism*, **11**:509–7.

Golab, S. (1992). Differentiation of the physical development in children and youth in relation to the socioeconomic and health status on the example of longitudinal studies in Nowa Huta, Poland. *Acta Medica Auxologica*, **24**:189–96.

Golden, M. H. N. (1988). The role of individual nutrient deficiencies in growth retardation of children as exemplified by zinc and protein. In *Linear Growth Retardation in Less Developed Countries*, vol. 14, ed. J. C. Waterlow, pp. 143–63. New York: Raven Press.

Golden, M. H. N. (1998). Catch-up weight-gain. In *The Cambridge Encyclopedia of Human Growth and Development*, eds. S. J. Ulijaszek, F. E. Johnston and M. A. Preece, pp. 348–50. Cambridge, U.K.: Cambridge University Press.

Golden, P. L., MacCagnan, T. J. and Pardridge, W. M. (1997). Human blood–brain barrier leptin receptor. *Journal of Clinical Investigation*, **99**:14–18.

Goldenberg, R. L., Cliver, S. P., Cutter, G. R., Hoffman, H. J., Cassady, G., Davis, R. O. and Nelson, K. G. (1991). Black–white differences in newborn anthropometric measurements. *Obstetrics and Gynaecology*, **78**:782–8.

Goldenberg, R. L., Hoffman, H. J., Cliver, S. P., Cutter, G. R., Nelson, K. G. and Copper, R. L. (1992). The influence of previous low birth weight on birth weight, gestational age, and anthropometric measurements in the current pregnancy. *Obstetrics and Gynecology*, **79**:276–80.

Goldstein, A. and Warren, R. (1962). Passage of caffeine into human gonadal and fetal tissue. *Biochemical Pharmacology*, **11**:166–8.

Goldstein, H. (1971). Factors influencing the height of seven year old children: Results from the National Child Development Study. *Human Biology*, **43**:92–111.

González Apraiz, A. and Rebato, E. (1995). La edad de menarquia en las niñas de la villa de Bilbao: Un estudio comparativo con otras poblaciones españolas (The age of menarche in females of the city of Bilbao: A comparative study with other Spanish populations). In *Avances en Antropologia Ecológica y Genética*, eds. J. L. Nieto and L. A. Moreno, pp. 153–9. Zaragoza, Spain: Universidad de Zaragoza.

Goodman, H., Schwartz, Y., Tai, L. and Gorin, E. (1990). Actions of growth hormone on adipose tissue: Possible involvement of autocrine and paracrine factors. *Acta Paediatrica Scandinavica*, **367**:132–6.

Goran, M. I., Driscoll, P., Johnson, R., Naby, T. R. and Hunter, G. R. (1996). Cross-calibration of body-composition techniques against dual-energy X-ray absorptiometry in young children. *American Journal of Clinical Nutrition*, **63**:299–305.

Gordon-Larsen, P., Zemel, B. S. and Johnston, F. E. (1997). Secular changes in stature, weight, fatness, overweight, and obesity in urban African American adolescents from the mid-1950s to the mid-1960s. *American Journal of Human Biology*, **9**:675–88.

Górny, S. (1977). *Trend sekularny w wysokoœci ciała poborowych w Polsce (Secular Changes in the Height of Conscripts in Poland)*, Sympozjum Skularne Zmiany w Stanie Fizyczym Populacji i Ocena Ich Znaczenia. Wrocław: Polish Academy of Science.

Gorski, R. A. (1990). Maturation of neural mechanisms and the pubertal process. In *Control of the Onset of Puberty*, eds. M. M. Grumbach, P. C. Sizonenko and M. L. Aubert, pp. 259–81. Baltimore, MD: Williams & Wilkins.

Gortmaker, S. L., Dietz, W. H., Sobol, A. M. and Wehler, C. A. (1987). Increasing pediatric obesity in the United States. *American Journal of Diseases of Children*, **141**:535–40.

Gortmaker, S. L., Must, A., Sobol, A. M., Peterson, K., Colditz, G. A. and Dietz, W. H. (1996). Television viewing as a cause of increasing obesity among children in the United States, 1986–1990. *Archives of Pediatrics and Adolescent Medicine*, **150**:356–62.

Graffar, M. (1956). Une méthode de classification sociale d'échantillons de population (A method for the social classification of population groups). *Courrier*, **6**:455–9.

Graffar, M. and Corbier, J. (1972). Contribution to the study of the influence of socio-economic conditions on the growth and development of the child. *Early Child Development and Care*, **1**:141–79.

Grahn, D. and Kratchman, J. (1963). Variation in neonatal death rate and birth weight in the United States and possible relations to environmental radiation, geology and altitude. *American Journal of Human Genetics*, **15**:329–52.

Graitcer, P. L. and Gentry, E. M. (1981). Measuring children: One reference for all. *Lancet*, i: 297–9.

Graves, A. B., Mortimer, J. A., Larson, E. B., Wenzlow, A., Bowden, J. D. and McCormick, W. C. (1996). Head circumference as a measure of cognitive reserve: Association with severity of impairment in Alzheimer's disease. *British Journal of Psychiatry*, **169**:86–92.

Gray, H. (1948). Predictions of adult stature. *Child Development*, **19**:167–75.

Green, A. A. and Mac Farlane, J. A. (1983). Method for the earlier recognition of abnormal stature. *Archives of Disease in Childhood*, **58**:535–7.

Green, H., Morikawa, M. and Nixon, T. (1985). A dual effector theory of growth hormone action. *Differentiation*, **29**:195–8.

Green, M. and Buckler, J. M. H. (2001). Prediction of ultimate stature of children delayed in puberty. *Acta Medica Auxologica*, **33**:13–18.

Greenberg, A. H., Najjar, S. and Blizzard, R. M. (1974). Effects of thyroid hormone on growth, differentiation and development. In *Handbook of Physiology*, Section 7, *Endocrinology*, vol. 3, *Thyroid*, eds. M. A. Greer and D. H. Solomon, pp. 377–89. Washington, D.C.: American Physiological Society.

Gregory, A., Doull, I., Pearce, N., Cheng, S., Leadbitter, P., Holgate, S. and Beasley, R. (1999). The relationship between anthropometric measurements at birth: Asthma and atopy in childhood. *Clinical and Experimental Allergy*, **29**:330–3.

Gregory, J. W., Greene, S. A., Thompson, J., Scrimgeour, C. M. and Rennie, M. J. (1992). Effects of oral testosterone undecanoate on growth, body composition, strength and energy expenditure of adolescent boys. *Clinical Endocrinology*, **37**:207–13.

Greksa, L. P., Spielvogel, H. and Caceres, E. (1985). Effect of altitude on the physical growth of upper-class children of European ancestry. *Annals of Human Biology*, **12**:225–32.

Grennert, L., Persson, P. H., Gennser, G. and Gullberg, B. (1980). Zygosity and intrauterine growth of twins. *Obstetrics and Gynecology*, **55**:684–7.

Greulich, W. W. and Pyle, S. I. (1950). *Radiographic Atlas of Skeletal Development of the Hand and Wrist*. Stanford, CA: Stanford University Press.

Greulich, W. W. and Pyle, S. I. (1959). *Radiographic Atlas of Skeletal Development of the Hand and Wrist*, 2nd edn. Stanford, CA: Stanford University Press.

Griggs, R. C., Kingston, W., Jozefowicz, R. F., Herr, B. E., Forbes, G. and Halliday, D. (1989). Effect of testosterone on muscle mass and muscle protein synthesis. *Journal of Applied Physiology*, **66**:498–503.

Gross, T., Sokol, R. J. and King, K. C. (1980). Obesity in pregnancy: Risks and outcome. *Obstetrics and Gynecology*, **56**:446–50.

Grossman, A., Kruseman, A. C., Perry, L., Tomkin, S., Schally, A. V., Coy, D. H., Rees, L. H., Comaru-Schally, A.-M. and Besser, G. M. (1982). New hypothalamic hormone, corticotropin-releasing factor, specifically stimulates the release of adrenocorticotropic hormone and cortisol in man. *Lancet*, i:921–2.

Grumbach, M. M. (1975). Onset of puberty. In *Puberty, Biologic and Social Components*, ed. S. R. Berenberg, pp. 1–21. Leiden, The Netherlands: Stenfert Krose.

Grumbach, M. M. and Kaplan, S. L. (1990). The neuroendocrinology of human puberty: An ontogenetic perspective. In *Control of the Onset of*

Puberty, eds. M. M. Grumbach, P. C. Sizonenko and M. L. Aubert, pp. 1–68. Baltimore, MD: Williams & Wilkins.

Grumbach, M. M. and Styne, D. M. (1998). Puberty: Ontogeny, neuroendocrinology, physiology, and disorders. In *Williams' Textbook of Endocrinology*, eds. J. D. Wilson, D. W. Foster, H. M. Kronenberg and P. R. Larsen, pp. 1509–1625. Philadelphia, PA: WB Saunders.

Grunfeld, C., Sherman, B. and Cavalieri, R. (1988). The acute effects of human growth hormone administration on thyroid function in normal men. *Journal of Clinical Endocrinology and Metabolism*, 67:1111–14.

Guendelman, S. and Abrams, B. (1995). Dietary intake among Mexican-American women: Generational differences and a comparison with white non-Hispanic women. *American Journal of Public Health*, 85:20–5.

Guendelman, S. and English, P. B. (1995). Effect of United States residence on birth outcomes among Mexican immigrants: An exploratory study. *American Journal of Epidemiology*, 142:S30–8.

Guihard-Costa, A. M., Grange, G., Larroche, J. C. and Papiernik, E. (1997). Sexual differences in anthropometric measurements in French newborns. *Biology of the Neonate*, 72:156–64.

Guillaume, M., Lapidus, L., Björntorp, P. and Lambert, A. (1997). Physical activity, obesity, and cardiovascular risk factors in children: The Belgian Luxembourg Child Study II. *Obesity Research*, 5:549–56.

Guillaume-Gentil, C., Assimacopoulos-Jeannet, F. and Jeanrenaud, B. (1993). Involvement of nonesterified fatty acid oxidation in glucocorticoid-induced peripheral insulin resistance *in vivo* in rats. *Diabetologia*, 36:899–906.

Guler, H. P., Zapf, J. and Froesch, E. R. (1987). Short-term metabolic effects of recombinant human insulin-like growth factor I in healthy adults. *New England Journal of Medicine*, 317:137–40.

Gunnell, D. J., Frankel, S. J., Nanchahal, K., Peters, T. J. and Smith, G. D. (1998). Childhood obesity and adult cardiovascular mortality: A 57-year follow-up study based on the Boyd Orr cohort. *American Journal of Clinical Nutrition*, 67:1111–18.

Guo, S., Chi, E., Wisemandle, W., Chumlea, W. C., Roche, A. F. and Siervogel, R. M. (1998a). Serial changes in blood pressure from childhood into young adulthood for females in relation to body mass index and maturational age. *American Journal of Human Biology*, 10:589–98.

Guo, S., Chumlea, W. C., Roche, A. F., Gardner, J. D. and Siervogel, R. M. (1994). The predictive value of childhood body mass index values for overweight at age 35 years. *American Journal of Clinical Nutrition*, 59:810–19.

Guo, S., Roche, A. F., Chumlea, W. C., Casey, P. H. and Moore, W. M. (1997a). Growth in weight, recumbent length, and head circumference for preterm low-birthweight infants during the first three years of life using gestation-adjusted ages. *Early Human Development*, 47:305–25.

Guo, S., Roche, A. F., Fomon, S. J., Nelson, S. E., Chumlea, W. C., Rogers, R. R., Baumgartner, R. N., Ziegler, E. E. and Siervogel, R. M. (1991). Reference data on gains in weight and length during the first two years of life. *Journal of Pediatrics*, 119: 355–62.

Guo, S., Roche, A. F. and Houtkooper, L. (1989). Fat-free mass in children and young adults predicted from bioelectric impedance and anthropometric variables. *American Journal of Clinical Nutrition*, 50:435–43.

Guo, S., Roche, A. F. and Moore, W. (1988). Reference data for head circumference and one-month increments from one to twelve months. *Journal of Pediatrics*, 113:490–4.

Guo, S., Siervogel, R. M., Roche, A. F. and Chumlea, W. C. (1992). Mathematical modelling of human growth: A comparative study. *American Journal of Human Biology*, 4:93–104.

Guo, S. S., Chumlea, W. C., Roche, A. F. and Siervogel, R. M. (1997b). Age- and maturity-related changes in body composition during adolescence into adulthood: The Fels Longitudinal Study. *International Journal of Obesity*, 21:1167–75.

Guo, S. S., Huang, C., Maynard, L. M., Demerath, E., Towne, B., Chumlea, W. C. and Siervogel, R. M. (2000a). Body mass index during childhood, adolescence and young adulthood in relation to adult overweight and adiposity: The Fels Longitudinal Study. *International Journal of Obesity*, **24**:1628–35.

Guo, S. S., Roche, A. F., Chumlea, W. C., Casey, P. H. and Moore, W. M. (1998*b*). Adjustments to the observed growth of preterm low-birth weight infants for application to infants who are small for gestational age at birth. *Acta Medica Auxologica*, **30**:71–87.

Guo, S. S., Roche, A. F., Chumlea, W. C., Johnson, C. L., Kuczmarski, R. J. and Curtin, R. (2000*b*). Statistical effects of varying sample sizes on the precision of percentile estimates. *American Journal of Human Biology*, **12**:64–74.

Guo, S. S., Wholihan, K., Roche, A. F., Chumlea, W. C. and Casey, P. H. (1996). Weight-for-length reference data for preterm, low birth weight infants. *Archives of Pediatric and Adolescent Medicine*, **150**:964–70.

Gutin, B., Islam, S., Manos, T., Cucuzzo, N., Smith, C. and Stachura, M. (1994). Relation of percentage of body fat and maximal aerobic capacity to risk factors for atherosclerosis and diabetes in black and white seven- to eleven-year-old children. *Journal of Pediatrics*, **125**:847–52.

Guyda, H. J. (1990). The adolescent growth spurt and its sex differences. In *Handbook of Human Growth and Developmental Biology*, vol. ll, Part B, eds. E. Meisami and P. S. Timiras, pp. 69–83. Boca Raton, FL: CRC Press.

Gworys, B. (1978). Zimany w budowie ciała młodzieży w wieku od 18 do 22 roku życia na przykładzie studentów I studentek Akademji Medycznej we Wrocławiu (Changes in body build in youths aged 18 to 22 years). *Materialy i Prace Antropologiczne*, **95**:81–106.

Haas, J. D., Frongillo, E. A., Stepic, C. D., Beard, J. L. and Hurtado, L. (1980). Altitude, ethnic, and sex difference in birth weight and length in Bolivia. *Human Biology*, **52**:459–77.

Habicht, J.-P., Martorell, R., Yarbrough, C., Malina, R. M. and Klein, R. E. (1974*a*). Height and weight standards. *Lancet*, **ii**:47.

Habicht, J.-P., Yarbrough, C., Martorell, R., Malina, R. M. and Klein, R. E. (1974*b*). Height and weight standards for preschool children: How relevant are ethnic differences in growth potential? *Lancet*, **i**:611–14.

Hadders-Algra, M. and Touwen, B. C. L. (1990). Body measurements, neurological and behavioral development in six-year-old children born preterm and/or small-for-gestational-age. *Early Human Development*, **22**:1–13.

Hadley, M. E. (1996). *Endocrinology*, 4th edn. Upper Saddle River, NJ: Prentice Hall.

Haeusler, G. and Frisch, H. (1994). Methods for evaluation of growth in Turner's syndrome: critical approach and review of the literature. *Acta Paediatrica*, **83**:309–14.

Haeusler, G., Schemper, M., Frisch, H., Blümel, P., Schmitt, K. and Plöchl, E. (1992). Spontaneous growth in Turner syndrome: Evidence for a minor pubertal growth spurt. *European Journal of Pediatrics*, **151**:283–7.

Hägg, U. and Taranger, J. (1991). Height and height velocity in early, average and late maturers followed to the age of 25: A prospective longitudinal study of Swedish urban children from birth to adulthood. *Annals of Human Biology*, **18**:47–56.

Hägg, U. and Taranger, J. (1992). Pubertal growth and maturity pattern in early and late maturers. *Swedish Dental Journal*, **16**:199–209.

Hahn, R. A. (1999). Why race is differentially classified on U.S. birth and infant death certificates: An examination of two hypotheses. *Epidemiology*, **10**:108–11.

Hakama, M. (1991). Screening. In *Oxford Textbook of Public Health. Applications in Public Health*, 2nd edn, vol. 3, eds. W. W. Holland, R. Detels and G. Knox, pp. 91–106. Oxford, U.K.: Oxford University Press.

Halaas, J. L., Gajiwala, K. S., Maffei, M., Cohen, S. L., Chait, B. T., Rabinowitz, D., Lallone, R. L., Burley, S. K. and Friedman, J. M. (1995). Weight-reducing effects of the plasma-protein encoded by the obese gene. *Science*, **269**:543–6.

Hales, C. N., Barker, D. J. P., Clark, P. M. S., Cox, L. J., Fall, C., Osmond, C. and Winter, P. D. (1991). Fetal and infant growth and impaired glucose tolerance at age 64. *British Medical Journal*, **303**:1019–22.

Hall, D. C. and Kaufmann, D. A. (1987). Effects of aerobic and strength conditioning on pregnancy outcomes. *American Journal of Obstetrics and Gynecology*, **157**:1199–203.

Hall, D. M. B. (2000). Growth monitoring. *Archives of Disease in Childhood*, **82**:10–15.

Hall, J. G. (1991). The relationship between karyotype and growth in Turner syndrome. In *Turner Syndrome: Growth Promoting Therapies*, eds. M. B. Ranke and R. G. Rosenfeld, pp. 9–13. Amsterdam, The Netherlands: Excerpta Medica.

Hall, K. and Sara, V. R. (1984). Somatomedin levels in childhood, adolescence and in adult life. *Clinics in Endocrinology and Metabolism*, **13**:91–112.

Hall, M. H. (1990). Definitions used in relation to gestational age. *Paediatric and Perinatal Epidemiology*, **4**:123–8.

Halmesmaki, E. (1988). Alcohol counseling of 85 pregnant problem drinkers: Effect on drinking and fetal outcome. *British Journal of Obstetrics*, **95**: 243–7.

Hambidge, K. M. (1986). Zinc deficiency in the weaning: How important? *Acta Paediatrica Scandinavica, Supplement*, **323**:52–8.

Hamill, P., Drizd, T., Johnson, C., Reed, R. and Roche, A. (1977). *NCHS growth Curves for Children, Birth–18 Years, United States*, Vital and Health Statistics, Series 11, no. 165. Washington, D.C.: U.S. Department of Health, Education, and Welfare.

Hamill, P. V. V., Drizd, T. A., Johnson, C. L., Reed, R. B., Roche, A. and Moore, W. M. (1979). Physical growth: National center for health statistics percentiles. *American Journal of Clinical Nutrition*, **32**:607–29.

Hamill, P. V. V., Johnston, F. E. and Lemeshow, S. (1972). *Height and Weight of Children: Socioeconomic status, United States*, Vital and Health Statistics, Series 11, no. 119. Washington, D.C.: U.S. Department of Health, Education, and Welfare.

Handler, A., Kistin, N., Davis, F. and Ferre, C. (1991). Cocaine use during pregnancy: Perinatal outcomes. *American Journal of Epidemiology*, **133**:818–25.

Hansen, J. D. I. (1965). Body composition and appraisal of nutriture. In *Human Body Composition: Approaches and Applications*, Symposium of the Society for the Study of Human Biology, vol. 7, ed. J. Brožek, pp. 255–63. Oxford, U.K.: Pergamon Press.

Hansman, C. (1970). Anthropometry and related data. In *Human Growth and Development*, ed. R. W. McCammon, pp. 101–54. Springfield, IL: Charles C. Thomas.

Harbison, R. D. and Mantilliplata, B. (1972). Prenatal toxicity, maternal distribution and placental transfer of tetrahydrocannabinol. *Journal of Pharmacology and Experimental Therapeutics*, **180**:446–53.

Hardy, J. B. and Mellits, E. D. (1972). Does maternal smoking during pregnancy have a long-term effect on the child? *Lancet*, **ii**:1332–6.

Harlan, W. R. (1980). Letter to the editor. *Journal of Pediatrics*, **96**:348.

Harlan, W. R., Grillow, G. P., Coroni-Huntley, J. and Leaverton, P. E. (1979). Secondary sex characteristics of boys 12 to 17 years of age: The U.S. Health Examination Survey. *Journal of Pediatrics*, **95**:293–7.

Harlan, W. R., Harlan, E. A. and Grillo, G. P. (1980). Secondary sex characteristics of girls 12 to 17 years of age: The U.S. Health Examination Survey. *Journal of Pediatrics*, **96**:1074–8.

Harris, E. F., Weinstein, S., Weinstein, L. and Poole, A. E. (1980). Predicting adult stature: A comparison of methodologies. *Annals of Human Biology*, **7**:225–34.

Harris, J. A., Vernon, P. A. and Boomsma, D. I. (1998). The heritability of testosterone: A study of Dutch adolescent twins and their parents. *Behavioral Genetics*, **28**:165–71.

Harris, J. E. (1982). Prenatal medical care and infant mortality. In *Economic Aspects of Health*, ed. V. R. Fuchs, pp. 15–52. Chicago, IL: University of Chicago Press.

Harrison, G. G., Branson, R. and Vaucher, Y. E. (1983). Association of maternal smoking with body composition of the newborn. *American Journal of Clinical Nutrition*, **38**:757–62.

Harrison, G. G., Udall, J. N. and Morrow, G. (1980). Maternal obesity, weight gain in pregnancy and infant birthweight. *American Journal of Obstetrics and Gynecology*, **136**:411–12.

Hartikainen-Sorri, A.-L. and Sorri, M. (1989). Occupational and socio-medical factors in preterm birth. *Obstetrics and Gynecology*, **74**:13–16.

Hartman, M. L., Iranmanesh, A., Thorner, M. O. and Veldhuis, J. D. (1993). Evaluation of pulsatile patterns of growth hormone release in humans: A brief review. *American Journal of Human Biology*, **5**:603–14.

Hartman, M. L., Veldhuis, J. D., Johnson, M. L., Lee, M. M., Alberti, K. G. M. M., Samojlik, E. and Thorner, M. O. (1992). Augmented growth hormone (GH) secretory burst frequency and amplitude mediate enhanced GH secretion during a two-day fast in normal men. *Journal of Clinical Endocrinology and Metabolism*, **74**:757–65.

Harvey, M. A. S., Smith, D. W. and Skinner, A. L. (1979). Infant growth standards in relation to parental stature. *Clinical Pediatrics*, **18**:602–13.

Haschke, F. (1983). Body composition of adolescent males. *Acta Paediatrica Scandinavica, Supplement*, **370**:1–23.

Haschke, F. (1989). Body composition during adolescence. In *Body Composition Measurements in Infants and Children*, Report of the 98th Ross Conference on Pediatric Research, eds. W. J. Klish, N. Kretchmer, pp. 76–83. Columbus, OH: Ross Laboratories.

Haschke, F., van't Hof, M. A. and The Euro-Growth Study Group. (2000). Euro-Growth references for length, weight, and body circumferences. *Journal of Pediatric Gastroenterology and Nutrition*, **31**:S14–S38.

Hasegawa, Y., Hasegawa, T., Takada, M. and Tsuchiya, Y. (1996). Plasma-free insulin-like growth factor I concentration in growth hormone deficiency in children and adolescents. *European Journal of Endocrinology*, **134**:184–9.

Hashimoto, N., Kawasaki, T., Kikuchi, T., Takahashi, H. and Uchiyama, M. (1995). Influence of parental obesity on the physical constitution of preschool children in Japan. *Acta Paediatrica Japonica*, **37**:150–3.

Hassan, H. M. S., Kohno, H., Kuromaru, R., Honda, S. and Ueda, K. (1996). Body composition, atherogenic risk factors and apolipoproteins following growth hormone treatment. *Acta Paediatrica*, **85**:899–901.

Hassing, J. M., Padmanabhan, V., Kelch, R. P., Brown, M. B., Olton, P. R., Sonstein, J. J., Foster, C. M. and Beitins, T. Z. (1990). Differential regulation of serum immunoreactive luteinizing hormone and bioactive follicle-stimulating hormone by testosterone in early pubertal boys. *Journal of Clinical Endocrinology and Metabolism*, **70**:1082–9.

Hassink, S. G., Sheslow, D. V., Lancey, E., Opentanova, I., Considine, R. V. and Caro, J. F. (1996). Serum leptin in children with obesity: Relationship to gender and development. *Pediatrics*, **98**:201–3.

Haste, F. M., Brooke, O. G., Anderson, H. R. and Bland, J. M. (1991). The effect of nutritional intake on outcome of pregnancy in smokers and non-smokers. *British Journal of Nutrition*, **65**:347–54.

Haste, F. M., Brooke, O. G., Anderson, H. R., Bland, J. M., Shaw, A., Griffin, J. and Peacock, J. L. (1990). Nutrient intakes during pregnancy: Observations on the influence of smoking and social class. *American Journal of Clinical Nutrition*, **51**:29–36.

Hata, T., Deter, R. L. and Hill, R. M. (1991). Reduction of soft tissue deposition in normal triplets. *Journal of Clinical Ultrasound*, **19**:541–5.

Hatch, M., Ji, B.-T., Shu, X. O. and Susser, M. (1997). Do standing, lifting, climbing, or long hours of work during pregnancy have an effect on fetal growth? *Epidemiology*, **8**:530–6.

Hauspie, R., Bergman, P., Bielicki, T. and Susanne, C. (1994). Genetic variance in the pattern of the growth curve for height: A longitudinal analysis of male twins. *Annals of Human Biology*, **21**:347–62.

Hauspie, R. C., Chrzastek-Spruch, H., Verleye, G., Kozlowska, M. A. and Susanne, C. (1996). Determinants of growth in body length from birth to 6 years of age: A longitudinal study of Lublin children. *American Journal of Human Biology*, **8**:21–9.

Hauspie, R. C., Das, S. A., Preece, M. A. and Tanner, J. M. (1980). A longitudinal study of the growth in height of boys and girls of West Bengal (India) aged six months to 20 years. *Annals of Human Biology*, **7**:429–41.

Hauspie, R. C. and Steendijk, R. (1993). Application of growth models in the analysis of pathological growth data: The case of hypophosphatemic vitamin D-resistant rickets. *American Journal of Human Biology*, **5**:181–92.

Hauspie, R. C., Vercauteren, M. and Susanne, C. (1997). Secular changes in growth and maturation: an update. *Acta Paediatrica*, **423**:20–7.

Hauspie, R. C. and Wachholder, A. (1986). Clinical standards for growth velocity in height of Belgian boys and girls, aged 2 to 18 years. *International Journal of Anthropology*, **1**:339–47.

Hauspie, R. C., Wachholder, A., Sand, E. A. and Susanne, C. (1992). Body length, body weight and head circumference in Belgian boys and girls aged 1–36 months: Sex difference and effect of socioeconomic status. *Acta Medica Auxologica* **24**:149–58.

Haworth, J. C., Ellestad-Sayed, J. J., King, J. and Dilling, L. A. (1980*a*). Fetal growth retardation in cigarette-smoking mothers is not due to decreased maternal food intake. *American Journal of Obstetrics and Gynecology*, **137**:719–23.

Haworth, J. C., Ellestad-Sayed, J. J., King, J. and Dilling, L. A. (1980*b*). Relation of maternal cigarette smoking, obesity and energy consumption to infant size. *American Journal of Obstetrics and Gynecology*, **138**:1185–9.

Hazum, E. and Conn, P. M. (1988). Molecular mechanism of gonadotropin releasing hormone (GnRH) action: The GnRH receptor. *Endocrine Reviews*, **9**:379–86.

He, Q. and Karlberg, J. (1999). Prediction of adult overweight during the pediatric years. *Pediatric Research*, **46**:697–703.

Hebel, J. R., Fox, N. L. and Sexton, M. (1988). Dose–response of birth weight to various measures of maternal smoking during pregnancy. *Journal of Clinical Epidemiology*, **41**:483–9.

Hediger, M. L., Overpeck, M. D., Kuczmarski, R. J., McGlynn, A., Maurer, K. R. and Davis, W. W. (1998). Muscularity and fatness of infants and young children born small- or large-for-gestational-age. *Pediatrics*, **102**:1–7.

Hediger, M. L., Overpeck, M. D., McGlynn, A., Kuczmarski, R. J., Maurer, K. R. and Davis, W. W. (1999). Growth and fatness at three to six years of age of children born small- or large-for-gestational age. *Pediatrics*, **104**:E33–8.

Hediger, M. L., Overpeck, M. D., Ruan, W. J. and Troendle, J. F. (2000). Early infant feeding and growth status of US-born infants and children aged 4–71 mo: Analyses from the third National Health and Nutrition Examination Survey, 1988–1994. *American Journal of Clinical Nutrition*, **72**:159–67.

Hediger, M. L., Scholl, T. O. and Salmon, R. W. (1989). Early weight gain in pregnant adolescents and fetal outcome. *American Journal of Human Biology*, **1**:665–72.

Heinig, M. J., Nommsen, L. A., Peerson, J. M., Lonnerdal, B. and Dewey, K. G. (1993). Energy and protein intakes of breast-fed and formula-fed infants during the first year of life and their association with growth velocity: The DARLING Study. *American Journal of Clinical Nutrition*, **58**:152–61.

Heinonen, S., Ryynänen, M. and Kirkinen, P. (1999). The effects on fetal development of high alpha-fetoprotein and maternal smoking. *American Journal of Public Health*, **89**:561–3.

Heinrichs, C., Munson, P. J., Counts, D. R., Cutler, G. B., Jr. and Baron, J. (1995). Patterns of human growth. *Science*, **268**:442–5.

Heitmann, B. L., Lissner, L., Sørensen, T. I. A. and Bengtsson, C. (1995). Dietary fat intake and weight gain in women genetically predisposed for obesity. *American Journal of Clinical Nutrition*, **61**:1213–17.

Helm, S. (1979). Skeletal maturity in Danish school children assessed by the TW-2 method. *American Journal of Physical Anthropology*, **51**:345–51.

Hemon, D., Berger, C. and Lazar, P. (1982). Maternal factors associated with small-for-dateness among twins. *Acta Geneticae Medicae et Gemellologiae*, **31**:241–5.

Henderson, G. I., Hoyumpa, A. M., Rothschild, M. D. and Schenker, S. (1980). Effect of ethanol and ethanol-induced hypothermia on protein synthesis in pregnant and fetal rats: Alcoholism. *Clinical and Experimental Research*, **4**:165–77.

Henderson, G. I., Turner, D., Patwardhan, R. V., Lumeng, L., Hoyumpa, A. M. and Schenker, S. (1981). Inhibition of placental valine uptake after acute and chronic maternal ethanol consumption. *Journal of Pharmacological and Experimental Therapeutics*, **216**:465–72.

Henkin, R. I. (1976). Trace metals in endocrinology. *Medical Clinics of North America*, **60**:779–97.

Henriksen, T. B., Hedegaard, M., Secher, N. J. and Wilcox, A. J. (1995). Standing at work and preterm delivery. *British Journal of Obstetrics and Gynaecology*, **102**:198–206.

Herman-Giddens, H. E., Slora, E. J., Hasemeier, C. M. and Wasserman, R. C. (1993). The prevalence of secondary sexual characteristics in young girls seen in office practice. *American Journal of Diseases of Children*, **147**:455.

Herman-Giddens, M. E. and Bourdony, C. J. (1995). *Assessment of Sexual Maturity Stages in Girls*. Elk Grove Village, IL: American Academy of Pediatrics.

Herman-Giddens, M. E., Slora, E. J., Wasserman, R. C., Bourdony, C. J., Bhapkar, M. V., Koch, G. G. and Haseimier, C. M. (1997). Secondary sexual characteristics and menses in young girls seen in office practice: A study from the Pediatric Research in Office Settings Network. *Pediatrics*, **99**:505–12.

Herman-Giddens, M. E., Wang, L. and Koch, G. (2001). Secondary sexual characteristics in boys: Estimates from the National Health and Nutrition Examination Survey III, 1988–1994. *Archives of Pediatrics and Adolescent Medicine*, **155**:1022–8.

Hermanussen, M. and Burmeister, J. (1989). Standards for the predictive accuracy of short term body height and lower leg length measurements on half annual growth rates. *Archives of Disease in Childhood*, **64**:259–63.

Hermanussen, M. and Burmeister, J. (1993). Children do not grow continuously but in spurts. *American Journal of Human Biology*, **5**:615–22.

Hermanussen, M., Gieger-Benoit, K., Burmeister, J. and Sippell, W. G. (1988). Periodical changes of short term velocity ('mini growth spurts') in human growth. *Annals of Human Biology*, **15**:103–9.

Hernesniemi, I., Zachmann, M. and Prader, A. (1974). Skinfold thickness in infancy and adolescence: A longitudinal correlation study in normal children. *Helvetica Paediatrica Acta*, **29**:523–30.

Herngreen, W. P., van Buuren, S., van Wieringen, J. C., Reerink, J. D., Verloove-Vanhorick, S. P. and Ruys, J. H. (1994). Growth in length and weight from birth to 2 years of a representative sample of Netherlands children (born 1988–89) related to socio–economic status and other background characteristics. *Annals of Human Biology*, **21**:449–63.

Hervey, G. R., Knibbs, A. V., Burkinshaw, L., Morgan, D. B., Jones, P. R. M., Chettle, D. R. and Vartsky, D. (1981). Effects of methandienone on the performance and body composition of men undergoing athletic training. *Clinical Science*, **60**:457–61.

Hewitt, D., Westropp, C. K. and Acheson, R. M. (1955). Oxford Child Health Survey: Effect of childish ailments on skeletal development. *British Journal of Preventive and Social Medicine*, **9**:179–86.

Heymsfield, S. B., McManus, C., Smith, J., Stevens, V. and Nixon, D. W. (1982). Anthropometric measurement of muscle mass: Revised equations for calculating bone-free arm muscle area. *American Journal of Clinical Nutrition*, **36**:680–90.

Heymsfield, S. B., Wang, J., Lichtman, S., Kamen, Y., Kehayias, J. and Pierson, R. N. J. (1989). Body composition in elderly subjects: A critical appraisal of clinical methodology. *American Journal of Clinical Nutrition*, **50**:1167–75.

Hickey, C. A., McNeal, S. F., Menefee, L. and Ivey, S. (1997). Prenatal weight gain within upper and lower recommended ranges: Effect on birth weight of Black and White infants. *Obstetrics and Gynecology*, **90**:489–94.

Hickey, M. S., Israel, R. G., Gardiner, S. N., Considine, R. V., McCammon, M. R., Tyndall, G. L., Houmard, J. A., Marks, R. H. L. and Caro, J. F. (1996). Gender differences in serum leptin levels in humans. *Biochemical and Molecular Medicine*, **59**:1–6.

Hilson, J. A., Rasmussen, K. M. and Kjolhede, C. L. (1997). Maternal obesity and breast-feeding success in a rural population of white women. *American Journal of Clinical Nutrition*, **66**:1371–8.

Himes, J. H. (1999a). Minimum time intervals for serial measurements of growth in recumbent length or stature of individual children. *Acta Paediatrica*, **88**:120–5.

Himes, J. H. (1999b). Maturation-related deviations and misclassification of stature and weight in adolescence. *American Journal of Human Biology*, **11**:499–504.

Himes, J. H. and Bouchard, C. (1989). Validity of anthropometry in classifying youths as obese. *International Journal of Obesity*, **13**:183–93.

Himes, J. H. and Dietz, W. H. (1994). Guidelines for overweight in adolescent preventive services: Recommendations from an expert committee. *American Journal of Clinical Nutrition*, **59**:307–16.

Himes, J. H., Roche, A. F. and Thissen, D. (1981). Parent-specific adjustments for assessment of recumbent length and stature. *Monographs in Paediatrics*, **13**:1–88.

Himes, J. H., Roche, A. F. and Webb, P. (1980). Fat areas as estimates of total body fat. *American Journal of Clinical Nutrition*, **33**:2093–100.

Hindmarsh, P., di Silvio, L., Pringle, P. J., Kurtz, A. B. and Brook, C. G. (1988*a*). Changes in serum insulin concentration during puberty and their relationship to growth hormone. *Clinical Endocrinology*, **28**:381–8.

Hindmarsh, P. C. (1998). Endocrinological regulation of post-natal growth. In *The Cambridge Encyclopedia of Human Growth and Development*, eds. S. J. Ulijaszek, F. E. Johnston and M. A. Preece, pp. 182–3. Cambridge, U.K.: Cambridge University Press.

Hindmarsh, P. C., Matthews, D. R., di Silvio, L., Kurtz, A. B. and Brook, C. G. (1988*b*). Relation between height velocity and fasting insulin concentrations. *Archives of Disease in Childhood*, **63**:665–6.

Hingson, R., Alpert, J., Day, N., Dooling, E., Kayne, H., Morelock, S., Oppenheimer, E. and Zuckerman, B. (1982). Effects of maternal drinking and marijuana use on fetal growth and development. *Pediatrics*, **70**:539–46.

Hitchcock, M. E., Gracey, M. and Owles, E. N. (1981). Growth of healthy breast-fed infants in the first six months. *Lancet*, **ii**:64–5.

Ho, K. Y., Evans, W. S., Blizzard, R. M., Veldhuis, J. D., Merriam, G. R., Samojlik, E., Furlanetto, R., Rogol, A. D., Kaiser, D. L. and Thorner, M. O. 1987). Effects of sex and age on the twenty-four hour profile of growth hormone secretion in man: Importance of endogenous estradiol concentrations. *Journal of Clinical Endocrinology and Metabolism*, **64**:51–8.

Ho, S. K. and Wu, P. K. (1975). Perinatal factors and neonatal morbidity in twin pregnancies. *American Journal of Obstetrics and Gynecology*, **122**:979–85.

Hochberg, Z. (1990). Growth hormone and IGF-I in growth-plate growth. *Growth and Growth Factors*, **5**:127–30.

Hochberg, Z., Hertz, P., Colin, V., Ish-Shalom, S., Yeshurun, D., Youdim, M. B. H. and Amit, T. (1992). The distal axis of growth hormone (GH) in nutritional disorders: GH-binding protein, insulin-like growth factor-I (IGF-1), and IGF-1

receptors in obesity and anorexia nervosa. *Metabolism: Clinical and Experimental*, **41**:106–12.

Hockelman, R. A., Kelly, J. and Zimmer, A. W. (1976). The reliability of maternal recall. *Clinical Pediatrics*, **15**:261–5.

Hoey, H., Cox, L. and Tanner, J. (1986). The age of menarche in Irish girls. *Irish Medical Journal*, **79**:283–5.

Hoey, H. M., Tanner, J. M. and Cox, L. A. (1987). Clinical growth standards for Irish children. *Acta Paediatrica Scandinavica, Supplement*, **338**:1–31.

Hoff, C., Wertelecki, W., Reyes, E., Zansky, S., Dutt, J., Stumpe, A., Till, D. and Butler, R. M. (1985). Maternal sociomedical characteristics and birth weights of firstborns. *Social Science and Medicine*, **21**:775–83.

Hoff, C., Wetelecki, W., Blackburn, W. R., Mendenhall, H., Wiseman, H. and Stumpe, A. (1986). Trend associations of smoking with maternal, fetal, and neonatal morbidity. *Obstetrics and Gynecology*, **68**:317–21.

Hoffmans, M. D. A. F., Obermann de Boer, G. L., Florack, E. I. M., van Kampen Donker, M. and Kromhout, D. (1988). Determinants of growth during early infancy. *Human Biology*, **60**:237–49.

Hoffstedt, J., Eriksson, P., Hellstrom, L., Rössner, S., Rydén, M. and Arner, P. (2000). Excessive fat accumulation is associated with the TNF_x-308 G/A promoter polymorphism in women but not in men. *Diabetologia*, **43**:117–20.

Hoffstedt, J., Poirier, O., Thorne, A., Lonnqvist, F., Herrmann, S. M., Cambien, F. and Arner, P. (1999). Polymorphism of the human β(3)-adrenoceptor gene forms a well-conserved haplotype that is associated with moderate obesity and altered receptor function. *Diabetes*, **48**:203–5.

Hofman, P. L., Cutfield, W. S., Robinson, E. M., Bergman, R. N., Menon, R. K., Sperling, M. A. and Gluckman, P. D. (1997). Insulin resistance in short children with intrauterine growth retardation. *Journal of Clinical Endocrinology and Metabolism*, **82**:402–6.

Högfeldt, P., Johansson, P. and Källén, B. (1991). Head circumference at birth and microcephaly. *Acta Medica Auxologica* **23**:227–36.

Hokken-Koelega, A. C., de Ridder, M. A., Lemmen, R. J., den Hartog, H., de Muinck Keizer-Schrama, S. M. and Drop, S. L. (1995). Children born small for gestational age: Do they catch up? *Pediatric Research*, 38:267–71.

Holl, R. W., Heinze, E., Seifert, M., Grabert, M. and Teller, W. M. (1994). Longitudinal analysis of somatic development in paediatric patients with IDDM: Genetic influences on height and weight. *Diabetologia*, 37:925–9.

Holm, V. A. (1995). Growth charts for Prader–Willi syndrome. In *Management of Prader–Willi Syndrome*, 2nd edn, eds. L. R. Greenwag and R. C. Alexander, pp. 335–8. New York: Springer-Verlag.

Holm, V. A. and Nugent, J. K. (1982). Growth in the Prader–Willi syndrome. *Birth Defects: Original Article Series*, 18:93–100.

Holmes, G. E., Miller, H. C., Hassanein, K., Lansky, S. B. and Goggin, J. E. (1977). Postnatal somatic growth in infants with atypical fetal growth patterns. *American Journal of Diseases of Children*, 131:1078–83.

Holmes, G. L. (1979). Evaluation and prognosis in nonorganic failure to thrive. *Southern Medical Journal*, 72:693–5.

Homer, C. and Ludwig, S. (1981). Categorization of etiology of failure to thrive. *American Journal of Diseases of Children*, 135:848–51.

Hong, Y., Pedersen, N., Brismar, K., Hall, K. and de Faire, U. (1996). Quantitative genetic analyses of insulin-like growth factor I (IGF-I), IGF-binding protein-1, and insulin levels in middle-aged and elderly twins. *Journal of Clinical Endocrinology and Metabolism*, 81:1791–7.

Hong, Y. L., Rice, T., Gagnon, J., Després, J. P., Nadeau, A., Pérusse, L., Bouchard, C., Leon, A. S., Skinner, J. S., Wilmore, J. H. and Rao, D. C. (1998). Familial clustering of insulin and abdominal visceral fat: The HERITAGE Family Study. *Journal of Clinical Endocrinology and Metabolism*, 83:4239–45.

Horning, M. G., Stratton, C., Nowlin, J., Wilson, A., Horning, E. C. and Hill, R. M. (1973). Placental transfer of drugs. In *Fetal Pharmacology*, ed. L. O. Boreus, pp. 355–71. New York: Raven Press.

Horon, I. L., Strobino, D. M. and MacDonald, H. M. (1983). Birth weights among infants born to adolescent and young adult women. *American Journal of Obstetrics and Gynecology*, 146:444–9.

Horowitz, M., Wishart, J. M., O'Loughlin, P. D., Morris, H. A., Need, A. G. and Nordin, B. E. (1992). Osteoporosis and Klinefelter's syndrome. *Clinical Endocrinology*, 36:113–18.

Horst, H. J., Bartsch, W. and Dirksen-Thedens, I. (1977). Plasma testosterone, sex hormone binding globulin capacity and per cent binding of testosterone and 5 alpha-dihydrotestosterone in prepubertal and adult males. *Journal of Clinical Endocrinology and Metabolism*, 45:522–7.

Horton, W. A., Hall, J. G., Scott, C. I., Pyeritz, R. E. and Rimoin, D. L. (1982). Growth curves for height for distrophic dysplasia, spondyloepiphyseal dysplasia congenita, and pseudoachondroplasia. *American Journal of Diseases of Children*, 136:316–19.

Horton, W. A., Rotter, J. I., Rimoin, D. L., Scott, C. I. and Hall, J. (1978). Standard growth curves for achondroplasia. *Journal of Pediatrics*, 93:435–8.

Howard, R. C., Burns, P. D. and Lichty, J. A. (1957). Arterial oxygen saturation and hematocrit values at birth. *American Journal of Diseases of Children*, 93:674–8.

Huang, Q. L., Rivest, R. and Richard, D. (1998). Effects of leptin on corticotropin-releasing factor (CRF) synthesis and CRF neuron activation in the paraventricular hypothalamic nucleus of obese (*ob/ob*) mice. *Endocrinology*, 139:1524–32.

Hube, F., Lietz, U., Igel, M., Jensen, P. B., Tornqvist, H., Joost, H. G. and Hauner, H. (1996). Difference in leptin mRNA levels between omental and subcutaneous abdominal adipose tissue from obese humans. *Hormone and Metabolic Research*, 28:690–3.

Hughes, J. M., Li, L., Chinn, S. and Rona, R. J. (1997). Trends in growth in England and Scotland, 1972 to 1994. *Archives of Disease in Childhood*, 76:182–9.

Hulanicka, B., Brajczewski, C., Jedlińska, T., Slawińska, T. and Waliszko, A. (1990). *Duze miasta – male miasta – wies: Wzrastanie dzieci w Polsce w 1988 roku (City – Town – Village: Growth of Children in Poland in 1988)*, Monographs of the Institute of Anthropology no. 7. Wrocław, Poland: Polish Academy of Sciences.

Hulanicka, B., Kolasa, E. and Waliszko, A. (1994). *Dziewczêta z Górnego Slaska: Spoleczne i Ekologiczne*

Uwarunkowania Dojrzewania (Girls in Upper Silesia: Social and Ecological Influences on Maturation), Monographs of the Institute of Anthropology no. 11. Wrocław, Poland: Polish Academy of Sciences.

Hulanicka, B. and Kotlarz, K. (1983). The final phase of growth in height. *Annals of Human Biology*, **10**:429–34.

Hulanicka, B. and Waliszko, A. (1991). Deceleration of age at menarche of girls in Poland. *Annals of Human Biology*, **18**:507–13.

Hulens, M., Beunen, G., Claessens, A. L., Lefevre, J., Thomis, M., Philippaerts, R., Borms, J., Vrijens, J., Lysens, R. and Vansant, G. (2001). Trends in BMI among Belgian children, adolescents and adults from 1969 to 1996. *International Journal of Obesity*, **25**:395–9.

Hulman, S., Kushner, H., Katz, S. and Falkner, B. (1998). Can cardiovascular risk be predicted by newborn, childhood, and adolescent body size? An examination of longitudinal data in urban African Americans. *Journal of Pediatrics*, **132**:90–7.

Hulsey, T. C., Levkoff, A. H. and Alexander, G. R. (1991). Birth weights of infants of black and white mothers without pregnancy complications. *American Journal of Obstetrics and Gynecology*, **164**:1299–1302.

Hunter, A. G. W., Hecht, J. T. and Scott, C. I., Jr. (1996). Standard weight for height curves in achondroplasia. *American Journal of Medical Genetics*, **62**:255–61.

Hunter, C. and Clegg, E. F. (1973). The effect of hypoxia on the caudal vertebrae of growing mice and rats. *Journal of Anatomy*, 227–44.

Hurgoiu, V. and Mihetiu, M. (1993). The skinfold thickness in preterm infants. *Early Human Development*, **33**:177–81.

Hussain, M. A., Schmitz, O., Mengel, A., Keller, A., Christiansen, J. S., Zapf, J. and Froesch, E. R. (1993). Insulin-like growth factor I stimulates lipid oxidation, reduces protein oxidation, and enhances insulin sensitivity in humans. *Journal of Clinical Investigation*, **92**:2249–56.

Hutchinson-Smith, B. (1973). Skinfold thickness in infancy in relation to birthweight. *Developmental Medicine and Child Neurology*, **15**:628–34.

Hyer, S. L., Rodin, D. A., Tobias, J. H., Leiper, A. and Nussey, S. S. (1992). Growth hormone deficiency during puberty reduces adult bone mineral density. *Archives of Disease in Childhood*, **67**:1472–4.

Hyman, I. and Dussault, G. (1996). The effect of acculturation on low birthweight in immigrant women. *Canadian Journal of Public Health*, **87**: 158–62.

Hyppönen, E., Kenward, M. G., Virtanen, S. M., Piitulainen, A., Virta-Autio, P., Tuomilehto, J., Knip, M., Akerblom, H. K. and The Childhood Diabetes in Finland (DiMe) Study Group. (1999). Infant feeding, early weight gain, and risk of type 1 diabetes. *Diabetes Care*, **22**:1961–5.

Ifft, D. L., Engstrom, J. L., Meier, P. P., Kavanaugh, K. and Yousef, C. K. (1989). Reliability of head circumference measurements for preterm infants. *Neonatal Network*, **8**:41–6.

Imaki, T., Shibasaki, T., Shizume, K., Masuda, A., Hotta, M., Kiyosawa, Y., Jibiki, K., Demura, H., Tsushima, T. and Ling, H. N. (1985). The effect of free fatty acids on growth hormone (GH)-releasing hormone-mediated GH secretion in man. *Journal of Clinical Endocrinology and Metabolism*, **60**:290–4.

Institute of Medicine. (1990). *Subcommittees on Nutritional Status and Weight Gain during Pregnancy and Dietary Intake and Nutrient Supplements during Pregnancy, Food and Nutrition Board: Nutrition during Pregnancy: Weight Gain: Nutrient Supplements.* Washington, DC: National Academy Press.

Iolascon, A. (1991). Growth in Cooley's patients: The Campania experience. *Acta Medica Auxologica*, **23**:37–41.

Iranmanesh, A., Lizarralde, G. and Veldhuis, J. (1991). Age and relative adiposity are specific negative determinants of the frequency and amplitude of growth hormone (GH) secretory bursts and the half-life of endogenous GH in healthy men. *Journal of Clinical Endocrinology and Metabolism*, **73**: 1081–8.

Isaacs, R. E., Gardner, D. G. and Baxter, J. D. (1987). Insulin regulation of rat growth hormone gene expression. *Endocrinology*, **120**:2022–8.

Isaksson, O. G. P., Edén, S. and Jansson, J.-O. (1985). Mode of action of pituitary growth hormone on target cells. *Annual Review of Physiology*, **47**:483–99.

Isaksson, O. G. P., Lindahl, A., Nilsson, A. and Isgaard, J. (1987). Mechanism of the stimulatory effect of growth hormone on longitudinal bone growth. *Endocrine Reviews*, **8**:426–38.

Iselius, L., Lindsten, J., Morton, N. E., Efendic, S., Cerasi, E., Haegermark, A. and Luft, R. (1982). Evidence for an autosomal recessive gene regulating the persistence of the insulin response to glucose in man. *Clinical Genetics*, **22**:180–94.

Isley, W. L., Underwood, L. E. and Clemmons, D. R. (1983). Dietary components that regulate serum somatomedin-C concentration in humans. *Journal of Clinical Investigation*, **71**:175–82.

Isselbacher, K. J. (1977). Metabolic and hepatic effects of alcohol. *New England Journal of Medicine*, **296**:612–16.

Iuliano-Burns, S., Mirwald, R. L. and Bailey, D. A. (2001). Timing and magnitude of peak height velocity and peak tissue velocities for early, average, and late maturing boys and girls. *American Journal of Human Biology*, **13**:1–8.

Jackson, A. and Pollock, M. (1976). Factor analysis and multivariate scaling of anthropometric variables for the assessment of body composition. *Medicine and Science in Sports and Exercise*, **8**:196–203.

Jackson, A. A. (1990). Protein requirements for catch-up growth. *Proceedings of the Nutrition Society*, **49**:507–16.

Jackson, M. J. (1989). Physiology of zinc: General aspects. In *Zinc in Human Biology*, ed. C. F. Mills, pp. 1–14. London: Springer-Verlag.

Jacob, R., Barrett, E., Plewe, G., Fagin, K. D. and Sherwin, R. S. (1989). Acute effects of insulin-like growth factor I on glucose and amino acid metabolism in the awake fasted rat. *Journal of Clinical Investigation*, **83**:1717–23.

Jacobson, S. W., Jacobson, J. L., Sokol, R. J., Martier, S. S., Ager, J. W. and Kaplan, M. G. (1991). Maternal recall of alcohol, cocaine, and marijuana use during pregnancy. *Neurotoxicology and Teratology*, **13**:535–40.

Jaeger, U. (1998). Secular trend in Germany. In *Secular Growth Changes in Europe*, ed. É. B. Bodzsár and C. Susanne, pp. 135–59. Budapest: Eötvös University Press.

Jakacki, R. I., Kelch, R. P., Sauder, S. E., Lloyd, J. S., Hopwood, N. J. and Marshall, T. C. (1982). Pulsatile secretion of luteinizing hormone in children. *Journal of Clinical Endocrinology and Metabolism*, **55**:453–8.

Janz, K. F., Golden, J. C., Hansen, J. R. and Mahoney, L. T. (1992). Heart rate monitoring of physical activity in children and adolescents: The Muscatine Study. *Pediatrics*, **89**:256–61.

Jaworski, A. A. (1974). New premature weight chart for hospital use. *Clinica Pediatrica*, **13**:513–16.

Jelliffe, D. B. and Jelliffe, E. F. P. (1974). Universal growth standards for preschool children. *Lancet*, **ii**:47.

Jeniček, M. and Demirjian, A. (1974). Age at menarche in urban French-Canadian girls. *Annals of Human Biology*, **2**:339–46.

Jensen, G. M. and Moore, L. G. (1997). The effect of high altitude and other risk factors on birthweight: Independent or interactive effects? *American Journal of Public Health*, **87**:1003–7.

Jensen, O. H. and Foss, O. P. (1981). Smoking in pregnancy: Effects on the birth weight and on thiocyanate concentration in mother and baby. *Acta Obstetrica et Gynecologica Scandinavica*, **60**:177–81.

Jenss, R. M. and Bayley, N. (1937). A mathematical method for studying the growth of a child. *Human Biology*, **9**:556–63.

Jéquier, E. and Tappy, L. (1999). Regulation of body weight in humans. *Physiological Reviews*, **79**:451–80.

Ji, C.-Y., Ohsawa, S. and Kasai, N. (1995). Secular changes in the stature, weight and age at maximum growth increments of urban Chinese girls from 1950 to 1985. *American Journal of Human Biology*, **7**:473–84.

Jiang, X., Srinivasan, S. R., Bao, W. and Berenson, G. S. (1993). Association of fasting insulin with longitudinal changes in blood pressure in children and adolescents: The Bogalusa Heart Study. *American Journal of Hypertension*, **6**:564–9.

Job, J.-C. and Rolland, A. (1986). Histoire naturelle des retards de croissance début intra utérin: Croissance pubertaire et taille adulte (Natural history of growth retardation beginning *in utero*: pubescent growth and adult stature). *Archives Françaises de Pédiatrie*, **43**:301–6.

Johansson, A. G., Burman, P., Westermark, K. and Ljunghall, S. (1992). The bone mineral density in acquired growth hormone deficiency correlates with circulating levels of insulin-like growth factor-I. *Journal of Internal Medicine*, **232**:447–52.

Johnson, M. S., Huang, T. T.-K., Figueroa-Colon, R., Dwyer, J. H. and Goran, M. I. (2001). Influence of leptin on changes in body fat during growth in African American and White children. *Obesity Research*, **9**:593–8.

Johnson, S. L. and Birch, L. L. (1994). Parents' and children's adiposity and eating style. *Pediatrics*, **94**:653–61.

Johnston, F. E. (1985). Validity of triceps skinfold and relative weight as measures of adolescent obesity. *Journal of Adolescent Health Care*, **6**:185–90.

Jones, D. Y., Nesheim, M. C. and Habicht, J.-P. (1985). Influences in child growth associated with poverty in the 1970s: An examination of HANES I and HANES II cross-sectional US national surveys. *American Journal of Clinical Nutrition*, **42**:714–34.

Jones, J. S., Newman, R. B. and Miller, M. C. (1991). Cross-sectional analysis of triplet birth weight. *American Journal of Obstetrics and Gynecology*, **164**:135–40.

Jones, K. L. and Smith, D. W. (1975). The fetal alcohol syndrome. *Teratology*, **12**:1–10.

Jones, M. A., Hitchen, P. J. and Stratton, G. (2000). The importance of considering biological maturity when assessing physical fitness measures in girls and boys aged 10 to 16 years. *Annals of Human Biology*, **27**:57–65.

Jørgensen, J. O. L., Pedersen, S., Laurberg, P., Weeke, J., Skakkebaer, N. E. and Christiansen, J. S. (1989). Effects of growth hormone therapy on thyroid function of growth hormone-deficient adults with and without concomitant thyroxine-substituted central hypothyroidism. *Journal of Clinical Endocrinology and Metabolism*, **69**:1127–32.

Joss, E. E., Temperli, R. and Mullis, P. E. (1992). Adult height in constitutionally tall stature: Accuracy of five different height prediction methods. *Archives of Disease in Childhood*, **67**:1357–62.

Juul, A., Fisker, S., Scheike, T., Hertel, T., Müller, J., Ørskov, H. and Skakkebæk, N. E. (2000). Serum levels of growth hormone binding protein in children with normal and precocious puberty: Relation to age, gender, body composition and gonadal steroids. *Clinical Endocrinology*, **52**:165–72.

Juul, A., Flyvbjerg, A., Frystyk, J., Muller, J. and Skakkebaek, N. E. (1996). Serum concentrations of free and total insulin-like growth factor I, IGF binding proteins 1 and 3 and IGFBP-3 protease activity in boys with normal or precocious puberty. *Clinical Endocrinology*, **44**:515–23.

Kahn, A. D. and Harrison, G. G. (1989). Relationship of minor illnesses to growth velocity in well-nourished children. *FASEB Journal*, **4**:A1054.

Kaminski, M., Rumeau-Rouquette, C. and Schwartz, D. (1978). Alcohol consumption in pregnant women and the outcome of pregnancy. *Alcoholism, Clinical and Experimental Research*, **224**:155–63.

Kandall, S., Albin, S., Lowinson, J., Berle, B., Eidelman, A. I. and Gartner, L. M. (1976). Differential effects of maternal heroin and methadone use on birthweight. *Pediatrics*, **58**:681–5.

Kandel, D. B. and Udry, J. R. (1999). Prenatal effects of maternal smoking on daughters' smoking: Nicotine or testosterone exposure? *American Journal of Public Health*, **89**:1377–83.

Kanehisa, H., Ikegawa, S., Tsunoda, N. and Fukunaga, T. (1995). Strength and cross-sectional areas of reciprocal muscle groups in the upper arm and thigh during adolescence. *International Journal of Sports Medicine*, **16**:54–60.

Kantero, R.-L., Tiisala, R. and Bäckström, L. (1971). Predictability of individual height and weight growth from cross-sectional distance curves from birth to 10 years: A mixed longitudinal study. *Acta Paediatrica Scandinavica, Supplement*, **220**:33–41.

Kao, P., Matheny, A. J. and Lang, C. (1994). Insulin-like growth factor-I comparisons in healthy twin children. *Journal of Clinical Endocrinology and Metabolism*, **78**:310–12.

Kapen, S., Boyar, R. M., Hellman, L. and Weitzman, E. D. (1975). Twenty-four-hour patterns of luteinizing hormone secretion in humans: Ontogenetic and sexual considerations. *Progress in Brain Research*, **42**:103–13.

Kaplowitz, P. B., d'Ercole, A. J., Van Wyk, J. J. and Underwood, L. E. (1982). Plasma somatomedin-C during the first year of life. *Journal of Pediatrics,* **100**:932–4.

Kaprio, J., Rimpela, A., Winter, T., Viken, R. J., Rimpela, M. and Rose, R. J. (1995). Common genetic influences on BMI and age at menarche. *Human Biology,* **67**:739–53.

Karlberg, J. (1987). Oscillation of human growth during the first three years of life. In *Modelling of Human Growth: With Special Reference to the Assessment of Longitudinal Growth Standards,* ed. J. Karlberg, pp. 138–50. Göteborg, Sweden: Kallered.

Karlberg, J. and Albertsson-Wikland, K. (1995). Growth in full-term small-for-gestational-age infants: From birth to final height. *Pediatric Research,* **38**:733–9.

Karlberg, J., Albertsson-Wikland, K., Naeraa, R. W., Rongen-Westerlaken, C. and Wit, J.-M. (1993). Reference values for spontaneous growth in Turner girls and its use in estimating treatment effects. In *Basic and Clinical Approach to Turner Syndrome,* eds. I. Hibi and K. Takano, pp. 83–92. Amsterdam, The Netherlands: Excerpta Medica.

Karlberg, J. and Fryer, J. G. (1990). A method for adjustment of final height for midparental height for Swedish children. *Acta Paediatricia Scandinavica,* **79**:468–9.

Karlberg, P., Niklasson, A., Ericson, A., Fryer, J. G., Hunt, R. G., Lawrence, C. J. and Munford, A. G. (1985). A methodology for evaluating size at birth. *Acta Paediatrica Scandinavica,* **319**:26–37.

Karlberg, P., Taranger, J., Engstrom, I., Karlberg, J., Landström, T., Lichtenstein, H., Lindstrom, B. and Svenn-Berg-Redegren, I. (1976). The somatic development of children in a Swedish urban community: A prospective longitudinal study. I. Physical growth from birth to 16 years and longitudinal outcome of the study during the same age period. *Acta Paediatrica Scandinavica, Supplement,* **258**:7–76.

Karniski, W., Blair, C. and Vitucci, J. S. (1987). The illusion of catch-up growth in premature infants: Use of the growth index and age correction. *American Journal of Diseases of Children,* **141**:520–6.

Kashiwazaki, H., Suzuki, T. and Takemoto, T. (1988). Altitude and reproduction of the Japanese in Bolivia. *Human Biology,* **60**:831–45.

Katakami, H., Downs, T. R. and Frohman, L. A. (1986). Decreased hypothalamic growth hormone–releasing hormone content and pituitary responsiveness in hypothyroidism. *Journal of Clinical Investigation,* **77**:1704–11.

Kato, N. and Takaishi, M. (1999). A longitudinal study of the physical growth of Japanese infants. *Annals of Human Biology,* **26**:353–63.

Kato, S., Ashizawa, K. and Satoh, K. (1998). An examination of the definition 'final height' for practical use. *Annals of Human Biology,* **25**:263–70.

Katz, S. H., Hediger, M. L., Zemel, B. S. and Parks, J. S. (1985). Adrenal androgens, body fat and advanced skeletal age in puberty: New evidence for the relations of adrenarche and gonadarche in males. *Human Biology,* **57**:401–13.

Katzman, D. K., Bachrach, L. K., Carter, D. R. and Marcus, R. (1991). Clinical and anthropometric correlates of bone mineral acquisition in healthy adolescent girls. *Journal of Clinical Endocrinology and Metabolism,* **73**:1332–9.

Katzmarzyk, P. T., Malina, R. M. and Beunen, G. P. (1997). The contribution of biological maturation to the strength and motor fitness of children. *Annals of Human Biology,* **24**:493–505.

Katzmarzyk, P. T., Malina, R. M., Pérusse, L., Rice, T., Province, M. A., Rao, D. C. and Bouchard, C. (2000). Familial resemblance in fatness and fat distribution. *American Journal of Human Biology,* **12**:395–404.

Kavanaugh, K., Engström, J. L. and Meier, P. P. (1990). How reliable are scales for weighing preterm infants? *Neonatal Network,* **9**:29–32.

Kaye, K., Elkind, L., Goldberg, D. and Tytun, A. (1989). Birth outcomes for infants of drug abusing mothers. *New York State Journal of Medicine,* **89**:256–61.

Keenan, B. S., Richards, G. E., Ponder, S. W., Dallas, J. S., Nagamani, M. and Smith, E. R. (1993). Androgen-stimulated pubertal growth: The effects of testosterone and dihydrotestosterone on growth hormone and insulin-like growth factor-I in the treatment of short stature and delayed puberty.

Journal of Clinical Endocrinology and Metabolism, **76**:996–1001.

Keita, S. O. Y. and Boyce, A. J. (2001). Using the term "race" in human biology research. *American Journal of Human Biology*, **13**:569–75.

Keith, L., Ellis, R., Berger, G. S., Depp, R., Filstead, W., Hatcher, R. and Keith, D. M. (1980). The Northwestern University Multihospital Twin Study. I. A description of 588 twin pregnancies and associated pregnancy loss, 1971 to 1975. *American Journal of Obstetrics and Gynecology*, **138**: 781–7.

Keith, L., MacGregor, S., Friedell, S., Rosner, M., Chasnoff, I. J. and Sciarra, J. J. (1989). Substance abuse in pregnant women: Recent experience at the Perinatal Center for Chemical Dependence of Northwestern Hospital. *Obstetrics and Gynecology*, **73**:715–20.

Kelijman, M. and Frohman, L. (1988). Enhanced growth hormone (GH) responsiveness to GH-releasing hormone after dietary manipulation in obese and non-obese subjects. *Journal of Clinical Endocrinology and Metabolism*, **66**:489–94.

Kelly, P., Sullivan, D., Bartsch, M., Gracey, M. and Ridout, S. (1984). Evolution of obesity in young people in Busselton, Western Australia. *Medical Journal of Australia*, **141**:97–9.

Kelly, P. J., Hopper, J. L., Macaskill, G. T., Pocock, N. A., Sambrook, P. N. and Eisman, J. A. (1991). Genetic factors in bone turnover. *Journal of Clinical Endocrinology and Metabolism*, **72**:808–14.

Kempe, A., Wise, P. H., Barkan, S. E., Sappenfield, W. M., Sachs, B., Gortmaker, S. L., Sobol, A. M., First, L. R., Pursley, D., Rinehart, H., Kotelchuck, M., Cole, F. S., Gunter, N. and Stockbauer, J. W. (1992). Clinical determinants of the racial disparity in very low birth weight. *New England Journal of Medicine*, **327**:969–73.

Keppel, K. G. and Taffel, S. M. (1987). Maternal smoking and weight gain in relation to birth weight and the risk of fetal and infant death. In *Smoking and Reproductive Health*, ed. M. J. Rosenberg, pp. 80–5. Littleton, MA: PSG Publishing.

Kerr, A., Jr. (1943). Weight gain in pregnancy and its relation to weight of infants and to length of labor.

American Journal of Obstetrics and Gynecology, **45**:950–60.

Kerrigan, J. R. and Rogol, A. D. (1992). The impact of gonadal steroid hormone action on growth hormone secretion during childhood and adolescence. *Endocrine Reviews*, **13**:281–98.

Khamis, H. J. and Guo, S. (1993). Improvement in the Roche–Wainer–Thissen stature prediction model: A comparative study. *American Journal of Human Biology*, **5**:669–79.

Khamis, H. J. and Roche, A. F. (1994). Predicting adult stature without using skeletal age: The Khamis–Roche method. *Pediatrics*, **94**:504–7.

Khamis, H. J. and Roche, A. F. (1995). Growth outcome of "normal" short children who are retarded in skeletal maturation. *Journal of Pediatric Endocrinology and Metabolism*, **8**:85–96.

Kiess, W., Reich, A., Meyer, K., Glasow, A., Deutscher, J., Klammt, J., Yang, Y., Müller, G. and Kratzsch, J. (1999). A role for leptin in sexual maturation and puberty? *Hormone Research, Supplement*, **51**:55–63.

Kimm, S. Y., Obarzanck, E., Barton, B. A., Aston, C. E., Similo, S. L., Morrison, J. A., Sabry, Z. I., Schreiber, G. B. and McMahon, R. P. (1996). Race, socioeconomic status, and obesity in 9- to 10-year old girls: The NHLBI Growth and Health Study. *Annals of Epidemiology*, **6**:266–75.

Kimura, K. (1981). Skeletal maturity in twins. *Journal of the Anthropological Society of Nippon*, **89**:457–78.

King, G. L., Goodman, D., Buzney, S., Moses, A. and Kahn, C. R. (1985). Receptors and growth-promoting effects of insulin and insulin-like growth factors on cells from bovine retinal capillaries and aorta. *Journal of Clinical Investigation*, **75**:1028–36.

Kirchengast, S., Gruber, D., Sator, M. and Huber, J. (1998). Impact of the age at menarche on adult body composition in healthy pre- and postmenopausal women. *American Journal of Physical Anthropology*, **105**:9–20.

Kirkinen, P., Jouppila, P., Koivula, A., Vuori, J. and Puukka, M. (1983). The effect of caffeine on placental and fetal blood flow in human pregnancy. *American Journal of Obstetrics and Gynecology*, **147**:939–42.

Kirkland, J. L., Gibbs, A. R., Kirkland, R. T. and Clayton, G. W. (1981). Height predictions in girls with idiopathic precocious puberty by the Bayley–Pinneau method. *Pediatrics*, **68**:251–2.

Kirkland, R. T. (1990). Failure to thrive. In *Principles and Practice of Pediatrics*, eds. F. A. Oski, C. D. de Angelis, R. D. Feigin, J. B. Warshaw, pp. 969–72. Philadelphia, PA: Lippincott.

Kistin, N., Benton, D., Rao, S. and Sullivan, M. (1990). Breast-feeding rates among black urban low-income women: Effect of prenatal education. *Pediatrics*, **86**:741–6.

Kistin, N., Handler, A., Davis, F. and Ferre, C. (1996). Cocaine and cigarettes: A comparison of risk. *Paediatric and Perinatal Epidemiology*, **10**:269–78.

Klebanoff, M. A., Levine, R. J., Clemens, J. D., der Simonian, R. and Wilkins, D. G. (1998*a*). Serum cotinine concentration and self-reported smoking during pregnancy. *American Journal of Epidemiology*, **148**:259–62.

Klebanoff, M. A., Mednick, B. R., Schulsinger, C., Secher, N. J. and Shiono, P. H. (1998*b*). Father's effect on infant birth weight. *American Journal of Obstetrics and Gynecology*, **178**:1022–6.

Klebanoff, M. A., Meirik, O. and Berendes, H. W. (1989). Second-generation consequences of small-for-dates birth. *Pediatrics*, **84**:343–7.

Klebanoff, M. A., Shiono, P. H. and Rhoads, G. G. (1990). Outcomes of pregnancy in a national sample of resident physicians. *Australian Paediatric Journal*, **19**:157–61.

Klebanoff, M. A. and Yip, R. (1987). Influence of maternal birth weight on rate of fetal growth and duration of gestation. *Journal of Pediatrics*, **111**:287–92.

Klein, K. O., Baron, J., Colli, M. J., McDonnell, D. P. and Cutler, G. B., Jr. (1994). Estrogen levels in childhood determined by an ultrasensitive recombinant cell bioassay. *Journal of Clinical Investigation*, **94**:2475–80.

Klein, K. O., Martha, P. M., Blizzard, R. M., Herbst, T. and Rogol, A. D. (1996). A longitudinal assessment of hormonal and physical alterations during normal puberty in boys. II. Estrogen levels as determined by an ultrasensitive bioassay. *Journal of Clinical Endocrinology and Metabolism*, **81**:3203–7.

Kleinman, J. C. and Kessel, S. S. (1987). Racial differences in low birth weight: Trends and risk factors. *New England Journal of Medicine*, **317**:749–53.

Kleinman, J. C. and Madans, J. H. (1985). The effects of maternal smoking, physical stature, and educational attainment on the incidence of low birth weight. *American Journal of Epidemiology*, **121**:843–55.

Klesges, R. C., Shelton, M. L. and Klesges, L. M. (1993). The effects of television on metabolic rate: Potential implications for childhood obesity. *Pediatrics*, **91**:281–6.

Kline, J., Stein, Z. and Hutzler, M. (1987). Cigarettes, alcohol, and marijuana: Varying associations with birth weight. *International Journal of Epidemiology*, **16**:44–51.

Knobil, E. (1980). The neuroendocrine control of the menstrual cycle. *Recent Progress in Hormone Research*, **36**:53–88.

Knobil, E. (1990). The GnRH pulse generator. *American Journal of Obstetrics and Gynecology*, **163**:1721–7.

Knudtzon, J., Waaler, P. E., Skjaerven, R., Solberg, L. K. and Steen, J. (1989). Growth data for Norwegian children 0–4 years and new clinical growth charts. In *Perspectives in the Science of Growth and Development*, ed. J. M. Tanner, pp. 149–54. London: Smith-Gordon.

Kobyliansky, E., Livshits, G. and Otremsky, I. (1987). Sibling similarity in development of covariation among physical traits in early childhood. *American Journal of Physical Anthropology*, **72**:77–87.

Koch, H. (1966). *Twins and Twin Relations*. Chicago, IL: University of Chicago Press.

Kolaczynski, J. W., Nyce, M. R., Considine, R. V., Boden, G., Nolan, J. J., Henry, R. L., Mudaliar, S. R., Olefsky, J. M. and Caro, J. F. (1996). Acute and chronic effect of insulin on leptin production in humans: Studies *in vivo* and *in vitro*. *Diabetes*, **45**:699–701.

Kolodziej, H., Szklarska, A. and Malina, R. M. (2001). Young adult height of offspring born to rural-to-urban migrant parents and urban-born parents. *American Journal of Human Biology*, **13**:30–4.

Kopczynska-Sikorska, J. and Miesowicz, I. (1980). Predicting adult stature for normal short Polish boys by Roche, Wainer and Thissen method. *Acta Medica Auxologica*, **12**:89–92.

Kopelman, P. G., Noonan, K., Goulton, R. and Forrest, A. J. (1985). Impaired growth hormone response to growth hormone releasing factor and insulin-hypoglycaemia in obesity. *Clinical Endocrinology*, 23:87–94.

Kostyo, J. L. (1968). Rapid effects of growth hormone on amino acid transport and protein synthesis. *Annals of the New York Academy of Sciences*, 148:389–407.

Kotani, K., Nishida, M., Yamashita, S., Funahashi, T., Fujioka, S., Tokunaga, K., Ishikawa, K., Tarui, S. and Matsuzawa, Y. (1997). Two decades of annual medical examinations in Japanese obese children: Do obese children grow into obese adults? *International Journal of Obesity*, 21:912–21.

Kotelchuck, M. (ed.) (1980). *Nonorganic Failure to Thrive: The Status of International and Environmental Etiologic Theories*. Greenwich, CT: Jai Press.

Kotelchuck, M. (1994). The adequacy of prenatal care utilization index: Its U.S. distribution and association with low birthweight. *American Journal of Public Health*, 84:1486–9.

Kouchi, M., Mukherjee, D. and Roche, A. F. (1985a). Curve fitting for growth in weight during infancy with relationships to adult status, and familial associations of the estimated parameters. *Human Biology*, 57:245–65.

Kouchi, M., Roche, A. and Mukherjee, D. (1985b). Growth in recumbent length during infancy with relationships to adult status and familial associations of the estimated parameters of fitted curves. *Human Biology*, 57:449–72.

Koupilová, I., Leon, D. A., McKeigue, P. M. and Lithell, H. O. (1999). Is the effect of low birth weight on cardiovascular mortality mediated through high blood pressure? *Journal of Hypertension*, 17:19–25.

Koziel, S. (2001). Relationships among tempo of maturation, midparent height, and growth in height of adolescent boys and girls. *American Journal of Human Biology*, 13:15–22.

Koziel, S., Jankowska, E. A. and Boznański, A. (2001). Birth weight and selected health parameters among 14-year old Polish adolescents. *Acta Medica Auxologica*, 33:31–7.

Kozlowski, L., Herman, C. and Frecker, R. (1980).

What researchers make of what cigarette smokers say: Filtering smokers' hot air. *Lancet*, i:699–700.

Kracier, J., Sheppard, M. S., Luke, J., Lussier, B., Moor, B. C. and Cowan, J. S. (1988). Effect of withdrawal of somatostatin and (GH)-releasing factor on GH release *in vitro*. *Endocrinology*, 122:1810–15.

Kramer, M. S. (1981). Do breast-feeding and delayed introduction of solid foods protect against subsequent obesity? *Journal of Pediatrics*, 98:883–7.

Kramer, M. S., Barr, R. G., Leduc, D. G., Boisjoly, C., McVey-White, L. and Pless, I. B. (1985a). Determinants of weight and adiposity in the first year of life. *Journal of Pediatrics*, 106:10–14.

Kramer, M. S., Barr, R. G., Leduc, D. G., Boisjoly, C. and Pless, I. B. (1983). Maternal psychological determinants of infant obesity: Development and testing of two new instruments. *Journal of Chronic Diseases*, 36:329–35.

Kramer, M. S., Barr, R. G., Leduc, D. G., Boisjoly, C. and Pless, I. B. (1985b). Infant determinants of childhood weight and adiposity. *Journal of Pediatrics*, 107:104–7.

Kramer, M. S., McLean, F. H., Boyd, M. E. and Usher, R. H. (1988). The validity of gestational age estimation by menstrual dating in term, preterm and postterm gestations. *Journal of the American Medical Association*, 260:3306–8.

Kramer, M. S., Platt, R., Yang, H., Joseph, K. S., Wen, S. W., Morin, L. and Usher, R. H. (1998). Secular trends in preterm birth: A hospital-based cohort study. *Journal of the American Medical Association*, 280:1849–54.

Kramer, M. S., Platt, R., Yang, H., McNamara, H. and Usher, R. H. (1999). Are all growth-restricted newborns created equal? *Pediatrics*, 103:599–602.

Kratzsch, J., Dehmel, B., Pulzer, F., Keller, E., Englaro, P., Blum, W. F. and Wabitsch, M. (1997). Increased serum GHBP levels in obese pubertal children and adolescents: Relationship to body composition, leptin and indicators of metabolic disturbances. *International Journal of Obesity*, 21:1130–6.

Krebs, N. F., Hambidge, K. M. and Walravens, P. A. (1984). Increased food intake of young children receiving a zinc supplement. *American Journal of Diseases of Childhood*, 138:270–3.

Krebs, N. F., Reidinger, C. J., Robertson, A. D. and Hambidge, K. M. (1994). Growth and intakes of energy and zinc in infants fed human milk. *Journal of Pediatrics*, **124**:32–9.

Krieger, I. (1973). Endocrines and nutrition in psychosocial deprivation in the U.S.A.: Comparison with growth failure due to malnutrition on an organic basis. In *Endocrine Aspects of Malnutrition: Marasmus, Kwashiorkor and Psychosocial Deprivation*, eds. L. I. Gardner and P. Amacher, pp. 129–62. Santa Ynez, CA: Kroc Foundation.

Kristiansson, B. and Fällström, S. P. (1981). Infants with low rate of weight gain. II. A study of environmental factors. *Acta Paediatrica Scandinavica*, **70**:663–8.

Kristiansson, B. and Fällström, S. P.(1987). Growth at the age of 4 years subsequent to early failure to thrive. *Child Abuse and Neglect*, **11**:35–40.

Krogman, W. M. (1970). *Growth of Head, Face, Trunk, and Limbs in Philadelphia White and Negro Children of Elementary and High School Age*, Monographs of the Society for Research in Child Development no. 35. Chicago, IL: University of Chicago Press.

Kromeyer, K., Hauspie, R. C. and Susanne, C. (1997). Socioeconomic factors and growth during childhood and early adolescence in Jena children. *Annals of Human Biology*, **24**:343–53.

Kromeyer-Hauschild, K. and Jaeger, U. (2000). Growth studies in Jena, Germany: Changes in sitting height, biacromial, and bicristal breadth in the past decenniums. *American Journal of Human Biology*, **12**:646–54.

Krueger, R. H. (1969). Some long-term effects of severe malnutrition in early life. *Lancet*, **ii**:514–17.

Kuczmarski, R. J., Ogden, C. L., Guo, S. S., Grummer-Strawn, L. M., Flegal, K. M., Mei, Z., Wei, R., Curtin, L. R., Roche, A. F. and Johnson, C. L. (2000). CDC growth charts: United States, 2000. *Advance Data, Vital and Health Statistics*, **314**:1–28.

Kuczmarski, R. J., Ogden, C. L., Guo, S. S., Grummer-Strawn, L. M., Flegal, K. M., Mei, Z., Wei, R., Curtin, L. R., Roche, A. F. and Johnson, C. L. (2002). CDC growth charts: United States, 2000. *Vital and Health Statistics National Center for*

Health Statistics. Washington, D.C.: Department of Health and Nutrition Services.

Kuh, D. and Wadsworth, M. (1989). Parental height: Childhood environment and subsequent adult height in a national birth cohort. *International Journal of Epidemiology*, **18**:663–8.

Kuhnert, B. R., Kuhnert, P. M., Debanne, S. and Williams, T. G. (1987*a*). The relationship between cadmium, zinc, and birth weight in pregnant women who smoke. *American Journal of Obstetrics and Gynecology*, **157**:1247–51.

Kuhnert, P. M., Kuhnert, B. R., Erhard, P., Brashear, W. T., Groh-Wargo, S. L. and Webster, S. (1987*b*). The effect of smoking on placental and fetal zinc status. *American Journal of Obstetrics and Gynecology*, **157**:1241–6.

Kulik-Rechberger, B., Rechberger, T. and Jakimiuk, A. J. (1999). Leptin correlates with the skinfold thickness in prepubertal and pubertal girls. *Acta Paediatrica*, **88**:103–4.

Kulin, H. E., Bwibo, N., Mutie, D. and Santer, S. J. (1982). The effect of chronic childhood malnutrition on pubertal growth and development. *American Journal of Clinical Nutrition*, **36**:527–36.

Kulin, H. E., Grumbach, M. M. and Kaplan, S. L. (1972). Gonadal–hypothalamic interaction in prepubertal and pubertal man: Effect of clomiphene citrate on urinary follicle stimulating hormone and luteinizing hormone and plasma testosterone. *Pediatric Research*, **6**:162–71.

Kuno, A., Akiyama, M., Yanagihara, T. and Hata, T. (1999). Comparison of fetal growth in singleton, twin, and triplet pregnancies. *Human Reproduction*, **14**:1352–60.

Kuroki, Y., Kurosawa, K. and Imaizumi, K. (1995). Growth patterns in children with Down syndrome: From birth to 15 years of age. In *Physical and Motor Development in Mental Retardation*, eds. A. Vermeer and W. E. Davis, pp. 159–67. Basel, Switzerland: Karger.

Kuromaru, R., Kohno, H., Ueyama, N., Hassan, H. M. S., Honda, S. and Hara, T. (1998). Long-term prospective study of body composition and lipid profiles during and after growth hormone (GH) treatment in children with GH deficiency:

Gender-specific metabolic effects. *Journal of Clinical Endocrinology and Metabolism*, **83**:3890–6.

Kuzma, J. W. and Sokol, R. J. (1982). Maternal drinking behaviour and decreased intrauterine growth. *Alcoholism, Clinical and Experimental Research*, **6**:396–402.

La Franchi, S., Hanna, C. E. and Mandel, S. H. (1991). Constitutional delay of growth: Expected versus final adult height. *Pediatrics*, **87**:82–7.

Labrie, F., Belanger, A., Cusan, L. and Candas, B. (1997). Physiological changes in dehydroepiandrosterone are not reflected by serum levels of active androgens and estrogens but of their metabolites: Intracrinology. *Journal of Clinical Endocrinology and Metabolism*, **82**:2403–9.

Lacey, K. A. and Parkin, J. M. (1974*a*). The normal short child: A community study of children in Newcastle-upon-Tyne. *Archives of Disease in Childhood*, **49**:417–24.

Lacey, K. A. and Parkin, J. M. (1974*b*). Causes of short stature: A community study of children in Newcastle-upon-Tyne. *Lancet*, **i**:42–5.

Laitinen, J., Power, C. and Järvelin, M.-R. (2001). Family social class, maternal body mass index, childhood body mass index, and age at menarche as predictors of adult obesity. *American Journal of Clinical Nutrition*, **74**:287–94.

Lampl, M. (1993). Evidence of saltatory growth in infancy. *American Journal of Human Biology*, **5**:641–52.

Lampl, M., Ashizawa, K., Kawabata, M. and Johnson, M. L. (1998). An example of variation and pattern in saltation and stasis growth dynamics. *Annals of Human Biology*, **25**:203–9.

Lampl, M. and Johnson, M. (1999). Wrinkles induced by the use of smoothing procedures applied to serial growth data. *Annals of Human Biology*, **26**:494–6.

Lampl, M. and Johnson, M. L. (1993). A case study in daily growth during adolescence: A single spurt or changes in the dynamics of saltatory growth? *Annals of Human Biology*, **20**:595–603.

Lang, J. M., Lieberman, E. and Cohen, A. (1996). A comparison of risk factors for preterm labor and term small-for-gestational-age birth. *Epidemiology*, **7**:369–76.

Langhoff-Roos, J., Lindmark, G., Gutstavson, K. H., Gebre-Medhin, M. and Meirik, O. (1987). Relative effect of parental birth weight on infant birth weight at term. *Clinical Genetics*, **32**:240–8.

Langinvainio, H., Koskenvuo, M., Kaprio, J. and Sistonen, P. (1984). Finnish twins reared apart. II. Validation of zygosity, environmental dissimilarity and weight and height. *Acta Geneticae Medicae et Gemellologiae*, **33**:251–8.

Laor, A., Stevenson, D. K., Shemer, J., Gale, R. and Seidman, D. S. (1997). Size at birth, maternal nutritional status in pregnancy, and blood pressure at age 17: Population based analysis. *British Medical Journal*, **315**:449–53.

Lapillonne, A., Braillon, P., Claris, O., Chatelain, P. G., Delmas, P. D. and Salle, B. L. (1997). Body composition in appropriate and in small for gestational age infants. *Acta Paediatrica*, **86**:196–200.

Large, D. M. and Anderson, D. C. (1979). Twenty-four hour profiles of circulating androgens and oestrogens in male puberty with and without gynaecomastia. *Clinical Endocrinology*, **11**:505–21.

Large, V., Hellström, L., Reynisdottir, S., Lönnqvist, F., Eriksson, P., Lannfelt, L. and Arner, P. (1997). Human beta-2 adrenoceptor gene polymorphisms are highly frequent in obesity and associate with altered adipocyte beta-2 adrenoceptor function. *Journal of Clinical Investigation*, **100**:3005–13.

Largo, R. H., Gasser, T., Prader, A., Stûetzle, W. and Huber, P. J. (1978). Analysis of the adolescent growth spurt using smoothing spline functions. *Annals of Human Biology*, **5**:421–34.

Largo, R. H. and Prader, A. (1983*a*). Pubertal development in Swiss boys. *Helvetica Paediatrica Acta*, **38**:211–28.

Largo, R. H. and Prader, A. (1983*b*). Pubertal development in Swiss girls. *Helvetica Paediatrica Acta*, **38**:229–43.

Laron, Z. (1999). The essential role of IGF-I: Lessons from the long-term study and treatment of children and adults with Laron syndrome. *Journal of Clinical Endocrinology and Metabolism*, **84**:4397–404.

Laron, Z., Anin, S., Klipper-Aurbach, Y. and Klinger, B. (1992). Effects of insulin-like growth factor-I on linear growth, head circumference and body fat in patients with Laron-type dwarfism. *Lancet*, 339:1258–61.

Laron, Z. and Klinger, B. (2000). Comparison of the growth-promoting effects of insulin-like growth factor I and growth hormone in the early years of life. *Acta Paediatrica*, 89:38–41.

Larroque, B., Kaminski, M., Lelong, N., Subtil, D. and Dehaene, P. (1993). Effects on birth weight of alcohol and caffeine consumption during pregnancy. *American Journal of Epidemiology*, 137:941–50.

Larsson, G., Bohlin, A. B. and Tunnel, R. (1985). Prospective study of children exposed to variable amounts of alcohol in utero. *Archives of Disease in Childhood*, 60:315–21.

Łaska-Mierzejewska, T. and Luczak, E. (1993). *Biologiczne mierniki sytuacji społeczno ekonomicznej ludnoœci wiejskiej w Polsce w latach 1967, 1977, 1987 (Biological Indicators of Socioeconomic Situation of the Rural Population in Poland in 1967, 1977 and 1987)*, Monographs of the Institute of Anthropology no. 10. Wrocław, Poland: Polish Academy of Sciences.

Lasker, G. W. and Mascie-Taylor, C. G. N. (1996). Influence of social class on the correlation of stature of adult children with that of their mothers and fathers. *Journal of Biosocial Science*, 18:117–22.

Lauer, R. M., Obarzanek, E., Hunsberger, S. A., van Horn, L., Hartmuller, V. W., Barton, B. A., Stevens, V. J., Kwiterovich, P. O., Jr., Franklin, F. A., Jr., Kimm, S. Y. S., Lasser, N. L. and Simons-Morton, D. G. (2000). Efficacy and safety of lowering dietary intake of total fat, saturated fat, and cholesterol in children with elevated LDL cholesterol: The Dietary Intervention Study in Children 1–6. *American Journal of Clinical Nutrition*, 72:S1332–42.

Launer, L. J., Hofman, A. and Grobbee, D. E. (1993). Relation between birth weight and blood pressure: Longitudinal study of infants and children. *British Medical Journal*, 307:1451–4.

Launer, L. J., Villar, J., Kestler, E. and de Onis, M. (1990). The effect of maternal work on fetal growth and duration of pregnancy: A prospective study. *British Journal of Obstetrics and Gynaecology*, 97:62–70.

Laursen, E. M., Molgaard, C., Michaelsen, K. F., Koch, C. and Müller, J. (1999). Bone mineral status in 134 patients with cystic fibrosis. *Archives of Disease in Childhood*, 81:235–40.

Law, C. M., Barker, D. J. P., Bull, A. R. and Osmond, C. (1991). Maternal and fetal influences on blood pressure. *Archives of Disease in Childhood*, 66:1291–5.

Law, C. M., de Swiet, M., Osmond, C., Fayers, P. M., Barker, D. J. P., Cruddas, A. M. and Fall, C. H. D. (1993). Initiation of hypertension *in utero* and its amplification throughout life. *British Medical Journal*, 306:24–7.

Law, C. M. and Shiell, A. W. (1996). Is blood pressure inversely related to birth weight? The strength of evidence from a systematic review of the literature. *Journal of Hypertension*, 14:935–41.

Lawrence, C., Fryer, J. G., Karlberg, P., Niklasson, A. and Ericson, A. (1989). Modelling of reference values for size at birth. *Acta Paediatrica Scandinavica, Supplement*, 350:55–69.

Lazarus, R., Baur, L., Webb, K. and Blyth, F. (1996). Body mass index in screening for adiposity in children and adolescents: Systematic evaluation using receiver operating characteristic curves. *American Journal of Clinical Nutrition*, 63:500–6.

Le Marchand, L., Kolonel, L., Earle, M. E. and Mi, M.-P. (1988). Body size at different periods of life and breast cancer risk. *American Journal of Epidemiology*, 128:137–52.

Lecomte, E., Herbeth, B., Nicaud, V., Rakotovao, R., Artur, Y. and Tiret, L. (1997). Segregation analysis of fat mass and fat-free mass with age-and sex-dependent effects: The Stanislas Family Study. *Genetic Epidemiology*, 14:51–62.

Lee, P. A. (1980). Independence of seasonal variation of growth from temperature change. *Growth*, 44:54–7.

Lee, P. A., Plotnick, L. P., Migeon, C. J. and Kowarski, A. A. (1978). Integrated concentrations of follicle stimulating hormone and puberty. *Journal of Clinical Endocrinology and Metabolism*, 46:488–90.

Leeson, P., Whincup, P. H., Cook, D. G., Donald, A. E., Papacosta, O., Lucas, A. and Deanfield, J. E. (1997). Flow mediated dilation in 9–11-year-old children: The influence of intrauterine and childhood factors. *Circulation*, 96:2233–8.

Lefevre, J., Beunen, G., Steens, G., Claessens, A. and Renson, R. (1990). Motor performance during adolescence and age thirty as related to age at peak height velocity. *Annals of Human Biology*, 17:423–35.

Lefevre, M., Lovejoy, J. C., de Felice, S. M., Keener, J. W., Bray, G. A., Ryan, D. H., Hwang, D. H. and Greenway, F. L. (2000). Common apolipoprotein A-IV variants are associated with differences in body mass index levels and percentage body fat. *International Journal of Obesity*, 24:945–53.

Leff, M., Orleans, M., Haverkamp, A. D., Barón, A. E., Alderman, B. W. and Freedman, W. L. (1992). The association of maternal low birthweight and infant low birthweight in a racially mixed population. *Paediatric and Perinatal Epidemiology*, 6:51–61.

Leger, J., Levy-Marchal, C., Bloch, J., Pinet, A., Chevenne, D., Porquet, D., Collin, D. and Czernichow, P. (1997). Reduced final height and indications for insulin resistance in 20 year olds born small for gestational age: Regional cohort study. *British Medical Journal*, 315:341–7.

Leger, J., Limoni, C., Collin, D. and Czernichow, P. (1998). Prediction factors in the determination of final height in subjects born small for gestational age. *Pediatric Research*, 43:808–12.

Lehingue, Y., Remontet, L., Muñoz, F. and Mamelle, N. (1998). Birth ponderal index and body mass index reference curves in a large population. *American Journal of Human Biology*, 10:327–40.

Lenko, H.L. (1979). Prediction of adult height with various methods in Finnish children. *Acta Paediatrica Scandinavica*, 68:85.

Lenko, H. L., Mäenpää, J. and Perheentupa, J. (1982). Acceleration of delayed growth with fluoxymesterone. *Acta Paediatrica Scandinavica*, 71:929–36.

Lenko, H. L., Perheentupa, J. and Soderholm, A. (1979). Growth in Turner's syndrome: Spontaneous and fluoxymesterone stimulated. *Acta Paediatrica Scandinavica, Supplement*, 227:57–63.

Leon, D. A., Koupilova, I., Lithell, H. O., Berglund, L., Mohsen, R., Vagero, D., Lithell, U. B. and McKeigue, P. M. (1996). Failure to realize growth potential *in utero* and adult obesity in relation to blood pressure in 50 year old Swedish men. *British Medical Journal*, 312: 401–6.

Leon, D. A., Lithell, H. O., Vågerö, D., Koupilova, I., Mohsen, R., Berglund, L., Lithell, U.-B. and

McKeigue, P. M. (1998). Reduced fetal growth rate and increased risk of death from ischaemic heart disease: Cohort study of 15 000 Swedish men and women born 1915–1929. *British Medical Journal*, 317:241–5.

Leonard, W. R., de Walt, K. M., Stansbury, J. P. and McCaston, M. K. (1995). Growth differences between children of highland and coastal Ecuador. *American Journal of Physical Anthropology*, 98: 47–57.

Lepercq, J., Lahlou, N., Timsit, J., Girard, J. and Mouzon, S. H. (1999). Macrosomia revisited: Ponderal index and leptin delineate subtypes of fetal overgrowth. *American Journal of Obstetrics and Gynecology*, 181:621–5.

Leslie, R. D., Lo, S., Millward, B. A., Honour, J. and Pyke, D. A. (1991). Decreased growth velocity before IDDM onset. *Diabetes*, 40:211–16.

Letarte, J. (1985). The hypothyroid infant: Clinical aspects. *Pediatric Adolescent Endocrinology*, 14:117.

Leung, A. K. C., Robson, L. M. and Fagan, J. E. (1993). Assessment of the child with failure to thrive. *American Family Physician*, 48:1432–8.

Levy, D., Garrison, R., Savage, D. D., Kannel, W. B. and Castelli, W. P. (1990). Prognostic implications of echocardiographically determined left ventricular mass in the Framingham Heart Study. *New England Journal of Medicine*, 322:1561–6.

Lhotská, L., Bláha, P., Vignerová, J., Roth, Z. and Prokopec, M. (1993). *5th Nationwide Anthropological Survey of Children and Adolescents 1991 (Czech Republic): Anthropometric Characteristics*. Prague: National Institute of Public Health.

Lhotská, L., Bláha, P., Vignerová, J., Roth, Z. and Prokopec, M. (1995). *5th Nationwide Anthropological Survey of Children and Adolescents 1991 (Czech Republic): Evaluation of Parents' Questionnaire*. Prague: National Institute of Public Health.

Li, C. Q., Windsor, R. A., Perkins, L., Goldenberg, R. L. and Lowe, J. B. (1993). The impact on infant birth weight and gestational age of cotinine-validated smoking reduction during pregnancy. *Journal of the American Medical Association*, 269:1519–24.

Li, J., Park, W. J. and Roche, A. F. (1998). Decanalization of weight and stature during childhood and adolescence. *American Journal of Human Biology*, **10**:351–9.

Lichty, J. A., Ting, R. Y., Bruns, P. D. and Dyar, E. (1957). Studies of babies born at high altitude. I. Relation of altitude to birthweight. *American Journal of Diseases in Childhood*, **93**:666–78.

Licinio, J., Mantzoros, C., Negrao, A. B., Cizza, G., Wong, M.-L., Bongiorno, P. B., Chrousos, G. P., Karp, B., Allen, C., Flier, J. S. and Gold, P. W. (1997). Human leptin levels are pulsatile and inversely related to pituitary–adrenal function. *Nature Medicine*, **3**:575–9.

Lieberman, E., Gremy, I., Lang, J. M. and Cohen, A. P. (1994). Low birthweight at term and the timing of fetal exposure to maternal smoking. *American Journal of Public Health*, **84**:1127–31.

Lieberman, E., Ryan, K. J., Monson, R. R. and Schoenbaum, S. C. (1987). Risk factors accounting for racial differences in the rate of premature birth. *New England Journal of Medicine*, **317**:743–8.

Lifschitz, M. H., Wilson, G. S., Smith, E. O. and Desmond, M. M. (1983). Fetal and postnatal growth of children born to narcotic-dependent women. *Journal of Pediatrics*, **102**:686–91.

Lifschitz, M. H., Wilson, G. S., Smith, E. O. and Desmond, M. M. (1985). Factors affecting head growth and intellectual function in children of drug addicts. *Pediatrics*, **75**:269–74.

Lifshitz, F. and Cervantes, C. D. (1990). Short stature. In *Pediatric Endocrinology*, 2nd edn, ed. F. Lifshitz, pp. 3–36. New York: Marcel Dekker.

Lifshitz, F. and Moses, N. (1986). Nutritional growth retardation. In *Human Growth: A Comprehensive Treatise*, 2nd edn, vol. 2, *Postnatal Growth*, eds. F. Falkner and J. M. Tanner, pp. 117–32. New York: Plenum Press.

Lindgren, G. (1976). Height, weight and menarche in Swedish urban school children in relation to socio-economic and regional factors. *Annals of Human Biology*, **3**:501–28.

Lindgren, G. (1978). Growth of schoolchildren with early, average and late ages of peak height velocity. *Annals of Human Biology*, **5**:253–67.

Lindgren, G. (1994). Aspects of research on socioeconomic conditions and growth based on Swedish data. *Humanbiologica Budapestiensis*, **25**:5–135.

Lindgren, G. (1996). Pubertal stages 1980 of Stockholm schoolchildren. *Acta Paediatrica*, **85**:1365–7.

Lindgren, G. (1998). Secular growth changes in Sweden. In *Secular Growth Changes in Europe*, eds. É. B. Bodzsár and C. Susanne, pp. 319–33. Budapest: Eötvös University Press.

Lindgren, G., Strandell, A., Cole, T., Healy, M. and Tanner, J. M. (1995). Swedish population reference standards for height, weight and body mass index attained at 6 to 16 years (girls) or 19 years (boys). *Acta Paediatrica*, **84**:1019–28.

Lindgren, G. W. and Strandell, A. (1986). *Fysisk utveckling och halsa: En analys av halsokortsuppifter for grundskoleever fodda 1967 (Physical Growth and Health: An Analysis of Health Chart Data of Elementary Schoolchildren Born in 1967)*. Stockholm: Institute of Education, Department of Educational Research.

Lindley, A. A., Gray, R. H., Herman, A. A. and Becker, S. (2000). Maternal cigarette smoking during pregnancy and infant ponderal index at birth in the Swedish Medical Birth Register, 1991–1992. *American Journal of Public Health*, **90**:420–3.

Lindsay, C. A., Thomas, A. J. and Catalano, P. M. (1997). The effect of smoking tobacco on neonatal body composition. *American Journal of Obstetrics and Gynecology*, **177**:1124–8.

Lindsay, R., Feldkamp, M., Harris, D., Robertson, J. and Rallison, M. (1994). Utah Growth Study: Growth standards and the prevalence of growth hormone deficiency. *Journal of Pediatrics*, **125**: 29–35.

Lindsay, R. S., Dabelea, D., Roumain, J., Hanson, R. L., Bennett, P. H. and Knowler, W. C. (2000). Type 2 diabetes and low birth weight: The role of paternal inheritance in the association of low birth weight and diabetes. *Diabetes*, **49**:445–9.

Link, K., Blizzard, R. M., Evans, W. S., Kaiser, D. L., Parker, M. W. and Rogol, A. D. (1986). The effect of androgens on the pulsatile release and the twenty-four-hour mean concentration of growth hormone in peripubertal males. *Journal of Clinical Endocrinology and Metabolism*, **62**:159–64.

Linn, S., Schoenbaum, S. C., Monson, R. R., Rosner, R., Stubblefield, P. C. and Ryan, K. J. (1983). The association of marijuana use with outcome of pregnancy. *American Journal of Public Health*, **73**:1161–4.

Linn, S., Schoenbaum, S. C., Rosner, B., Stubblefield, P. G. and Ryan, K. J. (1982). No association between coffee consumption and adverse outcomes of pregnancy. *New England Journal of Medicine*, **306**:141–5.

Lissau-Lund-Sørensen, I. and Sørensen, T. I. A. (1992). Prospective study of the influence of social factors in childhood on risk of overweight in young adulthood. *International Journal of Obesity*, **16**:169–75.

Lithell, H. O., McKeigue, P. M., Berglund, L., Mohsen, R., Lithell, U. B. and Leon, D. A. (1996). Relation of size at birth to non-insulin dependent diabetes and insulin concentrations in men aged 50–60 years. *British Medical Journal*, **312**:406–10.

Little, B. B., Snell, L. M., Klein, V. R. and Gilstrap, L. C. III, (1989). Cocaine abuse during pregnancy: Maternal and fetal implications. *Obstetrics and Gynecology*, **73**:157–60.

Little, R. E. (1986). Birthweight and gestational age: Mothers' estimates compared with state and hospital records. *American Journal of Public Health*, **76**:1350–1.

Little, R.E. (1987). Mother's and father's birthweight as predictors of infant birthweight. *Paediatric and Perinatal Epidemiology*, **1**:19–31.

Little, R. E. and Sing, C. F. (1987). Genetic and environmental influences on human birth weight. *American Journal of Human Genetics*, **40**:512–16.

Livshits, G. (1986). Growth and development of bodyweight, height and head circumference during the first two years of life: Quantitative genetic aspects. *Annals of Human Biology*, **13**:387–96.

Livson, N. and McNeill, D. (1962). The accuracy of recalled age of menarche. *Human Biology*, **34**:218–21.

Ljung, B. O., Fischbein, S. and Lindgren, G. (1977). Comparison of growth in twins and singleton controls of matched age followed longitudinally from 10 to 18 years. *Annals of Human Biology*, **4**:405–15.

Lloyd, J. K., Martel, J. K., Rollings, N., Andon, M. B., Kulin, H. E., Demers, L. M., Eggli, D. F., Kieselhorst, K. and Chinchilli, V. M. (1996). The effect of calcium supplementation and Tanner stage on bone density, content and area in teenage women. *Osteoporosis International*, **6**:276–3.

Loche, S., Cappa, M., Borrelli, P., Fædda, A., Crino, A., Cella, S. G., Corda, R., Muller, E. E. and Pintor, C. (1987). Reduced growth hormone response to growth-hormone-releasing hormone in children with simple obesity: Evidence for somatomedin-C mediated inhibition. *Clinical Endocrinology*, **27**:145–53.

Loesch, D. Z., Huggins, R. M. and Stokes, K. M. (1999). Relationship of birth weight and length with growth in height and body diameters from 5 years of age to maturity. *American Journal of Human Biology*, **11**:772–8.

Lohman, T. G., Boileau, R. A. and Massey, B. H. (1975). Prediction of lean body mass in young boys from skinfold thickness and body weight. *Human Biology*, **47**:245–62.

Lohman, T. G., Roche, A. F. and Martorell, R. (1988). *Anthropometric Standardization Reference Manual*. Champaign, IL.: Human Kinetics.

Longcope, C., Pratt, L. H., Schneider, S. H. and Fineberg, S. E. (1978). Aromatization of androgens by muscle and adipose tissue *in vivo*. *Journal of Clinical Endocrinology and Metabolism*, **46**:146–52.

Longo, L. D. P. (1977). The biological effects of carbon monoxide on the pregnant woman, fetus and newborn infant. *American Journal of Obstetrics and Gynecology*, **129**:69–103.

Lönnqvist, F., Arner, P., Nordfors, L. and Schalling, M. (1995). Over-expression of the obese (*ob*) gene in human obese subjects. *Nature Medicine*, **1**: 950–3.

Lönnqvist, F., Nordfors, L. and Schalling, M. (1999). Leptin and its potential role in human obesity. *Journal of Internal Medicine*, **245**: 643–52.

Low, W. D., Chan, S. T., Chang, K. S. F. and Lee, M. M. C. (1964). Skeletal maturation of Southern Chinese children. *Child Development*, **35**:1313–16.

Lowe, W. L., Jr. (1991). Biological actions of the insulin-like growth factors. In *Insulin-Like Growth Factors: Molecular and Cellular Aspects*, ed. D. Le Roith, pp. 49–85. Boca Raton, FL: CRC Press.

Lu, P. W., Briody, J. N., Ogle, G. D., Morley, K., Humphries, I. R. J., Allen, J., Howmangiles, R.,

Sillence, D. and Cowell, C. T. (1994). Bone mineral density of total body, spine, and femoral neck in children and young adults: A cross-sectional and longitudinal study. *Journal of Bone and Mineral Research*, **9**:1451–8.

Lubicka, D. (1994). Zimany wysokosci i ciezaru ciała u blizniat po 18 roku życia (Changes in height and body mass among twins over 18). *Przeglad Antropologiczny*, **57**:11–21.

Luciano, A., Bressan, F. and Zoppi, G. (1997). Body mass index reference curves for children aged 3–19 years from Verona, Italy. *European Journal of Clinical Nutrition*, **51**:6–10.

Luckert, B. (1996). Glucocorticoid-induced osteoporosis. In *Osteoporosis*, ed. R. Marcus, D. Feddman, J. Kelsey, pp. 801–20. San Diego, CA: Academic Press.

Luke, B. (1982). Letter to the Editor. *New England Journal of Medicine*, **306**:1549.

Luke, B., Mamelle, N., Keith, L., Muñoz, F., Minogue, J., Papiernik, E. and Johnson, T. R. B. (1995). The association between occupational factors and preterm birth: A United States nurses' study. *American Journal of Obstetrics and Gynecology*, **173**:849–62.

Luke, B., Minogue, J., Witter, F. R., Keith, L. G. and Johnson, T. R. B. (1993). The ideal twin pregnancy: Patterns of weight gain, discordancy, and length of gestation. *American Journal of Obstetrics and Gynecology*, **169**:588–97.

Luke, B., Witter, F. R., Abbey, H., Feng, T., Namnoum, A. B., Paige, D. M. and Johnson, T. R. B. (1991). Gestational age-specific birthweights of twins versus singletons. *Acta Geneticae Medicae et Gemellologiae*, **40**:69–76.

Lumey, L. H. (1992). Decreased birthweights in infants after maternal *in utero* exposure to the Dutch famine of 1944–1945. *Paediatric and Perinatal Epidemiology*, **6**:240–53.

Luna, A. M., Wilson, D. M., Wibbelsmaf, C. J., Brown, R. C., Nagashima, R. J., Hintz, R. L. and Rosenfeld, R. G. (1983). Somatomedins in adolescence: A cross-sectional study of the effect of puberty on plasma insulin-like growth factor I and II levels. *Journal of Clinical Endocrinology and Metabolism*, **57**:258–71.

Luo, Z. C., Albertsson-Wickland, K. and Karlberg, J. (1998). Target height as predicted by parental heights in a population-based study. *Pediatric Research*, **44**:563–71.

Luukkaa, V., Pesonen, U., Huhtaniemi, I., Lehtonen, A., Tilvis, R., Tuomilehto, T., Koulu, M. and Huupponen, R. (1998). Inverse correlation between serum testosterone and leptin in men. *Journal of Clinical Endocrinology and Metabolism*, **83**:3243–6.

Lyon, A. J. (1984). Factors influencing breastfeeding. *Acta Paediatrica Scandinavica*, **73**:268–70.

Lyon, A. J., Preece, M. A. and Grant, D. B. (1985). Growth curve for girls with Turner syndrome. *Archives of Disease in Childhood*, **60**:932–5.

MacArthur, C. and Knox, E. (1988). Smoking in pregnancy: Effects of stopping at different stages. *British Journal of Obstetrics and Gynaecology*, **95**:551–5.

MacDonald, R. S. (2000). The role of zinc in growth and cell proliferation. *Journal of Nutrition, Supplement*, **130**:S1500–8.

MacGregor, S. N., Keith, L. G., Chasnoff, I. J., Rosner, M. A., Chisum, G. M., Shaw, P. and Minogue, J. P. (1987). Cocaine use during pregnancy: Adverse perinatal outcome. *American Journal of Obstetrics and Gynecology*, **157**:686–90.

MacLean, H. E., Chu, S., Joske, F., Warne, G. L. and Zajac, J. D. (1995). Androgen receptor binding studies on heterozygotes in a family with androgen insensitivity syndrome. *Biochemical and Molecular Medicine*, **55**:31–7.

MacLean, W. C. and Graham, G. G. (1980). The effect of energy intake on nitrogen content of weight gained by recovering malnourished infants. *American Journal of Clinical Nutrition*, **33**:903–9.

MacMahon, B. (1973). *Age at Menarche, United States*, Vital and Health Statistics, Series 11, no. 133. Washington, D.C.: U.S. Department of Health, Education, and Welfare.

MacMahon, B., Alpert, M. and Salber, E. (1965). Infant weight and parental smoking habits. *American Journal of Epidemiology*, **82**:247–61.

Maes, M., van de Weghe, M., du Caju, M., Ernould, C., Bourguignon, J.-P. and Massa, G. (1997). A valuable improvement of adult height prediction methods in short normal children. *Hormone Research*, **48**: 184–90.

Maffei, M., Halaas, J., Ravussin, E., Pratley, R. E., Lee, G. H., Zhang, Y., Fei, H., Kim, S., Lallone, S., Ranganathan, S., Kern, P. A. and Friedman, J. M. (1995). Leptin levels in human and rodent: Measurement of plasma leptin and *ob* RNA in obese and weight-reduced subjects. *Nature Medicine*, **1**:1155–61.

Maffeis, C., Micciolo, R., Must, A., Zaffanello, M. and Pinelli, L. (1994). Parental and perinatal factors associated with childhood obesity in North-East Italy. *International Journal of Obesity*, **18**:301–5.

Maffeis, C., Pinelli, L. and Schutz, Y. (1996). Fat intake and adiposity in 8- to 11-year-old obese children. *International Journal of Obesity*, **20**:170–4.

Mafulli, M. (1996). Children in sport: Towards the year 2000. *Sports Exercise and Injury*, **63**:96–106.

Magarey, A. M. and Boulton, J. (1997). The Adelaide Nutrition Study 5: Differences in energy, nutrient and food intake at ages 11, 13 and 15 years according to fathers' occupation and parents' education level. *Australian Journal of Nutritional Dietetics*, **54**:15–23.

Magarey, A. M., Boulton, T. J. C., Chatterton, B. E., Schultz, C. and Nordin, B. E. C. (1999). Familial and environmental influences on bone growth from 11–17 years. *Acta Paediatrica*, **88**:1204–10.

Magnus, P. (1984). Causes of variation in birth weight: A study of offspring of twins. *Clinical Genetics*, **25**:15–24.

Magnus, P., Berg, K., Bjerkedal, T. and Nance, W. E. (1984). Parental determinants of birth weight. *Clinical Genetics*, **26**:397–405.

Mahoney, C. P. (1987). Evaluating the child with short stature. *Pediatric Clinics of North America*, **34**:825–49.

Malina, R. M. (1990). Research on secular trends in auxology. *Anthropologischer Anzeiger*, **48**:209–27.

Malina, R. M. (1994). Physical growth and biological maturation of young athletes. *Exercise and Sports Science Reviews*, **22**:389–433.

Malina, R. M. (1995). Issues in normal growth and maturation. *Current Opinion Endocrinology and Diabetes*, **2**:83–90.

Malina, R. M., Hamill, P. V. V. and Lemeshow, S. (1973). *Selected Body Measurements of Children 6–11*

Years, United States, Vital and Health Statistics, Series 11, no. 123. Washington, D.C.: U.S. Department of Health, Education, and Welfare.

Malina, R. M. and Katzmarzyk, P. T. (1999). Validity of the body mass index as an indicator of the risk and presence of overweight in adolescents. *American Journal of Clinical Nutrition*, **70**:S131–6.

Malina, R. M., Katzmarzyk, P. T. and Beunen, G. P. (1999). Relation between birth weight at term and growth rate, skeletal age, and cortical bone at 6–11 years. *American Journal of Human Biology*, **11**:505–11.

Mallmin, H., Ljunghall, S., Persson, I. and Bergstrom, R. (1994). Risk factors for fractures of the distal forearm: A population-based case-control study. *Osteoporosis International*, **4**:298–304.

Mamelle, N., Laumon, B. and Lazar, P. (1984). Prematurity and occupational activity during pregnancy. *American Journal of Epidemiology*, **119**:309–22.

Manchester, K. L. (1972). The effect of insulin on protein synthesis in diabetes. *Diabetes*, **21**:447–52.

Manglik, S., Cobanov, B., Flores, G., Nadjafi, R. and Tayek, J. A. (1998). Serum insulin but not leptin is associated with spontaneous and growth hormone (GH)-releasing hormone-stimulated GH secretion in normal volunteers with and without weight loss. *Metabolism: Clinical and Experimental*, **47**:1127–33.

Mantzoros, C. S., Flier, J. S. and Rogol, A. D. (1997). A longitudinal assessment of hormonal and physical alterations during normal puberty in boys. V. Rising leptin levels may signal the onset of puberty. *Journal of Clinical Endocrinology and Metabolism*, **82**:1066–70.

Marbury, M. C., Linn, S., Monson, R., Schoenbaum, S., Stubblefield, P. G. and Ryan, K. J. (1983). The association of alcohol consumption with outcome of pregnancy. *American Journal of Public Health*, **73**:1165–8.

Marbury, M. C., Linn, S., Monson, R. R., Wegman, D. H., Schoenbaum, S. C., Stubblefield, P. G. and Ryan, K. J. (1984). Work and pregnancy. *Journal of Occupational Medicine*, **26**:415–21.

Mardones-Santander, F., Salazar, G., Rosso, P. and Villarroel, L. (1998). Maternal body composition near term and birth weight. *Obstetrics and Gynecology*, **91**:873–7.

Markowitz, J., Grancher, K., Rossa, J., Aiges, H. and Daum, F. (1993). Growth failure in pediatric inflammatory bowel disease. *Journal of Pediatric Gastroenterology and Nutrition*, **16**:373–80.

Marks, K. H., Maisels, M. J., Moore, E., Gifford, K. and Friedman, Z. (1979). Head growth in sick premature infants: A longitudinal study. *Journal of Pediatrics*, **94**:282–5.

Marouka, K., Yagi, M., Akazawa, K., Kinukawa, N., Ueda, K. and Nose, Y. (1998). Risk factors for low birthweight in Japanese infants. *Acta Paediatrica*, **87**:304–9.

Marshall, J. A., Grunwald, G. K., Donahoo, W. T., Scarbro, S. and Shetterly, S. M. (2000). Percent body fat and lean mass explain the gender difference in leptin: Analysis and interpretation of leptin in Hispanic and non-Hispanic white adults. *Obesity Research*, **8**:543–52.

Marshall, W. A. (1971). Evaluation of growth rate in height over periods less than one year. *Archives of Disease in Childhood*, **46**:414–20.

Marshall, W. A. (1975). The relationship of variations in children's growth rates to seasonal climatic variations. *Annals of Human Biology*, **2**:243–350.

Marshall, W. A. and Tanner, J. M. (1969). Variations in pattern of pubertal changes in girls. *Archives of Disease in Childhood*, **44**:291–303.

Marshall, W. A. and Tanner, J. M. (1970). Variation in the pattern of pubertal changes in boys. *Archives of Disease in Childhood*, **45**:13–23.

Marshall, W. A. and Tanner, J. M. (1986). Puberty. In *Human Growth: A Comprehensive Treatise*, 2nd edn, vol. 2, *Postnatal Growth*, ed. F. Falkner and J. M. Tanner, pp. 171–209. New York: Plenum Press.

Martha, P. M., Jr., Gorman, K. M., Blizzard, R. M., Rogol, A. D. and Veldhuis, J. D. (1992*a*). Endogenous growth hormone secretion and clearance rates in normal boys, as determined by deconvolution analysis: Relationship to age, pubertal status, and body mass. *Journal of Clinical Endocrinology and Metabolism*, **74**:336–44.

Martha, P. M., Jr., Reiter, E. O., Davila, N., Shaw, M. A., Holcombe, J. H. and Baumann, G. (1992*b*). The role of body mass in the response to growth hormone therapy. *Journal of Clinical Endocrinology and Metabolism*, **75**:1470–3.

Martha, P. M., Jr., Rogol, A. D., Blizzard, R. M., Shaw, M. A. and Baumann, G. (1991). Growth hormone-binding protein activity is inversely related to 24-hour growth hormone release in normal boys. *Journal of Clinical Endocrinology and Metabolism*, **73**:175–81.

Martha, P. M., Jr., Rogol, A. D., Carlsson, L. M., Gesundheit, N. and Blizzard, R. M. (1993). A longitudinal assessment of hormonal and physical alterations during normal puberty in boys. I. Serum growth hormone-binding protein. *Journal of Clinical Endocrinology and Metabolism*, **77**:452–7.

Martha, P. M., Jr., Rogol, A. D., Veldhuis, J. D., Kerrigan, J. R., Goodman, D. W. and Blizzard, R. M. (1989). Alterations in the pulsatile properties of circulating growth hormone concentrations during puberty in boys. *Journal of Clinical Endocrinology and Metabolism*, **69**:563–70.

Martin, M. M., Martin, A. L. A. and Mossman, K. L. (1986). Testosterone treatment of constitutional delay in growth and development: Effect of dose on predicted versus definitive height. *Acta Endocrinologica, Supplement*, **279**:147–52.

Martin, T. R. and Bracken, M. B. (1986). Association of low birthweight with passive smoke exposure in pregnancy. *American Journal of Epidemiology*, **124**:633–42.

Martin, T. R. and Bracken, M. B. (1987). The association between low birthweight and caffeine consumption during pregnancy. *American Journal of Epidemiology*, **126**:813–21.

Martine, T., Claessens, A. L., Vlietinck, R. and Marchal, G. (1997). Accuracy of anthropometric estimation of muscle cross-sectional area of the arm in males. *American Journal of Human Biology*, **9**:73–86.

Martinez, F. D., Wright, A. L. and Taussig, L. M. (1994). The effect of paternal smoking on the birthweight of newborns whose mothers did not smoke. *American Journal of Public Health*, **84**:1489–91.

Martin-Hernández, T., Gálvez, M. D., Cuadro, A. T. and Herrera-Justiniano, E. (1996). Growth hormone secretion in normal prepubertal children: Importance of relations between endogenous secretion, pulsatility and body mass. *Clinical Endocrinology*, **44**: 327–34.

Martorell, R., Mendoza, F. S., Castillo, R. O., Pawson, I. G. and Budge, C. C. (1987). Short and plump physique of Mexican-American children. *American Journal of Physical Anthropology*, **73**:475–87.

Martorell, R. and Shekar, M. (1994). Growth faltering rates in Berkeley, Guatemala and Tamil Nadu: Implications for growth monitoring programs. *Food and Nutrition Bulletin*, **15**:185–91.

Martuzzi Veronesi, F. and Gueresi, P. (1994). Trend in menarcheal age and socioeconomic influence in Bologna (Northern Italy). *Annals of Human Biology*, **21**:187–96.

Martyn, C. N., Barker, D. J. P., Jespersen, S., Greenwald, S., Osmond, C. and Berry, C. (1995*a*). Growth *in utero*, adult blood pressure, and arterial compliance. *British Heart Journal*, **73**:116–21.

Martyn, C. N., Gale, C. R., Jespersen, S. and Sherriff, S. B. (1998). Impaired fetal growth and atherosclerosis of carotid and peripheral arteries. *Lancet*, **352**: 173–8.

Martyn, C. N. and Greenwald, S. E. (1997). Impaired synthesis of elastin in walls of aorta and large conduit arteries during early development as an initiating event in pathogenesis of systemic hypertension. *Lancet*, **350**:953–5.

Martyn, C. N., Meade, T. W., Stirling, Y. and Barker, D. J. P. (1995*b*). Plasma concentrations of fibrinogen and factor VII in adult life and their relation to intra-uterine growth. *British Journal of Haematology*, **89**:142–6.

Marubini, E., Resele, L. F. and Barghini, G. (1971). A comparative fitting of the Gompertz and logistic functions to longitudinal height data during adolescence in girls. *Human Biology*, **43**:237–52.

Marubini, E., Resele, L. F., Tanner, J. M. and Whitehouse, R. H. (1972). The fit of Gompertz and logistic curves to longitudinal data during adolescence on height, sitting height and biacromial diameter in boys and girls of the Harpenden Growth Study. *Human Biology*, **44**:511–24.

Massa, G., Bouillon, R. and Vanderschueren-Lodeweyckx, M. (1992*a*). Serum levels of growth hormone-binding protein and

insulin-like growth factor-I during puberty. *Clinical Endocrinology*, **37**:175–80.

Massa, G., de Zegher, F. and Vanderschueren-Lodeweyckx, M. (1992*b*). Serum growth hormone-binding proteins in the human fetus and infant. *Pediatric Research*, **32**:69–72.

Massa, G., Vanderschueren-Lodeweyckx, M. and Malvaux, P. (1990). Linear growth in patients with Turner syndrome: Influence of spontaneous puberty and parental height. *European Journal of Pediatrics*, **149**:246–50.

Matanoski, G., Kanchanaraksa, S., Lantry, D. and Chang, Y. (1995). Characteristics of nonsmoking women in NHANES I and NHANES I epidemiologic follow-up study with exposure to spouses who smoke. *American Journal of Epidemiology*, **142**:149–57.

Matkovic, V., Fontana, D., Tominac, C., Goel, P. and Chesnut, C. H. III (1990). Factors that influence peak bone mass formation: A study of calcium balance and the inheritance of bone mass in adolescent females. *American Journal of Clinical Nutrition*, **52**:878–88.

Matkovic, V., Ilich, J. Z., Skugor, M., Badenhop, N. E., Goel, P., Clairmont, A., Klisovic, D., Nahhas, R. W. and Landoll, J. D. (1997). Leptin is inversely related to age at menarche in human females. *Journal of Clinical Endocrinology and Metabolism*, **82**:3239–45.

Matsudo, S. M. M. and Matsudo, V. K. R. (1994). Self-assessment and physician assessment of sexual maturation in Brazilian boys and girls: Concordance and reproducibility. *American Journal of Human Biology*, **6**:451–5.

Matsui, T. and Yamaguchi, M. (1995). Zinc modulation of insulin-like growth factor's effect in osteoblastic MC3T3-E1 cells. *Peptides*, **16**:1063–8.

Matsukura, S., Taminato, T., Kitano, N., Seino, Y., Hamada, H., Uchihashi, M., Nakajima, H. and Hirata, Y. (1984). Effects of environmental tobacco smoke on urinary cotinine excretion in nonsmokers: Evidence for passive smoking. *New England Journal of Medicine*, **311**:828–32.

Matsumoto, K. (1982). Secular acceleration of growth in height in Japanese and its social background. *Annals of Human Biology*, **9**:399–410.

Matsuoka, H., Fors, H., Bosaeus, I., Rosberg, S., Albertsson-Wikland, K. and Bjarnason, R. (1999*a*).

Changes in body composition and leptin levels during growth hormone (GH) treatment in short children with various GH secretory capacities. *European Journal of Endocrinology*, **140**:35–42.

Matsuoka, H., Sato, K., Sugihara, S. and Murata, M. (1999*b*). Bone maturation reflects the secular trend in growth. *Hormone Research*, **52**:125–30.

Matthes, J. W. A., Lewis, P. A., Davies, D. P. and Bethel, J. A. (1994). Relation between birth weight at term and systolic blood pressure in adolescence. *British Medical Journal*, **308**:1074–7.

Mauras, N., Haymond, M. W., Darmaun, D., Vieira, N. E., Abrams, S. A. and Yergey, A. L. (1994). Calcium and protein kinetics in prepubertal boys: Positive effects of testosterone. *Journal of Clinical Investigation*, **93**:1014–19.

Mauras, N., Rogol, A. D. and Veldhuis, J. D. (1990). Increased hGH production rate after low-dose estrogen therapy in prepubertal girls with Turner's syndrome. *Pediatric Research*, **28**:626–30.

Maynard, L. M., Guo, S. S., Chumlea, W. C., Roche, A. F., Wisemandle, W. A., Zeller, C. M., Towne, B. and Siervogel, R. M. (1998). Total-body and regional bone mineral content and areal bone mineral density in children aged 8–18 years: The Fels Longitudinal Study. *American Journal of Clinical Nutrition*, **68**:1111–17.

Maynard, L. M., Wisemandle, W. A., Roche, A. F., Chumlea, W. C., Guo, S. S. and Siervogel, R. M. (2001). Childhood body composition in relation to body mass index: The Fels Longitudinal Study. *Pediatrics*, **107**:344–50.

Mazess, R. B. and Barden, H. S. (1988). Measurement of bone by dual-photon absorptiometry (DPA) and dual-energy X-ray absorptiometry (DEXA). *Annales Chirurgiae et Gynaecologiae*, **77**:197–203.

Mazzanti, L., Nizzoli, G., Tassinari, D., Bergamaschi, R., Magnani, C., Chiumello, G. and Cacciari, E. (1994). Spontaneous growth and pubertal development in Turner's syndrome with different karyotypes. *Acta Paediatrica*, **83**:299–304.

McCammon, R. W. (1970). *Human Growth and Development*. Springfield, IL: Charles C. Thomas.

McCance, D. R., Pettitt, D. J., Hanson, R. L., Jacobsson, L. T. H., Knowler, W. C. and Bennett, P. H. (1994). Birth weight and non-insulin dependent diabetes: Thrifty genotype, thrifty phenotype, or surviving small baby genotype? *British Medical Journal*, **308**:942–5.

McClung, J. (1969). *Effects of High Altitude on Human Birth*. Cambridge, MA: Harvard University Press.

McCullough, R. E., Reeves, J. T. and Liljegren, R. L. (1977). Fetal growth retardation and increased infant mortality at high altitude. *Archives of Environmental Health*, **32**:36–40.

McDermott, J. M., Drews, C. D., Adams, M. M., Hill, H. A., Berg, C. J. and McCarthy, B. J. (1999). Does inadequate prenatal care contribute to growth retardation among second-born African-American babies? *American Journal of Epidemiology*, **150**: 706–13.

McDonald, A. and Goldfine, J. (1988). Glucocorticoid regulation of insulin receptor gene transcription in IM-9 cultured lymphocytes. *Journal of Clinical Investigation*, **81**:499–504.

McDonald, A. D., Armstrong, B. C. and Sloan, M. (1992). Cigarette, alcohol, and coffee consumption and prematurity. *American Journal of Public Health*, **82**:87–90.

McIntyre, D. D., Bloom, S. L., Casey, B. M. and Leveno, K. J. (1999). Birth weight in relation to morbidity and mortality among newborn infants. *New England Journal of Medicine*, **340**:1234–8.

McKusick, V. A. (1972). *Heritable Disorders of Connective Tissue*, 4th edn. St. Louis, MO: Mosby.

McMurray, R. G., Harrell, J. S., Deng, S., Bradley, C. B., Cox, L. M. and Bangdiwala, S. I. (2000). The influence of physical activity, socioeconomic status, and ethnicity on the weight status of adolescents. *Obesity Research*, **8**:130–9.

McNaughton, S. A., Shepherd, R. W., Greer, R. G., Cleghorn, G. J. and Thomas, B. J. (2000). Nutritional status of children with cystic fibrosis measured by total body potassium as a marker of body cell mass: Lack of sensitivity of anthropometric measures. *Journal of Pediatrics*, **136**:188–94.

McPherson, R. S., Nichaman, M. Z., Kohl, H. W., Reed, D. B. and Labarthe, D. R. (1990). Intake and food sources of dietary fat among schoolchildren in The Woodlands, Texas. *Pediatrics*, **86**:520–6.

Melander, O., Mattiasson, I., Maršál, K., Groop, L. and Hulthén, U. L. (1999). Heredity for hypertension influences intra-uterine growth and the relation between fetal growth and adult blood pressure. *Journal of Hypertension*, 17:1557–61.

Melmed, S. and Slanina, S. N. (1985). Insulin suppresses triiodothyronine-induced growth hormone secretion by GH3 rat pituitary cells. *Endocrinology*, 117:532–7.

Mendlewicz, J., Linkowski, P., Kerkofs, M., Leproult, R., Copinschi, G. and Van Cauter, E. (1999). Genetic control of 24-hour growth hormone secretion in man: A twin study. *Journal of Clinical Endocrinology and Metabolism*, 84:856–62.

Mendoza, F. S. and Castillo, R. O. (1986). Growth abnormalities in Mexican-American children in the United States: The National Health and Nutrition Examination Survey. *Nutrition Research*, 6:1247–57.

Mertz, W., Tsui, J. C., Judd, J. T., Reiser, S., Hallfrisch, J., Morris, E. R., Steele, P. D. and Lashley, E. (1991). What are people really eating? The relation between energy intake derived from estimated diet records and intake determined to maintain body weight. *American Journal of Clinical Nutrition*, 54:191–255.

Mesa, M. S., Sanchez-Andres, A., Marrodan, M. D., Martin, J. and Fuster, V. (1996). Body composition of rural and urban children from the Central Region of Spain. *Annals of Human Biology*, 23:203–12.

Metzger, D. L. and Kerrigan, J. R. (1994). Estrogen receptor blockade with tamoxifen diminishes growth hormone secretion in boys: Evidence for a stimulatory role of endogenous estrogens during male adolescence. *Journal of Clinical Endocrinology and Metabolism*, 79:513–18.

Meyer, B. A. and Daling, J. R. (1985). Activity level of mother's usual occupation and low infant birth weight. *Journal of Occupational Medicine*, 27:841–6.

Michaelsen, K. F. (1997). Nutrition and growth during infancy: The Copenhagen Cohort Study. *Acta Paediatrica, Supplement*, 420:1–36.

Michaelsen, K. F., Petersen, S., Greisen, G. and Thomsen, B. L. (1994). Weight, length, head circumference, and growth velocity in a longitudinal study of Danish infants. *Danish Medical Bulletin*, 41:577–85.

Michaut, E., Niang, I. and Dan, V. (1972). La maturation osseuse pendant la période pubertaire (Skeletal maturation during the pubescent period). *Annales de Radiologie*, 15:767–79.

Milani, S., Benso, L., Vannelli, S., Larizza, D., Volta, C. and Bernasconi, S. (1994). Turner syndrome: Italian growth charts from infancy to adulthood. *Humanbiologica Budapestiensis*, 25:91–8.

Miller, F. J. W., Billewicz, W. Z. and Thomson, A. M. (1972). Growth from birth to adult life of 442 Newcastle-upon-Tyne children. *British Journal of Preventive and Social Medicine*, 26:224–30.

Miller, F. J. W., Court, S. D. M., Knox, E. G. and Brandon, S. (1974). *The School Years in Newcastle-upon-Tyne 1952–1962, being a Further Contribution to the Study of a Thousand Families*. London: Oxford University Press.

Miller, H. C., Hassaniun, K. and Hensleigh, P. A. (1976). Fetal growth retardation in relation to maternal smoking and weight gain in pregnancy. *American Journal of Obstetrics and Gynecology*, 125:55–60.

Miller, H. C. and Jekel, J. F. (1987a). Incidence of low birth weight infants born to mothers with multiple risk factors. *Yale Journal of Biology and Medicine*, 60:397–404.

Miller, H. C. and Jekel, J. F. (1987b). The effect of race on the incidence of low birth weight: persistence of effect after controlling for socioeconomic, educational, marital, and risk status. *Yale Journal of Biology and Medicine*, 60:221–32.

Miller, H. S., Lesser, K. B. and Reed, K. L. (1996). Adolescence and very low birth weight infants: A disproportionate association. *Obstetrics and Gynecology*, 87:83–8.

Miller, J. D., Tannenbaum, G. S., Colle, E. and Guyda, H. J. (1982). Daytime pulsatile growth hormone secretion during childhood and adolescence. *Journal of Clinical Endocrinology and Metabolism*, 55:989–94.

Mills, J. L., Graubard, B. I., Harley, E. E., Rhoads, G. G. and Berendes, H. W. (1984). Maternal alcohol consumption and birth weight: How much drinking during pregnancy is safe? *Journal of American Medical Association*, 252:1875–9.

Mills, J. L., Shiono, P. H., Shapiro, L. R., Crawford, P. B. and Rhoads, G. G. (1986). Early growth predicts timing of puberty in boys: Results of a 14-year nutrition and growth study. *Journal of Pediatrics*, **109**:543–7.

Milman, A. E., de Moor, P. and Luckens, F. D. W. (1951). Relation of purified pituitary growth hormone and insulin in regulation of nitrogen balance. *American Journal of Physiology*, **166**:354–63.

Milner, R. D. G. and Richards, B. (1974). An analysis of birthweight by gestational age of infants born in England and Wales, 1967 to 1971. *Journal of Obstetrics and Gynaecology of the British Commonwealth*, **81**:956–67.

Mirwald, R. L. and Bailey, D. A. (1986). *Maximal Aerobic Power*. London, Ontario, Canada: Sport Dynamics.

Mirwald, R. L. and Bailey, D. A. (1997). Seasonal height velocity variation in boys and girls. *American Journal of Human Biology*, **9**:709–15.

Mitchell, B. D., Kammerer, C. M., Mahaney, M. C., Blangero, J., Comuzzie, A. G., Atwood, L. D., Haffner, S. M., Stern, M. P. and MacCluer, J. W. (1996). Genetic analysis of the IRS: Pleiotropic effects of genes influencing insulin levels on lipoprotein and obesity measures. *Arteriosclerosis Thrombosis and Vascular Biology*, **16**:281–8.

Mitchell, M. C. and Lerner, E. (1989). A comparison of pregnancy outcome in overweight and normal weight women. *Journal of the American College of Nutrition*, **8**:617–24.

Mitchell, W. G., Gorrell, R. W. and Greenberg, R. A. (1980). Failure-to-thrive: A study in a primary care setting – epidemiology and follow-up. *Pediatrics*, **65**:971–7.

Mizutani, T., Nishikawa, Y., Adachi, H., Enomoto, T., Ikegami, H., Kurachi, H., Nomura, T. and Miyake, A. (1994). Identification of estrogen receptor in human adipose tissue. *Journal of Clinical Endocrinology and Metabolism*, **78**:950–4.

Moar, V. A. and Ounsted, M. K. (1982). Growth in the first year of life: How early can one predict size at twelve months among small-for-dates and large-for-dates babies? *Early Human Development*, **6**:65–9.

Mochizuki, M., Maruo, T., Masuko, K. and Ohtsu, T. (1984). Effects of smoking on fetoplacental-maternal system during pregnancy. *American Journal of Obstetrics and Gynecology*, **149**:413–20.

Moiraghi-Ruggenini, A., Cavallo, F., Zotti, C., Delcorno, G., Bonazzi, M. C., Primatesta, P., Gullotti, A., Olivieri, R., Casuccio, A., Casuccio, N., Vanini, G. C., Fusco, A., Gagliardi, L. and Tibaldi, C. (1992). Studio multicentrico su abitudini alcooliche a gravidanza (A multicenter study of alcohol intakes during pregnancy). *L'Igiene moderna*, **98**:137–74.

Molinari, L., Gasser, T., Largo, R. and Prader, A. (1995). Child–adult correlations for anthropometric measurements. In *Essays in Auxology 88*, eds. R. Hauspie, G. Lindgren and F. Falkner, pp. 164–77. Welwyn Garden City, U.K.: Castlemead Publications.

Molinari, L., Largo, R. H. and Prader, A. (1980). Analysis of the growth spurt at age seven (mid-growth spurt). *Helvetica Paediatrica Acta*, **35**:325–34.

Moll, G. W., Jr., Rosenfield, R. L. and Fang, V. S. (1986). Administration of low-dose estrogen rapidly and directly stimulates growth hormone production. *American Journal of Diseases of Childhood*, **140**:124–7.

Moll, P. P., Burns, T. L. and Lauer, R. M. (1991). The genetic and environmental sources of body mass index variability: The Muscatine Ponderosity Family Study. *American Journal of Human Genetics*, **49**:1243–55.

Montague, C. T., Prins, J. B., Sanders, L., Digby, J. E. and O'Rahilly, S. (1997). Depot- and sex-specific differences in human leptin mRNA expression: Implication for the control of regional fat distribution. *Diabetes*, **46**:342–7.

Moore, L. G., Jahnigen, D., Rounds, S. S., Reeves, J. T. and Grover, R. F. (1982*a*). Maternal hyperventilation helps preserve arterial oxygenation during high altitude pregnancy. *Journal of Applied Physiology: Respiratory, Environmental and Exercise Physiology*, **52**:690–4.

Moore, L. G., Rounds, S. S., Jahnigen, D., Grover, R. F. and Reeves, J. T. (1982*b*). Infant birth weight is related to maternal arterial oxygenation at high altitude. *Journal of Applied Physiology: Respiratory, Environmental and Exercise Physiology*, **52**:695–9.

Mora, S., Weber, G., Marenzi, K., Signorini, E., Rovelli, R., Proverbio, M. C. and Chiumello, G. (1999).

Longitudinal changes of bone density and bone resorption in hyperthyroid girls during treatment. *Journal of Bone and Mineral Research*, **14**:1971–7.

Moreno, L. A., Fleta, J. and Mur, L. (1998). Television watching and fatness in children. *Journal of the American Medical Association* **280**:1230–1.

Moreno, L. A., Sarria, A., Rodriguez, G. and Bueno, M. (2000). Trends in body mass index and overweight prevalence among children and adolescents in the region of Aragon (Spain) from 1985 to 1995. *International Journal of Obesity*, **24**:925–31.

Morishima, A. (1996). Aromatase deficiency in the male: The physiologic role of estrogens. *Clinical Endocrinology*, **44**:65–74.

Morishima, A., Grumbach, M. M., Simpson, E. R., Fisher, C. and Qin, K. (1995). Aromatase deficiency in male and female siblings caused by a novel mutation and the physiological role of estrogens. *Journal of Clinical Endocrinology and Metabolism*, **80**:3689–98.

Morley, R., Lister, G., Leeson Payne, C. and Lucas, A. (1994). Size at birth and later blood pressure. *Archives of Disease in Childhood*, **70**:536–7.

Morris, C. A., Demsey, S. A., Leonard, C. O., Dilts, C. and Blackburn, B. L. (1988). Natural history of Williams syndrome: Physical characteristics. *Journal of Pediatrics*, **113**:318–26.

Morrison, J. A., Barton, B. A., Obarzanek, E., Crawford, P. B., Guo, S. S. and Schreiber, G. B. (2001). Racial differences in the sums of skinfolds and percentage of body fat estimated from impedance in black and white girls, 9 to 19 years of age: The National Heart, Lung, and Blood Institute Growth and Health Study. *Obesity Research*, **9**:297–305.

Morrison, J. A., Sprecher, D., McMahon, R. P., Simon, J., Schreiber, G. B. and Khoury, P. R. (1996). Obesity and high-density lipoprotein cholesterol in black and white 9- and 10-year-old girls: The National Heart, Lung, and Blood Institute Growth and Health Study. *Metabolism: Clinical and Experimental*, **45**:469–74.

Morrison, N. A., Yeoman, R., Kelly, P. J. and Eisman, J. A. (1992). Contribution of *trans*-acting factor alleles to normal physiological variability: Vitamin D receptor gene polymorphisms and circulating osteocalcin. *Proceedings of the National Academy of Sciences (U.S.A.)*, **89**:6665–9.

Mosier, H. D., Dearden, L. C., Jansons, R. A. and Hill, R. R. (1977). Growth hormone, somatomedin and cartilage sulfation in failure of catch-up growth after propylthiouracil-induced hypothyroidism in the rat. *Endocrinology*, **100**:1644–51.

Mosier, H. D. and Jansons, R. A. (1976). Growth hormone during catch-up growth and failure of catch-up growth in rats. *Endocrinology*, **98**:214–19.

Mueller, W. and Reid, R. (1979). A multivariate analysis of fatness and relative fat patterning. *American Journal of Physical Anthropology*, **50**:199–208.

Mueller, W. H. (1976). Parent–child correlations for stature and weight among school aged children: A review of 24 studies. *Human Biology*, **48**:379–97.

Muñoz-Torres, M., Jodar, E., Quesada, M. and Escobar-Jiminez, F. (1995). Bone mass in androgen–insensitivity syndrome: Response to hormonal replacement therapy. *Calcified Tissue International*, **57**:94–6.

Muramatsu, S., Sato, Y., Miyao, M., Muramatsu, T. and Ito, A. (1990). A longitudinal study of obesity in Japan: Relationship of body habitus between at birth and at age 17. *International Journal of Obesity*, **14**:39–45.

Murata, M. (1993). Japanese specific bone age standard on the TW2. *Clinical Pediatric Endocrinology, Supplement*, **3**:35–41.

Murata, M. (2000). Secular trends in growth and changes in eating patterns of Japanese children. *American Journal of Clinical Nutrition, Supplement*, **72**:S1379–83.

Murphy, N. J., Bulkow, L. R., Schraer, C. D. and Lanier, A. P. (1993). Prevalence of diabetes mellitus in pregnancy among Yup'ik Eskimos, 1987–1988. *Diabetes Care*, **16**:315–17.

Murray, J. and Kippax, S. (1978). Children's social behavior in three towns with differing television expererience. *Journal of Communication*, **18**: 19–29.

Murray, J. L. and Bernfield, M. (1988). The differential effect of prenatal care on the incidence of low birth weight among blacks and whites in a prepaid health care plan. *New England Journal of Medicine*, **319**:1385–91.

Must, A., Jacques, P. F., Dallal, G. E., Bajema, C. J. and Dietz, W. H. (1992). Long-term morbidity and mortality of overweight adolescents: A follow-up of the Harvard Growth Study of 1922 to 1935. *New England Journal of Medicine*, 327:1350–5.

Naeye, R. L. (1981). Influence of maternal cigarette smoking during pregnancy on fetal and childhood growth. *Obstetrics and Gynecology*, 57:18–21.

Naeye, R. L., Blanc, W., Leblanc, W. and Khatamee, M. A. (1973). Fetal complications of maternal heroin addiction: Abnormal growth, infections, and episodes of stress. *Journal of Pediatrics*, 83:1055–61.

Nagy, T. R., Gower, B. A., Trowbridge, C. A., Dezenberg, C., Shewchuk, R. M. and Goran, M. I. (1997). Effects of gender, ethnicity, body composition, and fat distribution on serum leptin concentrations in children. *Journal of Clinical Endocrinology and Metabolism*, 82:2148–52.

Najjar, M. F. and Rowland, M. (1987). *Anthropometric Reference Data and Prevalence of Overweight: United States, 1976–80*, Vital and Health Statistics, Series 11, no. 238. Washington, D.C.: Department of Health and Human Services.

Nakao–Hayashi, J., Ito, H., Kanayasu, T., Morita, I. and Murota, S. (1992). Stimulatory effects of insulin and insulin-like growth factor 1 on migration and tube formation by vascular endothelial cells. *Atherosclerosis*, 92:141–9.

National Heart Lung and Blood Institute, United States (1998). Clinical guidelines on the identification, evaluation, and treatment of overweight and obesity in adults: The Evidence Report. *Obesity Research*, 6:S51–209.

Neggers, Y., Goldenberg, R. L., Cliver, S. P., Hoffman, H. J. and Cutter, G. R. (1995). The relationship between maternal and neonatal anthropometric measurements in term newborns. *Obstetrics and Gynecology*, 85:192–6.

Neinstein, L. S. (1982). Adolescent self-assessment of sexual maturation. *Clinica Pediatrica*, 21:482–4.

Németh, Á. and Eiben, O. G. (1997). Secular growth changes in Budapest in the 20th century. *Acta Medica Auxologica* 29:5–12.

Ness, R. (1991). Adiposity and age at menarche in Hispanic women. *American Journal of Human Biology*, 3:41–7.

Neville, A. M. and O'Hare, M. J. (1982). Hyperaldosteronism and related syndromes of mineralocorticoid excess. In *The Human Adrenal Cortex: Pathology and Biology – An Integrated Approach*, eds. A. M. Neville and M. J. O'Hare, pp. 202–41. Berlin: Springer-Verlag.

New Zealand Department of Health (1971). *Physical Development of New Zealand schoolchildren 1969*, Special Report no. 38. Wellington, New Zealand: Health Services Research Unit, Department of Health.

Neyzi, O., Bundak, R., Molzan, J., Günöz, H., Darendeliler, F. and Saka, N. (1993). Estimation of annual height velocity based on short-versus long-term measurements. *Acta Paediatrica*, 82:239–44.

Nicholson, A. B. and Hanley, C. (1953). Indices of physiological maturity: Derivation and interrelationships. *Child Development*, 24:3–38.

Nicklas, T. A., Webber, L. S., Koschak, M. L. and Berenson, G. S. (1992). Nutrient adequacy of low fat intakes for children: The Bogalusa Heart Study. *Pediatrics*, 89:221–8.

Nicklas, T. A., Webber, L. S., Srinivasan, S. and Berenson, G. S. (1993). Secular trends in dietary intakes and cardiovascular risk factors of 10-year-old children: The Bogalusa Heart Study (1973–1988). *American Journal of Clinical Nutrition*, 57:930–7.

Nicoloff, J. T. (1978). Thyroid hormone transport and metabolism; pathophysiologic implications. In *The Thyroid*, eds. S. C. Werner and S. H. Ingbar, pp. 88–99. New York: Harper & Row.

Nielsen, J. (1982). Effects of growth hormone, prolactin, and placental lactogen on insulin content and release, and deoxyribonucleic acid synthesis in cultured pancreatic islets. *Endocrinology*, 110:600–6.

Nieto, F. J., Szklo, M. and Comstock, G. W. (1992). Childhood weight and growth rate as predictors of adult mortality. *American Journal of Epidemiology*, 136:201–13.

Nilsen, S. T., Sagen, N., Kim, H. C. and Bergsjø, R. (1984). Smoking, hemoglobin levels, and birth weights in normal pregnancies. *American Journal of Obstetrics and Gynecology*, 148:752–8.

Nilsson, A., Isgaard, J. and Isaksson, O. G. P. (1996). Role of growth hormone in postnatal growth and metabolism. In *Pediatrics and Perinatology. The Scientific Basis*, 2nd edn, eds. P. D. Gluckman and M. A. Heymann, pp. 309–13. London: Edward Arnold.

Nimrod, A. and Ryan, K. J. (1975). Aromatization of androgens by human abdominal and breast fat tissue. *Journal of Clinical Endocrinology and Metabolism*, **40**:367–72.

Nishi, Y. (1990). Hereditary growth hormone deficiency. In *Pediatric Endocrinology*, 2nd edn, ed. F. Lifshitz, pp. 27–42. New York: Marcel Dekker.

Nordström, M. L. and Cnattingius, S. (1994). Smoking habits and birthweights in two successive births in Sweden. *Early Human Development*, **37**:195–204.

Norgan, N. and Ferro-Luzzi, A. (1982). Weight–height indices as estimators of fatness in men. *Human Nutrition: Clinical Nutrition*, **36C**:363–72.

Novak, L. P. (1963). Age and sex differences in body density and creatinine excretion of high school children. *Annals of the New York Academy of Sciences*, **110**:545–77.

Novak, L. P., Tauxe, W. N. and Orvis, A. L. (1973). Estimation of total body potassium in normal adolescents by whole-body counting: Age and sex differences. *Medicine and Science in Sports and Exercise*, **5**:147–55.

Nylind, B., Schele, R. and Klas, L. (1978). Changes in male exercise performance and anthropometric variables between the age of 19 and 30. *European Journal of Applied Physiology and Occupational Physiology*, **38**:145–50.

Oakley, J. R., Parsons, R. J. and Whitelaw, A. G. L. (1977). Standards for skinfold thickness in British newborn infants. *Archives of Disease in Childhood*, **52**:287–90.

Oates, R. K., Peacock, A. and Forrest, D. (1985). Long-term effects of nonorganic failure to thrive. *Pediatrics*, **75**:36–40.

Obarzanek, E., Schreiber, G. B., Crawford, P. B., Goldman, S. R., Barrier, P. M., Frederick, M. M. and Lakatos, E. (1994). Energy intake and physical activity in relation to indexes of body fat: The National Heart, Lung, and Blood Institute Growth and Health Study. *American Journal of Clinical Nutrition*, **60**:15–22.

Ogden, C. L., Troiano, R. P., Briefel, R. R., Kuczmarski, R. J., Flegal, K. M. and Johnson, C. L. (1997). Prevalence of overweight among preschool children in the United States, 1971 through 1994. *Pediatrics*, **99**:E1–7.

Ogle, G. D., Allen, J. R., Humphries, I. R. J., Lu, P. W., Briody, J. N., Morley, K., Howmangiles, R. and Cowell, C. T. (1995). Body-composition assessment by dual-energy X-ray absorptiometry in subjects aged 4–26 y. *American Journal of Clinical Nutrition*, **61**:746–53.

Ogston, S. A. and Parry, G. J. (1992). Results: strategy of analysis and analysis of pregnancy outcome. *International Journal of Epidemiology*, **21**:S45–87.

Olivier, G. and Devigne, G. (1983). Biology and social structure. *Journal of Biosocial Science*, **15**:379–89.

Olsen, D. and Ferin, M. (1987). Corticotropin-releasing hormone inhibits gonadotropin secretion in ovariectomized Rhesus monkey. *Journal of Clinical Endocrinology and Metabolism*, **65**:262–7.

Olsen, J., Pereira, A., Da, C. and Olsen, S. F. (1991). Does maternal tobacco smoking modify the effect of alcohol on fetal growth? *American Journal of Public Health*, **81**:69–73.

Olsen, J., Rachootin, P. and Schiodt, A. V. (1983). Alcohol use, conception time, and birthweight. *Journal of Epidemiology and Community Health*, **37**:63–5.

Onat, T. (1995). Validation of methods for predicting adult stature in Turkish girls. *American Journal of Human Biology*, **7**:757–67.

Onat, T. (1997). Socioeconomic differences in growth of metacarpal II during adolescence in females: Interactions with sexual and skeletal maturity. *American Journal of Human Biology*, **9**:439–48.

Ong, K. K. L., Ahmed, M. L. and Dunger, D. B. (1999*a*). The role of leptin in human growth and puberty. *Acta Paediatricia Scandinavica*, **88**:95–8.

Ong, K. K. L., Ahmed, M. L., Sherriff, A., Woods, K. A., Watts, A., Golding, J., The ALSPAC Study Team and Dunger, D. B. (1999*b*). Cord blood leptin is associated with size at birth and predicts infancy weight gain in humans. *Journal of Clinical Endocrinology and Metabolism*, **84**:1145–8.

Ortega, R. M., Requejo, A. M., Andrés, P., Lopez-Sobaler, A. M., Redondo, R. and González-Fernández, M. (1995). Relationship between diet composition and body mass index in a group of Spanish adolescents. *British Journal of Nutrition*, **74**:765–73.

Oryszczyn, M.-P., Annesi-Maesano, I., Campagna, D., Sahuquillo, J., Huel, G. and Kauffmann, F. (1999). Head circumference at birth and maternal factors related to cord blood total IgE. *Clinical and Experimental Allergy*, **29**:334–41.

Osmond, C., Barker, D. J. P., Winter, P. D., Fall, C. H. D. and Simmonds, S. J. (1993). Early growth and death from cardiovascular disease in women. *British Medical Journal*, **307**:1519–24.

O'Sullivan, J. B., Gellis, S. S., Tenny, B. O. and Mahan, C. M. (1965). Aspects of birth weight and its influencing variables. *American Journal of Obstetrics and Gynecology*, **92**:1023–9.

Otsuki, K. (1956). Family study on stature, body weight, and chest circumference in elementary school pupils with special regard to the parent–child correlation. *Jinruigaku Jinruiidengaku, Taishitsugaku Ronbunsho*, **26**:51–71. (In Japanese)

Ottosson, M., Vikman-Adolfsson, K., Enerback, S., Elander, A., Björntorp, P. and Edén, S. (1995). Growth hormone inhibits lipoprotein lipase activity in human adipose tissue. *Journal of Clinical Endocrinology and Metabolism*, **80**:936–41.

Ottosson, M., Vikman-Adolfsson, K., Enerback, S., Olivecrona, G. and Björntorp, P. (1994). The effects of cortisol on the regulation of lipoprotein lipase activity in human adipose tissue. *Journal of Clinical Endocrinology and Metabolism*, **79**:820–5.

Ounsted, M., Moar, V. and Scott, A. (1982). Growth in the first four years. IV. Correlations with parental measures in small-for-dates and large-for-dates babies. *Early Human Development*, **7**:357–66.

Ounsted, M. and Moar, V. A. (1986). Proportionality changes in the first year of life: The influence of weight for gestational age at birth. *Acta Paediatrica Scandinavica*, **75**:811–18.

Ounsted, M., Moar, V. A. and Scott, A. (1986). Growth and proportionality in early childhood. I. Within-population variations. *Early Human Development*, **13**:27–36.

Overpeck, M. D., Hediger, M. L., Zhang, J., Trumble, A. C. and Klebanoff, M. A. (1999). Birth weight for gestational age of Mexican American infants born in the United States. *Obstetrics and Gynecology*, **93**:943–7.

Owen, G. M., Garry, P. J., Seymoure, R. D., Harrison, G. G. and Acosta, P. B. (1981). Nutrition studies with White Mountain Apache preschool children in 1976 and 1969. *American Journal of Clinical Nutrition*, **34**:266–77.

Owen, P., Donnet, M. L., Ogston, S. A., Christie, A. D., Howie, P. and Patel, N. (1996). Standards for ultrasound fetal growth velocity. *British Journal of Obstetrics and Gynaecology*, **103**:60–9.

Padez, C. and Johnston, F. (1999). Secular trends in male adult height 1904–1996 in relation to place of residence and parent's educational level in Portugal. *Annals of Human Biology*, **26**:287–98.

Palmer, C. G. S., Cronk, C., Pueschel, S. M., Wisniewski, K. R., Laxova, R., Crocker, A. C. and Pauli, R. M. (1992). Head circumference of children with Down syndrome (0–36 months). *American Journal of Medical Genetics*, **42**:61–7.

Palmert, M. R., Hayden, D. L., Mansfield, M. J., Crigler, J. F., Jr., Crowley, W. F., Jr., Chandler, D. W. and Boepple, P. A. (2001). The longitudinal study of adrenal maturation during gonadal suppression: Evidence that adrenarche is a gradual process. *Journal of Clinical Endocrinology and Metabolism*, **86**:4536–42.

Palmert, M. R., Mansfield, M. J., Crowley, W. F., Jr., Crigler, J. F., Jr., Crawford, J. D. and Boepple, P. A. (1999). Is obesity an outcome of gonadotropin-releasing hormone agonist administration? Analysis of growth and body composition in 110 patients with central precocious puberty. *Journal of Clinical Endocrinology and Metabolism*, **84**:4480–8.

Paneth, N. S. (1995). The problem of low birth weight. *Future of Children*, **5**:19–34.

Pang, S. (1984). Premature adrenarche. In *Adrenal Diseases in Childhood: Pediatric Adolescent Endocrinology*, eds. M. I. New and L. S. Levine, pp. 173–84. New York: Karger.

Pankau, R., Partsch, C.-J., Gosch, A., Oppermann, H. C. and Wessel, A. (1992). Statural growth in Williams–Beuren syndrome. *European Journal of Pediatrics*, **151**:751–5.

Pařizková, J. (1976). Growth and growth velocity of lean body mass and fat in adolescent boys. *Pediatric Research*, **10**:647–50.

Părizková, J. and Berdychová, J. (1977). The impact of ecological factors on somatic and motor development of preschool children: Growth and development physique. *Symposia Biologica Hungarica*, **20**:235–42.

Park, J. H. Y., Grandjean, C. J. and Vanderhoof, J. A. (1989). Effects of pure zinc deficiency on glucose tolerance: Role of corticosterone. *Nutrition Research*, **9**:183–93.

Park, W. J., Li, J. and Roche, A. F. (1997). The decanalization of weight, recumbent length, and head circumference during infancy. *American Journal of Human Biology*, **9**:689–98.

Parker, D. C., Jadd, H. J., Rossman, L. G. and Yen, S. S. C. (1975). Pubertal sleep–wake patterns of episodic LH, FSH and testosterone release in twin boys. *Journal of Clinical Endocrinology and Metabolism*, **40**:1099–109.

Parker, J. D. and Schoendorf, K. C. (1992). Influence of paternal characteristics on the risk of low birth weight. *American Journal of Epidemiology*, **136**:399–407.

Parker, M. W., Johanson, A. J., Rogol, A. D., Kaiser, D. L. and Blizzard, R. M. (1984). Effect of testosterone on somatomedin C concentrations in prepubertal boys. *Journal of Clinical Endocrinology and Metabolism*, **58**:87–90.

Parks, J. S., Brown, M. R., Siervogel, R. M. and Towne, B. (2000). Variation in the estrogen receptor and gene for the timing of pubertal growth. *Acta Medica Auxologica*, **32**:43.

Partington, M. W. and Roberts, N. (1969). The heights and weights of Indian and Eskimo school children on James Bay and Hudson Bay. *Canadian Medical Association Journal*, **100**:502–9.

Pasquet, P. and Ducros, A. (1980). Variation of age at menarche in France. In *Man and His Environment*, eds. L. P. Vidyarthi, I. P. Singh and S. C. Tiwari, pp. 241–50. New Delhi: Concept Publishing.

Passaro, K. T., Little, R. E., Savitz, D. A., Noss, J. and ALSPAC Study Team (1996). The effect of maternal drinking before conception and in early pregnancy on infant birthweight. *Epidemiology*, **7**:377–83.

Pastrakuljic, A., Derewlany, L. O. and Koren, G. (1999). Maternal cocaine use and cigarette smoking in pregnancy in relation to amino acid transport and fetal growth. *Placenta*, **20**:499–512.

Patterson, M. L., Stern, S., Crawford, P. B., McMahon, R. P., Similo, S. L., Schreiber, G. B., Morrison, J. A. and Waclawiw, M. A. (1997). Sociodemographic factors and obesity in preadolescent black and white girls: NHLBI's Growth and Health Study. *Journal of the National Medical Association*, **89**:594–600.

Patterson, R. M. and Pouliot, R. N. (1987). Neonatal morphometrics and perinatal outcome: Who is growth retarded? *American Journal of Obstetrics and Gynecology*, **157**:691–3.

Paz, I., Seidman, D. S., Danon, Y. L., Laor, A., Stevenson, D. K. and Gale, R. (1993). Are children born small for gestation age at increased risk of short stature? *American Journal of Diseases of Childhood*, **147**:337–9.

Peacock, J. L., Bland, M. J. and Anderson, H. R. (1991). Effects on birthweight of alcohol and caffeine consumption in smoking women. *Journal of Epidemiology and Community Health*, **45**:159–63.

Pecquery, R., Leneveu, M.-C. and Giudicelli, Y. (1988). Influence of androgenic status on the a_2/β-adrenergic control of lipolysis in isolated fat cells: Predominant a_2-antilipolytic response in testosterone treated castrated hamsters. *Endocrinology*, **122**:2590–8.

Pedersen, N. L., Friberg, L., Floderus-Myrhed, B., McClearn, G. E. and Plomin, R. (1984). Swedish early separated twins: Identification and characterization. *Acta Geneticae Medicae et Gemellologiae*, **33**:243–50.

Pedersen, S. B., Hansen, P. S., Lund, S., Andersen, P. H., Odgaard, A. and Richelsen, B. (1996). Identification of oestrogen receptors and oestrogen receptor mRNA in human adipose tissue. *European Journal of Clinical Investigation*, **26**:262–9.

Pelletier, D. L., Frongillo, E. A. and Habicht, J.-P. (1993). Epidemiologic evidence for a potentiating effect of malnutrition on child mortality. *American Journal of Public Health*, **83**:1130–3.

Pelz, L., Timm, D., Eyermann, E., Hinkel, G. K., Kirchner, M. and Verron, G. (1982). Body height in Turner's syndrome. *Clinical Genetics*, **22**:62–6.

Peoples-Sheps, M. D., Siegel, E., Suchindran, C. M., Origasa, H., Ware, A. and Barakat, A. (1991). Characteristics of maternal employment during pregnancy: Effects on low birthweight. *American Journal of Public Health*, **81**:1007–12.

Permutt, S. and Farhi, L. (1969). Tissue hypoxia and carbon monoxide. In *Effects of Chronic Exposure to*

Low Levels of Carbon Monoxide on Human Health, Behavior and Performance, ed. Committee on Effects of Atmospheric Contaminants on Human Health and Welfare, pp. 18–24. Washington, D.C.: National Academy of Science and National Academy of Engineering (United States).

Persson, I., Ahlsson, F., Ewald, U., Tuvemo, T., Qingyuan, M., von Rosen, D. and Proos, L. (1999). Influence of perinatal factors on the onset of puberty in boys and girls: Implications for interpretation of link with risk of long term diseases. *American Journal of Epidemiology*, **150**:747–55.

Persson, L. A. (1985). Infant feeding and growth: A longitudinal study in three Swedish communities. *Annals of Human Biology*, **12**:41–52.

Persson, P. H., Grennert, L., Gennser, G. and Kullanders, S. (1978). A study of smoking and pregnancy with special reference to fetal growth. *Acta Obstetrica et Gynecologica Scandinavica*, **78**:33–9.

Pérusse, L. and Bouchard, C. (1994). Genetics of energy intake and food preferences. In *The Genetics of Obesity*, ed. C. Bouchard, pp. 125–34. Boca Raton, FL: CRC Press.

Pérusse, L., Chagnon, Y. C., Weisnagel, S. J., Rankinen, T., Snyder, E., Sands, J. and Bouchard, C. (2001). The human obesity gene map: The 2000 update. *Obesity Research*, **9**:135–68.

Pérusse, L., Leblanc, C., Tremblay, A., Allard, C., Thériault, G., Landry, F., Talbot, J. and Bouchard, C. (1987). Familial aggregation in physical fitness, coronary heart disease risk factors and pulmonary function measurements. *Preventive Medicine*, **16**:607–15.

Peters, T., Golding, J. and Butler, N. R. (1983). Plus ça change: Predictors of birthweight in two national studies. *British Journal of Obstetrics and Gynaecology*, **90**:1040–5.

Petersen, S., Gotfredsen, A. and Knudsen, F. U. (1988). Lean body mass in small for gestational age and appropriate for gestational age infants. *Journal of Pediatrics*, **113**:886–9.

Petersen, S., Gotfredsen, A. and Knudsen, F. U. (1989). Total body bone mineral in light-for-gestational-age infants and appropriate-for-gestational-age infants. *Acta Paediatrica Scandinavica*, **78**:886–9.

Peterson, K. E., Rathbun, J. M. and Herrera, M. A.

(1985). Growth rate analysis in failure to thrive: Treatment and research. In *New Directions in Failure to Thrive: Implications for Research and Practice*, ed. D. Drotar, pp. 157–76. New York: Plenum Press.

Petitti, D. B. and Coleman, C. (1990). Cocaine and the risk of low birth weight. *American Journal of Public Health*, **80**:25–8.

Petrie, L., Buskin, J. N. and Chesters, J. K. (1996). Zinc and the initiation of myoblast differentiation. *Journal of Nutritional Biochemistry*, **7**:670–6.

Pettigrew, A. R., Logan, R. W. and Wiliocks, J. (1977). Smoking in pregnancy: Effects on birthweight and on cyanide and thiocyanate levels in mother and baby. *British Journal of Obstetrics and Gynaecology*, **84**:31–4.

Pettit, D. J., Bennett, P. H., William, C., Knowler, H., Baird, R. and Aleck, K. A. (1985). Gestational diabetes mellitus and impaired glucose tolerance during pregnancy: Long-term effects on obesity and glucose tolerance in the offspring. *Diabetes, Supplement*, **34**:119–22.

Peyritz, R. E. (1997). Marfan syndrome and other disorders of fibrillin. In *Emery and Rimoin's Principles and Practice of Medical Genetics*, 3rd edn, vol. 1, eds. Rimoin, D. L., Connor, J. M. and Peyritz, R. E., pp. 1027–66. New York: Churchill Livingstone.

Pfeilschifter, J., Scheodt-Nave, C., Leidig-Bruckner, G., Woitge, H. W., Blum, W. F., Wuster, C., Haack, D. and Ziegler, R. (1996). Relationship between circulating insulin-like growth factor components and sex hormones in a population-based sample of 50- to 80-year-old men and women. *Journal of Clinical Endocrinology and Metabolism*, **81**:2534–40.

Pharoah, P. O. D., Stevenson, C. J. and West, C. R. (1998). Association of blood pressure in adolescence with birthweight. *Archives of Disease in Childhood*, **79**:F114–18.

Phebus, C. K., Gloninger, M. F. and Maciak, B. J. (1984). Growth patterns by age and sex in children with sickle cell disease. *Journal of Pediatrics*, **105**:28–33.

Philipp, K., Pateisky, N. and Endler, M. (1984). Effects of smoking on uteroplacental blood flow. *Gynecologic and Obstetric Investigation*, **17**:179–82.

Phillips, D. I. W., Barker, D. J. P., Hales, C. N., Hirst, S. and Osmond, C. (1994*a*). Thinness at birth and insulin resistance in adult life. *Diabetologia*, **37**:150–4.

Phillips, D. I. W., Borthwich, A. C., Stein, C. and Taylor, R. (1996). Fetal growth and insulin resistance in adult life: Relationship between glycogen synthase activity in adult skeletal muscle and birth weight. *Diabetic Medicine*, **13**:325–9.

Phillips, D. I. W., Hirst, S., Clark, P. M. S., Hales, C. N. and Osmond, C. (1994b). Fetal growth and insulin secretion in adult life. *Diabetologia*, **37**:592–6.

Phillips, D. I. W. and Young, J. B. (2000). Birth weight, climate at birth and the risk of obesity in adult life. *International Journal of Obesity*, **24**:281–7.

Phillips, J. A. III, Hjelle, P. H., Seeburg, P. H. and Zachmann, M. (1981). Molecular basis for familial isolated growth hormone deficiency. *Proceedings of the National Academy of Sciences (U.S.A.)*, **78**:6372–5.

Phillips, K. and Matheny, A. P. Jr. (1990). Quantitative genetic analysis of longitudinal trends in height: Preliminary results from the Louisville Twin Study. *Acta Geneticae Medicae et Gemellologiae*, **39**:143–63.

Pierce, J. P., Dwyer, T., di Giusto, E., Carpenter, T., Hannam, C., Amin, A., Yong, C., Sarfaty, G., Shaw, J. and Burke, N. (1987). Cotinine validation of self-reported smoking in commercially run community survey. *Journal of Chronic Diseases*, **40**:689–95.

Pietrobelli, A., Faith, M. S., Allison, D. B., Gallagher, D., Chiumello, G. and Heymsfield, S. B. (1998). Body mass index as a measure of adiposity among children and adolescents: A validation study. *Journal of Pediatrics*, **132**:204–10.

Piro, E., Pennino, C., Cammarata, M., Corsello, G., Grenci, A., Givdice, C., Morabito, M., Piccione, M. and Giuffré, L. (1990). Growth charts of Down syndrome in Sicily: Evaluation of 382 children 0–14 years of age. *American Journal of Medical Genetics*, **7**:66–70.

Pocock, N. A., Eisman, J. A., Hopper, J. H., Yeates, M. G., Sambrook, P. N. and Eberl, S. (1987). Genetic determinants of bone mass in adults. *Journal of Clinical Investigation*, **80**:706–10.

Podar, T., Onkamo, P., Forsén, T., Karvonen, M., Tuomilehto-Wolf, E. and Tuomilehto, J. (1999). Neonatal anthropometric measurements and risk of childhood-onset type 1 diabetes. *Diabetes Care*, **22**:2092–4.

Poggi, C., Le Marchand-Brustel, Y., Zapf, J., Froesch, E. R. and Freychet, P. (1979). Effects and binding of insulin-like growth factor I in the isolated soleus muscle of lean and obese mice: Comparison with insulin. *Endocrinology*, **105**:723–30.

Polychronakos, C., Abu-Srair, H. and Guyda, H. J. (1988). Transient growth deceleration in normal short children: A potential source of bias in growth studies. *European Journal of Pediatrics*, **147**:582–3.

Pombo, M., Herrera-Justiniano, E., Considine, R. V., Hermida, R. C., Gálvez, M. J., Martin, T., Barreiro, J., Casanueva, F. F. and Diéguez, C. (1997). Nocturnal rise of leptin in normal prepubertal and pubertal children and in patients with perinatal stalk-transection syndrome. *Journal of Clinical Endocrinology and Metabolism*, **82**:2751–4.

Pomerance, H. H. and Krall, J. M. (1981). Linear regression to approximate longitudinal growth curves: Revised standards for velocity of weight and length in infants. *Pediatric Research*, **15**:1390–5.

Poosha, P. V. R., Byard, P. J., Satyanarayana, M., Rice, J. P. and Rao, D. C. (1984). Family resemblance for craniofacial measurements in Velantie Brahmins from Andhra Pradesh, India. *American Journal of Physical Anthropology*, **65**:15–22.

Potau, N., Ibanez, L., Rique, S. and Carrascosa, A. (1997). Pubertal changes in insulin secretion and peripheral insulin sensitivity. *Hormone Research*, **48**:219–26.

Power, C., Lake, J. K. and Cole, T. J. (1997). Measurement and long-term health risks of child and adolescent fatness. *International Journal of Obesity*, **21**:507–26.

Prader, A. (1978). Catch-up growth. *Postgraduate Medical Journal*, **54**:133–43.

Prader, A. and Budliger, H. (1977). Korpermasse, Wachstumsgeschwindigkeit und Knochenalter gesunder Kinder in den ersten zwolf Jahren; Longitudinale Wachstumsstudie Zurich (Body measurements, growth velocity and bone age of healthy children up to 12 years of age: Longitudinal Growth Study Zurich). *Helvetica Paediatrica Acta, Supplementum*, **37**:1–44.

Prader, A., Hernesniemi, I. and Zachman, M. (1976). Skinfold thickness in infancy and adolescence: A longitudinal correlation study in normal children. *Pediatric Adolescent Endocrinology*, 1:84–8.

Prader, A., Largo, R. H., Molinari, L. and Issler, C. (1989). Physical growth of Swiss children from birth to 20 years of age: First Zurich Longitudinal Study of Growth and Development. *Helvetica Paediatrica Acta, Supplementum*, 52:1–125.

Prader, A., Tanner, J. M. and von Harnack, G. A. (1963). Catch-up growth following illness or starvation: An example of developmental canalization in man. *Journal of Pediatrics*, 62:646–59.

Prado, C. (1984). Secular change in menarche in women in Madrid. *Annals of Human Biology*, 11:165–6.

Prahl-Andersen, B., Kowalski, C. J. and Heydendael, P. H. J. M. (eds.) (1979). *A Mixed-Longitudinal Interdisciplinary Study of Growth and Development*. London: Academic Press.

Prahl-Andersen, B. and Roede, M. J. (1979). The measurement of skeletal and dental maturity. In *A Mixed-Longitudinal Interdisciplinary Study of Growth and Development*, eds. B. Prahl-Andersen, C. J. Kowalski and P. H. J. M. Heydendael, pp. 491–519. London: Academic Press.

Prasad, A. S., Schulert, A. R., Miale, A., Farid, Z. and Sandstead, H. H. (1963). Zinc and iron deficiencies in male subjects with dwarfism and hypogonadism but without ancylostomiasis, schistosomiasis or severe anemia. *American Journal of Clinical Nutrition*, 12:437–44.

Preece, M. A. (1986). Prepubertal and pubertal endocrinology. In *Human Growth: A Comprehensive Treatise*, 2nd edn, vol. 2, *Postnatal Growth*, eds. F. Falkner and J. M. Tanner, pp. 211–24. New York: Plenum Press.

Preece, M. A. (1988). Prediction of adult height: Methods and problems. *Acta Paediatrica Scandinavica*, 347:4–11.

Preece, M. A. and Baines, M. J. (1978). A new family of mathematical models describing the human growth curve. *Annals of Human Biology*, 5:1–24.

Preece, M. A., Cameron, N., Baines-Preece, J. C. and Silman, R. (1980). Auxological and serum hormonal changes during puberty. In *Pathophysiology of Puberty*, eds. E. Cacciari and A. Prader, pp. 79–88. London: Academic Press.

Preece, M. A., Cameron, N., Donmall, M. C., Dunger, D. B., Holder, A. T., Baines-Preece, J., Seth, J., Sharp, G. and Taylor, A. M. (1984). The endocrinology of male puberty. In *Human Growth and Development*, eds. J. Borms, R. Hauspie, A. Sand, C. Susanne and M. Hebbelinck, pp. 23–37. New York: Plenum Press.

Preece, M. A. and Holder, A. T. (1982). The somatomedins: A family of serum growth factors. In *Recent Advances in Endocrinology and Metabolism*, vol. 2, ed. J. L. H. O'Riordan, pp. 47–72. Edinburgh, U.K.: Churchill Livingstone.

Presl, J., Horejsi, J., Stroufova, A. and Herzmann, J. (1976). Sexual maturation in girls and the development of estrogen-induced gonadotropic hormone release. *Annales de Biologie Animale, Biochimie, Biophysique*, 16:377–83.

Price, D. A., Shalet, S. M. and Clayton, P. E. (1988). Management of idiopathic growth hormone deficient patients during puberty. *Acta Paediatrica Scandinavica*, 347:44–51.

Price, R. A., Cadoret, R. J., Stunkard, A. J. and Troughton, E. (1987). Genetic contributions to human fatness: An adoption study. *American Journal of Psychiatry*, 144:1003–8.

Price, R. A., Ness, R. and Laskarzewski, P. (1990). Common major gene inheritance of extreme overweight. *Human Biology*, 62:747–65.

Price, R. A., Ness, R. and Sørensen, T. I. A. (1991). Changes in commingled body mass index distributions associated with secular trends in overweight among Danish young men. *American Journal of Epidemiology*, 133:501–10.

Price, T. M., O'Brien, S. N., Welter, B. H., George, R., Anandjiwala, J. and Kilgore, M. (1998). Estrogen regulation of adipose tissue lipoprotein lipase: Possible mechanism of body fat distribution. *American Journal of Obstetrics and Gynecology*, 178:101–7.

Prokopec, M. (1977). Nomogramm zur Bestimmung der mittleren Höhe von Eltern sowie der Korrection der Höhe von Kindern nach der Höhe ihrer Eltern (Nomogram for the definition of mid-parent stature and the correction of the statures of children for mid-parent stature). *Ärztliche Jugendkunde*, 68: 77–80.

Prokopec, M. (1984). Secular trends in body size and proportions, and their biological meaning. *Studies in Human Ecology*, **6**:37–61.

Prokopec, M. and Bellisle, F. (1992). Body mass index variations from birth to adulthood in Czech youths. *Acta Medica Auxologica* **24**:87–93.

Proos, L. A. (1993). Anthropometry in adolescence: Secular trends, adoption, ethnic and environmental differences. *Hormone Research, Supplement*, **39**:18–24.

Proos, L. A., Hofvander, Y. and Tuvemo, T. (1991). Menarcheal age and growth pattern of Indian girls adopted in Sweden. II. Catch-up growth and final height. *Indian Journal of Pediatrics*, **58**:105–14.

Province, M. A., Arnqvist, P., Keller, J., Higgins, M. and Rao, D. C. (1990). Strong evidence for a major gene for obesity in the large, unselected, total Community Health Study of Tecumseh. *American Journal of Human Genetics*, **47**:A143.

Prue, D. M., Martin, J. E. and Hume, A. S. (1980). An initial evaluation of thiocyanate as a biochemical index of smoking exposure. *Behavior Therapy*, **11**:368–79.

Pryor, J. (1992). Physical and development status of preschool small-for-gestational-age children: A comparative study. *Journal of Paediatrics and Child Health*, **28**:38–44.

Pryor, J. (1996). The identification and long-term effects of fetal growth restriction. *British Journal of Obstetrics and Gynaecology*, **103**:1116–22.

Pucci, E., Chiovato, L. and Pinchera, A. (2000). Thyroid and lipid metabolism. *International Journal of Obesity*, **24**:S109–12.

Pugliese, M. T., Weyman-Daum, M. W., Moses, N. and Lifshitz, F. (1987). Parental health beliefs as a cause of non-organic failure to thrive. *Pediatrics*, **80**:175–82.

Qin, T., Shohoji, T. and Sumiya, T. (1996). Relationship between adult stature and timing of the pubertal growth spurt. *American Journal of Human Biology*, **8**:417–26.

Querec, L. J. (1980). *Comparability of Reporting between the Birth Certificate and the National Natality Survey*, Vital and Health Statistics, Series 2, no. 83. Washington, D.C.: U.S. Department of Health, Education, and Welfare.

Quivers, E. S., Driscoll, D. J., Garvey, C. D., Harris, A. M., Harrison, J., Huse, D. M., Murtaugh, P. and Weidman, W. H. (1992). Variability in response to a low-fat, low-cholesterol diet in children with elevated low-density lipoprotein cholesterol levels. *Pediatrics*, **89**:925–9.

Rabinowicz, T. (1986). The differentiated maturation of the human cerebral cortex. In *Human Growth: A Comprehensive Treatise*, 2nd edn, vol. 2, *Postnatal Growth*, eds. F. Falkner and J. M. Tanner, pp. 385–410. New York: Plenum Press.

Rabkin, C. S., Anderson, H. R., Bland, J. M., Brooke, O. G., Chamberlain, G. and Peacock, J. L. (1990). Maternal activity and birth weight: A prospective, population-based study. *American Journal of Epidemiology*, **131**:522–31.

Raine, T., Powell, S. and Krohn, M. A. (1994). The risk of repeating low birth weight and the role of prenatal care. *Obstetrics and Gynecology*, **84**:485–9.

Ramakrishnan, U., Martorell, R., Schroeder, D. G. and Flores, R. (1999). Role of intergenerational effects on linear growth. *Journal of Nutrition Supplement*, **129**: S544–9.

Ramer, C. M. and Lodge, A. (1975). Neonatal addiction: A two year study. I. Clinical and developmental characteristics of infants of mothers on methadone maintenance. *Addictive Diseases*, **2**:227–34.

Ramirez, G., Grimes, R. M., Annegers, J. F., Davis, B. R. and Slater, C. H. (1990). Occupational physical activity and other risk factors for preterm birth among US Army primigravidas. *American Journal of Public Health*, **80**:728–30.

Ramsay, J. O., Bock, R. D. and Gasser, T. (1995). Comparison of height acceleration curves in the Fels, Zurich, and Berkeley growth data. *Annals of Human Biology*, **22**:413–26.

Ramsay, M., Gisel, E. G. and Boutry, M. (1993). Non-organic failure to thrive: Growth failure secondary to feeding-skills disorder. *Developmental Medicine and Child Neurology*, **35**:285–97.

Ramsey, J. J., Kemnitz, J. W., Colman, R. J., Cunningham, D. and Swick, A. G. (1998). Different central and peripheral responses to leptin in Rhesus monkeys: Brain transport may be limited. *Journal of Clinical Endocrinology and Metabolism*, **83**:3230–5.

Ranke, M. B., Chavez-Meyer, H., Blank, B., Frisch, H. and Haeusler, G. (1991). Spontaneous growth and bone age development in Turner syndrome: Results of a multicentric study 1990. In *Turner Syndrome: Growth Promoting Therapies*, eds. M. B. Ranke and R. G. Rosenfeld, pp. 101–6. Amsterdam, The Netherlands: Elsevier Science.

Ranke, M. B., Heidemann, P., Knupfer, C., Enders, H., Schmaltz, A. A. and Bierich, J. R. (1988). Noonan syndrome: Growth and clinical manifestations in 144 cases. *European Journal of Pediatrics*, 148:220–7.

Rantakallio, P. (1978). The effect of maternal smoking on birth weight and the subsequent health of the child. *Early Human Development*, 2:371–82.

Rantakallio, P. (1983). A follow-up study up to the age of 14 of children whose mothers smoked during pregnancy. *Acta Paediatrica Scandinavica*, 72:747–53.

Rantakallio, P., Läärä, E., Koiranen, M. and Sarpola, A. (1995). Maternal build and pregnancy outcome. *Journal of Clinical Endocrinology and Metabolism*, 48:199–207.

Rao, D. C. and Morton, N. E. (1974). Path analysis of family resemblance in the presence of gene environment interaction. *American Journal of Human Genetics*, 26:767–72.

Rappaport, R. (1988). Endocrine control of growth. In *Linear Growth Retardation in Less Developed Countries*, ed. J. C. Waterlow, pp. 109–26. New York: Raven Press.

Rarick, G. L. and Seefeldt, V. (1974). Observations from longitudinal data on growth in stature and sitting height of children with Down's syndrome. *Journal of Mental Deficiency Research*, 18:63–78.

Ratcliffe, S. G., Axworthy, D. and Ginsborg, A. (1979). The Edinburgh study of growth and development in children with sex chromosome abnormalities. *Birth Defects: Original Article Series*, 15:243–60.

Rauch, F., Klein, K., Allolio, B. and Schönau, E. (1999). Age at menarche and cortical bone geometry in premenopausal women. *Bone*, 25:69–73.

Ravelli, G. P., Stein, Z. A. and Susser, M. W. (1976). Obesity in young men after famine exposure *in utero* and early infancy. *New England Journal of Medicine*, 295:349–53.

Rawlings, D. J., Cooke, R. J., McCormick, K., Griffin, I. J., Faulkner, K., Wells, J. C. K., Smith, J. S. and Robinson, S. J. (1999). Body composition of preterm infants during infancy. *Archives of Disease in Childhood*, 80:F188–91.

Reading, R., Raybould, S. and Jarvis, S. (1993). Deprivation, low birth weight, and children's height: Comparison between rural and urban areas. *British Medical Journal*, 307:1458–62.

Rebuffé-Scrive, M., Brönnegård, M., Nilsson, A., Eldh, J., Gustafsson, J.-A. and Björntorp, P. (1990). Steroid hormone receptors in human adipose tissues. *Journal of Clinical Endocrinology and Metabolism*, 71:1215–9.

Rebuffé-Scrive, M., Lundholm, K. and Björntorp, P. (1985). Glucocorticoid hormone binding to human adipose tissue. *European Journal of Clinical Investigation*, 15:267–72.

Rechler, M. M. (1993). Insulin-like growth factor binding proteins. *Vitamins and Hormones*, 47:1–114.

Reif, S., Beler, B., Villa, Y. and Spirer, Z. (1995). Long-term follow-up and outcome of infants with non-organic failure to thrive. *Israel Journal of Medical Sciences*, 31:483–9.

Reilly, J. J., Dorosty, A. R., Emmett, P. M. and The ALSPAC Study Team. (2000). Identification of the obese child: Adequacy of the body mass index for clinical practice and epidemiology. *International Journal of Obesity*, 24:1623–7.

Reinken, L., Stolley, H., Droese, W. and van Oost, G. (1979). Longitudinal growth data of development of weight, height, skinfold thickness, head-, chest- and abdominal circumferences in healthy children. *Klinische Pädiatrie*, 191:556–65.

Reinken, L. and van Oost, G. (1992). Longitudinale köperentwicklung gesunder Kinder von 0 bis 18 jahren: Körperlänge/-höhe, Köpergewicht und Wachstumgeschwindigkeit (Longitudinal study of the development of healthy children from zero to 18 years: body length/height, body weight and growth velocity). *Klinische Pädiatrie*, 204:129–33.

Reiter, E. O., Kulin, H. E. and Hamwood, S. M. (1974). The absence of positive feedback between estrogen and luteinizing hormone in sexually immature girls. *Pediatric Research*, 8:740–5.

Reiterer, E. E., Sudi, K. M., Mayer, A., Limbert-Zinterl, C., Stalzer-Brunner, C., Füger, G. and Borkenstein, M. H. (1999). Changes in leptin, insulin and body composition in obese children during a weight reduction program. *Journal of Pediatric Endocrinology and Metabolism*, 12:853–62.

Rensonnet, C., Kanen, F., Coremans, C., Ernould, C., Albert, A. and Bourguignon, J.-P. (1999). Pubertal growth as a determinant of adult height in boys with constitutional delay of growth and puberty. *Hormone Research*, 51:223–9.

Reynolds, E. L. (1943). Degree of kinship and pattern of ossification. *American Journal of Physical Anthropology*, 1:405–16.

Reynolds, E. L. and Sontag, L. W. (1944). Seasonal variations in weight, height, and appearance of ossification centers. *Journal of Pediatrics*, 24: 524–35.

Reynolds, E. L. and Wines, J. V. (1948). Individual differences in physical changes associated with adolescence in girls. *American Journal of Diseases in Children*, 75:329–50.

Reynolds, E. L. and Wines, J. V. (1951). Physical changes associated with adolescence in boys. *American Journal of Diseases of Children*, 82:529–47.

Rhoads, G. G. and Kagan, A. (1983). The relation of coronary disease, stroke, and mortality to weight in youth and middle age. *Lancet*, i:492–5.

Rice, T., Borecki, I. B., Bouchard, C. and Rao, D. C. (1993a). Segregation analysis of body mass index in an unselected French–Canadian sample: The Quebec Family Study. *Obesity Research*, 1:288–94.

Rice, T., Borecki, I. B., Bouchard, C. and Rao, D. C. (1993b). Segregation analysis of fat mass and other body composition measures derived from underwater weighing. *American Journal of Human Genetics*, 52:967–73.

Richard, D., Huang, Q. and Timofeeva, E. (2000). The corticotropin-releasing hormone system in the regulation of energy balance in obesity. *International Journal of Obesity*, 24:S36–9.

Rich-Edwards, J. W., Colditz, G. A., Stampfer, M. J., Willett, W. C., Gillman, M. W., Hennekens, C. H., Speizer, F. E. and Manson, J. E. (1999). Birthweight and the risk for type 2 diabetes mellitus in adult women. *Annals of Internal Medicine*, 130:278–84.

Rich-Edwards, J. W., Manson, J. E., Stampfer, M. J., Colditz, G. A., Willett, W. C., Rosner, B., Speizer, F. E. and Hennekens, C. H. (1995b). Height and the

risk of cardiovascular disease in women. *American Journal of Epidemiology*, 142:909–17.

Rich-Edwards, J. W., Stampfer, M., Manson, J., Rosner, B., Colditz, G. A., Willett, W. C. and Speizer, F. E. (1995a). Birthweight, breastfeeding and the risk of coronary heart disease in the Nurses' Health Study. *American Journal of Epidemiology, Supplement*, 141:78.

Richman, R. A. and Kirsch, L. R. (1988). Testosterone treatment in adolescent boys with constitutional delay in growth and development. *New England Journal of Medicine*, 319:1563–7.

Richter, J. (1980). On the relationship between the number of sibs, rank in sibship and the weight development in children and juveniles. *Anthropologiai Közlemények*, 24:209–12.

Richter, J. (1981). Deutet sich das Ende der Sexualakzeleration an? (Is the end of the sexual acceleration in sight?). *Zeitschrift für die gesamte Hygiene und ihre Grenzgebiete*, 27:485–7.

Richter, J. (1989). Ergebnisse langfristiger Entwicklungsbeobachtungen bei Mädchen (Long-term observations of development in girls). *Sozialpädiatrie in Praxis und Klinik*, 11:650–7.

Richter, J. and Kern, G. (1980). Über Menarchebeziehungen zweier Generationen-Untersuchungen an Müttern und ihren Töchtern in Görlitz (On age at menarche across two generations: A study of mothers and their daughters in Görlitz). *Ärztliche Jugendkunde*, 71:13–20.

Rico, H., Revilla, M., Hernandez, E. R., Villa, L. F. and Delbuergo, M. A. (1992). Sex differences in the acquisition of total bone mineral mass peak assessed through dual-energy X-ray absorptiometry. *Calcified Tissue International*, 51:251–4.

Rico, H., Revilla, M., Villa, L. F., Hernandez, E. R., de Buergo, M. A. and Villa, M. (1993). Body composition in children and Tanner's Stages: A study with dual-energy X-ray absorptiometry. *Metabolism: Clinical and Experimental*, 42:967–70.

Riedel, M., Hoeft, B., Blum, W. F., von zur Muhlen, A. and Brabant, G. (1995). Pulsatile growth hormone secretion in normal-weight and obese men: Differential metabolic regulation during energy restriction. *Metabolism: Clinical and Experimental*, 44:606–10.

Rimm, E. B., Stampfer, M. J., Giovannucci, E., Ascherio, A., Spiegelman, D., Colditz, G. A. and Willett, W. C. (1995). Body size and fat distribution as predictors of coronary heart disease among middle-aged and older U.S. men. *American Journal of Epidemiology*, **141**:1117–27.

Rimoin, D. L. (1976). Hereditary forms of growth hormone-deficiency and resistance. *Birth Defects*, **12**:15–29.

Rizza, R. A., Mandarino, L. J. and Gerich, J. E. (1982). Effects of growth hormone on insulin action in man. *Diabetes*, **31**:663–9.

Rizzoni, G., Basso, T. and Setari, M. (1984). Growth in children with chronic renal failure on conservative treatment. *Kidney International*, **26**:52–8.

Robbins, J. and Rall, J. E. (1983). The iodine containing hormones. In *Hormones in Blood*, 3rd edn, vol. 4, eds. C. H. Gray and V. H. T. James, pp. 219–65. New York: Academic Press.

Roberts, C. L. and Lancaster, P. (1999). National birthweight percentiles by gestational age for twins born in Australia. *Journal of Paediatrics and Child Health*, **35**:278–82.

Roberts, D. F., Rozner, L. M. and Swan, A. V. (1971). Age at menarche, physique, and environment in industrial north-east England. *Acta Paediatrica Scandinavica*, **60**:158–64.

Roberts, N., Hogg, D., Whitehouse, G. H. and Dangerfield, P. (1998). Quantitative analysis of diurnal variation in volume and water content of lumbar intervertebral discs. *Clinical Anatomy*, **11**:1–8.

Robinson, T. N., Hammer, L. D., Killen, J. D., Kraemer, H. C., Wilson, D. M., Hayward, C. and Taylor, C. B. (1993). Does television viewing increase obesity and reduce physical activity? Cross-sectional and longitudinal analyses among adolescent girls. *Pediatrics*, **91**:273–80.

Roche, A. F. (1965). The stature of mongols. *Journal of Mental Deficiency Research*, **9**:131–45.

Roche, A. F. (1968). Sex-associated differences in skeletal maturity. *Acta Anatomica*, **71**:321–40.

Roche, A. F. (1974). Differential timing of maximum length increments among bones within individuals. *Human Biology*, **46**:145–57.

Roche, A. F. (1980). The measurement of skeletal maturation. In *Human Physical Growth and Maturation: Methodologies and Factors*, eds. F. E. Johnston, A. F. Roche and C. Susanne, pp. 61–82. New York: Plenum Press.

Roche, A. F. (1989). The final phase of growth in stature. *Growth Genetics and Hormones*, **5**:4–6.

Roche, A. F. (1992). *Growth, Maturation and Body Composition: The Fels Longitudinal Study 1929–1991*. Cambridge, U.K.: Cambridge University Press.

Roche, A. F., Chumlea, W. C. and Thissen, D. (1988). *Assessing the Skeletal Maturity of the Hand–Wrist: Fels Method*. Springfield, IL: Charles C. Thomas.

Roche, A. F. and Davila, G. H. (1972). Late adolescent growth in stature. *Pediatrics*, **50**:874–80.

Roche, A. F. and Guo, S. S. (1998). Selection of variables, timing of examinations, and retention. In *Clinical Trials in Infant Nutrition*, eds. J. A. Perman and J. Rey, pp. 67–83. Philadelphia, PA: Lippincott–Raven Press.

Roche, A. F., Guo, S. and Moore, W. M. (1989*a*). Weight and recumbent length from 1 to 12 mo of age: Reference data for 1-mo increments. *American Journal of Clinical Nutrition*, **49**:599–607.

Roche, A. F., Guo, S., Siervogel, R. M., Khamis, H. J. and Chandra, R. K. (1993). Growth comparison of breast-fed and formula-fed infants. *Canadian Journal of Public Health*, **84**:132–5.

Roche, A. F., Guo, S. and Yeung, D. L. (1989*b*). Monthly growth increments from a longitudinal study of Canadian infants. *American Journal of Human Biology*, **1**:271–9.

Roche, A. F., Guo, S. S., Wholihan, K. and Casey, P. H. (1997). Reference data for head circumference-for-length in preterm low-birth-weight infants. *Archives of Pediatrics and Adolescent Medicine*, **151**:50–7.

Roche, A. F., Heymsfield, S. B. and Lohman, T. G. (1996). *Human Body Composition*. Champaign, IL: Human Kinetics.

Roche, A. F. and Himes, J. H. (1980). Incremental growth charts. *American Journal of Clinical Nutrition*, **33**:2041–52.

Roche, A. F., Mukherjee, D., Guo, S. and Moore, W. M. (1987). Head circumference reference data: Birth to 18 years. *Pediatrics*, **79**:706–12.

Roche, A. F., Mukherjee, D. and Guo, S. S. (1986). Head circumference growth patterns: Birth to 18 years. *Human Biology*, 58:893–906.

Roche, A. F., Roberts, J. and Hamill, P. (1975a). *Skeletal Maturity of Children 6–11 Years: Racial, Geographic Area, and Socioeconomic Differentials, United States*, Vital and Health Statistics, Series 11, no. 149. Washington, D.C.: Department of Health, Education, and Welfare.

Roche, A. F., Roberts, J. and Hamill, P. V. V. (1976). *Skeletal Maturity of Youths 12–17 Years: United States*, Vital and Health Statistics, Series 11, no. 160. Washington, D.C.: U.S. Department of Health, Education, and Welfare.

Roche, A. F., Roberts, J. and Hamill, P. W. (1978). *Skeletal maturity of Youths 12–17 Years: Racial, Geographic Area, and Socioeconomic Differentials, United States 1966–1970*, Vital and Health Statistics, Series 11, no. 167. Washington, D.C.: Department of Health, Education, and Welfare.

Roche, A. F., Siervogel, R. M., Chumlea, W. C., Reed, R. B., Valadian, I., Eichorn, D. and McCammon, R. W. (1982). *Serial Changes in Subcutaneous Fat Thicknesses of Children and Adults*, Monographs in Pediatrics no. 17. Basel, Switzerland: Karger.

Roche, A. F., Siervogel, R. M., Chumlea, W. C. and Webb, P. (1981). Grading body fatness from limited anthropometric data. *American Journal of Clinical Nutrition*, 34:2831–8.

Roche, A. F., Wainer, H. and Thissen, D. (1975b). *Predicting Adult Stature for Individuals*, Monographs in Paediatrics no. 3. Basel, Switzerland: Karger.

Roche, A. F., Wellens, R., Attie, K. M. and Siervogel, R. M. (1995). The timing of sexual maturation in a group of U.S. white youths. *Journal of Pediatric Endocrinology and Metabolism*, 8:11–18.

Rochiccioli, P., Pienkowski, C. and Tauber, M. T. (1991). Spontaneous growth in Turner syndrome: A study of 61 cases. In *Turner Syndrome: Growth Promoting Therapies*, Proceedings of a Workshop on Turner Syndrome, Frankfurtam Main, eds. M. B. Ranke and R. G. Rosenfeld, pp. 107–12. Amsterdam, The Netherlands: Excerpta Medica.

Rochiccioli, P., Tauber, M., Moisan, V. and Pienkowski, C. (1989). Investigation of growth hormone secretion in patients with intrauterine growth retardation. *Acta Paediatrica Scandinavica*, 349:42–6.

Rodrigues, S., Robinson, E. J., Kramer, M. S. and Gray-Donald, K. (2000). High rates of infant macrosomia: A comparison of a Canadian native and a non-native population. *Journal of Nutrition*, 130:806–12.

Rodriguez-Soriano, J., Vallo, V. and Martul, P. (1992). Growth in renal insufficiency. In *Human Growth: Basic and Clinical Aspects*, eds. M. Hernandez and J. Argente, pp. 111–16. Amsterdam, The Netherlands: Elsevier Science.

Roede, M. J. (1990). The secular trend in the Netherlands: The Third Nation-wide Growth Study. *Ärztliche Jugendkunde*, 81:330–6.

Roede, M. J. and van Wieringen, J. C. (1985). Growth Diagrams 1980: Netherlands Third Nation-wide Survey. *Tijdeschrift voor Sociale Gezondheidszorg*, 63:1–34.

Roemmich, J. N., Clark, P. A., Mai, V., Berr, S. S., Weltman, A., Veldhuis, J. D. and Rogol, A. D. (1998). Alterations in growth and body composition during puberty. III. Influence of maturation, gender, body composition, fat distribution, aerobic fitness, and energy expenditure on nocturnal growth hormone release. *Journal of Clinical Endocrinology and Metabolism*, 83:1440–7.

Rogers, A. (1984). The effects of illness on growth during childhood and adolescence. PhD thesis, University of Oxford, U.K.

Rogol, A. D. (1989). Growth hormone: Physiology, therapeutic use, and potential for abuse. *Exercise and Sport Sciences Reviews*, 17:353–77.

Rohner-Jeanrenaud, F. (2000). Hormonal regulation of energy partitioning. *International Journal of Obesity*, 24:S2–7.

Rohner-Jeanrenaud, F., Walker, C. D., Greco-Perotto, R. and Jeanrenaud, B. (1989). Central corticotropin-releasing factor administration prevents the excessive body weight gain of genetically obese (*fa/fa*) rats. *Endocrinology*, 124:733–9.

Rolland-Cachera, M.-F. and Bellisle, F. (1986). No correlation between adiposity and food intake: Why are working-class children fatter? *American Journal of Clinical Nutrition*, 44:779–87.

Rolland-Cachera, M.-F., Bellisle, F. and Sempé, M. (1989). The prediction in boys and girls of the weight/height^2 index and various skinfold measurements in adults: A two-decade follow-up study. *International Journal of Obesity*, 13: 305–11.

Rolland-Cachera, M.-F., Cole, T. J., Sempé, M., Tichet, J., Rossignol, C. and Charraud, A. (1991). Body mass index variations: Centiles from birth to 87 years. *European Journal of Clinical Nutrition*, 45:13–21.

Rolland-Cachera, M.-F., Deheeger, M., Akrout, M. and Bellisle, F. (1995). Influence of macronutrients on adiposity development: A follow-up study of nutrition and growth from 10 months to 8 years of age. *International Journal of Obesity*, 19:573–8.

Rolland-Cachera, M.-F., Deheeger, M. and Bellisle, F. (1996). Nutrient balance and android body fat distribution: Why not a role for protein? *American Journal of Clinical Nutrition*, 64:663–4.

Rolland-Cachera, M.-F., Deheeger, M., Bellisle, F., Sempé, M., Guilloud-Bataille, M. and Patois, E. (1984). Adiposity rebound in children: A simple indicator for predicting obesity. *American Journal of Clinical Nutrition*, 39:129–35.

Rolland-Cachera, M.-F., Deheeger, M., Guilloud-Bataille, M., Avons, P., Patois, E. and Sempé, M. (1987). Tracking the development of adiposity from one month of age to adulthood. *Annals of Human Biology*, 14:219–29.

Rolland-Cachera, M.-F., Sempé, M., Guillod-Bataille, M., Patois, E., Péquignot-Guggenbuhl, F. and Fautrat, V. (1982). Adiposity indices in children. *American Journal of Clinical Nutrition*, 36:178–84.

Rona, R. J. (1981). Genetic and environmental factors in the control of growth in childhood. *British Medical Bulletin*, 37:265–72.

Rona, R. J. and Chinn, S. (1982). National Study of Health and Growth: Social and family factors and obesity in primary school children. *Annals of Human Biology*, 9:137–45.

Rona, R. J. and Chinn, S. (1986). National Study of Health and Growth: Social and biological factors associated with height of children from ethnic groups living in England. *Annals of Human Biology*, 13:453–71.

Rona, R. J. and Chinn, S. (1987). National Study of Health and Growth: Social and biological factors associated with weight-for-height and triceps skinfold of children from ethnic groups in England. *Annals of Human Biology*, 14:231–48.

Rona, R. J. and Chinn, S. (1995). Genetic and environmental influences on growth. *Journal of Medical Screening*, 5:133–9.

Rona, R. J., Chinn, S. and du Ve Florey, C. (1985). Exposure to cigarette smoking and children's growth. *International Journal of Epidemiology*, 14:402–9.

Rona, R. J., Swan, A. V. and Altman, D. G. (1978). Social factors and height of primary schoolchildren in England and Scotland. *Journal of Epidemiology and Community Health*, 32:147–54.

Rongen-Westerlaken, C., Corel, L., van den Broeck, J., Massa, G., Karlberg, J., Albertsson-Wikland, K., Naeraa, R. W. and Wit, J. M. (1997). Dutch and Swedish study groups for GH treatment: Reference values for height, height velocity and weight in Turner's syndrome. *Acta Paediatrica*, 86:937–42.

Rönnemaa, T., Knip, T., Lautala, P., Viikari, J., Uhari, M., Leino, A., Kaprio, E. A., Salo, M. K., Dahl, M., Nuutinen, E. M., Pesonen, E., Pietikäinen, M. and Åkerblom, H. K. (1991). Serum insulin and other cardiovascular risk indicators in children, adolescents and young adults. *Annals of Medicine*, 23:67–72.

Room, M. R. (1977). Measurement and distribution of drinking patterns and problems in general populations. In *Alcohol Related Disabilities*, eds. G. Edwards, M. M. Gross, M. Keller, J. Moser and R. Room, pp. 61–87. Geneva, Switzerland: World Health Organization.

Root, A. W., Duckett, G., Sweetland, M. and Reiter, E. O. (1979). Effects of zinc deficiency upon pituitary function in sexually mature and immature male rats. *Journal of Nutrition*, 109:958–64.

Rose, S. R., Municchi, G., Barnes, K. M., Kamp, G. A., Uriarte, M. M., Ross, J. L., Cassorla, F. and Cutler, G. B., Jr. (1991). Spontaneous growth hormone secretion increases during puberty in normal girls and boys. *Journal of Clinical Endocrinology and Metabolism*, 69:428–35.

Rosell, S. and Belfrage, E. (1979). Blood circulation in adipose tissue. *Physiological Reviews*, 59:1078–1104.

Rosen, C. J. and Adler, R. A. (1992). Longitudinal changes in lumbar bone density among thyrotoxic patients after attainment of euthyroidism. *Journal of Clinical Endocrinology and Metabolism*, **75**:1531–4.

Rosen, T., Bosaeus, I., Tolli, J., Lindstedt, G. and Bengtsson, B. A. (1993). Increased body fat mass and decreased extracellular fluid volume in adults with growth hormone deficiency. *Clinical Endocrinology*, **38**:63–71.

Rosen, T. S. and Johnson, H. L. (1982). Children of methadone maintained mothers: Follow-up to 18 months of age. *Journal of Pediatrics*, **101**:192–6.

Rosenbaum, M. and Gertner, J. M. (1989). Metabolic clearance rates of synthetic human growth hormone in children, adult women, and adult men. *Journal of Clinical Endocrinology and Metabolism*, **69**:820–4.

Rosenbaum, M., Nicolson, M., Hirsch, J., Heymsfield, S. B., Gallagher, D., Chu, F. and Leibel, R. L. (1996). Effects of gender, body composition, and menopause on plasma concentrations of leptin. *Journal of Clinical Endocrinology and Metabolism*, **81**:3424–7.

Rosenberg, M. (1988). Birth weights in three Norwegian cities,1860–1984: Secular trends and influencing factors. *Annals of Human Biology*, **15**:275–88.

Rosenberg, S. N., Verzo, B., Engström, J. L., Kavanaugh, K. and Meier, P. P. (1992). Reliability of length measurements for preterm infants. *Neonatal Network*, **11**:23–7.

Rosenfeld, R. G. (1982). Evaluation of growth and maturation in adolescence. *Pediatrics in Review*, **4**:175–83.

Rosenfield, R. I., Furlanetto, R. and Bock, D. (1983). Relationship of somatomedin-C concentrations to pubertal changes. *Journal of Pediatrics*, **103**:723–8.

Rosett, H. L., Weiner, L., Lee, A., Zuckerman, B., Dooling, E. and Oppenheimer, E. (1983). Patterns of alcohol consumption and fetal development. *Obstetrics and Gynecology*, **61**:539–46.

Rosett, H. L., Weiner, L., Zukerman, B., McKinlay, S. and Edelin, K. C. (1980). Reduction of alcohol consumption during pregnancy with benefits to the newborn. *Alcoholism, Clinical and Experimental Research*, **4**:178–84.

Ross, J. G. and Pate, R. R. (1987). The National Children and Youth Fitness Study II: A summary of findings. *Journal of Physical Education, Recreation, and Dance*, **58**:51–6.

Ross, J. L., Cassorla, F. G., Skerda, M. C., Valk, I. M., Loriaux, D. L. and Cutler, G. B. (1983). A preliminary study of the effect of estrogen dose on growth in Turner's syndrome. *New England Journal of Medicine*, **309**:1104–6.

Rostand, A., Kaminski, M., Lelong, N., Dehaene, P., Delestret, I., Klein-Bertrand, C., Querleu, D. and Crepin, G. (1990). Alcohol use in pregnancy, craniofacial features, and fetal growth. *Journal of Epidemiology and Community Health*, **44**:302–6.

Roth, H. P. and Kirchgessner, M. (1997). Course of concentration changes of growth hormone, IGF-1, insulin and C-peptide in serum, pituitary and liver of zinc-deficient rats. *Journal of Animal Physiology and Animal Nutrition*, **77**: 91–101.

Roti, E., Minelli, R. and Salvi, M. (2000). Thyroid hormone metabolism in obesity. *International Journal of Obesity*, **24**:S113–15.

Rubin, D. H., Krasilnikoff, P. A., Leventhal, J. M., Weile, B. and Berget, A. (1986). Effect of passive smoking on birthweight. *Lancet*, **ii**:405–17.

Rubin, K., Schirduan, V., Gendreau, P., Sarfarazi, M., Mendola, R. and Dalsky, G. (1993). Predictors of axial and peripheral bone mineral density in healthy children and adolescents, with special attention to the role of puberty. *Journal of Pediatrics*, **123**:863–70.

Rudd, B. T., Rayner, P. H. and Thomas, P. H. (1986). Observations on the role of GH/IGF-1 and sex hormone-binding globulin (SHBG) in the pubertal development of growth hormone- deficient (GHD) children. *Acta Endocrinologica*, **279**:164–9.

Rudolph, A. J. (1985). Failure to thrive in the perinatal period. *Acta Paediatrica Scandinavica*, **319**:55–61.

Rummler, S. and Woit, I. (1990). Zur postnatalen Entwicklung von Kindern adipöser Mutter (The postnatal growth of the children of obese mothers). *Zeitschrift für Geburtshilfe und Perinatologie*, **194**:254–9.

Rush, D. (1974). Examination of the relationship between birthweight, cigarette smoking during pregnancy and maternal weight gain. *Journal of Obstetrics and Gynaecology of the British Commonwealth*, **81**:746–52.

Rush, D. and Cassano, P. (1983). Relationship between cigarette smoking and social class to birthweight and perinatal mortality among all births in Britain 5–11 April 1970. *Journal of Epidemiology and Community Health*, 37:249–55.

Russell, L. B., Badgett, S. K. and Saylors, C. L. (1960). Comparison of the effects of acute, continuous and fractionated irradiation during embryonic development. In *Immediate and Low Level Effects of Ionizing Irradiation*, ed. A. F. Buzzati-Traverso, pp. 343–59. London: Taylor & Francis.

Russo, E. G. and Zaccagni, L. (1993). Growth curves for Italian children from birth to 3 years: A sample from North Italy (Faenza, Ravenna). *Acta Medica Auxologica* 25:143–51.

Ruther, N. M. and Richman, C. L. (1993). The relationship between mothers' eating restraint and their children's attitudes and behaviors. *Bulletin of the Psychosomatic Society*, 31:217–20.

Rutledge, M. M., Clark, J., Woodruff, C., Krause, G. and Flynn, M. A. (1976). A longitudinal study of total body potassium in normal breastfed and bottle-fed infants. *Pediatric Research*, 10:114–17.

Ruxton, C. H. S., Reilly, J. J. and Kirk, T. R. (1999). Body composition of healthy 7- and 8-year-old children and a comparison with the 'reference child'. *International Journal of Obesity*, 23:1276–81.

Ruz, M. O., Castillo-Durán, C., Lara, X., Codoceo, J., Rebolledo, A. and Atalah, E. (1997). A 14-month zinc supplementation trial in apparently healthy Chilean preschool children. *American Journal of Clinical Nutrition*, 66:1406–13.

Ryan, A. S. (1997). The resurgence of breastfeeding in the United States. *Pediatrics*, 99:4–12.

Ryan, A. S., Martinez, G. A. and Roche, A. F. (1990). An evaluation of the associations between socioeconomic status and the growth of Mexican-American children: Data from the Hispanic Health and Nutrition Examination Survey (HHANES 1982–1984). *American Journal of Clinical Nutrition Supplement*, 51:944–52.

Ryan, L., Ehrlich, S. and Finnegan, L. (1987). Cocaine abuse in pregnancy: Effects on the fetus and newborn. *Neurotoxicology and Teratology*, 9:295–9.

Rydhström, H. (1992). Gestational duration and birth weight for twins related to fetal sex. *Gynecologic and Obstetric Investigation*, 33:90–3.

Saad, M. F., Riad-Gabriel, M. G., Khan, A., Sharma, A., Michael, R., Jinagouda, S. D., Boyadjian, R. and Steil, G. M. (1998). Diurnal and ultradian rhythmicity of plasma leptin: Effects of gender and adiposity. *Journal of Clinical Endocrinology and Metabolism*, 83:453–9.

Sack, J. (1996). The thyroid gland. In *Pediatrics and Perinatology: The Scientific Basis*, eds. P. D. Gluckman and M. A. Heymann, pp. 478–85. London: Edward Arnold.

Saggese, G., Baroncelli, G. I., Bertelloni, S. and Barsanti, S. (1996). The effect of long-term growth hormone (GH) treatment on bone mineral density in children with GH deficiency: Role of GH in the attainment of peak bone mass. *Journal of Clinical Endocrinology and Metabolism*, 81:3077–83.

Saggese, G., Bertelloni, S. and Baroncelli, G. I. (1990). Bone mineralization and calciotropic hormones in children with hyperthyroidism: Effects of methimazole therapy. *Journal of Endocrinological Investigation*, 13:587–92.

Sagone, A. L., Lawrence, T. and Balcerzak, S. P. (1973). Effect of smoking on tissue oxygen supply. *Blood*, 41:845–51.

Salardi, S., Tonioli, S., Tassoni, P., Tellarini, M., Mazzanti, L. and Cacciari, E. (1987). Growth and growth factors in diabetes mellitus. *Archives of Disease in Childhood*, 62:57–62.

Salmenperä, L., Perheentupa, J. and Siimes, M. A. (1985). Exclusively breast-fed healthy infants grow slower than reference infants. *Pediatric Research*, 19:307–12.

Saloman, F., Cuneo, R. C., Hesp, R. and Sonksen, P. H. (1989). The effects of treatment with recombinant human growth hormone on body composition and metabolism in adults with growth hormone deficiency. *New England Journal of Medicine*, 321:1797–1803.

Samuelsen, S. O., Magnus, P. and Bakketeig, L. S. (1998). Birth weight and mortality in childhood in Norway. *American Journal of Epidemiology*, 148:983–91.

Sanders, M., Allen, M., Alexander, G. R., Yankowitz, J., Graeber, J., Johnson, T. R. B. and Repka, M. X.

(1991). Gestational age assessment in preterm neonates weighing less than 1500 grams. *Pediatrics*, 88:542–6.

Sann, L., Darre, E., Lasne, Y., Bourgeois, J. and Bethenod, M. (1986). Effects of prematurity and dysmaturity on growth at age 5 years. *Journal of Pediatrics*, 109:681–6.

Sann, L., Durand, M., Picard, J., Lasne, Y. and Bethenod, M. (1988). Arm fat and muscle areas in infancy. *Archives of Disease in Childhood*, 63:256–60.

Sanna, E. and Soro, M. R. (2000). Anthropometric changes in urban Sardinian children 7 to 10 years between 1975–1976 and 1996. *American Journal of Human Biology*, 12:782–91.

Santamaria, F., Slavatore, D. and Greco, L. (1991). Growth patterns in cystic fibrosis. *Acta Medica Auxologica*, 23:91–8.

Sarria, A., Selles, H., Cañedo, L., Fleta, J., Blasco, M. J. and Bueno, M. (1987). Un autotest como método de cuantificación de la actividad física en adolescentes (Self-assessment as a method to quantify physical activity in adolescents). *Nutrition in Clinical Practice*, 7:41–8.

Sasano, H., Uzuki, M., Sawai, T., Nagura, H., Matsunaga, G., Kashimoto, O. and Harada, N. (1997). Aromatase in human bone tissue. *Journal of Bone and Mineral Research*, 12:1416–23.

Satgé, P., Mattei, J. F. and Dan, V. O. (1970). Avenir somatique des enfants atteints de kwashiorkor (Future somatic growth of infants with kwashiorkor). *Annales de Pédiatrie*, 17:368–81.

Savage, D. C. L., Forsyth, C. C., McCafferty, E. and Cameron, J. (1975). The excretion of individual adrenocortical steroids during normal childhood and adolescence. *Acta Endocrinologica*, 79:551–67.

Schaefer, F., Georgi, M., Wuhl, E. and Scharer, K. (1998). Body mass index and percentage fat mass in healthy German schoolchildren and adolescents. *International Journal of Obesity*, 22:461–9.

Schaefer, O. (1970). Pre-and post-natal growth acceleration and increased sugar consumption in Canadian Eskimos. *Canadian Medical Association Journal*, 103:1059–68.

Scheiwiller, E., Guler, H. P., Merryweather, J., Scandella, C., Maerki, W., Zapf, J. and Froesch, E. R. (1986). Growth restoration of insulin–deficient diabetic rats by recombinant human insulin–like growth factor I. *Nature*, 323:169–71.

Schell, L. M., Naamon, P. B. N., Relethford, J. H. and Hook, E. B. (1987). Effect on birthweight of mother's daily cigarette consumption during pregnancy and of trimesters of cigarette use. *American Journal of Physical Anthropology*, 72:251.

Scheven, B. A. and Milne, J. S. (1998). Dehydroepiandrosterone (DHEA) and DHEA-S interact with 1,25-dihydroxyvitamin D3 (1,25(OH)2D3) to stimulate human osteoblastic cell differentiation. *Life Sciences*, 62:59–68.

Schibler, D., Brook, C. G. D., Kind, H. P., Zachmann, M. and Prader, A. (1974). Growth and body proportions in 54 boys and men with Klinefelter's syndrome. *Helvetica Paediatrica Acta*, 29:325–33.

Schlesinger, L., Arévalo, M., Arredondo, S., Diaz, M., Lönnerdal, B. and Stekel, A. (1992). Effect of a zinc-fortified formula on immunocompetence and growth of malnourished infants. *American Journal of Clinical Nutrition*, 56:491–8.

Schlüter, K., Funfack, W., Pachaly, J. and Weber, B. (1976). Development of subcutaneous fat in infancy. *European Journal of Pediatrics*, 123:255–67.

Schmid, C., Ernst, M., Zapf, J. and Froesch, E. R. (1989). Release of insulin–like growth factor carrier proteins by osteoblasts: Stimulation by estradiol and growth hormone. *Biochemical and Biophysical Research Communications*, 160:788–94.

Schofield, P. W., Logroscino, G., Andrews, H. F., Albert, S. and Stern, Y. (1997). An association between head circumference and Alzheimer's disease in a population-based study of aging and dementia. *Neurology*, 49:30–7.

Schofield, P. W., Mosesson, R., Stern, Y. and Mayeux, R. (1995). The age at onset of Alzheimer's disease and an intracranial area measurement: A relationship. *Archives of Neurology*, 52:95–8.

Scholl, T. O., Hediger, M. L., Schall, J. I., Ances, I. G. and Smithe, W. K. (1995). Gestational weight gain, pregnancy outcome, and postpartum weight retention. *Obstetrics and Gynecology*, 86:423–7.

Scholl, T. O., Karp, R. J., Theophano, J. and Decker, E. (1987). Ethnic differences in growth and nutritional status: A study of poor schoolchildren in southern New Jersey. *Public Health Reports*, 102:278–83.

Schorah, C. J., Zemroch, P. J., Sheppard, S. and Smithells, R. W. (1978). Leucocyte ascorbic acid and pregnancy. *British Journal of Nutrition*, 39:139–49.

Schubring, C., Siebler, T., Englaro, P., Blum, W. F., Kratzsch, J., Triep, K. and Kiess, W. (1998). Rapid decline of serum leptin levels in healthy neonates after birth. *European Journal of Pediatrics*, 157:263–4.

Schumacher, M. C., Hasstedt, S. J., Hunt, S. C., Williams, R. R. and Elbein, S. C. (1992). Major gene effects for insulin levels in familial NIDDM pedigree. *Diabetes*, 41:416–23.

Schutte, J. E., Lilljeqvist, R. E. and Johnson, R. L. (1983). Growth of lowland native children of European ancestry during sojourn at high altitude (3200 m). *American Journal of Physical Anthropology*, 61:221–6.

Schwander, J. C., Hauri, C., Zapf, J. and Froesch, E. R. (1983). Synthesis and secretion of insulin-like growth factor and its binding protein by the perfused rat liver: Dependence on growth hormone status. *Endocrinology*, 113:297–305.

Schwartz, A. G. and Amos, H. (1968). Insulin dependence of cells in primary culture: Influence on ribosome integrity. *Nature*, 219:1366–7.

Schwartz, H. L. (1983). Effect of thyroid hormone on growth and development. In *Molecular Basis of Thyroid Hormone Action*, eds. J. H. Oppenheimer and H. H. Samuels, pp. 413–44. New York: Academic Press.

Schwartz, M. W., Peskind, E., Raskind, M., Boyko, E. J. and Porte, D., Jr. (1996a). Cerebrospinal fluid leptin levels: Relationship to plasma levels and to adiposity in humans. *Nature Medicine*, 2:589–91.

Schwartz, M. W., Seeley, R. J., Campfield, L. A., Burn, P. and Baskin, D. G. (1996b). Identification of targets for leptin action in rat hypothalamus. *Journal of Clinical Investigation*, 98:1101–6.

Scott, A., Moar, V. and Ounsted, M. (1981). The relative contribution of different maternal factors in small-for-gestational-age pregnancies. *European Journal of Obstetrics, Gynecology, and Reproductive Biology*, 12:157–65.

Scott, A., Moar, V. and Ounsted, M. (1982a). Growth

in the first four years. I. The relative effects of gender and weight for gestational age at birth. *Early Human Development*, 7:17–28.

Scott, A., Moar, V. and Ounsted, M. (1982b). The relative contribution of different maternal factors in large-for-gestational age pregnancies. *European Journal of Obstetrics, Gynecology, and Reproductive Biology*, 13:269–77.

Scott, J. A., Landers, M. C. G., Hughes, R. M. and Binns, C. W. (2001). Factors associated with breastfeeding at discharge and duration of breastfeeding. *Journal of Paediatric Child Health*, 37:254–61.

Scribner, R. and Dwyer, J. H. (1989). Acculturation and low birth weight among Latinos in the Hispanic HANES. *American Journal of Public Health*, 79:1263–7.

Secher, N. J., Kaern, J. and Hansen, P. (1985). Intrauterine growth in twin pregnancies: Prediction of fetal growth retardation. *Obstetrics and Gynecology*, 66:63–8.

Seeman, E., Hopper, J. L., Young, N. R., Formica, C., Goss, P. and Tsalamandris, C. (1996). Do genetic factors explain associations between muscle strength, lean mass, and bone density? A twin study. *American Journal of Physiology, Endocrinology and Metabolism*, 33:E320–7.

Segal, K. R., Landt, M. and Klein, S. (1996). Relationship between insulin sensitivity and plasma leptin concentration in lean and obese men. *Diabetes*, 45:988–91.

Seidman, D. S., Ever-Hadani, P., Stevenson, D. K., Slater, P. E., Harlap, S. and Gale, R. (1988). Birth order and birth weight re-examined. *Obstetrics and Gynecology*, 72:158–62.

Seidman, D. S., Gale, R., Stevenson, D. K., Laor, A., Bettane, P. A. and Danon, Y. L. (1993). Is the association between birthweight and height attainment independent of the confounding effect of ethnic and socioeconomic factors? *Israel Journal of Medical Sciences*, 29:772–6.

Seidman, D. S., Laor, A., Gale, R., Stevenson, D. K. and Danon, Y. L. (1991a). A longitudinal study of birth weight and being overweight in late adolescence. *American Journal of Diseases of Children*, 145:782–5.

Seidman, D. S., Laor, A., Gale, R., Stevenson, D. K., Mashiach, S. and Danon, Y. L. (1991*b*). Birth weight, current body weight and blood pressure in late adolescence. *British Medical Journal*, **302**: 1235–7.

Seidman, D. S., Slater, P. E., Ever-Hadani, P. and Gale, R. (1987). Accuracy of mother's recall of birth weight and gestational age. *British Journal of Obstetrics and Gynaecology*, **94**:731–5.

Selvin, S. and Janerich, D. T. (1971). Four factors influencing birthweight. *British Journal of Preventive and Social Medicine*, **25**:12–20.

Sempé, M., Bondallaz, C. H. and Limoni, C. (1996). Growth curves in untreated Ullrich–Turner syndrome: French reference standards 1–22 years. *European Journal of Pediatrics*, **155**:862–9.

Sempé, M., Pédron, G. and Roy-Pernot, M. P. (1979). *Auxologie Méthode et Séquences*. Paris: Laboratoire Théraplix.

Sentipal, J. M., Wardlaw, G. M., Mahan, J. and Matkovac, V. (1991). Influence of calcium intake and growth indexes on vertebral bone mineral density in young females. *American Journal of Clinical Nutrition*, **54**:425–8.

Sexton, M. and Hebel, J. R. (1984). A clinical trial of change in maternal smoking and its effect on birth weight. *Journal of the American Medical Association*, **251**:911–15.

Sforza, C., Grassi, G. P., Dugnani, S., Mauro, F. and Ferrario, V. F. (1999). Regional differences in anthropometric variables in children from Lombardy, Italy. *Acta Medica Auxologica*, **31**:143–54.

Shaheen, S. O., Sterne, J. A. C., Montgomery, S. M. and Azima, H. (1999). Birth weight, body mass index and asthma in young adults. *Thorax*, **54**:396–402.

Shannon, B., Peacock, J. and Brown, M. (1991). Body fatness, television viewing and calorie-intake of a sample of Pennsylvania sixth grade children. *Journal of Nutrition Education*, **23**:262–8.

Sharma, K., Considine, R. V., Michael, B., Dunn, S. R., Weisberg, L., Kurnik, B., Kurnik, P., O'Connor, J., Sinha, M. and Caro, J. F. (1997). Plasma leptin is partly cleared by the kidney and is elevated in hemodialysis patients. *Kidney International*, **51**:1980–5.

Sharp, P. S., Mohan, V., Maneschi, F., Vitelli, F., Cloke, H. R., Burrin, J. M. and Kohner, E. M. (1987). Changes in plasma growth hormone in diabetic and nondiabetic subjects during the glucose clamp. *Metabolism: Clinical and Experimental*, **36**:71–5.

Shaw, J. L. and Bassett, C. A. L. (1967). The effects of varying oxygen concentrations on osteogenesis and embryonic cartilage *in vitro*. *Journal of Bone and Joint Surgery*, **49A**:73–80.

Shea, S., Basch, C. E., Stein, A. D., Contento, I. R., Irigoyen, M. and Zybert, P. (1993). Is there a relationship between dietary fat and stature or growth in children 3 to 5 years of age. *Pediatrics*, **92**:579–86.

Shephard, R. J., Lavallée, H., LaBarre, R., Rajic, M., Jéquier, J.-C. and Volle, M. (1984). Body dimensions of Québecois children. *Annals of Human Biology*, **11**:243–52.

Shepherd, R. W., Holt, T. L., Greer, R., Cleghorn, G. J. and Thomas, B. J. (1989). Total body potassium in cystic fibrosis. *Journal of Pediatric Gastroenterology and Nutrition*, **9**:200–5.

Shiono, P. H., Klebanoff, M. A., Graubard, B. I., Berendes, H. W. and Rhoads, G. G. (1986*a*). Birth weight among women of different ethnic groups. *Journal of the American Medical Association*, **255**:48–52.

Shiono, P. H., Klebanoff, M. A. and Rhoads, G. G. (1986*b*). Smoking and drinking during pregnancy: Their effects on preterm birth. *Journal of the American Medical Association*, **255**:82–4.

Shirai, K., Shinomiya, M., Saito, Y., Umezono, T., Takahashi, K. and Yoshida, S. (1990). Incidence of childhood obesity over the last 10 years in Japan. *Diabetes Research and Clinical Practice*, **10**:S65–70.

Showstack, J. A., Budetti, P. P. and Minkler, D. (1984). Factors associated with birthweight: An exploration of the roles of prenatal care and length of gestation. *American Journal of Public Health*, **74**:1003–8.

Shu, X. O., Hatch, M. C., Mills, J., Clemens, J. and Susser, M. (1995). Maternal smoking, alcohol drinking, caffeine consumption, and fetal growth: Results from a prospective study. *Epidemiology*, **6**:115–20.

Shuttleworth, F. K. (1939). *The Physical and Mental Growth of Girls and Boys age Six to 19 in Relation to*

Age at Maximum Growth, Monographs of the Society for Research in Child Development, no. 4. Chicago, IL: University of Chicago Press.

Siegers, C. P., Jungblut, J. R., Klink, F. and Oberheuser, F. (1983). Effect of smoking on cadmium and lead concentrations in human amniotic fluid. *Toxicology Letters*, **19**:327–31.

Siervogel, R. M., Roche, A. F., Guo, S., Mukherjee, D. and Chumlea, W. C. (1991). Patterns of change in weight/stature^2 from 2 to18 years: Findings from long-term serial data for children in the Fels Longitudinal Growth Study. *International Journal of Obesity*, **15**:479–85.

Silbergeld, A., Lazar, L., Erster, B., Keret, R., Tepper, R. and Laron, Z. (1989). Serum growth hormone binding protein activity in healthy neonates, children and young adults: Correlation with age, height and weight. *Clinical Endocrinology*, **31**:295–303.

Silbermann, M. (1983). Hormones and cartilage. In *Cartilage*, vol. 2, ed. B. K. Hall, pp. 327–68. New York: Academic Press.

Silva, P. A. and Crosado, B. (1985). The growth and development of twins compared with singletons at ages 9 and 11. *Australian Paediatric Journal*, **21**:265–7.

Simell, O., Niinikoski, H., Rönnemaa, T., Lapinleimu, H., Routi, T., Lagström, H., Salo, P., Jokinen, E. and Viikari, J. (2000). Special Turku Coronary Risk Factor Intervention Project for babies (STRIP). *American Journal of Clinical Nutrition Supplement*, **72**:1316–31.

Simmons, R. A., Flozak, A. S. and Ogata, E. S. (1993). The effect of insulin, and insulin–like growth factor 1 on glucose transport in normal and small for gestional age fetal rats. *Endocrinology*, **133**:1361–8.

Simmons, R. A., Gounis, A. S., Bangalore, S. A. and Ogata, E. S. (1985). Intrauterine growth retardation: Fetal glucose transport is diminished in lung but spared in brain. *Pediatric Research*, **31**:59–63.

Simon, N. V., Deter, R. L., Hassinger, K. K., Levisky, J. S., Stefos, T. and Shearer, D. M. (1989). Evaluation of fetal growth by ultrasonography in twin pregnancy: A comparison between individual and cross-sectional growth curve standards. *Journal of Clinical Ultrasound*, **17**:633–40.

Simpson, E. R., Zhao, Y., Agarwal, V. R., Michael, M. D., Bulun, S. E., Hinshelwood, M. M.,

Graham-Lorence, S. G., Sun, T., Fisher, C. R., Qin, K. and Mendelson, C. R. (1997). Aromatase expression in health and disease. *Recent Progress in Hormone Research*, **52**:185–214.

Singer, L. (1986). Long-term hospitalization of failure-to-thrive infants: Developmental outcome at three years. *Child Abuse and Neglect*, **10**:479–86.

Singer, L. (1987). Long-term hospitalization of nonorganic failure-to-thrive infants: Patient characteristics and hospital course. *Developmental and Behavioral Pediatrics*, **8**:25–31.

Singh, G. K. and Yu, S. M. (1994). Birthweight differentials among Asian Americans. *American Journal of Public Health*, **84**:1444–9.

Singleton, A., Patois, E., Pédroñ, G. and Roy, M.-P. (1975). Croissance de la taille du segment supérieur et du diamètre biiliaque chez la fille après l'apparition des premières règles (Growth in height of the superior segment and of the bi-iliac diameter in females after menarche). *Archives Françaises de Pédiatrie*, **32**:859–70.

Sinha, M. K., Ohannesian, J. P., Heiman, M. L., Kriauciunas, A., Stephens, T. W., Magosin, S., Marco, C. and Caro, J. F. (1996). Nocturnal rise of leptin in lean, obese, and non–insulin- dependent diabetes mellitus subjects. *Journal of Clinical Investigation*, **97**:1344–7.

Sinha, Y. N., Wilkins, J. N., Selby, F. and Vanderlaan, W. P. (1973). Pituitary and serum growth hormone during undernutrition and catch–up growth in young rats. *Endocrinology*, **92**:1768–71.

Sizonenko, P. C. and Paunier, L. C. (1975). Correlation of plasma dehydroepiandrosterone, testosterone, FSH, and LH with stages of puberty and bone age in normal boys and girls and in patients with Addison's disease or hypogonadism or with premature or late adrenarche. *Journal of Clinical Endocrinology and Metabolism*, **41**:894–904.

Sklad, M. (1975). Über die körperliche Entwicklung von Zwillingen (On the growth of body size in twins). *Acta Medica Auxologica* **7**:49–61.

Skuse, D. H. (1985). Non-organic failure to thrive: A reappraisal. *Archives of Disease in Childhood*, **60**:173–8.

Slaughter, M. H., Lohman, T. G. and Boileau, R. A. (1978). Relationship of anthropometric dimensions to lean body mass in children. *Annals of Human Biology*, 5:469–82.

Slaughter, M. H., Lohman, T. G., Boileau, R. A., Horswill, C. A., Stillman, R. J., van Loan, M. D. and Bemben, D. A. (1988). Skinfold equations for estimation of body fatness in children and youth. *Human Biology*, 60:709–23.

Slemenda, C. W., Reister, T. K., Hui, S. L., Miller, J. Z., Christian, J. C. and Johnston, C. C., Jr. (1994). Influences on skeletal mineralization in children and adolescents: Evidence for varying effects of sexual maturation and physical activity. *Journal of Pediatrics*, 125:201–7.

Smith, C. P., Archibald, H. R., Thomas, J. M., Tarn, A. C., Williams, A. J. K., Gale, E. A. M. and Savage, M. O. (1988). Basal and stimulated insulin levels rise with advancing puberty. *Clinical Endocrinology*, 28:7–14.

Smith, D. M., Nance, W. E., Kang, K. W., Christian, J. C. and Johnston, C. C., Jr. (1973). Genetic factors in determining bone mass. *Journal of Clinical Investigation*, 52:2800–8.

Smith, D. W., Truog, W., Rogers, J. E., Greitzer, L. J., Skinner, A. L., McCann, J. J. and Haney, M. A. S. (1976). Shifting linear growth during infancy: Illustration of genetic factors in growth from fetal life through infancy. *Journal of Pediatrics*, 89:225–30.

Smith, E. P., Boyd, J., Frank, G. R., Takahashi, H., Cohen, R. M., Specker, B., Williams, T. C., LuBahn, D. B. and Korach, K. S. (1994). Estrogen resistance caused by a mutation in the estrogen-receptor gene in a man. *New England Journal of Medicine*, 331:1056–61.

Smith, I. E., Coles, C. D., Lancaster, J., Fernhoff, P. M. and Falek, A. (1986). The effects of volume and duration of prenatal ethanol exposure on neonatal physical and behavioral development. *Neurobehavioral Toxicology and Teratology*, 8:375–81.

Socol, M. L., Manning, F. A., Murata, Y. and Druzin, M. L. (1982). Maternal smoking causes fetal hypoxia: Experimental evidence. *American Journal of Obstetrics and Gynecology*, 142:214–18.

Sokol, R. J., Miller, S. I. and Reed, G. (1980). Alcohol abuse during pregnancy: An epidemiologic study. *Alcoholism, Clinical and Experimental Research*, 4:135–45.

Sørensen, H. T., Sabroe, S., Rothman, K. J., Gillman, M., Fischer, P. and Sørensen, T. I. A. (1997). Relation between weight and length at birth and body mass index in young adulthood: Cohort study. *British Medical Journal*, 315:1137.

Sørensen, T. I. A., Holst, C. and Stunkard, A. J. (1992a). Childhood body mass index: Genetic and familial environmental influences assessed in a longitudinal adoption study. *International Journal of Obesity*, 16:705–14.

Sørensen, T. I. A., Holst, C., Stunkard, A. J. and Theil, L. (1992b). Correlations of body mass index of adult adoptees and their biological relatives. *International Journal of Obesity*, 16:227–36.

Sørensen, T. I. A., Price, R. A., Stunkard, A. J. and Schulsinger, F. (1989). Genetics of obesity in adult adoptees and their biological siblings. *British Medical Journal*, 298: 87–90.

Sørensen, T. I. A. and Sonne-Holm, S. (1988). Risk in childhood of development of severe adult obesity: Retrospective, population-based case-cohort study. *American Journal of Epidemiology*, 127:104–13.

Sorva, R., Tolppanen, E.-M., Lankinen, S. and Perheentupa, J. (1989). Growth evaluation: Parent and child specific height standards. *Archives of Disease in Childhood*, 64:1483–7.

Spady, D. W., Atrens, M. A. and Szymanski, W. A. (1986). Effects of mothers' smoking on their infants' body composition as determined by total body potassium. *Pediatric Research*, 20:716–19.

Specker, B. L., Beck, A., Kalkwarf, H. and Ho, M. (1997). Randomized trial of varying mineral intake on total body bone mineral accretion during the first year of life. *Pediatrics*, 99:E12–9.

Specker, B. L., Mulligan, L. and Ho, M. (1999). Longitudinal study of calcium intake, physical activity, and bone mineral content in infants 6–18 months of age. *Journal of Bone and Mineral Research*, 14:569–76.

Spencer, E. M. (1991). *Modern Concepts of Insulin-Like Growth Factors*. New York: Elsevier.

Spencer, N. J., Bambang, S., Logan, S. and Gill, L. (1999). Socioeconomic status and birth weight: Comparison of an area-based measure with the Registrar General's social class. *Journal of Epidemiology and Community Health*, 53:495–8.

Sprauve, M. E., Lindsay, M. K., Drews-Botsch, C. D. and Graves, W. (1999). Racial patterns in the effects of tobacco use on fetal growth. *American Journal of Obstetrics and Gynecology*, **181**:S22–7.

Spurr, G. B. (1988). Effects of chronic energy deficiency on stature, work capacity and productivity. In *Chronic Energy Deficiency: Consequences and Related Issues*, eds. B. Schürch and N. S. Scrimshaw, pp. 95–134. Lausanne, Switzerland: International Dietary Energy Consultancy Group.

Stanhope, R., Ackland, F. M., Hamill, G., Clayton, J., Jones, J. and Preece, M. A. (1989). Physiological growth hormone secretion and response to growth hormone treatment in children with short stature and intrauterine growth retardation. *Acta Paediatrica Scandinavica, Supplement*, **349**:47–52.

Stanhope, R., Albanese, A., Hindmarsh, P. and Brook, C. G. D. (1992). The effects of growth hormone therapy on spontaneous sexual development. *Hormone Research, Supplement*, **38**:9–13.

Stanhope, R., Pringle, P. J. and Brook, C. G. D. (1988). The mechanism of the adolescent growth spurt induced by low dose pulsatile GnRH treatment. *Clinical Endocrinology*, **28**:83–91.

Stanik, S., Dornfeld, L. P., Maxwell, M. H., Viosca, S. P. and Korenman, S. G. (1981). The effect of weight loss on reproductive hormones in obese men. *Journal of Clinical Endocrinology and Metabolism*, **53**: 828–32.

Starbird, E. H. (1991). Comparison of influences in breastfeeding initiation of firstborn children, 1960–69 vs 1970–79. *Social Science and Medicine*, **33**:627–34.

Stein, Z. A. and Susser, M. (1984). Intrauterine growth retardation: Epidemiological issues and public health significance. *Seminars in Perinatology*, **8**:5–14.

Steinberger, J., Moorehead, C., Katch, V. and Rocchini, A. P. (1995). Relationship between insulin resistance and abnormal lipid profile in obese adolescents. *Journal of Pediatrics*, **126**:690–5.

Stephen, A. and Wald, N. (1990). Trends in individual consumption of dietary fat in the United States, 1920–1984. *American Journal of Clinical Nutrition*, **52**:457–69.

Sternfeld, B., Quesenberry, C. P., Eskenazi, B. and Newman, L. A. (1995). Exercise during pregnancy and pregnancy outcome. *Medicine and Science in Sports and Exercise*, **27**:634–40.

Stettler, N., Tershakovec, A. M., Zemel, B. S., Leonard, M. B., Boston, R. C., Katz, S. H. and Stallings, V. A. (2000). Early risk factors for increased adiposity: A cohort study of African American subjects followed from birth to young adulthood. *American Journal of Clinical Nutrition*, **72**:378–83.

Stevens, C. A., Hennekam, R. C. M. and Blackburn, B. L. (1990). Growth in the Rubinstein–Taybi syndrome. *American Journal of Medical Genetics, Supplement*, **6**:51–5.

Stevens-Simon, C., McAnarney, E. R. and Roghmann, K. J. (1993). Adolescent gestational weight gain and birth weight. *Pediatrics*, **92**:805–9.

Stinson, S. (1982). The effect of high altitude on the growth of children of high socioeconomic status in Bolivia. *American Journal of Physical Anthropology*, **59**:61–71.

Stolz, H. R. and Stolz, L. M. (1951). *Somatic Development of Adolescent Boys: A Study of the Growth of Boys during the Second Decade of Life*. New York: Macmillan.

Strauss, M. E., Lessen-Firestone, J. K., Chavez, C. J. and Stryker, J. C. (1979). Children of methadone-treated women at five years of age. *Pharmacology, Biochemistry and Behavior Supplement*, **11**:3–6.

Strauss, R. S. and Knight, J. (1999). Influence of the home environment on the development of obesity in children. *Pediatrics*, **103**:E85–92.

Streissguth, A. P., Martin, D. C., Martin, J. C. and Barr, H. M. (1981). The Seattle Longitudinal Prospective Study on Alcohol and Pregnancy. *Neurobehavioral Toxicology and Teratology*, **3**:223–33.

Strickland, A. L. and Shearin, R. B. (1972). Diurnal height variation in children. *Journal of Pediatrics*, **80**:1023–5.

Strobel, A., Issad, T., Camoin, L., Ozata, M. and Strosberg, A. D. (1998). A leptin missense mutation associated with hypogonadism and morbid obesity. *Nature Genetics*, **18**:213–15.

Strobino, D. M., Ensminger, M. E., Kim, Y. J. and Nanda, J. (1995). Mechanisms for maternal age differences in birth weight. *American Journal of Epidemiology*, **142**:504–14.

Stuff, J. E. and Nichols, B. L. (1989). Nutrient intake and growth performance of older infants fed human milk. *Journal of Pediatrics*, **115**:959–68.

Stunkard, A. J., Berkowitz, R. I., Stallings, V. A. and Cater, J. R. (1999). Weights of parents and infants: Is there a relationship? *International Journal of Obesity*, **23**:159–62.

Stunkard, A. J., Harris, J. R., Pederson, N. L. and McClearn, G. E. (1990). The body mass index of twins who have been reared apart. *New England Journal of Medicine*, **322**:1483–7.

Styne, D. M. (1995). The physiology of puberty. In *Clinical Paediatric Endocrinology*, ed. C. G. D. Brook, pp. 234–52. Oxford, U. K.: Blackwell.

Sukanich, A. C., Rogers, K. D. and McDonald, H. M. (1986). Physical maturity and outcome of pregnancy in primiparas younger than 16 years of age. *Pediatrics*, **78**:31–6.

Sulaiman, N. D., Florey, C., Taylor, D. J. and Ogston, S. A. (1988). Alcohol consumption in Dundee primigravidas and its effect on outcome of pregnancy. *British Medical Journal*, **296**:1500–3.

Sullivan, P. G. (1983). Prediction of the pubertal growth spurt by measurements of standing height. *European Journal of Orthodontics*, **5**:189–97.

Sumiya, T., Nakahara, H. and Shohoji, T. (1999). Height and menarcheal age in Japanese girls. *Acta Medica Auxologica*, **31**:15–24.

Susanne, C. (1980). Interrelations between some social and familial factors and stature and weight of young Belgian male adults. *Human Biology*, **52**:702–9.

Susanne, C. and Vercauteren, M. (1997). Focus on physical and sexual maturation: The case of Belgium. *Acta Biologica Szegediensis*, **42**:287–97.

Suwa, S. (1992). Standards for growth and growth velocity in Turner's syndrome. *Acta Paediatrica Japonica*, **34**:206–21.

Suwa, S. (1993). Standards for growth and growth velocity in Turner syndrome. In *Basic and Clinical Approach to Turner Syndrome*, eds. I. Hibi and K. Takano, pp. 69–76. Amsterdam, The Netherlands: Excerpta Medica.

Swanson, C. A., Jones, D. Y., Schatzkin, A., Brinton, L. A. and Ziegler, R. G. (1988). Breast cancer risk assessed by anthropometry in the NHANES I epidemiological follow-up study. *Cancer Research*, **48**:5363–7.

Szemik, M. (1980). Age at menarche and adult stature. *Studies in Human Ecology*, **4**:147–54.

Taffel, S. M. (1994). *Cesarean Delivery in the U.S., 1990*, Vital and Health Statistics, Series 21, no. 51. Washington, D.C.: Department of Health and Human Services.

Taittonen, L., Nuutinen, M., Turtinen, J. and Uhari, M. (1996). Prenatal and postnatal factors in predicting later blood pressure among children: Cardiovascular risk in young Finns. *Pediatric Research*, **40**:627–32.

Taitz, L. S. and Lukmanji, Z. (1981). Alterations in feeding patterns and rates of weight gain in South Yorkshire infants, 1971–1977. *Human Biology*, **53**:313–20.

Takahashi, E. (1984). Secular trend in milk consumption and growth in Japan. *Human Biology*, **56**:427–37.

Takahashi, Y., Minamitani, K., Kobayashi, Y., Minagawa, M., Yasuda, T. and Niimi, H. (1996). Spinal and femoral bone mass accumulation during normal adolescence: Comparison with female patients with sexual precocity and with hypogonadism. *Journal of Clinical Endocrinology and Metabolism*, **81**:1248–53.

Takaishi, M. (1995). Growth standards for Japanese children: An overview with special reference to secular change in growth. In *Essays on Auxology*, eds. R. Hauspie, G. Lindgren and F. Falkner, pp. 302–11. Welwyn Garden City, U.K.: Castlemead Publications.

Takaishi, M. and Kikuta, F. (1989). The changes of standing height in schoolgirls of a private school in Tokyo during the last 20 years. In *Perspectives in the Science of Growth and Development*, ed. J. M. Tanner, pp. 203–6. London: Smith-Gordon.

Takamura, K., Ohyama, S., Yamada, T. and Ishinishi, N. (1988). Changes in body proportions of Japanese medical students between 1961 and 1986. *American Journal of Physical Anthropology*, **77**:17–22.

Tambs, K., Moum, T., Eaves, L.J., Neale, M.C., Midthjell, K., Lund-Larsen, P.G. and Naess, S. (1992). Genetic and environmental contributions to the variance of body height in a sample of first and second degree relatives. *American Journal of Physical Anthropology*, **88**:285–94.

Tanaka, T. (1996). Postnatal growth. In *Pediatrics and Perinatology*, eds. P. D. Gluckman and M. A. Heymann, pp. 304–9. London: Edward Arnold.

Tanaka, T., Suwa, S., Yokoya, S. and Hibi, I. (1988). Analysis of linear growth during puberty. *Acta Paediatrica Scandinavica*, **347**:25–9.

Tanner, J. M. and Cameron, N. (1980). Investigation of the mid-growth spurt in height, weight and limb circumferences in single-year velocity data from the London 1966–67 growth survey. *Annals of Human Biology*, **7**:565–77.

Tanner, J. M. and Davies, P. S. W. (1985). Clinical longitudinal standards for height and height velocity for North American children. *Journal of Pediatrics*, **107**:317–29.

Tanner, J. M., Goldstein, H. and Whitehouse, R. H. (1970). Standards for children's height at ages 2–9 years allowing for height of parents. *Archives of Disease in Childhood*, **45**:755–62.

Tanner, J. M., Hayashi, T., Preece, M. A. and Cameron, N. (1982). Increase in length of leg relative to trunk in Japanese children and adults from 1957 to 1977: Comparison with British and with Japanese Americans. *Annals of Human Biology*, **9**:411–23.

Tanner, J. M., Healy, M. J. R., Lockhart, R. D., MacKenzie, J. D. and Whitehouse, R. H. (1956). Aberdeen Growth Study. I. The prediction of adult body measurements from measurements taken each year from birth to 5 years. *Archives of Disease in Childhood*, **31**:372–81.

Tanner, J. M. and Israelsohn, W. (1963). Parent–child correlation for body measurements of children between the ages one month and seven years. *Annals of Human Genetics*, **26**:245–59.

Tanner, J. M., Landt, K. W., Cameron, N., Carter, B. S. and Patel, J. (1983*a*). Prediction of adult height from height and bone age in childhood. *Archives of Disease in Childhood*, **58**:767–76.

Tanner, J. M., Lejarraga, H. and Cameron, N. (1975). The natural history of the Silver–Russell syndrome: A longitudinal study of thirty-nine cases. *Pediatric Research*, **9**:611–23.

Tanner, J. M., Oshman, D., Bahhage, F. and Healy, M. (1997). Tanner–Whitehouse bone age reference values for North American children. *Journal of Pediatrics*, **131**:34–40.

Tanner, J. M. and Whitehouse, R. H. (1975). Revised standards for triceps and subscapular skinfolds in British children. *Archives of Disease in Childhood*, **50**:142–5.

Tanner, J. M., Whitehouse, R. H., Cameron, N., Marshall, W. A., Healy, M. J. R. and Goldstein, H. (1983*b*). *Assessment of Skeletal Maturity and Prediction of Adult Height (TW2 Method)*, 2nd edn. London: Academic Press.

Tanner, J. M., Whitehouse, R. H., Marubini, E. and Resele, L. F. (1976). The adolescent growth spurt of boys and girls of the Harpenden Growth Study. *Annals of Human Biology*, **3**:109–26.

Tanner, J. M., Whitehouse, R. H. and Takaishi, M. (1966*a*). Standards from birth to maturity for height, weight, height velocity, and weight velocity: British children, 1965, Part I. *Archives of Disease in Childhood*, **41**:454–71.

Tanner, J. M., Whitehouse, R. H. and Takaishi, M. (1966*b*). Standards from birth to maturity for height, weight, height velocity, and weight velocity: British Children, 1965, Part II. *Archives of Disease in Childhood*, **41**:613–35.

Taranger, J. (1984). Ökande övervikt hos unga män: En medicinsk riskfaktor (Increasing overweight in young men: A medical risk factor). *Läkartidningen (The Medical Journal)*, **81**:1147. (Quoted by Lindgren, 1998.)

Taranger, J., Bruning, S., Claesson, I., Karlberg, P., Landström, T. and Lindström, B. (1976*a*). Skeletal development from birth to 7 years. *Acta Paediatrica Scandinavica, Supplement*, **258**:98–108.

Taranger, J., Lichtenstein, H. and Svennberg-Redegren, I. (1976*b*). Somatic pubertal development. *Acta Paediatrica Scandinavica, Supplement*, **258**:121–35.

Taylor, A. M., Thomson, A., Bruce-Morgan, C., Ahmed, M. L., Watts, A., Harris, D., Holly, J. M. P. and Dunger, D. B. (1999). The relationship between insulin, IGF-I and weight gain in cystic fibrosis. *Clinical Endocrinology*, **51**:659–65.

Taylor, S. J., Hird, K., Whincup, P. and Cook, D. (1998). Relation between birth weight and blood pressure is independent of maternal blood pressure. *British Medical Journal*, **317**:680.

Teitelman, A. M., Welch, L. S., Hellenbrand, K. G. and Bracken, M. B. (1990). The effects of maternal work activity on preterm birth and low birth weight. *American Journal of Epidemiology*, **131**:104–13.

Teixeira, P. J., Sardinha, L. B., Going, S. B. and Lohman, T. G. (2001). Total and regional fat and serum cardiovascular disease risk factors in lean and obese children and adolescents. *Obesity Research*, **9**:432–42.

Telama, R. and Yang, X. (2000). Decline of physical activity from youth to young adulthood in Finland. *Medicine and Science in Sports and Exercise*, **32**:1617–22.

Tennes, K. and Blackard, C. (1980). Maternal alcohol consumption, birth weight, and minor physical anomalies. *American Journal of Obstetrics and Gynecology*, **138**:774–80.

Tenovuo, A., Kero, P., Piekkala, P., Korvenranta, H., Sillanpaa, M. and Erkkola, R. (1987). Growth of 519 small for gestational age infants during the first two years of life. *Acta Paediatrica Scandinavica*, **76**:636–46.

Terada, H. and Hoshi, H. (1965a). Longitudinal study on the physical growth in Japanese: Growth in stature and body weight during the first three years of life. *Acta Anatomica Nippon*, **40**:166–77.

Terada, H. and Hoshi, H. (1965b). Longitudinal study on the physical growth in Japanese: Growth in chest and head circumferences during the first three years of life. *Acta Anatomica Nippon*, **40**:368–80.

Theintz, G., Buchs, B., Rizzoli, R., Slosman, D., Clavien, H., Sizonenko, P. C. and Bonjour, J.-P. (1992). Longitudinal monitoring of bone mass accumulation in healthy adolescents: Evidence for a marked reduction after 16 years of age at the levels of lumbar spine and femoral neck in female subjects. *Journal of Clinical Endocrinology and Metabolism*, **75**:1060–5.

Thieriot-Prévost, G., Boccara, J. F., Francoual, C., Badoual, J. and Job, J. C. (1988). Serum insulin-like growth factor 1 and serum growth–promoting activity during the first postnatal year in infants with intrauterine growth retardation. *Pediatric Research*, **24**:380–3.

Thomas, P. K. and Ferriman, D. G. (1957). Variation in facial and pubic hair growth in white women. *American Journal of Physical Anthropology*, **15**:171–80.

Thomas, P. W., Singhal, A., Hemmings-Kelly, M. and Serjeant, G. R. (2000). Height and weight reference curves for homozygous sickle cell disease. *Archives of Disease in Childhood*, **82**:204–8.

Thomis, M. A. I., Beunen, G. P., van Leemputte, M., Maes, H. H., Blimkie, C. J., Classens, A. L., Marchal, G., Willems, E. and Vlietinck, R. (1998). Inheritance of static and dynamic arm strength and some of its determinants. *Acta Physiologica Scandinavica*, **163**:59–71.

Thompson, R. G., Rodriguez, A., Kowarski, A., Migeon, C. J. and Blizzard, R. M. (1972). Integrated concentrations of growth hormone correlated with plasma testosterone and bone age in pre-adolescent and adolescent males. *Journal of Clinical Endocrinology and Metabolism*, **35**:344–7.

Thomsen, B. L., Ekström, C. T. and Sørensen, T. I. A. (1999). Development of the obesity epidemic in Denmark: Cohort, time and age effects among boys born 1930–1975. *International Journal of Obesity*, **23**:693–701.

Thomson, J. (1955). Observations on weight gain in infants. *Archives of Disease in Childhood*, **30**:322–7.

Thon, A., Heinze, E., Feilen, K. D., Holl, R. W., Schmidt, H., Koletzko, S., Wendel, U. and Nothjunge, J. (1992). Development of height and weight in children with diabetes mellitus: Report on two prospective multicentre studies, one cross-sectional, one longitudinal. *European Journal of Pediatrics*, **151**:258–62.

Thorner, M. O., Vance, M. L., Horvath, E. and Kovacs, K. (1992). The anterior pituitary. In *Williams' Textbook of Endocrinology*, 8th edn, eds. J. D. Wilson and D. W. Foster, pp. 221–310. Philadelphia, PA: WB Saunders.

Thorsdottir, I. and Birgisdottir, B. E. (1998). Different weight gain in women of normal weight before pregnancy: Postpartum weight and birth weight. *Obstetrics and Gynecology*, **92**:377–83.

Tibblin, G., Eriksson, M., Cnattingius, S. and Ekbom, A. (1995). High birthweight as a predictor of prostate cancer risk. *Epidemiology*, **6**:423–4.

Tiisala, R. and Kantero, R.-L. (1971). Some parent–child correlations for height, weight and skeletal age up to 10 years. *Acta Paediatrica Scandinavica, Supplement*, **220**:42–8.

Tilley, B. C., Barnes, A. B., Bergstralh, E., Labarthe, D., Noller, K. L., Colton, T. and Adam, E. (1985). A comparison of pregnancy history recall and medical records. *American Journal of Epidemiology*, **121**:269–81.

Tillmann, V., Thalange, N. K. S., Foster, P. J., Gill, M. S., Price, D. A. and Clayton, P. E. (1998). The relationship between stature, growth, and short-term changes in height and weight in normal prepubertal children. *Pediatric Research*, **44**:882–6.

Tiret, L., André, J. L., Ducimetière, P., Herbeth, B., Rakotovao, R., Guegen, R., Spyckerelle, Y. and Cambien, F. (1992). Segregation analysis of height-adjusted weight with generation- and age-dependent effects: The Nancy Family Study. *Genetic Epidemiology*, **9**:389–403.

Toledo, C., Alembik, Y., Aguirre Jaime, A. and Stoll, C. (1999). Growth curves of children with Down syndrome. *Annals of Genetics*, **42**:81–90.

Torrence, C. R., Horns, K. M. and East, C. (1995). Accuracy and precision of neonatal electronic incubator scales. *Neonatal Network*, **14**:35–41.

Towne, B., Guo, S., Roche, A. F. and Siervogel, R. M. (1993). Genetic analysis of patterns of growth in infant recumbent length. *Human Biology*, **65**: 977–89.

Towne, B., Parks, J. S., Brown, M. R., Siervogel, R. M. and Blangero, J. (2000). Effect of a luteinizing hormone β-subunit polymorphism on growth in stature. *Acta Medica Auxologica*, **32**:43–4.

Towne, B., Parks, J. S., Guo, S. S. and Siervogel, R. M. (1995). Quantitative genetic analysis of associations between pubertal growth pattern parameters. *American Journal of Human Genetics*, **57**:A173.

Towne, B., Siervogel, R. M., Parks, J. S., Brown, M. R., Roche, A. F. and Blangero, J. (1999). Genetic regulation of skeletal maturation from 3–15 years. *American Journal of Human Genetics*, **S65**:A401.

Travers, S. H., Jeffers, B. W., Bloch, C. A., Hill, J. O. and Eckel, R. H. (1995). Gender and Tanner stage differences in body composition and insulin sensitivity in early pubertal children. *Journal of Clinical Endocrinology and Metabolism*, **80**:172–8.

Trayhurn, P., Duncan, J. S. and Rayner, D. V. (1995). Acute cold-induced suppression of *ob* (obese) gene expression in white adipose tissue of mice: mediation by the sympathetic system. *Biochemical Journal*, **311**:729–33.

Trayhurn, P., Hoggard, N., Mercer, J. G. and Rayner, D. V. (1999). Leptin: Fundamental aspects. *International Journal of Obesity, Supplement*, **23**:22–8.

Tres, L. L., Smith, E. P., Van Wyk, J. J. and Kierszenbaum, A. L. (1986). Immunoreactive sites and accumulation of somatomedin-C in rat Sertoli–spermatogenic cell co-cultures. *Experimental Cell Research*, **16**:33–50.

Trichopoulos, D. (1986). Passive smoking, birthweight and estrogens. *Lancet*, **ii**:743.

Troiano, R. P., Flegal, K. M., Kuczmarski, R. J., Campbell, S. M. and Johnson, C. L. (1995). Overweight prevalence and trends for children and adolescents: The National Health and Nutrition Examination Surveys, 1963 to 1991. *Archives of Pediatrics and Adolescent Medicine*, **149**:1085–91.

Trudeau, F., Shephard, R. J., Arsenault, F. and Laurencelle, L. (2001). Changes in adiposity and body mass index from late childhood to adult life in the Trois-Rivières Study. *American Journal of Human Biology*, **13**:349–55.

Tsuzaki, S., Matsuo, N. and Osano, M. (1987). The physical growth of Japanese children from birth to 18 years of age: Cross-sectional percentile growth curve for height and weight. *Helvetica Paediatrica Acta*, **42**:111–19.

Tucker, L. A. (1986). The relationship of television viewing to physical fitness and obesity. *Adolescence*, **21**:797–806.

Tucker, L. A., Seljaas, G. T. and Hager, R. L. (1997). Body fat percentage of children varies according to their diet composition. *Journal of the American Dietetic Association*, **97**:981–6.

Tuddenham, R. D. and Snyder, M. M. (1954). *Physical Growth of Californian Boys and Girls from Birth to 18 years*, University of California Publications in Child Development. Berkeley, CA: University of California at Berkeley.

Tuvemo, T., Cnattingius, S. and Jonsson, B. (1999*a*). Prediction of male adult stature using anthropometric data at birth: A nationwide population-based study. *Pediatric Research*, **46**:491–5.

Tuvemo, T., Jonsson, B. and Persson, I. (1999*b*). Intellectual and physical performance and morbidity in relation to height in a cohort of 18-year-old Swedish conscripts. *Hormone Research*, **52**:186–91.

Twiesselmann, F. (1969). *Développement biométrique de l'enfant à l'adulte (Biometric Development from Infancy to Adulthood)*. Brussels: Presses Universitaires de Bruxelles.

Twisk, J., Kemper, C. G. and Snel, J. (1995). Tracking of cardiovascular risk factors in relation to lifestyle: The Amsterdam Growth Study. *HK Sport Science Monograph Series*, **6**:203–24.

Tzoumaka-Bakoula, C. (1993). Intrauterine growth retardation. In *Birth Risks*, vol. 31, ed. J. D. Baum, pp. 119–26. New York: Raven Press.

Uiterwaal, C. S. P. M., Anthony, S., Launer, L. J., Witteman, C. M., Trouwborst, A. M. W., Hofman, A. and Grobbee, D. E. (1997). Birth weight, growth, and blood pressure: An annual follow-up study of children ages 5 through 21 years. *Hypertension*, **30**:267–71.

Ulloa-Augirre, A., Blizzard, R. M., Garcia-Rubi, E., Rogol, A. D., Link, K., Christie, C. M., Johnson, M. L. and Veldhuis, J. D. (1990). Testosterone and oxandrolone, a non-aromatizable androgen, specifically amplify the mass and rate of growth hormone (GH) secreted per burst without altering GH secretory burst duration or frequency or the GH half-life. *Journal of Clinical Endocrinology and Metabolism*, **71**:846–54.

Underwood, L. E. (1996). Nutritional regulation of IGF-1 and IGFBPs. *Journal of Pediatric Endocrinology and Metabolism*, **9**:303–12.

Underwood, L. E., D'Ercole, A. J. and van Wyk, J. J. (1980). Somatomedin-C and assessment of growth. *Pediatric Clinics of North America*, **27**:771–82.

Underwood, L. E. and van Wyk, J. J. (1985). Normal and aberrant growth. In *Williams' Textbook of Endocrinology*, eds. J. D. Wilson and D. W. Foster, pp. 155–205. Philadelphia, PA: WB Saunders.

Unger, C., Weiser, J. K., McCullough, R. E., Keefer, S. and Moore, L. G. (1988). Altitude, low birth weight, and infant mortality in Colorado. *Journal of American Medical Association*, **259**:3427–32.

Uruena, M., Pantsioyou, S., Preece, M. A. and Stanhope, R. (1992). Is testosterone therapy for boys with constitutional delay of growth and puberty associated with impaired final height and suppression of the hypothalamo–pituitary–gonadal axis? *European Journal of Pediatrics*, **151**:15–18.

U.S. Department of Health and Human Services. (2000). *Healthy people 2010: Understanding and Improving Health*, 2nd edn. Washington D.C.: U.S. Government Printing Office.

U.S. Department of Health and Human Services. (2002). National Center for Health Statistics. http://www.cdc.gov/

Vågerö, D., Koupilová, I., Leon, D. A. and Lithell, U.-B. (1999). Social determinants of birthweight, ponderal index and gestational age in Sweden in the 1920s and the 1980s. *Acta Paediatrica*, **88**:445–53.

Valdez, R., Athens, M. A., Thompson, G. H., Bradshaw, B. S. and Stern, M. P. (1994). Birthweight and adult health outcomes in a biethnic population in the U.S.A. *Diabetologia*, **37**:624–31.

Valdez, R., Greenlund, K. J., Wattigney, W. A., Bao, W. and Berenson, G. S. (1996). Use of weight-for-height indices in children to predict adult overweight: The Bogalusa Heart Study. *International Journal of Obesity*, **20**:715–21.

Vallee, B. L. and Falchuk, K. H. (1981). Zinc and gene expression. *Philosophical Transactions of the Royal Society of London, Series B, Biological Sciences*, **294**:185–97.

van den Brandt, P., Dirx, M. J. M., Ronckers, C. M., van den Hoogen, P. and Goldbohm, R. A. (1997). Height, weight, weight change, and postmenopausal breast cancer risk: The Netherlands Cohort Study. *Cancer Causes and Control*, **8**:39–47.

van den Broeck, J., Brand, R., Massa, G., Herngreen, W. P. and Wit, J.-M. (2000). Length velocity acceleration at 9 months of age in a representative birth cohort of Dutch infants. *Journal of Pediatric Endocrinology and Metabolism*, **13**:45–54.

van der Kallen, C. J. H., Cantor, R. M., van Greevenbroek, M. M. J., Geurts, J. M. W., Bouwman, F. G., Aouizerat, B. E., Allayee, H., Buurman, W. A.,

Lusis, A. J., Rotter, J. I. and de Bruin, T. W. A. (2000). Genome scan for adiposity in Dutch dyslipidemic families reveals novel quantitative trait loci for leptin, body mass index and soluble tumor necrosis factor receptor superfamily 1A. *International Journal of Obesity*, **24**:1381–91.

van der Werff ten Bosch, J. J. and Bot, A. (1986). Growth hormone and androgen effects in the third decade. *Acta Endocrinologica, Supplement*, **279**:29–34.

van Duzen, J., Carter, J. P. and van der Zwagg, R. (1976). Protein and calorie malnutrition among Navajo Indian children: a follow-up. *American Journal of Clinical Nutrition*, **29**:657–62.

van Mechelen, W., Twisk, J. W. R., Post, G. B., Snel, J. and Kemper, H. C. G. (2000). Physical activity of young people: The Amsterdam Longitudinal Growth and Health Study. *Medicine and Science in Sports and Exercise*, **32**:1610–16.

van Venrooij-Ysselmuiden, M. E. (1978). Mixed longitudinal data on height, weight, limb circumferences and skinfold measurements of Dutch children. *Human Biology*, **50**:369–84.

van Venrooij-Ysselmuiden, M. E. and van Ipenburg, A. (1978). Mixed longitudinal data on skeletal age from a group of Dutch children living in Utrecht and surroundings. *Annals of Human Biology*, **5**:359–80.

van Wieringen, J. C. (1972). *Secular Changes of Growth: 1964–1966 Height and Weight Surveys in the Netherlands in Historical Perspective*. Leiden, The Netherlands: Netherlands Institute for Preventive Medicine TNO.

van Wieringen, J. C. (1986). Secular growth changes. In *Human Growth: A Comprehensive Treatise*, 2nd edn, vol. 3, *Methodology*, eds. F. Falkner and J. M. Tanner, pp. 307–31. New York: Plenum Press.

van Wieringen, J. C., Wafelbakker, F., Verbrugge, H. P. and De Haas, J. H. (1971). *Growth Diagrams 1965 Netherlands*. Groningen, The Netherlands: Wolters-Noordhoff.

van Wyk, J. J. (1984). The somatomedins: Biological actions and physiologic control mechanism. In *Hormonal Proteins and Peptides: Growth Factors*, vol. 12, ed. C. H. Li, pp. 81–125. Orlando, FL: Academic Press.

van Wyk, J. J., Graves, D. C., Casella, S. J. and Jacobs, S. (1985). Evidence from monoclonal antibody studies that insulin stimulates deoxyribonucleic acid synthesis through the type I somatomedin receptor. *Journal of Clinical Endocrinology and Metabolism*, **61**:639–43.

Vanderschueren-Lodeweyckx, M. (1993). The effect of simple obesity on growth and growth hormone. *Hormone Research*, **40**:23–30.

Vanhala, M. J., Vanhala, P. T., Keinänen-Kiukaanniemi, S. M., Kumpusale, E. A. and Takala, J. K. (1999). Relative weight gain and obesity as a child predict metabolic syndrome as an adult. *International Journal of Obesity*, **23**:656–69.

van't Hof, M. A., Haschke, F., Darvay, S. and The Euro-Growth Study Group. (2000a). Euro-Growth references on increments in length, weight, head and arm circumference during the first 3 years of life. *Journal of Pediatric Gastroenterology and Nutrition, Supplement*, **31**:S39–47.

van't Hof, M. A., Haschke, F. and The Euro-Growth Study Group (1997). Limitations of growth charts derived from longitudinal studies: The Euro-Growth Study. *International Journal of Sports Medicine*, **18**:S204–7.

van't Hof, M. A., Haschke, F. and The Euro-Growth Study Group (2000b). The Euro-Growth Study: Why, who, and how. *Journal of Pediatric Gastroenterology and Nutrition, Supplement*, **31**:S3–13.

van't Hof, M. A., Haschke, F. and The Euro-Growth Study Group (2000c). Euro-Growth references for body mass index and weight for length. *Journal of Pediatric Gastroenterology and Nutrition, Supplement*, **31**:S48–59.

Vaucher, Y. E., Harrison, G. G., Udall, J. N. and Morrow, G. III (1984). Skinfold thickness in North American infants 24–41 weeks gestation. *Human Biology*, **56**:713–31.

Veldhuis, J. D., Iranmanesh, A., Ho, K. K. Y., Waters, M. J., Johnson, M. L. and Lizarralde, G. (1991). Dual defects in pulsatile growth hormone secretion and clearance subserve the hyposomatotropism of obesity in man. *Journal of Clinical Endocrinology and Metabolism*, **72**:51–9.

Veldhuis, J. D. and Johnson, M. L. (1992). Deconvolution analysis of hormone data. *Methods in Enzymology*, **201**:539–75.

Ventura, S. J. and Martin, J. A. (1993). *Advance Report of Final Natality Statistics, 1991*, Monthly Vital Statistics Report 42, no. 3. Washington, D.C: U.S. Department of Health and Human Services.

Ventura, S. J., Martin, J. A., Curtin, S. C. and Mathews, T. J. (1998). *Advance Report of Final Natality Statistics, 1996*, Monthly Vital Statistics Report 46, no. 11. Washington, D.C.: U.S. Department of Health and Human Services.

Vercauteren, M. (1993). Croissance, facteurs socio-familiaux et évolution séculaire (Growth, socio-familial factors and secular changes). *Bulletin et Mémoires de la Société d'Anthropologie de Paris*, 5:85–92.

Vercauteren, M., Hauspie, R. C. and Susanne, C. (1998). Biometry of Belgian boys and girls: Changes since Quételet. In *Secular Growth Changes in Europe*, eds. É. B. Bodzsar and C. Susanne, pp. 47–63. Budapest: Eötvös University Press.

Vercauteren, M. and Susanne, C. (1985). The secular trend of height and menarche in Belgium: Are there any signs of a future stop? *European Journal of Pediatrics*, 144:306–9.

Verkerk, P. H. (1992). The impact of alcohol misclassification on the relationship between alcohol and pregnancy outcome. *International Journal of Epidemiology*, 21:33–43.

Vestbo, E., Damsgaard, E. M., Frøland, A. and Morgensen, C. E. (1996). Birth weight and cardiovascular risk factors in an epidemiological study. *Diabetologia*, 39:1598–1602.

Vicens-Calvet, E., Sureda, J., Blanco, M. G. and Pineda, C. (1977). Growth and maturity in diabetic children before puberty. In *Recent Progress in Pediatric Endocrinology*, Proceedings of the Serono Symposia, vol. 12, eds. G. Chiumello and Z. Laron, pp. 313–19. New York: Academic Press.

Victora, C. G., Kirkwood, B. R., Fuchs, S. C., Lombardi, C. and Barros, F. C. (1990). Is it possible to predict which diarrhoea episodes will lead to life-threatening dehydration? *International Journal of Epidemiology*, 19:736–42.

Vignerová, J. and Bláha, P. (1998). The growth of the Czech child during the past 40 years. In *Secular Growth Changes in Europe*, eds. É. B. Bodzsar and C.

Susanne, pp. 93–107. Budapest: Eötvös University Press.

Vignerová, J., Lhotská, L., Bláhá, P. and Roth, Z. (1997). Growth of the Czech child population 0–18 years compared to the World Health Organization growth reference. *American Journal of Human Biology*, 9:459–68.

Vignolo, M., Milani, S., Cerbello, G., Coroli, P., di Battista, E. and Aicardi, G. (1992). Fels, Greulich–Pyle, and Tanner–Whitehouse bone age assessments in a group of Italian children and adolescents. *American Journal of Human Biology*, 4:493–500.

Vignolo, M., Naselli, A., Magliano, P., di Battista, E., Aicardi, M. and Aicardi, G. (1999). Use of the new US90 standards for TW–RUS skeletal maturity scores in youths from the Italian population. *Hormone Research*, 51:168–72.

Vigouroux, E. (1990). Hormonal regulation of postnatal growth: Thyroid and growth hormones. In *Handbook of Human Growth and Developmental Biology*, vol. 2, part B, *Growth, Nutrition, and Metabolism*, eds. E. Meisami and P. S. Timiras, pp. 23–37. Boca Raton, FL: CRC Press.

Vijayakumar, M., Fall, C. H. D., Osmund, C. and Barker, D. J. P. (1995). Birthweight, weight at one year, and left ventricular mass in adult life. *British Heart Journal*, 73:363–7.

Vikman, K., Carlsson, B., Billig, H. and Edén, S. (1991). Expression and regulation of growth hormone (GH) receptor messenger ribonucleic acid (messenger RNA) in rat adipose tissue, adipocytes, and adipocyte precursor cells: GH regulation of GH receptor messenger mRNA. *Endocrinology*, 129:1155–61.

Villamor, E., Gofin, R. and Adler, B. (1998). Maternal anthropometry and pregnancy outcome among Jerusalem women. *Annals of Human Biology*, 25:331–43.

Villar, J., de Onis, M., Kestler, E., Bolanos, F., Cerezo, R. and Bernedes, H. (1990). The differential neonatal morbidity of the intrauterine growth retardation syndrome. *American Journal of Obstetrics and Gynecology*, 163:151–7.

Villarreal, S. F., Martorell, R. and Mendoza, F. (1989). Sexual maturation of Mexican-American adolescents. *American Journal of Human Biology*, 1:87–95.

Visness, C. M. and Kennedy, K. I. (1997). Maternal employment and breast-feeding: Findings from the 1988 National Maternal and Infant Health Survey. *American Journal of Public Health*, 87:945–50.

Vobecky, J. S., Vobecky, J. and Normand, L. (1995). Risk and benefit of low fat intake in childhood. *Annals of Nutrition and Metabolism*, 39:124–33.

Voigt, M., Schneider, K. T. M. and Jährig, K. (1996). Analyse des Geburtengutes des Jahrgangs 1992 der Bundesrepublik Deutschland. I. Neue Perzentilwerte für die Körpermaße von Neugeborenen (Analysis of birth weight of the year 1992 in the Republic of Germany. I. New percentiles for body mass of newborns). *Geburtshilfe und Frauenheilkunde*, 56:550–8.

Voigt, M., Schneider, K. T. M. and Jährig, K. (1997). Analyse des Geburtengutes des Jahrgangs 1992 der Bundesrepublik Deutschland. II. Mehrdimensionale Zusammenhänge zwischen Alter, Körpergewicht und Körperhöhe der Mutter und dem Geburtsgewicht (Analysis of birth weight of the year 1992 in the Republic of Germany. II. Multivariate relationships of age, body weight, and stature of the mother with birth weight). *Geburtshilfe und Frauenheilkunde*, 57: 246–55.

von Hinkel, G. K. and Schambach, H. (1980). Vorausberechnung der Endgröße bei hochwuchsigen Kindern (Prediction of end height in tall children). *Deutsche Gesundheitswesen*, 35:1670–2.

von Kries, R., Koletzko, B., Sauerwald, T., von Mutius, E., Barnert, D., Grunert, V. and von Voss, H. (1999). Breast feeding and obesity: Cross-sectional study. *British Medical Journal*, 319:147–50.

Voors, A. W., Harsha, D. W., Webber, L. S. and Berenson, G. S. (1981). Obesity and external sexual maturation: The Bogalusa Heart Study. *Preventive Medicine*, 10:50–61.

Voss, L. D., Bailey, B. J. R., Cumming, K., Wilkin, T. J. and Betts, P. R. (1990). The reliability of height measurement: The Wessex Growth Study. *Archives of Disease in Childhood*, 65:1340–4.

Voss, L. D., Mulligan, J., Betts, P. R. and Wilkin, T. J. (1992). Poor growth in school entrants as an index of organic disease: The Wessex Growth Study. *British Medical Journal*, 305:1400–2.

Voss, L. D., Wilkin, T. J., Bailey, B. J. R. and Betts, P. R. (1991). The reliability of height and height velocity in the assessment of growth: The Wessex Growth Study. *Archives of Disease in Childhood*, 66:833–7.

Waaler, P. E. (1983). Anthropometric studies in Norwegian children. *Acta Paediatrica Scandinavica, Supplement*, 308:1–41.

Wabitsch, M., Blum, W. F., Muche, R., Heinze, E., Haug, C., Mayer, H. and Teller, W. (1996). Insulin-like growth factors and their binding proteins before and after weight loss and their associations with hormonal and metabolic parameters in obese adolescent girls. *International Journal of Obesity*, 20:1073–80.

Wabitsch, M., Hauner, H., Heinze, E., Bockmann, A., Benz, R., Mayer, H. and Teller, W. (1995). Body fat distribution and steroid hormone concentrations in obese adolescent girls before and after weight reduction. *Journal of Clinical Endocrinology and Metabolism*, 80:3469–75.

Wagenknecht, L. E., Cutter, G. R., Haley, N. J., Sidney, S., Manolio, T. A., Hughes, G. H. and Jacobs, D. R. (1990). Racial differences in serum cotinine levels among smokers in the Coronary Artery Risk Development in (Young) Adults Study. *American Journal of Public Health*, 80:1053–6.

Wainer, H., Roche, A. F. and Bell, S. (1978). Predicting adult stature without skeletal age and without paternal data. *Pediatrics*, 61:569–72.

Wainwright, R. L. (1983). Change in observed birth weight associated with change in maternal cigarette smoking. *American Journal of Epidemiology*, 117: 668–75.

Wald, N. J., Boreham, J., Bailey, A., Ritchie, C., Haddow, J. E. and Knight, G. (1984). Urinary creatinine as marker of breathing other people's smoke. *Lancet*, i:230–1.

Wales, J. K. H. and Milner, R. D. G. (1987). Knemometry in assessment of linear growth. *Archives of Disease in Childhood*, 62:166–71.

Waliszko, A., Jedlińska, W., Kotlarz, K., Palus, D., Slawińska, T., Szmyd, A. and Szwedziňska, A. (1980). *Stan rozwoju fizycznego dzieci i młodziezy szkolnej*

(Physical Development of Schoolchildren and Youths), Monographs of the Institute of Anthropology no. 1. Wrocław, Poland: Polish Academy of Sciences.

Walker, B. R., McConnachie, A., Noon, J. P., Webb, D. J. and Watt, G. C. M. (1998). Contribution of parental blood pressures to association between low birth weight and adult high blood pressure: Cross-sectional study. *British Medical Journal*, 316:834–7.

Walker, J., Van Wyk, J. J. and Underwood, L. E. (1992). Stimulation of statural growth by recombinant insulin-like growth factor-I in a child with growth hormone insensitivity syndrome (Laron type). *Journal of Pediatrics*, 121:641–6.

Walker, M., Shaper, A. G., Phillips, A. N. and Cook, D. G. (1989). Short stature, lung function and risk of heart attack: The British Regional Heart Study. *International Journal of Epidemiology*, 18:602–6.

Walpole, I., Zubrick, S. and Pontré, J. (1990). Is there a fetal effect with low to moderate alcohol use before or during pregnancy? *Journal of Epidemiology and Community Health*, 44:297–301.

Walravens, P. A., Hambidge, K. M. and Koepfer, D. M. (1989). Zinc supplementation in infants with a nutritional pattern of failure to thrive: A double-blind, controlled study. *Pediatrics*, 83: 532–8.

Walravens, P. A., Krebs, N. F. and Hambidge, K. M. (1983). Linear growth of low income preschool children receiving a zinc supplement. *American Journal of Clinical Nutrition*, 38:195–201.

Walter, H. (1977). Socio-economic factors and human growth: Observations on schoolchildren from Bremen. Growth and development: physique. *Symposia Biologica Hungarica*, 20:49–62.

Walter, H., Fritz, M. and Welker, A. (1975). Untersuchungen zur sozialen Verteilung von Körperhöhe un Körpergewicht (Research on the social class distribution of body height and body weight). *Zeitschrift für Morphologie und Anthropologie*, 67:6–18.

Wang, Q., Bing, C., Al-Barazanji, K., Mossakowaska, D. E., Wang, X. M., McBay, D. L., Neville, W. A., Taddayon, M., Pickavance, L., Dryden, S., Thomas, M. E. A., McHale, M. T., Gloyer, I. S., Wilson, R., Buckingham, J. R. S., Arch, P., Trayhurn, P. and

Williams, G. (1997). Interactions between leptin and hypothalamic neuropeptide Y neurons in the control of food intake and energy homeostasis in the rat. *Diabetes*, 46:335–41.

Wang, X., Guyer, B. and Paige, D. M. (1994). Differences in gestational age-specific birthweight among Chinese, Japanese and White Americans. *International Journal of Epidemiology*, 23:119–28.

Wang, X., Zuckerman, B., Coffman, G. A. and Corwin, M. J. (1995). Familial aggregation of low birth weight among whites and blacks in the United States. *New England Journal of Medicine*, 333:1744–9.

Warner, J. T., Cowan, F. J., Dunstan, F. D. J. and Gregory, J. W. (1997). The validity of body mass index for the assessment of adiposity in children with disease states. *Annals of Human Biology*, 24:209–15.

Waterlow, J. C. (1981). Observations on the suckling's dilemma: A personal view. *Journal of Human Nutrition*, 35:85–98.

Waterlow, J. C. (1986). Metabolic adaptation of low intakes of energy and protein. *Annual Review of Nutrition*, 6:495–526.

Waterlow, J. C. (1988). Observations on the natural history of stunting. *Nestlé Nutrition Workshop Series*, 14:1–16.

Weber, G., Seidler, H., Wilfing, H. and Hauser, G. (1995). Secular change in height in Austria: An effect of population stratification? *Annals of Human Biology*, 22:277–88.

Weber, H. P., Kowalewski, S., Gilje, A., Mollering, M., Schnaufer, I. and Fink, H. (1976). Unterschiedliche Calorienzufuhr bei 75 "low birth weights": Einfluss auf Gewichtszunahme, Serumeiweiss, Blutzucker und Serumbilirubin (Variations in the caloric intakes of 75 "low birth weights": Influence upon weight gain, serum albumen, blood sugar and serum bilirubin). *European Journal of Pediatrics*, 122:207–16.

Weiner, G. and Milton, T. (1970). Demographic correlates of low birth weight. *Journal of Epidemiology*, 91:260–72.

Weinstein, R. S. and Haas, J. D. (1977). Early stress and later reproductive performance under conditions of malnutrition and high altitude hypoxia. *Medical Anthropology*, 1:25–54.

Welch, Q. B. (1970). Fitting growth and research data. *Growth*, **34**:293–312.

Wellens, R., Malina, R. M., Beunen, G. and Lefevre, J. (1990). Age at menarche in Flemish girls: Current status and secular changes in the 20th century. *Annals of Human Biology*, **17**:145–52.

Wellens, R., Malina, R. M., Roche, A. F., Chumlea, W. C., Guo, S. S. and Siervogel, R. M. (1992). Body size and fatness in young adults in relation to age at menarche. *American Journal of Human Biology*, **4**:783–7.

Węłon, Z. and Bielicki, T. (1971). Further investigations of parent–child similarity in stature, as assessed from longitudinal data. *Human Biology*, **43**:517–25.

Wen, S. W., Kramer, M. S. and Usher, R. H. (1995). Comparison of birth weight distributions between Chinese and Caucasian infants. *American Journal of Epidemiology*, **141**:1177–87.

Wenzel, A., Droschl, H. and Melson, B. (1984). Skeletal maturity in Austrian children assessed by the GP and the TW2 methods. *Annals of Human Biology*, **11**:173–7.

Wenzel, A. and Melsen, B. (1982). Skeletal maturity in 6–16-year-old Danish children assessed by the Tanner–Whitehouse-2 method. *Annals of Human Biology*, **9**:277–81.

Westwood, M., Kramer, M. S., Munz, D., Lovett, J. M. and Watters, G. V. (1983). Growth and development of full-term nonasphyxiated small-for-gestational age newborns: Follow-up through adolescence. *Pediatrics*, **71**:376–82.

Weyer, C., Pratley, R. E., Lindsay, R. S. and Tataranni, P. A. (2000). Relationship between birth weight and body composition, energy metabolism, and sympathetic nervous system activity later in life. *Obesity Research*, **8**:559–65.

Whincup, P. H., Cook, D. G., Adshead, F. A., Taylor, S. J. C., Walker, M. and Alberti, K. G. M. M. (1997). Childhood size is more strongly related than size at birth to glucose and insulin levels in 10–11-year-old children. *Diabetologia*, **40**:319–26.

Whincup, P. H., Cook, D. G. and Papacosta, O. (1992). Do maternal and intrauterine factors influence blood pressure in childhood? *Archives of Disease in Childhood*, **67**:1423–9.

Whitaker, R. C., Wright, J. A., Pepe, M. S., Seidel, K. D. and Dietz, W. H. (1997). Predicting obesity in young adulthood from childhood and parental obesity. *New England Journal of Medicine*, **337**:869–73.

Whitehead, R. G. (1983). Nutritional aspects of human lactation. *Lancet*, **i**:167–9.

Whitehead, R. G. and Paul, A. A. (1984). Growth charts and the assessment of infant feeding practices in the western world and in developing countries. *Early Human Development*, **9**:187–207.

Whitehead, R. G. and Paul, A. A. (1985). Human lactation, infant feeding, and growth: Secular trends. In *Nutritional Needs and Assessment of Normal Growth*, vol. 7, eds. M. Gracey and F. Falkner, pp. 85–122. New York: Raven Press.

Whitehouse, R. H., Tanner, J. M. and Healy, M. J. R. (1974). Diurnal variation in stature and sitting height in 12–14-year-old boys. *Annals of Human Biology*, **1**:103–6.

Whitelaw, A. G. L. (1971). The association of social class and sibling number with skinfold thickness in London schoolboys. *Human Biology*, **43**:414–20.

Whitelaw, A. G. L. (1977). Infant feeding and subcutaneous fat at birth and at one year. *Lancet*, **ii**:1098–9.

Wich, J. (1983). Body height correlations between parents and their children aged 4–6 years. *Materialy i Prace Anthropologiczne*, **103**:85–92.

Widhalm, K. and Schönegger, K. (1999). BMI: Does it really reflect body fat mass? *Journal of Pediatrics*, **134**:522–3.

Widjaja, A., Stratton, I. M., Horn, R., Holman, R. R., Turner, R. and Brabant, G. (1997). UKPD20: Plasma leptin, obesity, and plasma insulin in type 2 diabetic subjects. *Journal of Clinical Endocrinology and Metabolism*, **82**:654–7.

Wieghart, M., Hoover, J., Choe, S. H., McGrane, M. M., Rottman, F. M., Hanson, R. W. and Wagner, T. E. (1988). Genetic engineering of livestock: Transgenic pigs containing a chimeric bovine growth hormone (PEPCK/bGH) gene. *Journal of Animal Science, Supplement*, **66**: S266.

Wierman, M. E. and Crowley, W. F., Jr. (1986). Neuroendocrine control of the onset of puberty. In *Human Growth: A Comprehensive Treatise*, 2nd edn, vol. 2, *Postnatal Growth*, eds. F. Falkner and J. M. Tanner, pp. 225–41. New York: Plenum Press.

Wilcox, M., Gardosi, J., Mongelli, M., Ray, C. and Johnson, I. R. (1993). Birth weight from pregnancies dated by ultrasonography in a multicultural British population. *British Medical Journal*, **307**:588–91.

Wilcox, W. D., Gold, B. D. and Tuboku-Metzger, A. J. (1991). Maternal recall of infant birth weight. *Clinical Pediatrics*, **30**:509–10.

Wilding, J. P. H., Gilbey, S. G., Lambert, P. D., Ghatei, M. A. and Bloom, S. R. (1993). Increases in neuropeptide Y content and gene expression in the hypothalamus of rats treated with dexamethasone are prevented by insulin. *Neuroendocrinology*, **57**:581–7.

Williams, D. P., Going, S. B., Lohman, T. G., Harsha, D. W., Srinivasan, S. R., Webber, L. S. and Berenson, G. S. (1992). Body fatness and risk for elevated blood pressure, total cholesterol, and serum lipoprotein ratios in children and adolescents. *American Journal of Public Health*, **82**:358–63.

Williams, R. L., Cheyne, K. L., Houtkooper, L. K. and Lohman, T. G. (1989). Adolescent self-assessment of sexual maturation. *Journal of Adolescent Health Care*, **9**:480–2.

Williams, S., Davie, G. and Lam, F. (1999). Predicting BMI in young adults from childhood data using two approaches to modelling adiposity rebound. *International Journal of Obesity*, **23**:348–54.

Williams, T., Berelowitz, M., Joffe, S. N., Thorner, M. O., Rivier, J., Vale, W. and Frohman, L. A. (1984). Impaired growth hormone response to growth hormone releasing factor in obesity: A pituitary defect reversed with weight reduction. *New England Journal of Medicine*, **311**:1403–7.

Williams, T. and Handford, A. (1986). Television and other leisure activities. In *The Impact of Television: A Natural Experiment in Three Communities*, ed. T. M. Williams, pp. 143–213. Orlando, FL: Academic Press.

Wilson, D. M., Kraemer, H. C., Ritter, P. L. and Hammer, L. D. (1987). Growth curves and adult height estimation for adolescents. *American Journal of Diseases of Children*, **141**:565–70.

Wilson, J. and Jungner, G. (1968). *Principles and Practice of Screening for Disease*, Public Health Paper no. 34. Geneva, Switzerland: World Health Organization.

Wilson, R. S. (1979). Twin growth: Initial deficit, recovery, and trends in concordance from birth to nine years. *Annals of Human Biology*, **6**:205–20.

Wilson, R. S. (1986). Growth and development of human twins. In *Human Growth: A Comprehensive Treatise*, 2nd edn, vol. 3, Methodology, eds. F. Falkner and J. M. Tanner, pp. 197–211. New York: Plenum Press.

Windham, G. C., Fenster, L., Hopkins, B. and Swan, S. H. (1995). The association of moderate maternal and paternal alcohol consumption with birthweight and gestational age. *Epidemiology*, **6**:591–7.

Windsor, R. A., Morris, J., Cutter, G., Lowe, J., Higginbotham, J., Perkins, L. and Konkol, L. (1989). Sensitivity, specificity and predictive value of saliva thiocyanate among pregnant women. *Addictive Behaviors*, **14**:447–52.

Wingerd, J. (1970). The relation of growth from birth to 2 years to sex, parental size and other factors, using Rao's method of the transformed time scale. *Human Biology*, **42**:105–31.

Wingerd, J. and Schoen, E. J. (1974). Factors influencing length at birth and height at five years. *Pediatrics*, **53**:737–41.

Wingerd, J., Solomon, I. L. and Schoen, E. J. (1973). Parent-specific height standards for preadolescent children of three racial groups, with method for rapid determination. *Pediatrics*, **52**:555–60.

Winter, J. S. D., Faiman, C., Hobson, W. C., Prasad, A. V. and Reyes, F. I. (1975). Pituitary–gonadal relations in infancy. I. Patterns of serum gonadotropin concentrations from birth to four years of age in man and chimpanzee. *Journal of Clinical Endocrinology and Metabolism*, **40**:545–51.

Winter, K. (1962). Akzeleration: Nicht nur ein Problem des Jugendalters? (Acceleration: Not just a problem of youth?). *Deutsches Gesundheitwesen*, **17**:954–65.

Wisborg, K., Henriksen, T. B., Hedegaard, M. and Secher, N. J. (1996). Smoking during pregnancy and preterm birth. *British Journal of Obstetrics and Gynaecology*, **103**:800–5.

Wit, J.-M., Kalsbeek, E. J., Wijk-Hoek, J. M. and Leppink, G. J. (1987). Assessment of the usefulness of weekly knemometric measurements in growth studies. *Acta Paediatrica Scandinavica*, **76**:974–80.

Witter, F. R. and Luke, B. (1991). The effect of maternal height on birth weight and birth length. *Early Human Development*, **25**:181–6.

Wnuk-Lipinski, E. (1990). The Polish country profile: Economic crisis and inequalities in health. *Social Science and Medicine*, **81**:859–66.

Wohlert, M. (1989). *The Influence of Biological, Social and Organizational Conditions on Pregnancy and Delivery*. Aarhus, Denmark: Werks Offset.

Wollmann, H. A., Kirchner, T., Enders, H., Preece, M. A. and Ranke, M. B. (1995). Growth and symptoms in Silver–Russell syndrome: Review on the basis of 386 patients. *European Journal of Pediatrics*, **154**:958–68.

Wolthers, O. D., Heuck, C. and Skjaerbaek, C. (1999). Diurnal rhythm in serum leptin. *Journal of Pediatric Endocrinology and Metabolism*, **12**:863–6.

Woods, J. R., Plessinger, M. A. and Clark, K. E. (1987). Effect of cocaine on uterine blood flow and fetal oxygenation. *Journal of the American Medical Association*, **257**:957–61.

World Health Organization (1998). *Obesity: Preventing and Managing the Global Epidemic*, Report of a WHO consultation. Geneva, Switzerland: World Health Organization.

World Health Organization Working Group on Infant Growth (1994). *An Evaluation of Infant Growth: A Summary of Analyses Performed in Preparation for the WHO Expert Committee on Physical Status: The Use and Interpretation of Anthropometry*. Geneva, Switzerland: Nutrition Unit, World Health Organization.

World Health Organization Working Group on Infant Growth (1995). An evaluation of infant growth: The use and interpretation of anthropometry in infants. *Bulletin of the World Health Organization*, **73**: 165–74.

Wright, C., Avery, A., Epstein, M., Birks, E. and Croft, D. (1998*a*). New chart to evaluate weight faltering. *Archives of Disease in Childhood*, **78**:40–3.

Wright, C. M. (2000). Identification and management of failure to thrive: A community perspective. *Archives of Disease in Childhood*, **82**:5–9.

Wright, C. M., Callum, J., Birks, E. and Jarvis, S. (1998*b*). Effect of community-based management in failure to thrive: Randomized controlled trial. *British Medical Journal*, **317**:571–4.

Wright, C. M. and Cheetham, T. D. (1999). The strengths and limitations of parental heights as a predictor of attained height. *Archives of Disease in Childhood*, **81**:257–60.

Wright, C. M., Corbett, S. S. and Drewett, R. F. (1996). Sex differences in weight in infancy and the British 1990 national growth standards. *British Medical Journal*, **313**:513–14.

Wright, C. M., Edwards, A. G. K., Halse, P. C. and Waterston, A. J. R. (1991). Weight and failure to thrive in infancy. *Lancet*, **337**:365–6.

Wright, C. M., Matthews, J. N. S., Waterston, A. and Aynsley-Green, A. (1994). What is a normal rate of weight gain in infancy? *Acta Paediatrica*, **83**:351–6.

Wright, K., Dawson, J. P., Fallis, D., Vogt, E. and Lorch, V. (1993). New postnatal growth grids for very low birth weight infants. *Pediatrics*, **91**:922–6.

Wyshak, G. (1983). Secular changes in age at menarche in a sample of U.S. women. *Annals of Human Biology*, **10**:69–74.

Xu, B., Jarvelin, M.-R. and Pekkanen, J. (2000). Body build and atopy. *Journal of Allergy and Clinical Immunology*, **105**:393–4.

Xu, X., de Pergola, G. and Björntorp, P. (1990). The effects of androgens on the regulation of lipolysis in adipose precursor cells. *Endocrinology*, **126**:1229–34.

Yamaguchi, M. and Hashizume, M. (1994). Effect of beta-alanyl-L-histidinato zinc on protein components in osteoblastic MC3T3-EI cells: Increase in osteocalcin, insulin-like growth factor-I and transforming growth factor-beta. *Molecular and Cellular Biochemistry*, **136**:163–9.

Yamashita, S. and Melmed, S. (1986). Effect of insulin on rat anterior pituitary cells. *Diabetes*, **35**:440–7.

Yarbrough, D. E., Barrett-Connor, E., Kritz-Silverstein, D. and Wingard, D. L. (1998). Birth weight, adult weight, and girth as predictors of the metabolic syndrome in postmenopausal women: The Rancho Bernardo Study. *Diabetes Care*, **21**:1652–8.

Yeung, D. L. (1983). *Infant Nutrition: A Study of Feeding Practices and Growth from Birth to 18 Months*. Ottawa: Canadian Public Health Association.

Yip, R. (1987). Altitude and birth weight. *Journal of Pediatrics*, 111:869–76.

Yip, R., Binkin, N. J. and Trowbridge, F. L. (1988). Altitude and childhood growth. *Journal of Pediatrics*, 113:486–9.

Yip, R., Scanlon, K. and Trowbridge, F. (1993). Trends and patterns in height and weight status of low-income U.S. children. *Critical Reviews in Food Science and Nutrition*, 33:409–21.

Yoneyama, K., Nagata, H. and Sakamoto, Y. (1988). A comparison of height growth curves among girls with different ages of menarche. *Human Biology*, 60:33–41.

Young, N. R., Gordon Baker, H. W. and Guangda, L. (1993). Body composition and muscle strength in healthy men receiving testosterone enanthate for contraception. *Journal of Clinical Endocrinology and Metabolism*, 77:1028–32.

Younoszai, M. K., Kacic, A. and Haworth, J. C. (1968). Cigarette smoking during pregnancy: The effect upon the hematocrit and acid–base balance of the newborn infant. *Canadian Medical Association Journal*, 99:197–200.

Yu, W. H., Kimura, M., Walczewska, A., Karanth, S. and McCann, S. M. (1997). Role of leptin in hypothalamic–pituitary function. *Proceedings of the National Academy of Sciences (U.S.A.)*, 94: 1023–8.

Yudkin, P. L., Aboualfa, M., Eyre, J. A., Rodman, C. W. G. and Wilkinson, A. R. (1987). The influence of elective preterm delivery on birthweight and head circumference standards. *Archives of Disease in Childhood*, 62:24–9.

Yukawa, K., Uchino, C., Yamawaki, M. and Katayose, M. (1985). Longitudinal observation from six to seventeen years old on physical growth and obesity of female students in senior high school. *Japanese Journal of School Health*, 27:392–400.

Yuval, Y., Seidman, D. S., Achiron, R., Goldenberg, M., Alcalay, M., Mashiach, S. and Lipitz, S. (1995). Intrauterine growth of triplets as estimated from liveborn birth weight data. *Ultrasound in Obstetrics and Gynecology*, 6:345–8.

Zacharias, L. and Rand, W. M. (1986). Adolescent growth in weight and its relation to menarche in contemporary American girls. *Annals of Human Biology*, 13:369–86.

Zacharias, L., Wurtman, R. and Schatzoff, M. (1970). Sexual maturation in contemporary American girls. *American Journal of Obstetrics and Gynecology*, 108:833–46.

Zacharin, M. (2000). Use of androgens and oestrogens in adolescents: A review of hormone replacement treatment. *Journal of Pediatric Endocrinology and Metabolism*, 13:3–11.

Zachmann, M., Prader, A., Sobel, E. H., Crigler, J. F., Ritzen, E. M., Atares, M. and Ferrandez, A. (1986). Pubertal growth in patients with androgen insensitivity: Indirect evidence for the importance of estrogens in pubertal growth of girls. *Journal of Pediatrics*, 108:694–7.

Zachmann, M., Sobradillo, B., Frank, M., Frisch, H. and Prader, A. (1978). Bayley–Pinneau, Roche–Wainer–Thissen, and Tanner height predictions in normal children and in patients with various pathologic conditions. *Journal of Pediatrics*, 93:749–55.

Zadik, Z., Chalew, S. A., McCarter, R. J., Jr., Meistas, M. and Kowarski, A. A. (1985). The influence of age on the 24-hour integrated concentration of growth hormone in normal individuals. *Journal of Clinical Endocrinology and Metabolism*, 60:513–16.

Zakrzewska, K. E., Cusin, I., Sainsbury, A., Rohner-Jeanrenaud, F. and Jeanrenaud, B. (1997). Glucocorticoids as counterregulatory hormones of leptin toward an understanding of leptin resistance. *Diabetes*, 46:717–19.

Zamudio, S., Droma, T., Norkyel, K. Y., Acharya, G., Zamudio, J. A., Niermeyer, S. N. and Moore, L. G. (1993). Protection from intrauterine growth retardation in Tibetans at high altitude. *American Journal of Physical Anthropology*, 91:215–24.

Zanchetta, J. R., Plotkin, H. and Filgueira, M. L. A. (1995). Bone mass in children: Normative values for the 2–20-year-old population. *Bone, Supplement*, 16:393–9.

Zapf, J., Donath, M. Y., Froesch, E. R. and Schmid, C. (1998). Actions of administered IFG I in humans. *Clinical Pediatric Endocrinology*, 7:9–18.

Zapf, J., Schmid, C. H. and Froesch, E. R. (1984). Biological and immunological properties of insulin-like growth factors (IGF) I and II. *Clinics in Endocrinology and Metabolism*, 13:3–30.

Zapf, J., Walter, H. and Froesch, E. R. (1981). Radioimmunological determination of insulin-like growth factors I and II in normal subjects and in patients with growth disorders and extrapancreatic tumor hypoglycemia. *Journal of Clinical Investigation*, **68**:1321–30.

Zarén, B., Lindmark, G. and Gebre-Medhin, M. (1996). Maternal smoking and body composition of the newborn. *Acta Paediatrica*, **85**:213–19.

Zarjevski, N., Cusin, I., Vettor, R., Rohner-Jeanrenaud, F. and Jeanrenaud, B. (1993). Chronic intracerebroventricular neuropeptide-Y administration to normal rats mimics hormonal and metabolic changes of obesity. *Endocrinology*, **133**:1753–8.

Zarów, R. (1992). Adult stature prediction in girls according to different methods. *Acta Medica Auxologica* **24**:159–66.

Zeitler, P., Argente, J., Chowen-Breed, J. A., Clifton, D. K. and Steiner, R. A. (1990). Growth hormone releasing hormone messenger ribonucleic acid in the hypothalamus of the adult male rat is increased by testosterone. *Endocrinology*, **127**:1362–8.

Zelson, C. (1973). Infant of the addicted mother. *New England Journal of Medicine*, **288**:1393–5.

Zhang, J. and Bowes, W. A. (1995). Birth-weight-for-gestational- age patterns by race, sex, and parity in the United States population. *Obstetrics and Gynecology*, **86**:200–8.

Zhang, J., Peddada, S. D., Malina, R. M. and Rogol, A. D. (2000). Longitudinal assessment of hormonal and physical alterations during puberty in boys. VI. Modeling of growth velocity, mean growth hormone (GH mean), and serum testosterone (T) concentrations. *American Journal of Human Biology*, **12**:814–24.

Zive, M. M., McKay, H., Frank-Spohrer, G. C., Broyles, S. L., Nelson, J. A. and Nader, P. R. (1992). Infant-feeding practices and adiposity in 4-year-old Anglo- and Mexican-Americans. *American Journal of Clinical Nutrition*, **55**:1104–8.

Zumoff, B., Fukushima, D. K., Weitzman, E. D., Kream, J. and Hellman, L. (1974). The sex difference in plasma cortisol concentration in man. *Journal of Clinical Endocrinology and Metabolism*, **39**:805–8.

Zumoff, B. and Strain, G. W. (1994). A perspective on the hormonal abnormalities of obesity: Are they cause or effect? *Obesity Research*, **2**:56–67.

Zureik, M., Bonithon-Kopp, C., Lecomte, E., Siest, G. and Ducimetière, P. (1996). Weights at birth and in early infancy, systolic pressure, and left ventricular structure in subjects aged 8 to 24 years. *Hypertension*, **27**:339–45.

Index

Page references in italics refer to illustrations or tables.